Food, Pharmaceutical and Bioengineering Division 2016

Core Programming Area at the 2016 AIChE Annual Meeting

San Francisco, California, USA
13 - 18 November 2016

Volume 1 of 2

ISBN: 978-1-5108-3436-1

Printed from e-media with permission by:

Curran Associates, Inc.
57 Morehouse Lane
Red Hook, NY 12571

Some format issues inherent in the e-media version may also appear in this print version.

Printed by Curran Associates, Inc. (2017)

For permission requests, please contact AIChE
at the address below.

AIChE
120 Wall Street, FL 23
New York, NY 10005-4020

Phone: (800) 242-4363
Fax: (203) 775-5177

www.aiche.org

Additional copies of this publication are available from:

Curran Associates, Inc.
57 Morehouse Lane
Red Hook, NY 12571 USA
Phone: 845-758-0400
Fax: 845-758-2633
Email: curran@proceedings.com
Web: www.proceedings.com

TABLE OF CONTENTS

VOLUME 1

VOLUME 2

470501 Engineering Intelligently Designed Nano- and Microparticles to Control Interactions with the Immune System

Abstract:

Research Interests:

Nano- and micro-particulate carriers enable the site-specific delivery for controlled biological responses and can harness the intrinsic pathways by which the body responds to natural invaders. These particles are in size range which naturally associates with many innate immune cells, including antigen presenting cells (APCs), and can interface with these cells to deliver a stimulus. Through controlled design properties, engineered nano- and microparticle drug delivery vehicles have the potential to expand the breadth of many therapeutic approaches, impacting immunological outcomes through cell-specific targeted delivery. However, in many applications, such as mucosal vaccines or controlled-release lung depots, optimal particle properties have not yet been identified. Many particle design considerations have been established for systemic intravenous administration to create long-circulating drug delivery vehicles, yet less is known about particle design parameters critical to interfacing with and directing the immune system.

My research to date has focused on filling in these gaps by designing particles to engineer a controlled response with the immune system. Importantly, this research has ranged from particle synthesis, functionalization, and characterization to extensive evaluations in a range of *in vitro* and *in vivo* models. During my PhD, I investigated the role of particle properties to increase local residence times and alter the cellular fate of nano- and microparticles delivered to the lung. In these works, particle association in lung APC populations was characterized as a function of particle parameters, including size, surface charge, shape, and surface chemistry. This work demonstrated the potential use of cationic nanoparticles as pulmonary mucosal vaccines, which were shown to produce superior local and systemic immune responses, as well as PEGylated formulations to extend residence times in the airways. Research

during my postdoc has focused on designing particle therapeutics, again at the interface between particle formulations and interactions with blood cells. This work has evaluated particle properties of size, shape, modulus, and surface chemistry to understand interactions with blood cells and inflamed endothelial cells under physiological blood flow, specifically during inflammatory events. Additional projects funded independently under the UM Presidentâ€™s Postdoctoral Fellowship have evaluated controlled presentation of toll-like-receptor (TLR) adjuvant molecules on the surface of particle carriers to direct Th1 immune stimulation. Combined, these works demonstrate a potential for rationally designed particle therapeutics to interface with the immune system. Future research will continue to evaluate particle design features on controlled immune interactions, especially in non-intravenous routes of administration, towards the development of novel treatments for vaccinations, cancer, inflammation, and allergy.

Teaching Interests:

Transport, Mass and Energy Balances, Heat and Mass Transfer. I have experience guest lecturing for undergraduate level Chemical Process Analysis through the Mentored Teaching Award at North Carolina State University, as well as Heat and Mass Transfer at the University of Michigan.

456396 The Effect of Infusion Position on Convection-Enhanced Delivery of Anticancer Drugs to Remnant Brain Tumour after Surgery

Abstract:

Research Interests: biomedical engineering and bioprocessing, with a special focus on transport processes in biological and physiological systems.

Teaching Interests: mass transfer

Convection enhanced delivery is a promising method to overcome the blood-brain barrier and brain-tumour barrier to improve the drug delivery into tumour interior and enhance the treatment efficacy. In this study, a 3-D geometry of a remnant brain tumour and its normal holding tissue is reconstructed from MR images. A developed mathematical model is applied to investigate the effects of infusion position on the spatial distribution of carmustine and paclitaxel. Results show that infusion position is an important impact factor in CED treatments, and the effect of infusion position varies with the drug. For carmustine, most of the drugs accumulate around the infusion site to form a highly heterogeneous distribution in tumour. The heterogeneity and the spatial average concentration can be reduced by infusing carmustine into the cavity. Paclitaxel is capable to penetrate deep into the brain tumour tissue with the enhanced interstitial fluid flow. However, infusing paclitaxel in the downstream can reduce the homogeneity and spatial averaged concentration in the whole remnant tumour.

462670 Yeast Cell Factories: Construction of Platform Strains and Development of Synthetic Biology Tools

Abstract:

Research Interests:

Biological conversion of renewable feedstock into fuels and chemicals has been intensively investigated due to increasing concerns on sustainability and global climate change. Compared with its counterparts, *Saccharomyces cerevisiae*, the bakerâ€™s yeast, is more industrially relevant thanks to its well-studied genetic and physiological background, the availability of a large collection of genetic tools, the compatibility of high-density and large-scale fermentation, the resistance to phage infection, and the high tolerance against toxic inhibitors and products. Therefore, *S. cerevisiae* is one of the most popular cell factories and has been successfully used in the modern fermentation industry to produce a wide variety of products such as ethanol, organic acids, amino acids, enzymes, and therapeutic proteins.

My Ph.D. work was mainly focused on the construction of yeast cell factories for efficient and cost-effective production of biofuels and chemicals using synthetic biology and metabolic engineering approaches. Notably, a wide variety of products with industrial interests are derived from a few precursor metabolites. For example, lactate, 2,3-butanediol, and *iso*-butanol are derived from pyruvate, while *n*-butanol, polyhydroxybutyrate, isoprenoids, fatty acids, fatty alcohols, alkanes, and fatty acid ethyl esters are derived from acetyl-CoA. Therefore, we aim to develop platform yeast cell factories with enhanced supply of these precursor metabolites for efficient production of biofuels and chemicals. To achieve this, it generally involves host genome engineering such as the disruption of the competing pathways and biosynthetic pathway engineering such as the introduction of heterologous pathways. For example, in order to construct an acetyl-CoA overproducing yeast strain, the ethanol and glycerol formation routes were partially or fully blocked and alternative acetyl-CoA biosynthetic pathways, such as the pyruvate dehydrogenase (PDH) from a bacterium and the ATP-dependent citrate lyase (ACL) from an oleaginous yeast, were introduced. The higher efficiency and lower energy requirement of PDH and ACL, when combined with host genome engineering, increased the acetyl-CoA levels by more than three-fold and the production of acetyl-CoA derived product by more than 12-fold.

During my postdoc research, I switched my focus to the development of novel synthetic biology tools to facilitate the development of cell factories. Although a wide variety of cell factories have been constructed using the traditional strategies mentioned above, we are still facing the challenges of constructing efficient cell factories in a short period of time. Therefore, novel synthetic biology tools are demanded to revolutionize cell factory development. In terms of the biosynthetic pathway engineering, we constructed a series of plasmids with different copy numbers, which enabled the fine-tuning of gene expression levels and optimization of multi-gene biosynthetic pathways. In terms of genome engineering, we took advantage of the recently developed CRISPR/Cas9 system to achieve both gain of function (knock-in and gene activation) and loss of function (knock out and gene repression) in yeast. More importantly, we combined gene deletion, gene repression, and gene activation into a single system, in which a full spectrum of gene expression profiles (zero expression, down-regulation, and up-regulation) was achieved. Now we are performing genome-scale studies to develop a rapid and efficient strategy for the construction of cell factories and unveil the synergistic effects between gene deletion, gene repression, and gene activation in yeast.

Teaching Interests:

As a graduate student in the University of Illinois, I have the opportunity to work with and mentor many talented undergraduate and graduate students. Three semesters of TA in the department of Chemical and Biomolecular engineering have given me a better understanding on how to teach effectively in a big classroom, how to mentor students with various academic background, as well as the skills to communicate with students from a wide range of countries. Besides these formal teaching assignments, I have served as a mentor for undergraduate and junior graduate students in the lab, helping them to understand not only the synthetic biology principles and experimental techniques, but also the experimental design and problem-solving skills.

Keywords: Yeast cell factory; Synthetic biology; Genome engineering; CRISPR/Cas9; Plasmid copy number

464165 Engineering Immune Development By Recapitulating Tissue Microenvironments

Abstract:

Research Interests:

Bone marrow transplantation, also known as hematopoietic stem cell transplantation (HSCT), is a process in which healthy blood stem cells within the bone marrow of a donor are extracted and transferred into a recipient. The process is typically performed in patients with advanced blood-based diseases such as leukemia and multiple myeloma. Prior to a HSCT, the patient undergoes radiation and cytotoxic chemotherapy regimen to eradicate as much of the disease as possible. However, the process severely impairs the immune system as a side effect. The profound long-term post-transplant immunodeficiency often results in severe complications in patients such as infections, the development of graft-versus-host disease (GVHD) and the relapse of the underlying disease. Thus, while effective, HSCT continues to be a risky procedure. One way to alleviate the risk is to speed up the formation of new T- and B-cells of the immune system that are responsible in identifying and preventing such complications. Currently, certain proteins and cell-infusions are used in the clinic that can increase the absolute number of immune cells by inducing division of existing mature cells that already recognize certain pathogens. However, the generation of new immune cells that can identify unknown pathogens that the HSCT recipient would encounter is necessary to completely restore the immune system.

The immune itself system arises from blood stem cells, known as hematopoietic stem cells (HSCs). HSCs reside in 'niches,' which are sites within specific tissues in the body in which cells reside at steady-state. In humans, the HSC niche is contained in the bone marrow. Using signals presented in the bone marrow, the HSCs undergo cell differentiation and ultimately end up as T- or B-cells. This project seeks to replenish the T- and B-cells using a synthetic bone marrow that can capture transplanted HSCs after an HSCT and induce their differentiation into specific types of immune cells that prevent the common complications associated with HSCT. To reconstitute bone containing bone marrow in a mouse, a porous scaffold substrate and a specific bone-forming protein are sufficient to produce bone marrow in a pocket below the skin and mimic a healthy HSC niche environment. In this work alginate was used, a sugar-based polymer that can form a gel-like scaffold. The interweaving polymer chains created a large surface area to provide sufficient space for HSCs to engraft. By adding specific â€˜address labelsâ€™ in the form of biological signals it was possible to specify the differentiation of these HSCs into immune T- and B-cells. The synthetic marrow was able to rapidly induce the development of new T- and B-cells after a HSCT in a mouse model that rapidly restored the immune cell counts. It was observed that the diversity of the T-cell receptor, which recognizes pathogens, was significantly higher both in terms of gene rearrangement and the number of cells with a particular configuration in treated animals. Ongoing work seeks to induce the development of regulatory T-cells that are known to play a role in alleviating GVHD. To prevent the relapse of leukemia, the project will also investigate the combination of the synthetic marrow with a clinical vaccination regimen to induce the formation of T- and B-cells against specific targets that are known to the associated with leukemia. More broadly, there are a number of natural and induced events that can permanently or transiently weaken the immune system such as aging, autoimmune and infectious diseases. The technology development in this project could offer an option to restore the immune defenses in patients with a compromised immune system and to prevent life-threatening conditions.

Teaching Interests:

I am interested in Chemical Engineering and Bioengineering undergraduate as well as graduate level teaching in the areas of biomaterials, kinetics, thermodynamics.

469437 Deconstructing the Tumor Microenvironment and Its Contribution to Metastasis

Abstract:

Research Interests:

Despite aggressive surgical, radiological, and chemotherapeutic intervention, breast cancer recurrence remains a problem in up to 20% of patients. Most studies focus solely on tumors and tumor cells but often ignore the contribution of the normal tissues that surround and interact with tumor cells to disease progression. I am interested in combining my chemical and biomedical engineering expertise toward examining the mechanisms driving tumor recurrence and metastasis. I aim to study the relationship between tumor cells, normal tissues, and the immune system as a means of deconstructing the tumor microenvironment to understand how each component individually and as a whole contributes to metastasis and relapse after surgical and radiation treatment. The emphasis of my laboratory will be on developing bio-inspired ***in vitro*** models that recapitulate the tumor microenvironment by learning from ***in vivo*** mouse models. These studies will ultimately enable the design of optimized breast cancer therapeutics. ***I will build a program around analyzing the interplay between the physical, chemical, and biological cues that influence cancer metastasis and recurrence.*** Specifically, I will study (1) the collateral effect of radiation and surgery of normal tissues on tumor and immune cell migration patterns, (2) the molecular profiles of normal tissues wounded from therapy, and (3) the changes in the biomechanical properties of the tumor microenvironment and surrounding normal tissues following therapy. The idea that tissue response to therapeutic interventions may contribute to cancer recurrence is in its infancy. My research has the potential to challenge current paradigms about breast cancer treatment. My undergraduate and graduate studies provide me with the foundation to use biomimetic materials to study disease, and I will build on my strengths to design novel tumor microenvironment models in my independent research laboratory. My expertise has grown to encompass basic cancer biology, molecular imaging, biomechanical characterization of tissues, and ***in vivo*** experimental design through my postdoctoral work, which uniquely positions me to carry out research at the interface of engineering and medicine.

Successful Proposals:

K99/R00 Pathway to Independence Award, NIH/NCI; 2016-Present

Katherine McCormick Advanced Postdoctoral Fellowship; 2015-2016

T32 NIH NRSA Postdoctoral Fellowship in the Radiation Sciences; 2012-2013, 2015-2016

American Heart Association Founders Affiliate Predoctoral Fellowship; 2009-2011

Postdoctoral Project: "The Role of the Immune System in Modulating Breast Tumor Cell Migration following Radiation-Induced Changes in the Tissue Microenvironment"

Under the supervision of Edward Graves and Amato Giaccia, Department of Radiation Oncology, Stanford University

Ph.D. Dissertation: "Dual Antibody Functionalized Polyvinyl Alcohol and Alginate Hydrogels for Synergistic Endothelial Cell Adhesion"

Under the supervision of Debra Auguste, School of Engineering and Applied Sciences, Harvard University

Education:

Ph.D., Harvard University, Engineering Sciences, 2012

S.M., Harvard University, Engineering Sciences, 2007

S.B., Massachusetts Institute of Technology, Chemical Engineering, 2006

Selected Peer-Reviewed Publications:

1. **Rafat M**, Ali R, Graves EE. Imaging Radiation Response in Tumor and Normal Tissue. ***American Journal of Nuclear Medicine and Molecular Imaging***. 2015; 5(4): 317-332 (Review).
2. You JO*, **Rafat M***, Almeda D, Maldonado N, Guo P, Nabzdyk CS, Chun M, LoGerfo FW, Hutchinson JW, Pradhan-Nabzdyk LK, Auguste DT. pH-Responsive Scaffolds Generate a Pro-Healing Response. ***Biomaterials***. 2015; 57: 22-32. *Equal contribution. ***Highlighted in Materials Today. May 21, 2015***.
3. Vilalta M, **Rafat M**, Giaccia AJ, Graves EE. Recruitment of Circulating Breast Cancer Cells Is Stimulated by Radiotherapy. ***Cell Reports***. 2014; 8(2): 402-409.
4. **Rafat M**, Rotenstein L, Hu J, Auguste DT. Engineered Endothelial Cell Adhesion via VCAM1 and E-selectin Antibody-Presenting Alginate Hydrogels. ***Acta Biomaterialia***. 2012; 8(7): 2697-2703.
5. **Rafat M**, Rotenstein L, You JO, Auguste DT. Dual Functionalized PVA Hydrogels that Adhere Endothelial Cells Synergistically. ***Biomaterials***. 2012; 33(15): 3880-3886.

Teaching Interests:

Teaching Experience

I held two teaching fellow positions during graduate school, where I earned Certificates of Distinction in Teaching. I was a Teaching Fellow in 2007 for a graduate-level course in drug delivery that focused on engineering solutions for delivering drugs to specific sites in the body. In 2009, I served as the Head Teaching Fellow for an undergraduate-level introductory course in fluid mechanics and transport processes.

During my postdoctoral fellowship in 2014, I participated in the Stanford Pre-Collegiate Studies Program, where I lectured students in the Cancer Biology section of the Topics in Biosciences and Biotechnology module.

Mentoring Experience

I have extensive experience teaching and mentoring students in a research capacity. Through the National Science Foundation Research Experiences for Undergraduates (REU) program, I have mentored three students since 2007. I also advised three undergraduate students on their theses while at Harvard. My mentees contributed to co-authored publications and multiple conference presentations. At Stanford, I have supervised a summer veterinary student and a clinical research assistant who was interested in receiving wet lab training before attending medical school. I am currently supervising an undergraduate student on her own project. I emphasize critical thinking and challenge students to be active in the design, implementation, and analysis of their research. Mentoring students has been incredibly rewarding; I am eager to continue advising and instructing students as a junior faculty member.

Teaching Philosophy

My previous teaching experiences have given me a realistic sense of how to teach classes and mentor students in the future. As a professor, I will give students the necessary tools for developing analytical thinking skills and to solve problems beyond cookie-cutter solutions. I will structure classes using a combination of hands-on learning through laboratory projects as well as reading the current literature for students to gain an understanding of cutting-edge research. I will also expose students to problems with unknown solutions to challenge them to think critically and devise solutions creatively.

472132 Bacterial Biofilms: From the Built Environment to Human Diseases

Abstract:

Research Interests: Biofilms are communities of sessile microorganisms ubiquitously found in both natural and built environments, as well as numerous hosts including human. This mode of growth is preferred by bacteria for surviving in challenging environments because the unique structure of biofilms (e.g., extracellular polymeric substrates (EPS)) provide embedded bacteria with protection from harsh environmental factors such as antimicrobials. The enhanced resistance of biofilms against biocidal agents compared to free swimming bacteria (up to 1000 fold) is one of the leading causes of chronic infections in human and persistent biofouling in industrial settings. Although biofilm formation has become an important research topic in the past two decades, there is still a gap between fundamental knowledge obtained from experimental studies focusing on mono- or dual-species biofilms formed under controllable conditions and environmental studies based on multi-species biofilms in natural environments. This gap has severely hindered the mechanistic understanding of biofilm formation and the role of biofilms in human diseases, as well as the development of surfaces for biofilm prevention and removal. To control bacterial biofilms for public health, there is an urgent need to bridge the gap between experimental and environmental studies on biofilm formation and develop 'smart' antifouling surfaces.

Motivated by the significance of biofilms, I set my long-term career goal to control biofilm formation through interdisciplinary research and teaching. To prepare for my career, I pursued my Ph.D. work with a focus on biofilm control using chemically modified surfaces to direct bacterial adhesion. This well-defined surface system was used to study biofilm formation and the development of biofilm-related high antibiotic resistance. The obtained results together with the effects of surface topography on bacterial adhesion were used to engineer 'smart' antifouling surfaces via topographic surface modification. Engineering surface topography is a promising approach for biofilm control because it does not involve biocidal agents, and thus avoids the problem of antibiotic resistance. With my long-term goal to develop an academic career in the field of bacterial control, I extended my postdoctoral research to study biofilm formation and biofilm-associated high drug tolerance in built environments. People on average spend 90% of their time indoors. Surfaces in indoor environments such as air condition systems are all susceptible to colonization by microorganisms growing in biofilms, thus, it is critical to understand the physiology and virulence of biofilms formed by bacterial pathogens in the built environment, especially, in the hospital environment. Therefore, my postdoctoral research focuses on understanding the effects of environmental factors on bacterial biofilm formation and antibiotic resistance, so that the built environment can be better engineered for antifouling purposes. A well-defined surface system for studying the role of synergistic interaction between multispecies bacteria in biofilm formation and biofilm antibiotic resistance was also developed. The obtained results can help fill the gap between experimental and environmental studies on biofilm formation and design antifouling surfaces with enhanced and prolonged antifouling effects for environmental and medical applications.

In my faculty position, I plan to use my knowledge and skills of microbiology, engineering, material science, and chemistry to understand biofilm formation in built environments and the role of biofilms in environmental acquired chronic infections, especially those related to nosocomial infections. My doctoral and postdoctoral training has prepared me for building a successful independent research program that will make major contributions to the understanding and control of biofilms. Specifically, I will identify the microbial composition in biofilms formed in built environments and humans, as well as the transportation of microbes from environments to humans through human activities by following the 'footprints' of microbes. Meanwhile, the interaction between microbes from different species in the survival of surface attached communities under harsh environments is also a significant and interesting question to study. I also want to understand how biofilm pathogen and host interactions affect biofilm formation and biofilm antibiotic resistance in humans.

The results of environmental, ***in vitro***, or ***in vivo*** studies can allow my research group to have a more comprehensive understanding of the role of biofilms in the transportation of bacterial pathogens from built environments to humans and inspire novel and ecofriendly strategies for biofilm control to effectively interrupt this process. The results of these studies can inform future epidemiological work on cystic fibrosis and microbial exposures in vulnerable populations such as patients with compromised immunity and bedridden elderly, as well as inform local or domestic health departments on the microbial composition in biofilms formed in built environments. Additionally, given the interdisciplinary nature of biofilm research and strong medical relevance, there are exciting opportunities to build collaboration with other fields, such as biostatistics, microbiology, architecture science/engineering, ecology and evolution, and immunology to pursue large multidisciplinary grants. Results from these studies also have the potential to inform industry stakeholders in the development of innovative antifouling products. Beyond research studies, the biofilm related problems are also suitable for integrating into engineering courses and for outreach with local K-12 students and teachers to promote STEM education and talent development.

Teaching Interests: With training in Chemical Engineering, I am comfortable to teach all core courses in this discipline and I am interested in developing new courses related to nano- or micro- techniques for cell manipulation, microbes and human life, and techniques for exploring cell microenvironments.

475397 Structure-Guided Protein Engineering for Targeted Immunotherapy

Abstract:

Structure-Guided Protein Engineering for Targeted Immunotherapy

Jamie Berta Spangler

Postdoctoral Fellow, Stanford University School of Medicine

Research Interests:

The repertoire of naturally occurring proteins is finite and many molecules induce multiple confounding effects, limiting their efficacy as therapeutics. Recently, there has been a growing interest in manipulating existing proteins or engineering new proteins to address the deficiencies of molecules found in nature. Researchers have traditionally taken an unbiased approach to protein engineering, but as our knowledge of protein structure-function relationships advances, we have the exciting opportunity to apply molecular principles to guide engineering. Leveraging cutting-edge tools and unique expertise from structural biology and molecular engineering, I aim to implement a structure-based engineering approach to elucidate the determinants of protein function in order to inform the development of novel protein therapeutics. I am particularly interested in engineering immunological proteins such as antibodies and cytokines to bias immune activity for clinically relevant applications including cancer, infectious diseases, and autoimmune disorders. Harnessing molecular insights and new engineering platforms I have developed in my postdoctoral research, I am poised to pursue a wealth of translational directions that will expand the scope of protein activity to address biomedical challenges.

Successful Proposals:

Leukemia & Lymphoma Society Career Development Postdoctoral Fellowship, Repligen Koch Institute for Integrative Cancer Research Graduate Fellowship, National Defense Science and Engineering Graduate Fellowship

Postdoctoral Project:

"Engineering the interleukin-2 cytokine system to therapeutically modulate immune homeostasis."

Supervised by Prof. K. Christopher Garcia, Departments of Molecular & Cellular Physiology and Structural Biology, Stanford University School of Medicine

Ph.D. Dissertation:

"Characterization and informed design of down-regulating epidermal growth factor receptor antibodies."

Supervised by Prof. K. Dane Wittrup, Department of Biological Engineering, Massachusetts Institute of Technology

Research Experience:

I have cultivated a longstanding interest in understanding and manipulating the molecular determinants of protein function in health and disease and have consequently pursued training in a combination of fundamental biophysical and engineering approaches to biochemistry. As an undergraduate in Biomedical Engineering at Johns Hopkins University, I conducted research in Kalina Hristova's lab, elucidating the mechanistic basis for pathogenicity of mutations in fibroblast growth factor receptor-3 (FGFR3). This work equipped me with essential biochemical and biophysical tools and piqued my interest in studying protein interactions relevant to disease. As a Ph.D. student in Biological Engineering at Massachusetts Institute of Technology, I worked under the supervision of Prof. K. Dane Wittrup. In my thesis research, I used epidermal growth factor receptor (EGFR) as a model antigen to establish antibody-mediated receptor down-regulation as an effective therapeutic strategy and designed multispecific antibody fusions that utilized this strategy to inhibit tumor growth in mice. Through this work I developed a vast knowledge of cell biology and biochemistry and gained experience with antibody engineering and in vivo models of cancer. As a postdoctoral researcher in Professor K. Christopher Garcia's group at Stanford, I have adopted a structure-based approach to biochemistry, integrating new techniques in structural biology, cytokine engineering, and molecular and cellular immunology. My work has focused on interleukin-2 (IL- 2), a multifunctional cytokine that regulates immune homeostasis and that has been used therapeutically to treat diseases such as cancer and autoimmune disorders. The efficacy of IL-2 treatment has been limited by its concurrent promotion of both effector and regulatory immune cells, and I have implemented a multi-pronged approach toward engineering the IL-2 system to modulate the balance of its immunostimulatory versus immunosuppressive effects to achieve targeted therapy. As an independent investigator, I plan to combine my experience in both basic and applied biomedical sciences to build a translational research program at the interface of structural biology and molecular engineering.

Teaching Interests:

Throughout my academic research career, I have been fortunate to gain experience in teaching as well. As a teaching assistant for an introductory Biomedical Engineering course at Johns Hopkins University, I helped lead laboratory sessions and advised students on independent design projects, providing me with the opportunity to interactively train students in applying engineering and design approaches to solve medically relevant problems. As a Ph.D. student, I served as a teaching assistant for a graduate-level course in biomolecular kinetics and cellular dynamics at Massachusetts Institute of Technology. The course focused heavily on computer programming in MATLAB and I instructed several tutorials on this software and also led both lecture- and discussion-based recitation sessions.

In addition to classroom teaching, a formative experience in my pedagogical training has been the opportunity to mentor students. During my graduate career, I mentored six undergraduate students, many of whom have gone on to pursue biomedical careers: three are in top bioengineering Ph.D. programs and two are in medical school. I have also supervised three students and a technician during my postdoctoral tenure, and these opportunities to inspire the next generation of scientists have reinforced my desire to pursue a faculty career.

Future Directions:

The research program I will pursue in my own laboratory represents a culmination of the skills I have amassed through previous work, synthesizing engineering and design experience with structural and biophysical tools to create selective and efficacious molecular therapeutics for a broad range of immunological applications. The most prominent challenge to therapeutic development of cytokines is their pleiotropic nature, as stimulation of signaling on multiple immune cell subsets leads to dose-limiting toxicity and harmful off-target effects. By harnessing structural insights to isolate and selectively engineer the cellular activation properties of cytokines, I will construct a blueprint for biasing protein activity that can be tailored to a variety of disease objectives. In addition to affording unprecedented control over protein behavior, my research will deepen our understanding of the molecular mechanisms that underlie cellular responses and inform the design of candidate drugs for ready clinical translation. While initial work will focus on engineering antibodies to tune cytokine activity, my long term vision is to apply the modular engineering and therapeutic platforms I develop to other systems to allow for the direct manipulation of a diverse array of ligand-receptor interactions including hormone, growth factor, and membrane protein systems.

I recently devised structure-guided evolutionary strategies using yeast surface display methodology to establish two novel antibody-based platforms that modulate cytokine behavior: [1] Cytokine-targeted antibodies that allosterically disrupt cytokine-receptor interactions to bias activity; and [2] Agonistic antibody fragments that use a single binding site to bridge two receptor subunits and activate signaling in the absence of cytokine. Both platforms can be exploited to direct cytokine activity toward specific cell subsets based on surface receptor expression profiles, and their modularity allows for targeting of any protein interaction or cell population of interest to elicit either an immunostimulatory response (for treatment of cancer or chronic infection) or an immunosuppressive response (for engraftment or treatment of autoimmune disorders). In addition to building on platforms I have already established, I plan to develop new engineering platforms to isolate antibodies that act through alternative mechanisms such as modulation of internalization or molecular trafficking properties. The unique mechanisms recruited in these platforms will complement other activities inherent to the antibody scaffold such as competitive inhibition and Fc-mediated effector functions. Moreover, the stable and readily engineerable antibody construction will allow for development of bispecific molecules that synergistically combine antibodies with distinct modes of action, and structural feedback on molecules we develop will further inform design. My experience with mouse models of cancer and autoimmune disease and continued collaborations with leading T cell immunologists established during my postdoctoral tenure will facilitate pre-clinical evaluation and ultimate translation of engineered therapeutics. Through integration of expertise in molecular engineering and structural biology, I am uniquely positioned to make rapid progress in developing new approaches toward targeted therapy for immunomodulation and a host of other biomedical applications.

Selected Publications:

Spangler JB, Tomala J, Luca VC, Jude KM, Dong S, Ring AM, Votavova P, Pepper M, Kovar M, Garcia KC. Antibodies to interleukin-2 elicit selective T cell subset potentiation through distinct conformational mechanisms. ***Immunity.*** 2015;42(5):815-25. PMCID: 4439582.

Mitra S, Ring AM, Amarnath S, **Spangler JB**, Li P, Ju W, Fischer S, Oh J, Spolski R, Weiskopf K, Kohrt H, Foley JE, Rajagopalan S, Long EO, Fowler DH, Waldmann TA, Garcia KC, Leonard WJ. Interleukin-2 activity can be fine tuned with engineered receptor signaling clamps. ***Immunity.*** 2015;42(5):826-38. PMCID: 4560365.

Spangler JB, Moraga I, Mendoza JL, Garcia KC. Insights into cytokine-receptor interactions from cytokine engineering. ***Annu Rev Immunol.*** 2015;33:139-67. PMCID: 4445396.

Moraga I, **Spangler J**, Mendoza JL, Garcia KC. Multifarious determinants of cytokine receptor signaling specificity. ***Adv Immunol.*** 2014;121:1-39. PMCID: 4449261.

Spangler JB, Manzari MT, Rosalia EK, Chen TF, Wittrup KD. Triepitopic antibody fusions inhibit cetuximab-resistant BRAF and KRAS mutant tumors via EGFR signal repression. ***J Mol Biol.*** 2012;422(4):532-44. PMCID: 4041985.

Spangler JB, Neil JR, Abramovitch S, Yarden Y, White FM, Lauffenburger DA, Wittrup KD. Combination antibody treatment down-regulates epidermal growth factor receptor by inhibiting endosomal recycling. ***Proc Natl Acad Sci U S A.*** 2010;107(30):13252-7. PMCID: 2922117.

You M*, **Spangler J***, Li E, Han X, Ghosh P, Hristova K. Effect of pathogenic cysteine mutations on FGFR3 transmembrane domain dimerization in detergents and lipid bilayers. ***Biochemistry.*** 2007;46(39):11039-46.

475630 Enhancer-Mediated Regulation of Transcriptional Bursting

Abstract:

Research Interests:

My research interests lie in quantitative analysis of regulation in development, using both experimental and computational approaches. I completed my Ph.D. studies with Professor Stanislav Shvartsman in the department of Chemical and Biological engineering at Princeton University. Using a combination of genetic tools, high-throughput imaging, and mathematical modeling, I reconstructed developmental dynamics of the highly conserved ERK enzymatic activity from snapshots of fixed early *Drosophila* embryos. Reconstruction showed that the dynamic activation profile of ERK regulates the spatial and temporal extent of the target gene expression. Given that increasing number of studies report the link between changes in ERK kinetics and developmental abnormalities, my project contributed to establish a platform where quantitative regulation of enzyme activity can be studied in vivo.

As a postdoctoral researcher with Professor Michael Levine in the Lewis-Sigler Institute for Integrative Genomics at Princeton University, I applied quantitative image analysis and modeling techniques to explore another critical aspect in development: transcriptional regulation. Transcription is a critical yet a complex process that determines when and where a gene needs to be turned on and off to ensure normal development. While the importance of transcriptional regulation is well acknowledged, complex nature of the process makes it challenging to understand the comprehensive underlying mechanisms of transcriptional regulation.

During the past one year as a postdoctoral researcher, I showed that transcription occurs in a series of sequential stochastic bursts, and these bursts can be readily visualized in living fly embryos. Through quantitative image analysis,

I provided evidence that enhancers regulate transcription by modulating the frequency of these bursts, implying that the regulation of bursting frequency is a key parameter of gene control in development. I also showed that a shared enhancer can co-activate two linked reporter genes, arguing against the conventional enhancer-promoter looping models, which would express sequential activation of two reporter genes. The work has contributed to provide new insights on the enhancer-mediated gene regulation in quantitative manner.

Based on these results and my experience in applying quantitative techniques to biological systems, I will further explore the underlying mechanism of transcriptional bursting and the role of enhancer-promoter communication in transcriptional regulation. I will take advantage of the broad range of molecular cloning, optogenetics, and quantitative tools available in the *Drosophila* system. My initial projects will focus on (1) extracting the properties that comprise minimal unit of transcriptional burst, (2) probing the correlation between the cluster size of RNA polymerase II and the transcriptional burst size, and (3) testing the role of chromosomal loop domains in transcriptional regulation. Given that more and more mutations in the enhancer regions in the genome are found to cause developmental abnormalities, successful implementation of the proposed experiments will shed light on the nature of transcriptional bursts, the relationship between enhancers and bursts, and the effect of enhancer-mediated transcriptional regulation in development.

Teaching Interests:

I have served as a TA once for the undergraduate Chemical Reaction Engineering course and twice for the graduate level Chemical Reactor Engineering course, both of which are core classes for undergraduate and graduate students in Chemical and Biological engineering at Princeton University, respectively. I received the Air Products Assistant in Instruction Award in 2012, from the graduate Chemical Reactor Engineering class. I have mentored three undergraduate students (2 engineering and 1 molecular biology), and two rotation graduate students in molecular biology.

Selected Publications:

1. Fukaya T*, **Lim B***, and Levine M. 2016. Enhancer control of transcriptional bursting. *Cell.* 2016. dx.doi.org/10.1016/j.cell.2016.05.025 [epub ahead of print]

2. **Lim B**, Dsilva CJ, Levario TJ, Lu H, Schupbach T, Kevrekidis IG, Shvartsman SY. Dynamics of inductive ERK signaling in the *Drosophila* embryo. *Curr Biol.* 2015; 25: 1784-1790.

3. Dsilva CJ*, **Lim B***, Lu H, Singer A, Kevrekidis IG, Shvartsman SY. Temporal ordering and registration of images in studies of developmental dynamics. *Development.* 2015; 142: 1717-1724.

4. **Lim B**, Samper N, Lu H, Rushlow C, Jimenez G, Shvartsman SY. Kinetics of gene derepression by ERK signaling. *Proc Natl Acad Sci U S A.* 2013; 110: 10330-10335.

* Equal contribution

475682 Understanding and Controlling Protein Stability from Coarse-Grained Protein Models

Abstract:

Understanding and Controlling Protein Stability from Coarse-Grained Protein Models

Marco A. Blanco

National Institute of Standards and Technology, Gaithersburg, MD 20899

Institute for Bioscience and Biotechnology Research, University of Maryland, Rockville, MD 20850

Research Interests:

Protein-based therapeutics represents one of the fastest growing markets in pharmaceutical industry. The high efficacy and specificity of biotherapeutics make them valuable for targeting different type of illnesses. Nonetheless, most, if not all, proteins present a very wide range of behaviors that limit the development of commercially viable products. Small changes in the protein sequence and formulation conditions may cause conformational and physical instabilities that lead to the self-assembly of proteins (e.g., fluid phase separation, oligomerization, and irreversible aggregation). These stability concerns constitute the primary route for the formation of impurities in biotherapeutics, as well as may induce unfavorable immunologic responses in patients. The ability to accurately and unambiguously predict the behavior of proteins in solution will help control instability problems via an adequate design of proteins and selection of formulation conditions.

My research work has focused on understanding the stability of proteins in solution via simple but realistic coarse-grained computational models, which are combined with biomolecular and biophysical experiments (e.g., mutagenesis, chromatography, calorimetry, scattering) to provide both refinement and validation. By implementing advanced Monte Carlo and Molecular Dynamics techniques, these models allow one to evaluate the thermodynamic pathways by which proteins are destabilized, which might be otherwise challenging to probe due to the generally short-living nature of most of the key species in those processes. Thus, the resulting models provide powerful, physics-based tools to assess how different factors (e.g., pH, excipients, protein structure) affect those phenomena that cause instability in proteins such as phase separation, protein unfolding, and nonnative aggregation; moreover, they offer guidelines for rationally designing/selecting proteins with higher physical stability.

The Blanco laboratory will built on the use of both computational and experimental methods applied to topics within the general theme of protein stability, engineering, and self-assembly. A specific emphasis will be placed on developing new and improved theoretical models and experimental techniques to predict the conformational and solution stability of proteins and peptides, and thus to gain insights regarding how the local structure and surrounding molecular environment influence the underlying mechanisms associated to physical degradation of proteins. Such predicting methods will have direct application in fields such as biomedicine (e.g., elucidating critical stages during amyloidosis), drug discovery (e.g., rational design of high-affinity peptide-based drugs), drug development (e.g., identifying optimal formulation and manufacturing conditions for biopharmaceutics), and biomaterials (e.g., developing "smart" protein-based materials with controllable self-assembly properties). Since research in these fields encompasses various disciplines, there will be strong collaborations from both academy and pharmaceutical industry.

Teaching Interests:

In addition to my research career, I have also gained significant teaching experience. I lectured an undergraduate course at the Industrial University of Santander in Colombia about the implementation of statistical and quantum mechanical computational methods to thermodynamics and transport phenomena problems related to Chemical Engineering. I also TAed thermodynamics courses at both undergraduate and graduate level, as well as mentored undergraduate students and visiting scholars about application of molecular simulations to protein systems.

Given my background in protein science and chemical engineering, my teaching interests extend in undergraduate core chemical engineering courses that include thermodynamics, heat and mass transfer, chemical kinetics, and applied math. Furthermore, I would also like to develop and teach undergraduate- and graduate-level courses in topics such as protein engineering, statistical thermodynamics applied to chemical engineering, and computational and experimental methods for protein characterization.

Research Projects:

PhD Projects:

- "Measuring Protein-Protein interaction via Scattering Experiments from Low to High Protein Concentrations"

- "Evaluating the Effect of Solution Conditions and Protein Sequence on the Cluster Formation of γD-Crystallin"

- "A Novel Algorithm for Predicting Amyloidosis and Protein β-Aggregation"

- "Solution Behavior and Aggregation Propensity of Peptide-Polyacrylic Acid Conjugates"

Supervised by Christopher J. Roberts, Department of Chemical and Biomolecular Engineering, University of Delaware.

Postdoctoral Projects:

- "Effect of Specific Interactions on the Fluid Phase Behavior and Self-Assembly of Proteins and Nanocolloids"

- "Characterization of the Solution Structure and Phase Behavior of Monoclonal Antibodies via Coarse-Grained Models"

- "Effect of Electrostatic Interactions on the Oligomerization and Amyloid Formation of the Cardiomyopathy-Related Protein Transthyretin"

Supervised by Vincent K. Shen and Zvi Kelman, National Institute of Standards and Technology and Institute for Bioscience and Biotechnology Research, University of Maryland.

Selected Publications:

1. Marco A. Blanco, Erinc Sahin, Anne S. Robinson, and Christopher J. Roberts. "Coarse-grained model for colloidal protein interactions, B_{22}, and protein cluster formation". J. Phys. Chem. B, 117, 16013 (2013).

2. Marco A. Blanco, Tatiana Perevozchikova, Vincenzo Martorana, Mauro Manno, and Christopher J. Roberts. "Protein-protein interactions in dilute to concentrated solutions: a-Chymotrypsinogen in acidic conditions". J. Phys. Chem. B, 118, 5817 (2014).

3. Bradford Paik, Marco A. Blanco, Xinqiao Jia, Christopher J. Roberts, and Kristi Kiick. "Aggregation of poly(acrylic acid)-containing elastin-mimetic copolymers". Soft Matter, 11, 1839 (2015).

4. Marco A. Blanco and Vincent K. Shen. "Effect of the surface charge distribution on the fluid phase behavior of charged colloids and proteins". J. Chem. Phys. [submitted].

475765 Microbial Biosynthesis of Bioorthogonal Functionalities and Applications

Abstract:

Research Interests:

Natural products biosynthesized by microbes offer unique functionalities that can be used as chemical handles in bioorthogonal chemistry. Bioorthogonal chemistry have recently emerged as one of the most powerful tools in drug discovery and chemical biology. Despite the importance of the unique chemical functionalities involved in biorthogonal chemistry, the biosynthetic routes to bioorthogonal functionalities are not well understood. In my graduate career, we identified and characterized two biosynthetic pathways in microbes that lead to the generation of a bioorthogonal functionality-terminal alkyne through unprecedented pathways. These pathways were further exploited for tagging various natural products with terminal alkyne functionalities through biochemical pathway engineering. The natural product tagging strategy also enabled the development of an ***in situ*** natural products quantification platform based on bioorthogonal chemistry with the newly installed alkyne tag. Using this quantification platform as a screening method, we applied directed evolution strategy to improve the activity of a membrane-bound bifunctional desaturase/acetylenase. This study not only provides alkyne biosynthetic tools that are useful for chemical biology, but it also has applications in which enzymatic generation of terminal alkynes and ***in situ***bioorthorgonal chemical transformation are required or preferred.

I would like to continue studying microbial biosynthesis of structurally and functionally interesting chemical functionalities and find applications in chemical biology and drug development.

Teaching Interests:

Aside from my research, I also have teaching experience as a teaching assistant during my graduate life. I led discussion sections, prepared teaching materials and held office hours for a core course "Kinetics" for undergraduate majored in chemical engineering. I also spent two semesters organizing and preparing teaching materials for a seminar-based course "Berkeley lectures on energy: energy from biomass". I also mentored two graduate students, and two undergraduate students in my group.

Based on my own academic background, I would love to teach general chemical engineering (e.g. Kinetics, Thermodynamics) for undergraduate students. I would also like to develop a course "Biomolecular Engineering", which is a combination of lectures and lab sessions. The lectures would focus on the use of chemical engineering principles in the design of biologically-based processes (e.g. microbial fermentation, bioreactors, etc), and also cover topics in biotechnology such as enzyme kinetics, recombinant DNA technology, protein expression, protein engineering, synthetic biology, etc. Students would also have chances to practice basic biotechnical skills in lab sessions, such as molecular cloning, protein purification, enzyme kinetic assays, bioreactor-based fermentation, etc.

473004 Modulating Antigen-Specific T Cell Immunity with Biomaterials-Based Vaccine

Abstract:

Research Interests: Over the past several decades, biomaterials have been widely engineered for treating or diagnosing diseases such as cancers. My research experience and interests have been focused on using micro/nanofabrication of biomaterials for cancer therapies and diagnosis. In this poster session, I will highlight the major research directions I have been involved in during my graduate and postdoctoral research. The results of these efforts demonstrate my capabilities in using biomaterials for micro/nanofabrication, drug delivery, bio-sensing, and more recently, for vaccine design. Thus far my efforts have led to 21 papers (6 as first author) Building on this training, I will describe my plans to develop a distinct, independent research program at the interface of biomaterial, surface and immunology. Briefly, I plan to build a research lab studying immune detection and biomaterial vaccine for tissue inflammation and fibrosis.

I completed my PhD in Biomedical Engineering at Florida State University in 2013 under the supervision of Dr. Jingjiao Guan. My research was focused on developing non-conventional micro particles with well-defined shapes, diameters and components.[1, 2] These particles have been employed in biomedical applications such as drug delivery, cell tracking, and radiation therapy for cancer.[3-7] Upon completing my PhD, I worked as a postdoctoral researcher at Worcester Polytechnic Institute under the supervision of Dr. Ming Su, studying enhanced radiation therapy of cancer. Briefly, my research focused on modifying gold nanoparticle surface to enhance their uptake by cancer cells. Upon X radiation, X-ray photons in radiation therapy can generate photoelectrons and Auger electrons, which can cause ionization of water and formation of reactive free radicals. These free radicals diffuse through chain reactions in cells, and damage DNA in mitochondria and nuclei by extracting hydrogen atoms from ribose sugars, kill cancer cells by cleaving of the polynucleotide backbone in cancer cells. During this period of time, I gained valuable training in surface modification of biomaterials for radiation therapy cancer.[3, 5, 8, 9] In August 2014, I joined the University of Maryland College Park under the supervision of Dr. Christopher Jewell. My research in the Jewell lab has focused on exploiting biomaterials to generate specific, tunable immune responses.[10, 11] My research in this area has two directions: 1) investigating and understanding how biomaterials impact the structure of major secondary immune organs such as lymph nodes and spleens, and 2) developing biomaterial vaccines for cancer treatment.

Through training in three different labs, I have gained valuable training and experience in grant writing, scientific management and teaching. These experiences will allows me to build and run my future own labs.

References

[1] Wang ZB, **Zhang P**, Kirkland B, Liu YR, Guan JJ. Microcontact printing of polyelectrolytes on PEG using an unmodified PDMS stamp for micropatterning nanoparticles, DNA, proteins and cells. Soft Matter 2012;8:7630-7.

[2] **Zhang P**, Guan J. Fabrication of multilayered microparticles by integrating layer-by-layer assembly and microcontact printing. Small 2011;7:2998-3004.

[3] Wang CM, Sun A, Qiao Y, **Zhang P**, Ma LY, Su M. Cationic surface modification of gold nanoparticles for enhanced cellular uptake and X-ray radiation therapy. J Mater Chem B 2015;3:7372-6.

[4] **Zhang P**, Xia JF, Wang ZB, Guan JJ. Gold nanoparticle-packed microdisks for multiplex Raman labelling of cells. Nanoscale 2014;6:8762-8.

[5] **Zhang P**, Qiao Y, Xia J, Guan J, Ma L, Su M. Enhanced radiation therapy with multilayer microdisks containing radiosensitizing gold nanoparticles. ACS Appl Mater Interfaces 2015;7:4518-24.

[6] Kirkland B, Wang Z, **Zhang P**, Takebayashi S, Lenhert S, Gilbert DM, et al. Low-cost fabrication of centimetre-scale periodic arrays of single plasmid DNA molecules. Lab Chip 2013;13:3367-72.

[7] Liu Y, Kirkland B, Shirley J, Wang Z, **Zhang P**, Stembridge J, et al. Development of a single-cell array for large-scale DNA fluorescence in situ hybridization. Lab Chip 2013;13:1316-24.

[8] **Zhang P**, Qiao Y, Wang CM, Ma LY, Su M. Enhanced radiation therapy with internalized polyelectrolyte modified nanoparticles. Nanoscale 2014;6:10095-9.

[9] Qiao Y, **Zhang P**, Wang CM, Ma LY, Su M. Reducing X-Ray Induced Oxidative Damages in Fibroblasts with Graphene Oxide. Nanomaterials-Basel 2014;4:522-34.

[10] **Zhang P**, Chiu YC, Tostanoski LH, Jewell CM. Polyelectrolyte Multilayers Assembled Entirely from Immune Signals on Gold Nanoparticle Templates Promote Antigen-Specific T Cell Response. Acs Nano 2015;9:6465-77.

[11] Chakrabarti KR, Andorko JI, Whipple RA, **Zhang P**, Sooklal EL, Martin SS, et al. Lipid tethering of breast tumor cells enables real-time imaging of free-floating cell dynamics and drug response. Oncotarget 2016;7:10486-97. (*cover article*)

***Featured on cover of *Genetic Engineering and Biotechnology* News

Teaching Interests:

STATEMENT OF TEACHING PHILOSOPHY

With all of the roles that professors play, I consider teaching and mentoring be two of the most important ones. Through these two roles, professors deliver multiple levels of information to students, ranging from professional skills to problem-solving techniques, such as math and logic that can be employed from class room to daily life. With my past experiences as both a student and mentor, I have established skills and a philosophy (see below) that would allow me to train students at different levels (e.g. undergraduate and graduate). Based on these experiences and skills, I will expand my teaching and training activities to my research group, the department and the surrounding community. My teaching plan includes using a variety of teaching tools to deliver scientific knowledge to the students, which will help train another generation of outstanding scientists and engineers.

PAST EXPERIENCE

Teach students in accordance with their aptitude. During my Ph.D. study, I served as a teaching assistant (TA) from 2008 to 2011 in courses including *Transport Phenomena I (ECH 3204)*, *Mass and Energy Balances I (ECH 3203)*, *Separation Process (ECH3418)*, *Quantitative Anatomy and Systems Physiology (BME4403C)* and *Lab TA in Quantitative Anatomy and Systems Physiology (BME4403C)*. Besides routine and impersonal teaching jobs such as grading, other works including office hours and teaching in the lab allowed me to interact with students directly. During these interactions, my responsibility involved helping students find out what was wrong with their homework or experiment and guiding them to understand how and why their thought processes led to the incorrect answer. These works required me to understand how the students used the knowledge to finish their work. What I found interesting was that, while different students made similar mistakes in their assignments, the reasons and logic behind these mistakes were often completely different. For example, some students might do their homework or experiment carelessly and get the wrong answer, while others could make a similar mistake due to insufficient understanding of the knowledge. Through these interactions with students, I established my teaching philosophy to "teach students in accordance with their aptitude". This philosophy was further enforced in my other teaching and mentoring experiences.

Establish students with a vision and tangible future plan. I worked as a lecturer for students in a middle school in a rural area of China during summer 2006 and 2007. The goal of this work was to offer children from low income families a chance to improve their studies in Math, Chemistry, and English. What surprised me during these teachings activities was that some kids did not study hard simply because they lacked a vision or a visible future plan that was connected to their studying. For example, one student mentioned to me that he did not need to learn the knowledge since he was born to be labor worker, and that this idea was rooted in his mind by his parents. While this student's studying attitude changed when he learned how the knowledge could change his future, I questioned how long this change could last. The lesson I learned from this experience was that teaching efficiency can be improved if knowledge is delivered to students alongside a concrete vision so that they can link what they are learning with a tangible future plan.

Visible benefits are important, but so is inherent happiness during learning. I mentored students at different levels since 2008. These mentees included undergraduates from Florida State University, the University of Maryland College Park, and other universities such as the California Institute of Technology. The mentoring process included training students on micro/nanofabrication for different bio-applications (e.g. drug delivery, bio-sensing). In Maryland, I mentored a high school student from Wheaton High School in a program called "Enhance Participation in Research (EPR, supported by National Science Foundation)". In this one year program, the mentee selected a biomedical topic that she was interested in; in turn, the mentor provided feedback throughout the process, from collecting literature to finalizing the topic into a poster and report at the end of the year. In addition to mentoring students in experimental design and data analysis, I also directed an undergraduate in the University of Maryland College Park in writing proposals based on her research, which was funded with an ASPIRE fellowship from the university ($3000). In these mentoring processes, I found that the motivations for students in research mainly came from three sources: future career development, happiness in research, or both. For students driven by future career development (e.g. attending medical school), a project that can develop into a scientific paper is a very effective strategy; as for students driven by the inherent happiness of performing research, a topic that intrigues him/her is very important. For example, my Wheaton High School student was very interested in why her dog was slow and reluctant to move when it got old. While there was not enough literature reported, she learned knowledge by comparing the disease in her dog with Alzheimer's disease in humans. Through this learning process, the student developed a motivation to become a biomedical engineer in college. While both mentees learned a great deal of intellectual information on their respective subjects, they informed me that they gained a joy though the research.

TEACHING PHILOSOPHY

Teachings, in my point of view, can deliver two fundamental techniques to the students: 1) professional skills in one specific area, and 2) robust problem-solving techniques. For professional skills in chemical and biomedical engineering, the students can be trained with highly refined techniques such as generating micro-devices through photolithography. By obtaining these techniques, the students can start their careers in international companies such as Intel, Eli Lilly and Company, Novartis, and Merck & Co. In addition, while expertise in one specific area can help students to build their professional life, the problem-solving techniques can arm them with common sense skills that can be applied to almost all areas of life. Some of these problem-solving techniques include logic, math, and the fundamental laws of physics and biology. Based on my teaching and mentoring experiences, I would like to help students obtain these techniques through 3 specific teaching tools, as mentioned above: **1) teach students in accordance with their aptitude; 2) establish students with a vision and a tangible future plan; and 3) help students find happiness in learning.**

TEACHING AND MENTORING PLAN

My plan for training young scientists and engineers in my classes and research group will focus on three aspects: 1) Implement problem solving contents; 2) teach and help students according to their career goals and aptitudes; and 3) improve my own communication and teaching skills during interactions with students.

Teach in current and future courses. My (previous) training is in chemical and biomedical engineering. I was mostly interested in teaching transport phenomena, mass and balance transport, and physiology. However, I am enthusiastic about any courses that will allow me to interact with students. My teaching will focus on two aspects: 1) the fundamental sciences inherent to the courses, and 2) problem-solving techniques associated with these courses. In addition to conventional teaching in class, the student will be asked to design questions based on the courses for peers, structure a project toward specific problems associated with the course, and compare the potential solutions to the given problems. I will also try to develop new courses relevant to my research. For example, one course will be "Micro/Nano-Vaccines". This course will cover the necessary skills and techniques in nanotechnology and immunology and how these two areas interact for novel vaccine development. In this course, students will be introduced to conventional (i.e. emulsion) and non-conventional (i.e. top-down techniques, such as photolithography) methods for generating micro/nanoparticles; students will also learn about immunology and how the micro/nanoparticles modulate the immune system *in vivo*. At the end of this course, the student will give a presentation on one of the existing vaccines and propose how they will improve that vaccine with micro/nanotechnology.

Advise mentees' research and professional development. Advising my own research group is one of the most exciting responsibilities for me. I am eager to work with students who interact and learn in different ways. My advising activities will include at least two parts: 1) a regular portion to discuss each student's research project, and 2) a portion that discuss flexible topics that depends on the student's stage of development. The regular meeting will start with a discussion of research difficulties. Students will not be provided a direct answer; instead, their problems will be answered with other questions. For example, if the student cannot synthesize a polymer particles via emulsion, he/she will be challenged by a series of questions such as: What are the components used in the synthesis? What are the functions of each component? How do these components interact? What is the formulation in other literature? By asking these kinds of questions, the students will be expected to think on their own. The flexible part of the meeting will depend on the student's need. For example, the senior students may want discussions on how to apply for national meetings, fellowships, or even learn to write grants. The student's professional plan is also an important part of my advising, in which he/she will create a mutually agreed upon plan including action steps that are to be updated every 6 months.

Improve my own professional skills and technical expertise. Teaching and interacting with students will offer valuable opportunities to improve my own skills as a teacher and expand my technical expertise. For example, the "Micro/Nano-Vaccine" course will expose me to immunology that I have only partially studied in the past, and I will thus use this teaching experience to enrich my knowledge in this area. The experience of guiding students in lab will also allow me to learn how to mentor them effectively. Therefore, through these process, I will become even more effective and responsive to individuals with different backgrounds, research styles, and goals.

475794 Increasing the Scale and Rate of Metabolic Engineering through Systems Synthetic Biology

Abstract:

Research Interests:

Biological systems are fantastic platforms for solving myriad world issues, as they contain a universe of highly specific catalysts, self-replicate, undergo evolution to improve function, and integrate information about their surroundings to adjust their behavior. As we strive for ever-increasing yields of green fuels and chemicals, and as we attempt to move our designs from the laboratory to the real world (e.g. the microbiome), critical gaps are arising between the theoretical behavior of engineered organisms and the ways in which they behave in reality. To fill these gaps, my graduate and postdoctoral work have benchmarked an approach to metabolic engineering which fuses emerging systems-level understanding of processes occurring at the whole-cell/community level with high-throughput, iterative design. In this poster, I will highlight several of my recent efforts which demonstrate the promise of a systems synthetic biology approach to more quickly merge engineering goals with biological reality.

My graduate research focused on expediting the development cycle of engineered yeast, which is a workhorse organism for the production of diverse chemicals, yet can be time-consuming and unpredictable to engineer. In particular, tuning the rate of enzyme synthesis, improving enzyme catalytic properties, and streamlining the genetic background of the yeast host are each critical to optimizing a bioprocess, yet poor quantitative models and slow genetic tools often made these design cycles "shots in the dark". To improve enzyme synthesis, I quantified the rate-limiting potential of RNA secondary structure and nucleosome occupancy and in response developed computational techniques which exploit these phenomena to enable metabolic engineers to achieve desired enzyme levels in a single design cycle. In addition, because the design rules governing biocatalyst properties and host performance remain undiscovered, I further developed ***in vivo*** continuous evolution and RNA interference, respectively, for high-throughput optimization of these qualities in yeast, enabling such design rules to be inferred as the optimization proceeds. In each case, my innovative approaches expedited the design cycle by several orders of magnitude over the current state-of-the-art and provide a toolset for my future lab to use when exploring alternate behaviors, use locations, and genetic backgrounds of engineered microorganisms.

As one of my goals while an independent researcher, I aim to broaden the scope of metabolic engineering to include the microbial communities that live in the human gut, as this complex ecosystem plays critical roles in the extraction of nutrients, production of vitamins, and the progression of disease, yet whose engineering is currently regarded as a "grand challenge" in human health. The critical and growing importance of the gut microbiome was recently highlighted by the announcement of the National Microbiome Initiative this year. As this organ is, in its simplest sense, a packed bed reactor, chemical engineers are ideally poised to make significant contributions in this area. In my postdoctoral work, I am laying the foundation for engineering this system through the discovery of factors enabling specified residence time and gene expression levels in engineered probiotics using high-throughput experimental and computational approaches. Looking ahead, the techniques and design principles I have developed will enable my future lab to engineer "smart" probiotics which monitor the gut environment for signatures of disease and subsequently synthesize and deliver therapeutics in close proximity to affected areas. Additionally, integration with approaches developed during my graduate work will enable the construction of highly robust therapeutic strains which adjust their genetic content for improved function in response to changing environmental conditions. Third, these engineered strains will enable the elucidation and control of the three-dimensional structure of gut microbiota, enabling further insight and control over disease progression through modeling the currently unknown spatial reaction networks in the gut. It is clear through these examples that developing solutions for this and related microbial communities, as well as advancing microbial production of fuels and chemicals to the commercial scale, requires the use of high-throughput methods to both develop improved strains and also improve quantitative understanding of increasingly complex engineerable biological systems. My lab will be uniquely positioned at the forefront of the development and implementation of this systems synthetic biology approach to address pressing challenges in health, energy, and the environment.

Teaching Interests:

When I was a freshman in high school, a substitute teacher for geometry once asked the class if a student could introduce the material. To my astonishment, the class unanimously agreed that I should do it. It was an unforgettable experience to explain the surface areas and volumes of pyramids to my classmates at the same time I myself was learning the material – and I was pleasantly surprised to find that I enjoyed it. It was satisfying to see the eyes of my fellow students light up as difficult concepts came within their reach, and were then able to use that concept to solve problems. I reasoned that if it was this much fun to teach material "cold", it would be even more rewarding to teach things I know a lot about - and I was right! During graduate school, I gained an enjoyment for the puzzle of designing effective homework and quiz problems. I especially enjoyed leading office hours where I was able to interact face-to-face with my students and match their individual learning style. I am eager to broaden my impact as a professor by improving the engagement of diverse groups of students with chemical engineering and thereby provide a foundation upon which a variety of careers can be built.

Introductory Courses

My prior experience and interests make me ideally suited to teach mass and energy balances, kinetics, and transport. I utilize several strategies for improving student learning over the "information fire hose" format of the traditional lecture. First, problem-based learning engages students and makes often abstract concepts easier to grasp, as I found when I encouraged students to play with silly putty in preparation for a lab involving non-newtonian fluids. Second, peer-teaching formats such as "Think-Pair-Share" have been shown to be effective at enhancing conceptual understanding over traditional formats. I implement these techniques during lecture or recitation by allowing students solve problems in small groups, and then reconvening to discuss the challenges they encountered. Third, through exit surveys in which students briefly describe what they learned after each lecture, I simultaneously facilitate student reflection (critical to improving future recall) and gauge my teaching effectiveness. Fourth, learning how to study is something that students often get no formal training in, and is especially critical during introductory courses in order successfully transition students from high school to college. I teach metacognition (i.e. learning how to learn) through in-class activities, for example, where students write down the point they are most confused about and then form small groups to address each other's confusion. Finally, just as feedback is critical to the consistent performance of a chemical process, mid-semester course evaluations enable me to quickly adapt to the changing needs of my students by refining my teaching style, class format, and content. Through these efforts I provide a robust framework for my students to tackle any problems they face in their chosen field.

Advanced Courses

My expertise makes me well-suited for the development of a course on biochemical and metabolic engineering, including advanced enzyme kinetics, flux balance analysis, and the kinetics of cell growth. I can also develop a bioinformatics course for metabolic engineers which could be used either as a standalone class or as a module (e.g. in a computational methods or the metabolic engineering course mentioned above). This course would cover the principles and techniques of high-throughput sequencing analysis, RNA and protein structure prediction, and flux balance analysis. In these courses, I will use current literature to inspire discussion of underlying theories and techniques. Then, proposal writing will allow students to creatively combine what they have learned with the state-of-the-art in the field. Evaluation of these proposals will be facilitated through the formation of in-class review sections, allowing students to practice both project conception and critique. Taken together, I envision that these courses will provide guidance in areas which too often are left to be picked up on-the-fly after graduation.

Mentorship

My primary goals as an advisor are to develop in my mentees 1) a critical understanding of their field, 2) a framework to advance their field, and 3) the ability to effectively communicate their findings and goals with others. To do so, I will organize regular journal clubs to develop critical reading skills, teach new concepts, and serve as launch points for brainstorming sessions. I will provide my mentees with a model for effective lab management, including the keeping of electronic records (via a lab wiki) and centralized inventories. A wiki enables the creation of a core knowledge base, promoting efficient lab work and allowing new members to acclimate quickly. I will also strongly encourage mentorship of undergraduate researchers to increase focus on experimental design and develop teaching skills. In addition to authoring papers and presenting at conferences, teaching assistantships train clear communication and in this setting I will encourage my mentees to develop evaluation material and lead several lectures with my guidance. To combine experimental design and communication, my mentees will also be involved in grantwriting, which has the secondary benefit of providing familiarity with the various funding agencies and facilitating efficient acquisition of funding in the future. Throughout this process, roadblocks will naturally occur both in and out of the lab, and I view them as opportunities for discovering something new, scientifically and personally. I will re-evaluate shared goals with each mentee through yearly self-evaluations, and use individual development plans to achieve them. Through these efforts, I aim to guide mentees towards a career which is well-suited to their interests and expertise.

Commitment to Diversity and Inclusion

My research is framed by the principle that the most effective and novel solutions to problems arise by assembling a diverse collection of microbes and assessing each in an unbiased fashion. I embrace this principle in the classroom and lab by valuing diversity in those I teach, facilitating collaborations between individuals of different backgrounds, cultivating a growth mindset in my mentees, and rewarding alternative points of view during group discussions. I further aim to increase participation of underrepresented groups in my research through the creation of a day-long synthetic biology workshop connecting high school teachers in underprivileged areas to the concepts and techniques of engineering living systems, and providing digital materials for teaching these concepts their students. This workshop will be complemented by a program in which promising high school students which would not otherwise have the opportunity to do research are connected with a member of my lab to undertake a project over the summer. I aim to integrate these programs with existing outreach efforts to repair the leaky STEM pipeline and increase the diversity of perspectives available for solving the challenges of the future.

475827 Designing Novel Interfaces to Control Beneficial and Pathogenic Microbes

Abstract:

Research Interests: Bacteria that colonize on medically and environmentally relevant surfaces are subject to a variety of physical forces. For example, in potentially pathological cases, bacterial cells withstand shear stresses along blood vessels and electric discharges near medical implants as they develop infectious, mature multicellular aggregates known as biofilms. And in an environmentally relevant example, oil-degrading bacteria overcome complex interfacial energies to colonize fluid interfaces and establish films of bacteria at (oil-water) interfaces (FBI). My research program exploits fundamental knowledge of the physiological responses of bacteria to physical forces to investigate the development of new interfaces that will both help minimize the deleterious and optimize the beneficial effects of bacterial film formation. The outcomes of this research will inform new biomedical and technical strategies to: (1) eradicate multidrug-resistant (MDR) infections, (2) control microbial dynamics relevant to health and biotechnology, and (3) alleviate the ecological impacts of oil spills.

Research Experience: My research has encompassed interrelated areas ranging from food microbiology, to medical bacteriology and biotechnology. I have investigated how beneficial microbes can contribute to human health, as well as how bacteria are involved in many recalcitrant infections. Over the course of my graduate education, I focused on developing technologies targeting the resistance mechanisms of intrinsically quiescent bacterial cells, called persister cells. I studied how manipulating the electrophysiological properties of ***P. aeruginosa*** persisters can sensitize these cells to antibiotics and metallo-drug complexes previously demonstrated to be ineffective against persister cells. My current postdoctoral training builds upon my prior experience to decipher the interfacial properties of bacterial films.

I have gained expertise in characterizing 2- and 3-D bacterial films from physical and metabolic standpoints. Now, I am developing artificial microniches as functional microbial nanoculture systems using soft nanomaterials to advance the general understanding of microbial dynamics involved in complex biofilms. The microniches, when further advanced, will have applications for personalized, therapeutic medicine or for high-throughput screening of unculturable microbial species relevant to biotechnology and drug discovery. My research experience and relevant products are highlighted below:

Postdoctoral Training–Supervisors: Kathleen Stebe and Daeyeon Lee, Chemical and Biomolecular Engineering, University of Pennsylvania

Project 1: "*Artificial Microniches for High Throughput Study of Microbial Communities*."

1) ***Niepa THR, Hou L, Jiang H, Goulian M, Koo H, Stebe K, and Lee D*** (2016) Microbial Nanoculture as an Artificial Microniche (**Scientific reports, Accepted**).

2) ***Hann SD, Niepa THR, Stebe KJ, Lee D. (2016).*** One-step generation of cell-viable compartments via polyelectrolyte complexation in an aqueous two phase system (Submitted)

3) ***Kim D., Sengupta A., Niepa THR, Lee BH, Weljie A., Murata R., Stebe K.J., Lee D., Koo H.*** ***Candida albicans*** stimulate ***Streptococcus mutans*** microcolony development via cross-kindom biofilm-derived metabolites (In preparation).

Project 2: "*Physical and Metabolic Properties of Films of Bacteria at Fluid Interfaces (FBI)*."

4) ***Murphy D, Gemmell B, Vaccari L, Li C, Bacosa H, Evans M, Gemmell C, Harvey T, Jalali M, Niepa THR.*** (2016) An in-depth survey of the oil spill literature since 1968: Long term trends and changes since Deepwater Horizon (Submitted).

5) ***Niepa THR, Vaccari L, Leheny R, Goulian M, Lee D, and Stebe K*** (2016). Films of Bacteria at Interfaces (FBI) formation: A dynamic response to interfacial stress (In preparation)

Doctoral Training–Advisor: Dacheng Ren, Biomedical and Chemical Engineering, Syracuse University

PHD. Thesis: *Electrochemical Control of Bacterial Persister Cells*

6) ***Szkotak R, Niepa THR, Jawrani N, Gilbert JL, Jones MB, Ren D.*** Differential Gene Expression to Investigate the Effects of Low-level Electrochemical Currents on ***Bacillus subtilis.*** **AMB Express 1 (2011) 39.**

7) ***Niepa THR, Gilbert JL, and Ren D.*** Controlling ***Pseudomonas aeruginosa*** persister cells by weak electrochemical currents and synergistic effects with tobramycin. **Biomaterials 33(30) (2012) 7356-7365.**

8) ***Niepa THR, Snepenger LM, Hao Wang, Sivan S, Gilbert JL, Jones MB, and Ren D*** (2016) Sensitizing ***Pseudomonas aeruginosa*** cells to antibiotics by electrochemical disruption of membrane functions. **Biomaterials 74 (2016) 267-279.**

9) ***Niepa THR, Hao Wang, Dabrowiak JC, Gilbert JL, and Ren D*** (2015) Synergy between tobramycin and trivalent chromium ion in electrochemical control of ***Pseudomonas aeruginosa***. **Acta Biomaterialia 36 (2016) 286–295**.

10) ***Ren D., Zhang M., Niepa THR, and Gilbert J. (2014).*** Systems and Method for Controlling Bacterial Persister Cells with Weak Electric Currents. (**Awarded Patent** # 8 569 027).

11) ***Niepa THR, Wang H, Gilbert JL and Ren D.*** (2016) Electrochemical Control of ***Pseudomonas aeruginosa*** Persister and Biofilm Cells using TGON™ 805 in single and dual chamber systems (Submitted).

12) ***Niepa THR, Haveman G, Goyal P, and Ren D*** (2016) Electrochemical treatment of ***Pseudomonas aeruginosa***in a rabbit model of sinusitis (In preparation).

Teaching Interests: My vision is to promote active learning among organized team of students and to develop new methods to help them identify emerging research topics and design new solutions from a multidisciplinary approach. Thus, I am looking forward to teaching the core courses (including Transport, Thermodynamics, Reaction Engineering, Separation Processes) and specialty Bioengineering courses such as Biomaterials, Fields, Forces, and Flows in Biological Systems, Biochemical Engineering or Bioelectrochemistry.

Teaching Experience: Over the course of my academic and professional training, I have participated in and contributed to various teaching and research training experiences. These inspired my pursuit of an academic career so that I can continue to support future generations of scholars, especially those from disadvantaged backgrounds. As a research assistant, my main involvement in teaching was through guest lecturing in the areas of biotechnology, medical devices, and entrepreneurship. Lectures were structured within the course framework as taught by a faculty member, or based on my area of expertise in microbial biofilms. My efforts were well-received by students and also effective as indicated by positive reviews in independent evaluations and students' test performance. Another aspect of teaching entails training graduate and undergraduate students to acquire the research skills essential for their completion of their independent study. I helped four graduate and three undergraduate students to quickly become familiar with lab techniques and so independent researchers to contribute to the topic of bacterial electrophysiology. The success of my efforts led to co-authorship with two of my research mentees in scientific publications. Also as a doctoral student, my mentoring experiences reinforced my desire to become a faculty member. In the summers of 2012 and 2013, I coordinated the Syracuse University-Emerging Frontiers in Research and Innovation–Research Experience and Mentoring (EFRI–REM) program. I have pursued new mentoring opportunities during my fellowship at the University of Pennsylvania through the NSF Louis Stokes Alliance for Minority Participation (LSAMP) or through presentations at seminars and keynote addresses during STEM conferences. My commitment to engagement and teaching resulted in my receipt of the ***Appreciation Award for Dedication, Continuous Inspiration, and outstanding Contribution to Science and Engineering Students***from the Greater Philadelphia LSAMP.

Entrepreneurship: I have professional experience from starting up a small business that has both enhanced my creativity and increased my understanding of the challenges of and opportunities for technology transfer. In 2011, I successfully transferred a basic scientific idea developed in the lab into a business idea. I co-founded Helios Innovative Technology, Inc. (now PurpleSun, Inc.), which develops automated sterilization devices to prevent bacterial cross-contamination and fight hospital-acquired infections.

Founder of Helios Innovative Technology (now PurpleSun); Funding: 2011 New York State Plan Business Competition ($60,000), Angel Investor ($500,000).

13) Romo L., Garritano N., **Niepa THR.**(2013). Door Handle Sterilization System, WO Patent 2,013,025,894. (Patent Application).

Future Directions: As faculty member, I will strengthen and grow my research efforts while developing new, effective teaching programs. The goal of my research efforts is to elucidate the physicochemical mechanisms regulating microbial growth in various environments and, based on this knowledge, design sustainable means to control bacteria. My previous research on the electrophysiology of highly drug-tolerant bacterial cells opens the door for new approaches to cure chronic infections. Building on this foundation, my planned translational-research program would target the design of alternate ways to diagnose and eradicate bacteria using conductive-substrate interfaces and medical devices. Another research area I will pursue is the development of soft nanomaterials for biotechnological applications such as drug release and identification. By developing the means for selective diffusion through functional membranes, I propose to control physiological responses of microbes and direct their metabolic activity towards desired ends e.g., creation of new materials and bioproducts. Thus, the development of the functional membrane would allow the creation of functional nanocultures to encapsulate probiotics or unculturable cells as well as upstream bioprocess technologies to facilitate high-throughput screening and characterization of bioactive molecules.

Finally, I aim to elucidate the effects of interfacial properties of oil-water on the metabolic performance of bacteria in bioprocesses involving oil-water emulsions. I propose to characterize the mechanics and metabolic profiles of films of bacteria at interfaces subjected to strong interfacial energies, and elaborate on this understanding to model sustainability in bioremediation or prevent the biodeterioration of economically relevant products.

As a teacher, my vision is to promote active learning among organized team of students and to develop new methods to help them identify emerging research topics and design new solutions from a multidisciplinary approach. One of my goals as a faculty member is to translate my research ideas to medical and industrial technologies and encourage the development of an entrepreneurial spirit among my students so they are aware of this additional pathway through which research outcomes can benefit society.

475850 *Engineering Micro-Flows:Integrated Experimental-Computational Approach*

Abstract:

Research Interests:

Microfluidic systems have emerged as powerful tools for controllable manipulation of micro-scale flows. Ease of fabrication together with high-speed visualization prompt rapid growth in applications of microfluidic tools. Single-phase microfluidics, which is used to guide biological suspensions inside microchannels without addition of an immiscible suspending fluid, can be utilized for flow cytometry, rare cell diagnostic, growth of biofilm streamers and extensional rheometry. Multi-phase, i.e droplet, microfluidic platforms bring together the precise liquid handling capabilities of microfluidics with unprecedented control over size uniformity and the production rate of droplets. Droplet microfluidics are used for micro-particle synthesis, droplet-based digital polymerase chain reaction (ddPCR), droplet-based DNA sequencing, single-cell analysis and enzyme kinetic studies. Owing to the widespread use of microfluidics in various engineering applications, there is a critical demand for developing new microfluidic tools by relying on understanding the mechanisms of micro-scale flow. However, the fluid dynamics of suspensions and emulsions in complex geometries of microfluidics mean that experimental trials are the main approach to gain insight. The alternative path is utilizing both microfluidic experiments and mesoscale computational methods as interacting partners to address challenges of mixture flow at micro-scale.

A main thrust of my research will focus on developing novel platforms for micro-flow engineering enabled by microfluidic technology. I will employ micro-fabrication, microfluidics and microscopy techniques to accompany mesoscale simulations, in particular the lattice-Boltzmann Method (LBM). The insight provided by my integrated experimental-computational approach permits exploiting the inherent forces of the system, such as hydrodynamic and capillary forces, for developing easily deployable microfluidic devices. The outcome of my research will be applied to biomedical micro-filtration in single-phase microfluidics and on-chip liquid handling and droplet encapsulation systems for genomics, single-cell analysis and tissue engineering. Fundamental fluid dynamic questions which have hampered efficient integration of single-phase and multi-phase microfluidic, such as stochastic entrapment of components in each compartment, will be addressed through my efforts to develop microfluidic self-assembly and label-free sorting systems. On the computational front, I will apply the lattice-Boltzmann framework to micro-scale flow geometries to develop high quality computational models of mixture flow. By adding new modules to LBM using an object-oriented design and implementation framework, I will develop a software for micro-flow applications.

I have been working on my integrated microfluidics-LBM approach through my consistent experimental and computational research. The same line of work will be continued in my career as an assistant professor. I will use the capacity of single-phase microfluidic platforms to develop accurate on-chip micro-filtration tools for cancer diagnostic applications. The advantages of my proposed microfluidic filtration systems over other on-chip technologies, including the technology developed during my postdoctoral work, are smaller sample size and higher separation accuracy. I achieve this goal by utilizing the fluid dynamics of inertial mixture flow around obstacles in microchannels. I will also be focused on developing microchannel self-assembly systems for controlled encapsulation of components in microfluidic droplet generators. The inefficiency of the current on-chip technologies caused by stochastic entrapment will be resolved using my integrated experimental-computational approach. My third direction of research concerns with label-free sorting of droplets by exploiting hydrodynamic, capillary and magnetic forces. In addition to microfluidic experiments, the LBM simulations will assist to gain insight into the physics of micro-scale flow.

Teaching Interests:

I believe that modern engineering education should integrate computational methods throughout varied disciplines. Design tools such as AutoCAD and SolidWorks, simulation and scientific pogramming have a place in almost any course and should be introduced early and often during undergraduate education. With this in mind, I am interested and prepared to provide instruction for courses on the following topics, where I would develop a syllabus that included heavy integration of computational tools:

Undergraduate Level:

Physics and Applications of Microfluidics

Microfluidics Lab

Transport Phenomena

Applied Computational Methods in Science and Engineering

Graduate Level:

Advanced Fluid Dynamics

Rheology of Complex Fluids

Applied Microfluidics

Mesoscale Simulation Methods and Applications

475861 Engineering the Spatial Organization of Proteins for Applications in Synthetic Biology and Beyond

Abstract:

Engineering the Spatial Organization of Proteins for Applications

in Synthetic Biology and Beyond

Christopher Jakobson[1], Marilyn F. Slininger[1], Edward Y. Kim[1], Jeff E. Glasgow[2], Michael A. Asensio[3], Norma M. Morella[4], Emily C. Hartman[2], Alex Chien[5], Yiqun Chen[1], Elias Valdivia[4], Matthew B. Francis[2], Niall M. Mangan[6], and Danielle Tullman-Ercek[7]

[1]Dept. of Chemical Engineering, UC Berkeley

[2]Dept. of Chemistry, UC Berkeley

[3]Dept. of Bioengineering, UC Berkeley

[4]Dept. of Plant and Microbial Biology, UC Berkeley

[5]Biophysics Graduate Group, UC Berkeley

[6]Dept. of Applied Mathematics, U. of Washington

[7]Dept. of Chemical Engineering, Northwestern University

Research Interests:

Protein encapsulation, the storage of biologically active molecules inside protein containers to enhance cargo function, is ubiquitous in nature. My graduate research with Prof. Danielle Tullman-Ercek focused on two examples of this strategy: (i) bacterial microcompartments, subcellular organelles of *Salmonella*, *Yersinia*, and other pathogens, and (ii) virus-like particles, such as those based on bacteriophage MS2. By engineering these systems as platform technologies for synthetic biology, we also discover biological design principles that can guide the development of therapeutics and vaccines.

Bacterial microcompartments are protein-bound containers that house specialized metabolic pathways, allowing invading pathogens to proliferate in the host gut [1]. We have focused on engineering these structures to improve the performance of heterologous biosynthetic pathways encapsulated in the organelles. We have learned how to control organelle formation [2,3], control small molecule transport across the protein membrane [4], and localize heterologous cargo using a variety of cellular zip codes [5]. We have identified surprising plasticity in microcompartment-mediated metabolism and used mathematical models to understand microcompartment function in detail [6]. This toolbox allows us orchestrate the formation of custom subcellular organelles in bacteria and to encapsulate heterologous pathways in a well-controlled manner.

Using the bacteriophage MS2, another protein container, we have explored the possibility of using protein membranes to control the diffusion of small molecules. We have demonstrated that changes to the chemical character of the pores of the MS2 virus-like particle can influence the kinetic behavior of an encapsulated enzyme [7]. We have also devised a selection for MS2 virus-like particle assembly in order to comprehensively characterize the mutational landscape of the MS2 coat protein gene, and we have identified a single point mutant that mediates a stable conversion of the viral capsid structure from a 27 nm to a 17 nm icosahedron [8].

Our studies are but two examples of the growing intersection of synthetic biology and the study of host-pathogen interactions, demonstrating that the knowledge gained from engineering complex biological systems is broadly applicable to understanding their function and evolution. My independent research will focus on engineering probiotics to deplete the small molecules metabolized in microcompartment organelles, characterizing the mutational landscapes of complex systems such as viral capsids to understand their evolution, and designing spatial organization strategies to optimize the performance of heterologous biosynthetic pathways. My laboratory will combine chemical engineering expertise with experimental biology and mathematical modeling to take a unique approach to these questions.

Teaching Interests:

I have teaching experience on the instructional teams for the undergraduate Fluid Mechanics and Heat & Mass Transfer courses at Cornell University, as well as the graduate Protein Engineering course at UC Berkeley. In each of these courses, our focus as teachers was on student-driven learning in the form of guided problem-solving sessions, group projects, and facilitated discussion of primary literature [9]. I also serve as a scientific advisor to the K-12 Alliance professional development program for science teachers in California. I am part of a teaching team that leads a weeklong summer institute for teacher leaders from districts throughout the state, providing instruction on science content and modeling effective science pedagogy [10]. I will combine the lessons I have learned from teaching at the university and secondary levels to continue to improve my instruction of core chemical engineering courses.

As an independent investigator, I also plan to develop a course on Mathematical Methods in Biological Engineering for graduate and upper-level undergraduate students. The course will provide students with the analytical and numerical tools required to address chemical and biological phenomena at the molecular, organism, and population scales using models of chemical kinetics, cellular metabolism, and disease epidemiology. A key student outcome will be proficiency in the use of MATLAB as a numerical toolkit to address nonlinear dynamics in biological systems.

1. Jakobson and Tullman-Ercek, *PLoS Pathogens* (2016).

2. Kim*, **Jakobson***, and Tullman-Ercek, *PLoS ONE*(2014)

3. Jakobson, Chen, Slininger, Valdivia, Kim, and Tullman-Ercek, *in revision.*

4. Slininger, **Jakobson**, and Tullman-Ercek, *submitted.*

5. Jakobson, Kim, Slininger, Chien, and Tullman-Ercek, *J. Biol. Chem.*(2015)

6. Jakobson, Slininger, Tullman-Ercek, and Mangan, *in preparation.*

7. Glasgow, Asensio, **Jakobson**, Francis, and Tullman-Ercek, *ACS Synth. Biol.* (2015).

8. Asensio*, Morella*, **Jakobson**, Hartman, Glasgow, Sankaran, Zwart, and Tullman-Ercek, *submitted.*

*These authors contributed equally.

Manuscripts in preparation and submitted available upon request.

9. Sample syllabi, schedules, and course materials for Protein Engineering available upon request.

10. See: https://www.wested.org/service/k-12-alliance-science-content-institutes/

475964 Predicting the Aggregation Behaviour in Biopharmaceuticals

Abstract:

Profile

Sarah Hedberg

Imperial College London and Fujifilm Diosynth Biotechnologies

Academic advisers: Dr Daryl Williams and Dr Jerry Heng

PhD Thesis:

Predicting and Controlling Aggregation for mAbs During Bioprocessing, 2016

Research Interests:

My principal area of research is to develop an improved fundamental understanding of the role of protein phase behaviour that can be utilised in downstream bioprocess engineering. Potential research activities would include:

i) Studying the impact of protein aggregation in the downstream processing of biopharmaceuticals (primarily chromatography) and the formulation process;

(ii) The development and improvement of experimental techniques that can be used to predict protein aggregation and estimate the osmotic second virial coefficient (e.g. self-interaction chromatography and cross-interaction chromatography);

(iii) The development of robust crystallisation approaches for proteins using the second virial coefficient data;

(iv) To employ modelling tools for predictioning and control of these biopharmaceutical downstream operations.

My research has always been driven by the needs of the biopharmaceutical industry, with a current focus on controlling protein aggregation, which is critical in order to avoid immunogenicity issues in biopharmaceuticals. Problems with protein aggregation and other impurities would have the greatest impact on the downstream and formulation process. Previous research has shown that the second virial coefficient, describing the thermodynamics of the protein-protein interactions in solution, can be a useful tool for predicting the aggregation propensity of proteins. Automatic, high-throughput techniques requiring minimal proteins such as self-interaction chromatography, cross-interaction chromatography and dynamic light scattering have previously demonstrated to be very successful in determining the second virial coefficient or the interaction parameter that can be helpful in predicting protein aggregation.

My research work has always been in collaboration with industry and the industry partners for my current research are Fujifilm Diosynth Biotechnologies and JML Biotechnology Consulting. The research undertaken has later been directly implemented by these companies. I am planning to maintain these industrial links in my research and to work with these companies in my future research career.

My research vision is that we need to look closer at the thermodynamics governing the protein-protein interactions and approaching the problem from a physical chemical viewpoint. This may then help us gain a better understanding as well as being able to predict and control protein phase behaviour, protein stability and aggregation, protein crystallisation as well as performance for formulation and downstream operations.

Teaching Interests:

My current teaching interests lies within the biochemical engineering/ bioprocess engineering field. I am interested in promoting more biochemical engineering and biological engineering within the traditional chemical engineering teaching environment. I am especially hoping to teach and incorporate more of the downstream purification process and formulation process courses on biopharmaceuticals into the chemical engineering curriculum.

I have previously been part of designing, planning, teaching and marking a wide range of subjects in Chemical Engineering for the past 4 years, in my Department at Imperial College, but also at Tsinghua University, as a summer school course leader, and University College London, as a teaching associate. This teaching has ranged from fundamental chemistry, separation unit operations and chromatography to sustainable development (with a chemical engineering outlook). I have obtained a teaching recognition as an Associate Fellow of the Higher Education Academy (AFHEA) from the Higher Education Academy in the UK.

Publications

Hou, Y., Hedberg, S. & Schneider, I. C. *Differences in Adhesion and Protrusion Properties Correlate with Differences in Migration Speed under EGF Stimulation*. BMC Biophysics,2012, 5: 1-12.

Hedberg, S.H.M., Heng, J.Y.Y., Williams, D.R., and Liddell, J.M. *Self-Interaction Chromatography of MAbs: Accurate Measurement of Dead Volumes.* Pharmaceutical Research, 2015, 32(12): p. 3975-3985.

Hedberg, S.H.M., Heng, J.Y.Y., Williams, D.R., and Liddell, J.M. *Micro Scale Self-Interaction Chromatography of Proteins: A MAb Case-Study.* Journal of Chromatography A, 2016, 1434: p. 57-63.

Hedberg, S.H.M., Brown L., Heng, J.Y.Y., Williams, D.R., *Optimised Protein Immobilisation Reactions and*

Immobilisation Kinetics for Chromatography Media, submitted to Colloids and Surfaces B, 2016.

Hedberg, S.H.M., Heng, J.Y.Y., Williams, D.R., *Kinetics and Thermodynamics of Protein Aggregation*, in preparation 2016.

476183 Engineering the Plant Microbiome to Complement Host Phenotype

Abstract:

Engineering the plant microbiome to complement host phenotype

Plant roots are a rich source of carbon in an otherwise nutrient starved soil environment, and as a result, a diverse community of microbes is attracted to the plant. The complex network of beneficial and antagonistic interactions of community members, with each other and with the host, forms a microbiome that is generally a beneficial extension of the host plant. The plant microbiome can increase nutrient acquisition rates, increase resistance to pathogens, and alleviate environmental stresses such as drought. Understanding how the microbiome functions and interfaces with the host has applications in food and bioenergy agriculture, production of secondary metabolites from plants for commodity chemicals and pharmaceuticals, and will improve our general understanding of the host-microbiome relationship. My research goal is to engineer the microbiome to optimize plant growth for sustainable agriculture and biochemical production.

Research Interests:

My current research is focused on the bacterial microbiome of *Populus,* a genera of fast-growing trees cultivated as a second-generation biofuel feedstock. Using metabolic models, comparative genomics, and phenotype screening we showed that the extreme diversity within 19 *Pseudomonas fluorescens* strains isolated from *Populus* was driven by the metabolite profile of the host and correlated with spatial organization of the microbiome. To gain further insight into spatial aspects of cooperation and inhibition, we designed a novel patterning approach to study distance dependent interactions between bacterial strains. We are also studying the molecular mechanisms that lead to phenotypic changes in the host when colonized by individual bacteria or mixtures using phenotype measures, metabolomics, and transcriptomics. The results show that bacteria can behave synergistically to help optimize and modify plant growth phenotypes. To study community behavior, we measured microbiome response to host stresses using 16S sequencing and identified candidate bacterial taxa that may help alleviate host response to drought, light limitation, and toxin stresses.

My research group will build on the results described by studying how constructed communities of bacteria respond to the host and environmental stresses. Specifically, we will design communities of genome-sequenced bacterial isolates using individual phenotypes, phylogenetic relationships, and predicted genomic content and test the ability of the community to mimic the natural microbiome response to stress. Our approach will include quantitative analysis of bacterial strains and their interactions, and characterization of community response to host stress. Models of bacterial kinetics, metabolism, and interactions in culture conditions will be used to predict behavior on the host plant. Integrating quantitative experimental data with predictive models will help us understand how the microbiome assembles in nature, ultimately leading to the ability to engineer constructed communities of bacteria with specific and predictable functions.

Postdoctoral Project: Interactions between *Populus* and cultured representatives of its bacterial microbiome

Mentors: Dale A Pelletier and David J Weston, Oak Ridge National Laboratory, Biosciences Division

PhD Dissertation: Kinetics of vesicular stomatitis virus mRNA and genomes during infection

Mentor: John Yin, University of Wisconsin-Madison, Department of Chemical and Biological Engineering

Research Career

My research career began as an undergraduate where I studied gas hydrate formation at the Colorado School of Mines. I helped develop software for predicting hydrate formation in gas pipelines and tested predictions using a novel reactor in the laboratory. This valuable experience familiarized me with quantitative experiments and laboratory techniques, but more importantly convinced me to go to graduate school to continue my academic career. In graduate school I changed my focus to biological engineering, where I learned how to design experiments and control biological systems to make quantitative measurements of viral replication. I used the data I collected to build kinetic models of viral infection, and showed that viral mRNA production is independent from host environment and only depends on viral genome number. This suggests that transcription is robust and insensitive and helps to identify targets for anti-viral strategies for RNA viruses. While studying interactions between two organisms (viruses and cells), I became interested in complex ecological systems with multi-member interactions. I was fortunate to find a post-doctoral position as a part of the Plant-Microbe Interfaces project at Oak Ridge National Laboratory studying the microbiome of *Populus* trees. I have focused specifically on understanding bacterial genetic and phenotypic diversity in the system to help determine how interactions between microbiome members and the host lead to overall system function. In my next step I will study how multi-member systems can be engineered to perform specific functions.

Teaching Interests:

As I have moved through my academic career I have become more active in my role as a teacher and mentor. During my undergraduate studies I worked closely with many of my peers, and began to appreciate the different learning and teaching styles present in the group. I had the opportunity to informally tutor younger students in heat and mass transfer classes; in both cases their grades improved after we began working together. In graduate school I got my first opportunity to formally teach classes to students. As a teaching assistant for the introductory transport course, I enjoyed spending time developing lectures and example problems for students, and focused on how I could engage the entire class in a single topic. I also worked as a TA for the process control laboratory where I helped students run experiments and taught them to present their results in concise reports. As a post-doc I have worked with undergraduate summer students and new graduate students from chemistry, biochemistry, biological systems engineering and chemical engineering. Their unique backgrounds have helped me generalize my teaching methods. I have helped them learn how to work in the laboratory, analyze data, and present results. The students helped identify metabolic pathways, isolated bacteria that grow on *Populus* metabolites, identified genes that affect plant growth, and developed methods for studying root growth in microfluidic devices. Watching students succeed with my input has been extremely important to my development and has helped me appreciate the commitment of my teachers and mentors throughout my academic career.

476193 Microfluidic Line of Attack to Comprehend Biological Systems

Abstract:

Research Interests:

My research interests lie in developing microfluidic systems that have applications in cellular and other biological systems analysis as well as medical diagnostics and drug discovery. My primary goal is to investigate the cellular heterogeneity of primary cancer / patient-derived cells. Assessing cellular-level heterogeneity can provide a better understanding in a patient-specific manner how chemotherapy drugs interact with cancer cells so that personalized treatments can be designed. This goal will be extended to isolate individual cells of a desired phenotype and expand them for manufacturing of therapeutic cell lines. Single-cell isolation and expansion is a broad platform that can impact areas of tissue engineering, regenerative medicine, and disease therapies. Apart from cellular study, I would also like to utilize microfluidic systems to perform biological and biomedical assays with low sample/reagent volumes and parallelization capabilities. One of my targeted area will be the fundamental study of crystal nucleation and growth for different body-stones (crystals) in the gallbladder, kidney, and pancreas. My diverse research background—which includes experience with cell cultures, cell assays, and engineering microfluidic platforms for biological and chemical analysis—combined with training in novel experimental tools—such as advanced high-speed imaging techniques, fluorescence microscopy, confocal microscopy and scanning electron microscopy, particle image velocimetry, and microfabrication—puts me in a unique position to pursue active research in these areas.

In addition to my research expertise, my current position as Research Assistant Professor provides me hands-on experience with proposal writing for external funding along with interactions with prospective collaborators. These additional experiences put me a step ahead to propel my future research endeavors as a tenure-track faculty in the fields of cancer therapy and diagnostics.

Teaching Interests:

Apart from my research career, I have extensive teaching experiences as an instructor and teaching assistant at two different institutions. The first class that I taught as an instructor was 'Chemical Engineering Transport Laboratory' at Texas Tech University. Along with my teaching, I've also designed and developed two new experimental demonstrations that had been included in 'Unit Operation Laboratory' and 'Chemical Engineering Transport Laboratory' courses. The second-class as instructor was 'Introduction to Chemical Processes' a sophomore level course at Texas Tech. I have also held teaching duties as teaching assistant for undergrad fluid mechanics (Texas Tech), transport phenomena (University of North Dakota). I did guest lectures for two consecutive years for 'Introduction to Microsystems I' (senior and graduate level Electrical Engineering course) at Texas Tech. With respect to future teaching plans, I am capable of teaching both undergraduate and graduate level classes in Chemical Engineering. At the graduate level, I will offer new course 'Biomicrofabrication and Biomicrofluidics' that will emphasize microfabrication techniques, microscale flows and approaches for cellular analysis.

Proposal Contributions:NSF AIR-TT, NSF EAGER, and CPRIT.

Education:

Ph.D., Chemical Engineering, Texas Tech University, Lubbock, TX, 2013

M.S., Chemical Engineering, University of North Dakota, Grand Forks, ND, 2008

B.Sc., Chemical Engineering, Bangladesh University of Engineering & Technology, Dhaka, Bangladesh, 2004

Experience:

Research Assistant Professor, Department of Chemical Engineering, Texas Tech University, 03/2016-Present.

Post-doctoral Research Associate, Department of Chemical Engineering, Texas Tech University, 10/2014-03/2016.

Teaching (Instructor), Department of Chemical Engineering, Texas Tech University, 01/2014-08/2014.

Invention Disclosures:

1. Vanapalli, S. A., **Bithi, S. S.** and Sun, M. Methods and devices to control fluid volumes, reagent and particle concentration in arrays of microfluidic drops, US Patent Application 20140051062, 2014.
2. Vanapalli, S. A., **Bithi, S. S.** Microfluidic devices and methods of bioassays, PCT/US2016/036787, filed June 2016.

Publications:

1. **Bithi, S. S.** and Vanapalli, S. A. Collective Dynamics of Non-Coalescing and Coalescing Droplets in Microfluidic Parking Networks, ***Soft Matter***, 11, 5122-5132, (2015)
2. **Bithi, S. S.**, Wang, W. S., Sun, M., Blawzdziewicz, J. and Vanapalli, S. A. Coalescing drops in microfluidic parking networks: A multifunctional platform for drop-based microfluidics, ***Biomicrofluidics***, 8(3), 034118, (2014)
3. **Bithi, S. S.** and Vanapalli, S. A. Behavior of a train of droplets in a fluidic network with hydrodynamic traps, ***Biomicrofluidics***, 4, 044110, (2010) (**Selected as Research Highlight**)
4. Sun, M., **Bithi, S. S.** and Vanapalli, S. A. Microfluidic static droplet arrays with tuneable gradients in material composition, ***Lab on a Chip***, 11, 3949, (2011) (**Cover article**)
5. Maddala. J, Srinivasan, B., **Bithi, S. S.**, Vanapalli, S. A. and Rengaswamy, R. Design of a model-based feedback controller for droplet sorting and synchronization in a microfluidic loop, ***AIChE J***, 58, 2120, (2011)

476206 Biotechnological and Health Applications of Multiscale ME (Metabolism and protein Expression) Models

Abstract:

Research Interests:

- Novel systems biology algorithms to analyze Biomedical Big Data--massive and diverse datasets capturing complex biological interactions and are difficult to analyze using conventional data analysis methods.
- Multiscale models of metabolism, protein expression, and regulation, and multiscale optimization of bioprocesses.
- Antimicrobial strategies using systems biology models and multiple â€“omics data (genomics, transcriptomics, proteomics, epigenomics.
- Computer-aided microbe design for value-added chemicals, nutrients for animal feed, and therapeutic proteins.

Teaching Interests:

- 6 years of Teaching Assistant experience for Process Dynamics and Control
- Certification for completion of a 15-hour course on teaching university courses (at UCSD)
- I make appropriate use of active learning techniques, where student learning is facilitated using in-class clicker (real-time survey) questions on complex concepts that activate existing knowledge, identify misconceptions, and improve knowledge retention. Additional modern pedagogical methods will be used, including the inverted classroom and a balance of formative (ongoing or real-time) assessment and summative (exams, final

presentations, etc.) assessment.

Peer-Reviewed Journal Articles

1. E Brunk, KJ George, J Alonso-Guttierrez, M Thompson, E Baidoo, G Wang, **L Yang**, D McCloskey, T Baath, C Petzold, J Monk, EJ O'Brien, HG Martin, A Feist, P Adams, JD Keasling, BO Palsson, and TS Lee (2016) Characterizing Strain Variation in Engineered E. coli Using a Multi-omics Based Work ow. Cell Systems 2: 335-346.
2. **L Yang**, J Tan, EJ O'Brien, JM Monk, D Kim, HJ Li, P Charusanti, A Ebrahim, CJ Lloyd, JT Yurkovich, B Du, A Drager, A Thomas, Y Sun, MA Saunders, and BO Palsson (2015) A systems biology denition of the core proteome of metabolism and expression is consistent with high-throughput data. Proc Natl Acad Sci USA 112:10810-10815.
3. **L Yang**, S Srinivasan, R Mahadevan, and WR Cluett (2015) Characterizing metabolic pathway diversication in the context of perturbation size. Metab Eng 28: 14-122.
4. K Zhuang, **L Yang**, WR Cluett, and R Mahadevan (2013) Dynamic strain scanning optimization: an ecient strain design strategy for balanced yield, titer, and productivity. DySScO strategy for strain design. BMC Biotechnol 13:8.
5. **L Yang**, WR Cluett, and R Mahadevan (2011) EMILiO: a fast algorithm for genome-scale strain design. Metab Eng 13:272-281.
6. S Garg, **L Yang**, and R Mahadevan (2010) Thermodynamic analysis of regulation in metabolic networks using constraint-based modeling. BMC Research Notes, 3:125.
7. **L Yang**, R Mahadevan, and WR Cluett (2008) A bilevel optimization algorithm to identify enzymatic capacity constraints in metabolic networks. Comput Chem Eng, 32:2072-2085.

Submitted and pre-print articles

1. **L Yang**, D Ma, A Ebrahim, CJ Lloyd, MA Saunders, and BO Palsson. solveME: fast and reliable solution of nonlinear ME models. Under review.
2. **L Yang**, A Ebrahim, CJ Lloyd, MA Saunders, BO Palsson. dynamicME: Dynamic simulation and renement of integrated models of metabolism and protein expression. Under review.
3. **L Yang**, JT Yurkovich, CJ Lloyd, A Ebrahim, MA Saunders, BO Palsson. Principles of proteome allocation are revealed using proteomic data and genome-scale models. Submitted.
4. D Ma, **L Yang**, RMT Fleming, I Thiele, BO Palsson, MA Saunders (2016) Quadruple-precision solution of genome-scale models of Metabolism and macromolecular Expression. arXiv Preprint arXiv:1606.00054v1.

476212 Engineering Proteins and Metabolic Pathways for Biomedical, Bioenergy, and Biomaterial Applications

Abstract:

Research Interests:

Proteins play a key role in building, activating, and regulating what is essential in living organisms. From assembling pilin subunits on the *Staphylococcus aureus* cell surface, to generating cellulose microfibrils in *Arabidopsis* thaliana and converting carbon dioxide to methane in *Methanosarcina barkeri,* enzymes are the main actors in many biological processes. Therefore, understanding enzyme structure and function, and engineering enzyme stability and activity to generate novel bioengineering tools is crucial for solving many biomolecular engineering issues. Recent advances in molecular biology and protein engineering have created a platform to engineer enzymes as more efficient molecular "staplers" or "scissors" with higher activity and more specificity toward desired substrates. Furthermore, metabolic engineering methods are used to regulate enzymatic pathways for generating biofuel or other useful chemicals. My PhD and post-doctorate research has been focused on protein and metabolic engineering. Most of my research has evolved around controlling enzyme-mediated protein ligation and immobilization on the surface to generate site-specific protein assemblies, protein engineering using directed evolution and yeast surface display to create more efficient enzymes, as well as metabolic engineering to generate biofuel. As for my future plan, I intend to utilize recent genetic and protein engineering methods to create novel bioengineering tools with applications in bioenergy, biomedical, and biomaterial research.

Research Experience

Postdoctoral Researcher, University of Delaware

Projects: Engineering anaerobic methanogenic pathway to generate biofuels, Developing a light-activated enzyme.

Mentor: Dr. Wilfred Chen

PhD dissertation, University of Tennessee at Knoxville

Thesis: Engineering Sortase A for generating site-specific protein 3D assemblies

Advisor: Dr. Eric T. Boder

Teaching Interests:

Based on my training in teaching and research the Chemical and Biomolecular Engineering courses that I am readily able to teach include *Bioengineering Fundamentals, Transport Phenomena, and Mathematical Methods in Engineering.* I am also able to develop new biochemical engineering courses in undergraduate or graduate level. These courses will be focused on molecular and cell biology, protein engineering, as well as metabolic engineering.

476226 Towards Treatment of Neurodegenerative Disorders through Nanoparticle Mediated Enzyme Replacement Therapy

Abstract:

Research Interests:

Approximately 1 in 5,000 to 8,000 children are born annually with a lysosomal storage disease (LSD). Current treatment of LSDs, called enzyme replacement therapy (ERT), involves weekly to monthly intravenous (IV) infusions of missing endogenous enzymes. While ERT has proven effective in LSDs without central nervous system (CNS) involvement, the brain has remained untreatable due to the presence of the blood-brain barrier (BBB), which prevents passage of 98% of small molecule drugs, including enzymes, from the blood into the brain. Patients with neuropathic LSDs, which account for 50-70% of diseases in this category, present with severe CNS degeneration, ataxia, and premature death, with no treatment available on the market. The disease of focus, GM1 gangliosidosis, causes premature death between the ages of two and four with no hope available for parents of patients. Encapsulating enzyme into polymeric vesicles made up of polyethylene glycol and polylactic acid, called polymersomes, facilitates the transport of this missing enzyme into the brain through IV injections, effectively extending ERT therapy to the brain for the first time.

The missing enzyme in GM1 gangliosidosis, β-galactosidase (βgal) was loaded into polymersomes at a high efficiency, with 0.08 ± 0.03 mg βgal/mg polymersome, while maintaining its activity at 452 ± 82 a.u. Release of βgal was doubled over 6 hours when loaded polymersomes are placed into citrate buffer (pH=4.1) when compared to HEPES buffer (pH=7.4), indicating increased enzyme release expected in the lysosome. Attachment of ApoE was confirmed with an increase in polymersome diameter to 194.4 ± 3.5 nm and an increase in protein content from 0.06 ± 0.11 mg/mL to 1.03 ± 0.71 mg/mL. GM1SV3 cells, an immortalized cell line from felines with GM1 gangliosidosis, present low density lipoprotein receptors (LDLR) on the surface, as shown in confocal microscopy, indicating that they are a good model for ApoE mediated delivery through the BBB. ApoE increases the uptake of polymersomes into GM1/SV3 fibroblasts when compared to unlabeled control polymersomes.

My work highlights our ability to create polymersomes capable of protecting βgal in the blood stream, targeting the nanocarrier to the BBB, and after transcytosis, releasing βgal in the lysosome of neural cells. This combination of ERT and nanotechnology, demonstrating the capability of our carrier to transport enzymes to the brain while maintaining their activity, will create a paradigm shift in the treatment of CNS disease, providing treatment for currently untreatable and fatal diseases like GM1 gangliosidosis.

Teaching Interests:

As an instructor, it is my privilege to shape future engineers and motivate them to contribute to the direction of the field. I have demonstrated my ability to teach in both a classroom and laboratory setting through Teaching Assistant (TA) experience in Introduction to Chemical Engineering, Senior Design Lab, and Principles of Chemical Engineering at Auburn University. Though each experience was unique, I learned the universal importance of flexibility when teaching a course. It is necessary to constantly evolve in order to service the needs of your students as they progress and their understanding of the material matures.

I am passionate about inspiring students to excel in the chemical engineering field and excited to take on any chemical engineering course. Instruction of Thermodynamics, Principles of Chemical Engineering, Transport Phenomena, and Biomedical Engineering aligns closely with my strengths. My previous research experience developing novel nanocarriers for drug delivery applications and working in both the Chemical Engineering and Veterinary Medicine departments has prepared me for the integration of biomedical projects and assignments into any course, as well as the design of a Drug Delivery course. It is my goal as an educator to provide every student with the opportunity to engage in their own learning through flexible techniques. I want my students to leave the classes I teach with an understanding of the way in which the knowledge they have gained can contribute to the direction of the chemical engineering field.

476236 Microfluidic Technologies in High-Throughput Chemical Screens to Decipher Genetic Basis of Behavior and Development in *C. Elegans*

Abstract:

Research Interests:

The nematode *C. elegans* is of great interest to unravel fundamental biological processes common to the whole animal kingdom and to model human diseases. However, the large number of animals required per assay, complexity of protocols involved, and sensitivity to environmental perturbations render research on *C. elegans*time-consuming and labor-intensive, which slows down progress and limits the scope of addressable questions.

During my post-doc at Georgia Tech, I have exploited the possibilities of microfluidics to enable traditionally challenging types of screens on first larval stage (L1) and adult nematodes. The study of first larval stage *C. elegans*is particularly difficult because of the small size and high motility of the animals. I overcame this challenge by developing a microfluidic platform that uses nanoliter droplets of a reversible hydrogel to offer advanced manipulation of individual animals. This work paves the way for high-throughput screening of L1 nematodes with high-resolution imaging. This will facilitate the study of numerous post-embryonic developmental cellular events such as cell differentiation and cell migration, as well as important developmental decision-making, which specifically take place during the first larval stage.

I have also developed a microfluidic platform for combinatorial chemical screening of adult nematodes. Due to the exponential increase of combinations with the number of individual components, traditional technologies require a large amount of chemicals rendering the screens impractical. I solved this bottleneck by creating a droplet-on-demand platform that enables the preparation of precise mixtures of reagents while handling nanoliter volumes and the delivery of the stimuli to the animals in a robust way. Such a platform offers the possibility of performing combinatorial screens of rare compounds. It will enable deciphering intraspecies communication by screening combinations of pheromones that are difficult to extract or synthesize. It can also benefit drug screens to identify combinatorial therapeutic strategies.

In the future, I plan on building on my experience with *C. elegans* and microfluidics to pursue chemical screens with droplets. In addition, I intend to expand this approach to other model organisms, which will lead to an increase in throughput of assays by several orders of magnitude and open new exciting possibilities.

Teaching Interests:

During my PhD, I have taught undergrad and grad students, leading lab classes and solving-problem classes in addition to the usual TA duties (grading and helping students). I have interacted with students from a wide variety of backgrounds from which I have learned how to adapt to different learning styles. In addition, during my PhD and post-doc, I have mentored several undergrad and grad students on research projects as well.

I can teach Chemical Engineering core classes such as Numerical Methods Applied to Chemical Engineering, Mechanics of Fluids, Thermodynamics, in undergrad and I am willing to design new classes such as Microfluidics or Micro-Nano-Technologies at the graduate level.

476018 Biomaterials and Stem Cell-Based Therapeutics

Abstract:

Research Interests:
My research focuses on making stem cell and molecular therapies as an effective method to model and treat diseases. I believe that this goal can best be achieved by employing engineering fundamentals using biomaterials in stem cell research. At this interface between engineering and stem cell biology, my research leverages many disciplines including polymer synthesis, microfabrication technique, and gene/protein delivery. My goal is to run a multidisciplinary research program that focuses on three broad areas: 1) to fundamentally understand the mechanisms by which chemical and mechanical signals are sensed by stem cells under physiological and pathological conditions, 2) to design biomaterials that provide spatial and temporal control in regulating stem cell differentiation and assembly, and 3) to merge our understanding of stem cell signaling with our engineered stimuli-responsive biomaterials to a range of applications in regenerative medicine, drug delivery, and disease modelling. I envision that students within this research program will have strong collaborative interactions with other scientists, engineers, and clinicians, which will further enrich this training environment.

PhD Dissertation: "Engineering Vascular Networks Using Synthetic Matrices."
Advisor: Sharon Gerecht, Ph.D., Chemical and Biomolecular Engineering, The Johns Hopkins University.

Postdoctoral Project: "Targeted Anti-Thrombotic Therapeutics for Deep Vein Thrombosis."
Advisor: Elliot L. Chaikof, M.D., Ph.D., Department of Surgery, Harvard Medical School.

Research Experience:
My background as a chemical engineer, along with my training in biomaterials, stem cell biology and

vascular surgery lab, uniquely position myself to contribute substantial progress in the fields. During my doctoral work with Dr. Sharon Gerecht at the Johns Hopkins University, I investigated the mechanism of stem cell assembly into functional vascular networks. We developed a synthetic matrix, which properties can be tuned to provide temporal and spatial control for vascular morphogenesis that are useful for therapeutic angiogenesis and tissue regeneration. My postdoctoral training with Dr. Elliot L. Chaikof at Harvard Medical School has further broadened my research horizon in the area of drug delivery to target inflammation and thrombosis associated with vascular injury and stem cell transplantation. We engineered fusion proteins that selectively deliver anti-coagulant agents to activated platelets right at the site of injury, therefore inhibiting thrombosis without increasing the risk of systemic bleeding. I also led collaborations with engineers and surgical residents to evaluate anti-inflammatory and anti-thrombotic therapeutics in a clinically relevant model of deep vein thrombosis and islet transplantation.

Teaching Interests:
I believe that teaching is an integral part of a great research university, as rigorous education for both undergraduate and graduate students is a foundation for an innovative research. Throughout my academic training, I had the opportunity to TA "Senior Design Class for Chemical Engineering," as well as to design and teach advanced course in "Biotechnology for Regenerative Medicine." Additionally, I have also mentored and trained several undergraduate and master students in our research groups. As a faculty member, I am interested in teaching "Introduction to Chemical Engineering" and "Transport Phenomena" among other possible courses within the department. In addition to those classes, I aim to design graduate curricula that will introduce students to the evolving field of biomaterials, tissue engineering, and drug delivery.

Education:
B.S., 2007, Chemical and Biomolecular Engineering, University of Notre Dame.
Ph.D., 2012, Chemical and Biomolecular Engineering, Johns Hopkins University.

Selected Awards and Honors:
2007, *B.S. Magna Cum Laude*, University of Notre Dame.
2011, *U.S. New Investigator Travel Award*, International Society of Thrombosis and Haemostasis (ISTH).
2012, *Siebel Scholar, Class of 2012*, Siebel Scholar Foundation.
2014-2017, *JDRF Postdoctoral Fellowship,* Juvenile Diabetes Research Foundation.

Selected Publications (14 journal articles in total. Citation: 458. H-index: 10):

1. **Hanjaya-Putra, D.**, Yee, J., Ceci, D., Truitt, R., Yee, D., and Gerecht, S. (2010). Vascular endothelial growth factor and substrate mechanics regulate in vitro tubulogenesis of endothelial progenitor cells. *Journal of Cellular and Molecular Medicine*.
2. **Hanjaya-Putra, D.**, Bose, V., Shen, Y.-I., Yee, J., Khetan, S., Fox-Talbot, K., Steenbergen, C., Burdick, J.A., and Gerecht, S. (2011). Controlled activation of morphogenesis to generate a functional human microvasculature in a synthetic matrix. *Blood*.
3. **Hanjaya-Putra, D.**, Wong, K.T., Hirotsu, K., Khetan, S., Burdick, J.A., and Gerecht, S. (2012). Spatial control of cell-mediated degradation to regulate vasculogenesis and angiogenesis in hyaluronan hydrogels. *Biomaterials*.
4. **Hanjaya-Putra, D.**, Shen, Y.-I., Wilson, A., Fox-Talbot, K., Khetan, S., Burdick, J.A., Steenbergen, C., and Gerecht, S. (2013). Integration and Regression of Implanted Engineered Human Vascular Networks During Deep Wound Healing. *Stem Cells Translational Medicine*.
5. Kusuma, S., Shen, Y.-I., **Hanjaya-Putra, D.**, Mali, P., Cheng, L., and Gerecht, S. (2013). Self-organized vascular networks from human pluripotent stem cells in a synthetic matrix. *Proceedings of the National Academy of Sciences*.
6. Krishnamurthy, V.R., Sardar, M.Y.R., Ying, Y., Song, X., Haller, C., Dai, E., Wang, X., **Hanjaya-Putra, D**., Sun, L., Morikis, V., Simon, S.I., Woods, R.J., Cummings, R.D., Chaikof, E.L., (2015). Glycopeptide analogues of PSGL-1 inhibit P-selectin in vitro and in vivo. *Nature Communications*.
7. Naik, N., **Hanjaya-Putra, D.**, Haller C., Allen, M.G., Chaikof, E.L., (2015). Rapid homogenous endothelialization of high aspect ratio microvascular networks. *Biomedical Microdevices*.
8. **Hanjaya-Putra, D.**, Dai E., Haller C., Hagemeyer, C.E., Karlheinz, P., Chaikof, E.L., (2016). Targeted

Anti-Thrombotic Prophylaxis for Deep Vein Thrombosis. *Circulation (submitted)*.

Patents:

1. **Hanjaya-Putra, D.**, Vo E.T., Wanjare, M., Gerecht, S., "Smooth Muscle-Like Cells (SMLCs) Derived from Human Pluripotent Stem Cells," US 13/581,341. (2012).
2. **Hanjaya-Putra, D.**, Gerecht, S., "Hydrogel-Based Vascular Lineage Cell Growth and Uses," US 14/553,442. (2015).

476038 Engineering the Molecular Interactions for Biomedical Applications

Abstract:

Molecular Engineering for Biomedical Applications

Handan Acar

The Institute for Molecular Engineering, The University of Chicago, Chicago, IL 60637, USA

hacar@uchicago.edu

Research Interests:

Interest in peptides as potential drug candidates has naturally increased with recent progress in enhancing their structural and chemical stability, along with their ability to cross the cell membrane. Peptide-based therapeutics that target not only specific cells but also specific protein-protein interactions in the cytoplasm will herald a new era of "personalized medicine". In this approaching new era, chemical tools to engineer peptides into molecular probes will be invaluable. The combination of carefully selected and localized interactions will produce stable assemblies, which can be reversible, highly tunable, dynamic, and modular as required by the specific application. Effective approaches to the design of molecular therapeutics must be safe, sensitive, efficient, and rapidly adapted to new targets with minimal effort and expense. To enable new translational clinical technologies, my research will focus on molecular design that incorporates these aspects into the new materials.

I have extensive research experience in the design and synthesis of self-assembling peptide-based nanostructures, including the modulation of the fundamental driving forces of self-assembly. As a Ph.D student of Prof. Mustafa Ozgur Guler (Bilkent University, Ankara, Turkey), I designed and synthesized self-assembling peptide sequences to generate targeted structures in both organic solvents and water via the specific interactions of rationally placed amino acid sequences. My innovative work produced a novel bottom-up synthesis method for peptide template-directed self-assembled mineralization, generating one-dimensional inorganic nanostructures. I was awarded a patent for this technology.

I sought to broaden the translational focus of my research during my postdoctoral work at Institute for Molecular Engineering at the University of Chicago with Matthew Tirrell, PhD. Taking advantage of the university's rich mix of clinicians and translational scientists, I work closely with James LaBelle MD, PhD, a pediatric oncologist in the Pritzker School of Medicine whose research involves peptide therapeutic translation to refractory malignancies. This cutting-edge collaboration between myself, Dr. Tirrell and Dr. LaBelle led to the development of a self-assembled nanoparticle platform, comprised of enzymatically cleavable peptide amphiphiles, that carries therapeutic peptides into cells. This platform offers the unique ability to follow intracellular drug trafficking and enzymatic cleavage in real time. The broad-ranging application of this technology to clinical medicine, biochemical research, and peptide-based nanoparticle development led us to file a patent application through the University of Chicago.

Teaching Interests:

I have extensive teaching experience, starting as a Ph.D. student; I was the teaching assistant of three graduate courses in Bilkent University. As an independent teacher at university level, I designed and implemented the "Fundamentals of Thermodynamics for Mechanical Engineering" at Iowa State University (Spring 2014), which greatly assisted in my development as an instructor. At the end of the semester my students attained grades higher than the departmental average, a clear indication of my abilities as a teacher. Finally, I have mentored graduate and undergraduate students in my group as a postdoctoral researcher.

Future Direction:

My research lab will establish chemical methodologies to tune the supramolecular interactions of self-assembling molecules to surmount the key challenges for peptide-based therapeutics, diagnostics and delivery platforms. In self-assembly processes, the fundamental building blocks organize themselves into functional structures as driven by the energetics in the system. These behaviors typically emerge from supramolecular phenomena such as hydrogen bonding, hydrophobic effects, electrostatics, metal-ligand and π-π interactions, and van der Waals forces. Sufficient combinations of these rationally placed interactions can produce the reversible self-assembly of stable structures, with these structures presenting a highly tunable and modular platform for a variety of applications.

As an independent investigator, I will combine my experiences to build a lab with a strong focus on designing materials from a molecular perspective for translational clinical technologies. A specific emphasis will be placed on using an interdisciplinary approach from physical and synthetic chemistry for biomedical applications.

Selected Publications: (out of 14)

Acar, H.; Srivastava, S.; Chung E.J.; Scnorenberg, M.R.; Barett; LaBelle, J.L.; Tirrell, M.; *"Self-Assembling Peptide-Based Building Blocks in Medical Applications"*, accepted to Advanced Drug Delivery Reviews (2016).

Acar, H.; Cinar, S.; Thunga, M.; Kessler, M.R.; Hashemi, N.; Montazami, R.; 2014 "*Study of physically transient insulating materials for transient and biocompatible electronics*"; Advanced Functional

Materials; 24(26), 4135–4143 (2014).

Acar, H.; Genc, R.; Urel, M.; Erkal, T.S.; Dana, A.; Guler, M.O.; *Self assembled peptide nanofiber template one-dimensional gold nanostructures exhibiting resistive switching"*; Langmuir*;* 28, 16347-54 (2012).

Acar, H.; Garifullin, R.; Guler, M.O.; *Self-assembled template-directed synthesis of one-dimensional silica and titania nanostructures;* Langmuir*;* 27, 1079-84 (2011).

For more information, please visit: http://tirrell.ime.uchicago.edu/People/Handan_Acar.html

476042 Interfacing Cells and Materials for Advanced Delivery Systems

Abstract:

Research Interests:

The clinical translation of novel therapies and delivery systems is limited by their poor biological performance. Material-based delivery systems (e.g. nanoparticles) perform complex delivery functions such as on-demand or controlled release, but they are limited in performing basic biological functions (e.g. circulating, targeting tissues, penetrating barriers, and entering diseased cells). Cell-based therapies routinely perform these basic biological functions as they have evolved to interact with living systems, but they are not suitable for drug loading and controlled release. My research has led to the development of a number of hybrid delivery systems capable of performing both biological and synthetic functions [1-4]. These advanced delivery systems incorporate findings from my phenomenological studies [5-6], which have elucidated the key biophysical, biochemical, and material parameters that facilitate enhanced delivery.

Ph.D. Research:

My graduate research with Professor Samir Mitragotri at the University of California, Santa Barbara focused on improving the biological performance of nanoparticle delivery systems. During my Ph.D I developed three distinct blood-cell inspired hybrid systems that enhanced the in vivo delivery abilities of nanoparticles by utilizing and mimicking circulatory blood cells, including: (i) red blood cells [2-3], (ii) white blood cells [1], and (iii) platelets [4]. The first two examples utilize a strategy known as "cellular hitchhiking", which involves the attachment of polymeric particles to the surface of circulatory cells so as to transfer innate circulatory (e.g. long circulation of red blood cells) and targeting abilities (e.g. inflammation targeting of white blood cells) from cell to particle. The second strategy involves the design and application of synthetic platelets which incorporate the essential biophysical (elasticity [5] and shape [6]) and biochemical (surface biology) parameters of natural platelets into a single nanoparticle capable of performing hemostasis and preventing blood loss. During my Ph.D. I was an NSF GRFP Fellow and authored more than 25 peer-reviewed research publications in the field of drug delivery.

Postdoctoral Research:

My postdoctoral research with Professor Robert Langer at the Massachusetts Institute of Technology focuses on the development of oral drug delivery systems. I am currently expanding the broad theme of interfacing cells and materials [7] for the improved delivery of live-probiotics to the microbiome. Using a layer-by-layer encapsulation approach, I have shown that encapsulated probiotics exhibit enhanced: (i) protection against harsh gastrointestinal tract conditions, (ii) mucoadhesion and growth on intestines, and (iii) integration with the hostâ€™s microbiome in vivo as compared to non-encapsulated probiotics [8]. Another focus during my postdoctoral work has been on the clinical translation of controlled release polymeric delivery systems. I am currently working with clinicians in an ongoing clinical trial investigating the bioavailability of micronutrients delivered via a novel particle-based oral delivery system.

Future Research:

As a faculty member I will focus on two research areas:

Targeted nanoparticle delivery:

(a) Biologically-active nanoparticles for enhanced tissue targeting and penetration

(b) Determining organ-level and cell-specific distribution of nanoparticles

Microbiome delivery:

(a) Microbiome modulation via delivery of encapsulated probiotics and microbes

(b) Delivery strategies to maintain a healthy microbiome during antibiotic treatments

Teaching Interests:

My teaching philosophy places importance on integrating quantitative mathematical and rational/logical qualitative approaches. Emphasis will be on teaching and understanding the physical laws and fundamentals of chemical engineering to direct and guide quantitative mathematical approaches. As a graduate student I was a Teaching Assistant (TA) for core chemical engineering courses as well as bioengineering electives. As a TA I designed and presented lectures, held office hours, and presented MATLAB- and Mathematica-based approaches for computer-aided methods to solve partial differential equations. During my undergraduate years I served as a tutor for freshman and sophomore chemistry, physics, and mathematics courses.

References:

1. **Anselmo, A. C.**; Gilbert, J. B.; Kumar, S.; Gupta, V.; Cohen, R. E.; Rubner, M. F.; Mitragotri, S., Monocyte-Mediated Delivery of Polymeric Backpacks to Inflamed Tissues: A Generalized Strategy to Deliver Drugs to Treat Inflammation. Journal of Controlled Release 2015, 199, 29-36.
2. **Anselmo, A. C.**; Gupta, V.; Zern, B. J.; Pan, D.; Zakrewsky, M.; Muzykantov, V.; Mitragotri, S., Delivering Nanoparticles to Lungs While Avoiding Liver and Spleen through Adsorption on Red Blood Cells. ACS Nano 2013, 7, 11129-11137.
3. **Anselmo, A. C.**; Kumar, S.; Gupta, V.; Pearce, A. M.; Ragusa, A.; Muzykantov, V.; Mitragotri, S., Exploiting Shape, Cellular-Hitchhiking and Antibodies to Target Nanoparticles to Lung Endothelium: Synergy between Physical, Chemical and Biological Approaches. Biomaterials 2015, 68, 1-8.
4. **Anselmo, A. C.**; Modery-Pawlowski, C. L.; Menegatti, S.; Kumar, S.; Vogus, D. R.; Tian, L. L.; Chen, M.; Squires, T. M.; Sen Gupta, A.; Mitragotri, S., Platelet-Like Nanoparticles: Mimicking Shape, Flexibility, and Surface Biology of Platelets to Target Vascular Injuries. ACS Nano 2014, 8, 11243-11253.

5. **Anselmo, A. C.**; Zhang, M.; Kumar, S.; Vogus, D. R.; Menegatti, S.; Helgeson, M. E.; Mitragotri, S., Elasticity of Nanoparticles Influences Their Blood Circulation, Phagocytosis, Endocytosis, and Targeting. ACS Nano 2015, 9, 3169-3177.
6. *Kolhar, P.; ***Anselmo, A. C.**; Gupta, V.; Pant, K.; Prabhakarpandian, B.; Ruoslahti, E.; Mitragotri, S., Using Shape Effects to Target Antibody-Coated Nanoparticles to Lung and Brain Endothelium. Proceedings of the National Academy of Sciences 2013, 110, 10753-10758.
7. *Jaklenec, A.; ***Anselmo, A. C.**; Hong, J.; Vegas, A. J.; Kozminsky, M.; Langer, R.; Hammond, P. T.; Anderson, D. G., High Throughput Layer-by-Layer Films for Extracting Film Forming Parameters and Modulating Film Interactions with Cells. ACS Applied Materials Interfaces 2016, 8, 2255-2261.
8. **Anselmo, A. C.**, McHugh, K.J., Webster, J., Langer, R., Jaklenec, A. Layer-by-Layer Encapsulation of Probiotics for Delivery to the Microbiome. In revision since July 2016.

476061 Engineering Proteins for Magnetic Resonance Imaging at Molecular and Atomic Resolutions

Abstract:

Research Interests:

Imaging biological processes in intact tissue and cells with molecular specificity and high resolution presents a fundamental challenge to biomedical research. Magnetic resonance imaging (MRI), being non-invasive and not limited by tissue depth, enables whole-body imaging at sub-100µm resolution; however, the main bottleneck is the lack of compatible smart sensors that can recognize specific molecular events and accordingly output an MRI signal. On a different level of resolution, nuclear magnetic resonance (NMR) spectroscopy, a structural biology method compatible with living cells and tissues, allows the visualization of the movement of atoms inside a protein when it performs a biological task. Similarly, what prevents NMR from a wide application to the study of live cells and tissue is the lack of smart agents, which in this case, would restrict NMR signals only from those selective molecules, making atom-specific spectra a feasible goal. **I propose to engineer protein-based smart agents for the two applications that non-invasively image biomolecules at the molecular and atomic resolutions.**The growing genomic data from the sequencing of animals and microbes, together with the understanding and computational prediction of protein structures and functions, provides unique opportunities nowadays to repurpose and modify proteins for applications distinct from their native roles. Since these engineered proteins can be genetically encoded, they can leverage the large amount of genetic engineering methods developed over the decades to achieve functions that are difficult for synthetic agents, such as the precise localization in subcellular compartment, expression-level coupling with genes and circuits, and specificity in molecular recognition. The development of the new non-invasive molecular imaging methods could lead to various applications, such as the functional imaging of the brain in line with the BRAIN initiative, while the development of atomic-resolution imaging of biomacromolecules in live cells could lead to a new imaging paradigm and information that have not been previously accessed.

Postdoctoral Project: ***Genetically encoded gas nanostructures as acoustomagnetic reporters for noninvasive biological imaging.***

Under supervision of Mikhail G. Shapiro, California Institute of Technology

PhD Dissertation: ***Development of Solid-state NMR Spectroscopy for Membrane Proteins: Application to the Mercury Transporter MerF.***

Under supervision of Stanley J. Opella, University of California, San Diego (UCSD)

Research Experience:

My research has been at the interface between biology and physics. In the graduate school, I joined a laboratory that had work spanning from molecular biology, protein chemistry to quantum mechanics theory of the NMR pulse sequences and computational biology on protein structures. The overall goal of my research was to develop a general method for obtaining atomic-resolution structures of membrane proteins in their native lipid bilayer environment. I devoted a large portion of the efforts to the development of new solid-state NMR and computational methods. Besides, I also applied the new methods to solve the structure of a mercury transporter protein from microbial bioremediation pathways.

While the goal for a structural biologist is to understand a protein's native function, my current postdoctoral research in the realm of protein engineering aims to utilize such understanding to repurpose and generate new functions for the proteins. For example, we currently work on gas vesicles, a group of protein-only nanostructures whose native role is to regulate buoyancy of the microbes, and we repurpose them as a genetically encoded MRI reporter gene. During my postdoctoral training, I have also gained insights into the imaging aspects of magnetic resonance methods and the techniques for live animal imaging.

As faculty I would like to leverage my knowledge in protein engineering and magnetic resonance to develop new imaging methodology that can find utilities in various studies on live cells and animals. Just as how green fluorescent protein (GFP) changes the paradigm of optical imaging, I believe the engineering of protein-based probes can break through key bottlenecks in current non-invasive imaging methods and can potentially lead to new platforms.

Selected Publications:

Lu GJ, Tian Y, Vora N, Marassi FM, Opella SJ (2013) The structure of the mercury transporter MerF in phospholipid bilayers: a large conformational rearrangement results from N-terminal truncation. ***J. Am. Chem. Soc.*** 135(25):9299-9302.

Lu GJ, Opella SJ (2014) Mechanism of dilute-spin-exchange in solid-state NMR. ***J. Chem. Phys.*** 140(12): 124201.

Lu GJ, Opella SJ (2014) Resonance assignments of a membrane protein in phospholipid bilayers by combining multiple strategies of oriented sample solid-state NMR. ***J. Biomol. NMR*** 58(1): 69-81.

Lu GJ, Opella SJ (2013) Motion-adapted pulse sequences for oriented sample (OS) solid-state NMR of biopolymers. ***J. Chem. Phys.*** 139(8):084203.

Lu GJ, Park SH, Opella SJ (2012) Improved 1H amide resonance line narrowing in oriented sample solid-state NMR of membrane proteins in phospholipid bilayers. ***J. Magn. Reson.*** 220:54-61.

Lu GJ, Son WS, Opella SJ (2011) A general assignment method for oriented sample (OS) solid- state NMR of proteins based on the correlation of resonances through heteronuclear dipolar couplings in samples aligned parallel and perpendicular to the magnetic field. ***J. Magn. Reson.*** 209(2):195-206.

Lu GJ, Garen CR, Cherney MM, Cherney LT, Lee C, James MNG (2007) Expression, purification and preliminary X-ray analysis of the C-terminal domain of an arginine repressor protein from Mycobacterium tuberculosis. ***Acta Crystallogr Sect F Struct Biol Cryst Commun*** 63(Pt 11):936-939.

Teaching Interests:

At the graduate school, I TAed in analytical chemistry laboratory course, and presented weekly seminars as the only TA for a protein biochemistry course. At CalTech, I presented a 3-hour guest lecture in the Molecular Imaging course offered from chemical engineering department. Lastly, I mentored several undergraduate students during my Ph.D. and postdoctoral studies. Although my undergraduate background was mainly on biology, the exposure to multidisciplinary research in chemistry and physics during my Ph.D. and postdoctoral studies provided me confidence and enthusiasm to teach in various disciplines.

476128 Molecular Engineering for Cellular Imaging: From Fluorescence to Magnetic Resonance

Abstract:

Research Interests: Light-emitting proteins such as luciferase and GFP (green fluorescent protein) have revolutionized biological research by enabling genetically targeted labeling and optical imaging of cells and biomolecules. However, rapid advances in synthetic biology have imposed increasingly challenging requirements for improved imaging technologies that can enable the study of cellular function within several distinct contexts relevant to disease, healthcare, and industrial bioproduction. Addressing these requirements necessitates the development of smart molecular reporters for multiple imaging modalities, providing an exciting opportunity for biomolecular engineering. My research aims to meet this challenge through two distinct approaches involving **molecular engineering of: (1) reporter genes for magnetic resonance imaging (MRI) and (2) O_2-independent fluorescent reporters for optical imaging in low-O_2 conditions.**

Research Experience:

(1) Reporter genes for noninvasive cellular imaging using MRI (*Advisor: Mikhail G. Shapiro, Caltech***)**

MRI is a widely established modality for noninvasive imaging of intact animals and patients. However, the lack of sensitive molecular reporters for MRI (analogous to GFP and luciferase for optics), represents a critical missing capability for extending MRI to the study of cellular functions in whole organisms. In this regard, we recently introduced diffusion altering reporter genes that render cells visible in diffusion-weighted magnetic resonance images simply by increasing the rate of water exchange across the cell membrane[1]. This was achieved by overexpressing the human water channel, aquaporin, which results in robust MRI contrast without any deleterious effects on cell viability. Notably, aquaporin-based contrast is readily detectable at nanomolar levels of expression and in small (≈10%) subsets of aquaporin-labeled cells in mixed populations, which places aquaporin among the most sensitive reporter genes for MRI. Finally, we demonstrated the potential utility of aquaporin by imaging dynamic gene expression in a mouse brain tumor model.

(2) Engineering O_2-independent fluorescent reporter genes (*Advisor: Charles M. Schroeder, UIUC***)**

The significance of low O_2 tensions (hypoxia & anoxia) is illustrated by the critical role of hypoxia in regulating bacterial pathogenesis, antibiotic sensitivity, gut microbiome dynamics, and cancer chemoresistance. Consequently, live cell imaging in low-O_2 environments is of tremendous interest in biomedical science. However, widely prevalent GFP and luciferase-based reporters show O_2-dependent light emission, which precludes robust quantitative cell imaging. On the other hand, alternative O_2-independent reporters based on flavin-binding photosensory proteins (FbFPs) are limited by poor brightness, stability, toxicity, a small library size, and an incomplete understanding of their performance and properties as cellular reporters[2]. To this end, a key outcome of my research has been the characterization[3], expansion, and diversification of the O_2-independent FbFP fluorescent palette using directed evolution[4] and genome mining[5]. Notably, we engineered four new FbFP variants, which display improved brightness, photostability, small size (half the size of GFP), and a broad operational pH and thermal range (pH 3-11, up to 60°C).

Future Research Directions:

The transformative impact that optical reporters have had on biology can be attributed to the rich repertoire of cellular biosensors that have been developed based on foundational members of the GFP family. In contrast, there remains a critical dearth of genetically encoded MRI and optical sensors for cellular imaging in live animals and hypoxia respectively. As an independent investigator, I am interested in leveraging the previously described foundational classes of MRI and optical reporter genes to develop viable sensors for studying cellular function. Initial projects in the Mukherjee laboratory will have the following aims.

Aim 1: O_2-independent fluorescent biosensors. I propose to develop FbFP-based biosensors for imaging bacterial second messengers and cellular energy in physiologically realistic hypoxic models of bacterial infection, antibiotic resistance[6], and host-pathogen interactions. Concurrently, we will explore new approaches for further enhancing FbFP performance through improvements in protein folding, solubility, chromophore binding, and enriching cellular flavin content.

Aim 2: Functional MRI reporters. I propose to expand the library of diffusion altering reporter genes through the development, characterization, and application of pH and phosphorylation-sensitive aquaporins for imaging pH fluctuations and cell signaling events in preclinical animal models of infection, disease, and injury. Furthermore, I am interested in repurposing a recently described eukaryotic Mn^{2+} binding protein to develop a genetically encoded MRI-based sensor for calcium imaging *in vivo*.

In summary, I envision the proposed research to culminate in the development of improved imaging tools as well as platform technologies that will enable new approaches to study cellular function and gene expression in the context of realistic *in cellulo* and *in vivo* models.

Teaching Interests:

My teaching interests encompass topics in biomolecular engineering with a focus on imaging and therapy. Specifically, I am interested in developing and introducing a cross-disciplinary course on *Molecular*

Engineering for Imaging and Therapy that integrates fundamental principles from chemical kinetics, thermodynamics, and imaging along with techniques in molecular biology and protein engineering with an overall emphasis on the development of molecular and cell-based sensors and therapeutics. A central goal of my teaching philosophy is to help students identify exciting new avenues for innovation and apply principles from biomolecular engineering to brainstorm creative solutions. Through this initiative, I seek to emphasize the importance of cross-disciplinary approaches in biomolecular engineering while encouraging students to identify and appreciate the foundational engineering and scientific principles that bioengineering is built upon.

Teaching Experience:

In my PhD, I worked as a teaching assistant for four semesters during which time I extensively lectured courses in *Process Control*, *Biomolecular Engineering*, and *Bioenergy Technology* for which I received the campus nomination for "List of Teachers ranked Excellent". During my postdoc, I was a guest lecturer for topics on MRI contrast agents and relaxation theory as part of a course in *Molecular Imaging*. In addition, I have mentored three undergraduate researchers of whom two are currently pursuing graduate studies in chemical and biomolecular engineering.

Publications:

1. Mukherjee, A., et al. Diffusion altering reporter genes for magnetic resonance imaging. In revision (preprint available on bioarXiv)

2. Mukherjee, A. et al. Flavin-based fluorescent proteins: Emerging Paradigms in Biological Imaging. *Curr. Op. Biotech.*. (2014)

3. Mukherjee, A., et al. Characterization of flavin-based fluorescent proteins: an emerging class of powerful fluorescent probes. *PLOS ONE*. (2013)

4. Mukherjee, A., et al. Directed evolution of bright mutants of a flavin-binding fluorescent protein from *Pseudomonas putida*. *J. Biol. Eng.* (2012).

5. Mukherjee, A., et al. Engineering and characterization of LOV-based fluorescent reporter proteins in *C. reinhardtii* & *V. frigida*. *ACS Syn. Biol.*(2014)

6. Mohan, R.*, Mukherjee, A.*, et al. A Multiplexed Microfluidic Platform for Antibiotic Susceptibility Screening. *Biosens. Bioelectron*. (2013) (* co-first)

476130 Harnessing Diverse Microorganisms for Biochemical Production

Abstract:

Research Interests:

Biochemical production of compounds offers advantages over conventional synthesis due to the high specificity and diversity of biological catalysts, reactions occurring under labile conditions, and the use of renewable substrates. Metabolic engineering of microbes has primarily focused on modifying the model organisms *E. coli* and *S. cerevisiae* due to their fast and robust growth on inexpensive fermentation broths and the wealth of genetic tools available to them. Heterologous pathway addition in these platform hosts has had mixed success, often with natural producers reaching higher titers due to improper recombinant enzyme folding and regulation. Additionally, other traits should be considered before selecting a production host, such as substrate utilization, endogenous pathways, product toxicity, secretion of product or enzymes, growth temperature, susceptibility to contamination, pathogenicity of host, and growth environment. These factors have led to the development of alternative hosts for chemical production, including *C. glutamicum* (amino acid synthesis), *Synechocystis* sp. (CO_2 autotroph), *C. thermocellum* (consolidated biomass processing), and *B. subtilis* (GRAS certified, carotenoids, protein secretion), just to name a few. Further, advances in synthetic biology and â€“omics techniques have enabled the rapid modification and characterization of microbes. Our laboratory will use these methods and develop additional tools to harness the diversity of microorganisms for biochemical production. Specifically, we aim to 1) engineer organisms that are uniquely tolerant to reaction conditions, 2) use protein engineering and synthetic biology tools to enhance enzyme production and regulation in these hosts, and 3) discover novel pathways and organisms through bioprospecting to expand the range of chemicals to be synthesized and traits to be selected from.

Product toxicity and removal are two long-standing challenges to bioprocess engineering. Co-solvent culturing aims to alleviate both problems by stripping off products as they are generated, sequestering them from the organism and eliminating them from further conversion. We plan to use sustainable solvents and organisms that grow under these unique reaction conditions to produce compounds of interest, such as aldehydes and alkanes, that are highly toxic to cells and have high volatility. In addition to conventional metabolic engineering to increase biochemical production, we will develop new tools to enhance enzyme expression and create gene knockouts in non-model organisms. In particular, we will develop protein-folding reporters to assess and enhance heterologous enzyme stability in non-model hosts. Combinatorial libraries and metabolic flux analysis will be used to balance enzyme expression. CRISPR-based tools will be created for gene editing and chromosomal integration of pathways. Our lab will seek to not only modify organisms that have already been cultured, but to discover new hosts from environmental samples with unique characteristics to enhance biochemical production. Through sequencing and annotation of pools of organisms from specific environments, we hope to find novel enzymes for both production and remediation of compounds of interest. We aim to study the mechanisms that give rise to the unique traits in these organisms to further improve them for specific processes or transfer them to other species of interest. Through this work, we hope to create a set of design rules for utilizing microbes for chemical production as well as put together a pipeline of methods that is needed to systematically discover, characterize and engineer an organism of interest.

Teaching Interests:

I hope to use teaching of courses such as kinetics, separations, control theory and bioprocess engineering as well as outreach with high school students and teachers to share and further develop our study of biochemical engineering of microorganisms. Additionally, I plan to develop new courses focused around hand-on design of biological systems and implement team-based projects in classes I teach.

476135 Systems Biology and Systems Pharmacology Approaches to HIV Infection and TB/HIV Co-Infection

Abstract:

Systems biology and systems pharmacology approaches to HIV infection and TB/HIV co-infection

Elsje Pienaar, University of Michigan

Research Investigator

Research Interests:

I aim to establish a research group that uses computational approaches to study within-host dynamics and treatment of infectious diseases, such as tuberculosis (TB) and human immunodeficiency virus (HIV). My work will leverage my established collaborations with experts in animal models of HIV and TB-HIV co-infection. I will build multi-scale multi-compartment computational models of HIV infection to determine 1) the roles of cytokine and CD4+/CD8+ T cell imbalances, in the context of different CD4+ T cell levels, in driving immune-deficiency in HIV+ patients; 2) pharmacokinetics and viral dynamics in different physiological compartments which influence HIV treatment efficacy and the persistence of viral reservoirs, and 3) precision treatment options based on patient-specific data.

Ultimately, I will integrate this model with a host-scale model of tuberculosis (TB) infection and treatment to 1) study the natural progression of this globally devastating co-infection, and 2) evaluate different coordinated treatment strategies for TB and HIV.

Through my experience in infectious diseases (experimental and computational), systems biology, systems pharmacology, and productive collaborative experience with experimentalists, I am well positioned to achieve these goals.

Research Proposal Experience:

I have worked with current and past mentors to prepare successful NIH research proposals. I have submitted proposals for independent funding from the Simons Foundation. I am awaiting a decision on an NIH training grant (K01) and I am preparing an application for a PhRMA foundation research starter grant.

Teaching Interests:

My educational experiences include classroom teaching (guest lectures in chemical engineering and mathematical biology), student mentoring (graduate and undergraduate students in chemical engineering and microbiology) and formal training in pedagogy (Certificate in learning, instructing and knowledge (Linköping University, 2012)). I look forward to teaching any core Chemical Engineering courses, as well as developing specialty graduate courses in systems biology/systems pharmacology.

Academic Appointments

2015 – Present. **Research investigator** in Chemical Engineering, Microbiology and Immunology, University of Michigan, Ann Arbor, MI.

Grants

2011 Knut and Alice Wallenberg Travel Grant

2012 Microbiology and Immunology Postdoctoral Travel Grant

Honors and Awards

2014 BMES Innovation and Career Development Award

2010 Outstanding Doctoral Dissertation Award, UNL College of Engineering

2009 NIH-NIAID Scholarship Keystone Symposia

Education and Training

2012 – 2015 **Research fellow**. Chemical Engineering, Microbiology and Immunology, University of Michigan, Ann Arbor, MI. *Advisors: Dr. Jennifer Linderman and Dr. Denise Kirschner.*

2010 – 2012 **Postdoctoral fellow**. Medical Microbiology, Linköping University, Linköping, Sweden. *Advisor: Dr. Maria Lerm.*

2007 – 2010 **PhD**, Chemical and Biomolecular Engineering, Biological Sciences University of Nebraska-Lincoln, Lincoln, NE. *Advisor: Dr. Hendrik Viljoen.*

2005 – 2007 **MSc**, Chemical and Biomolecular Engineering, Biological Sciences University of Nebraska-Lincoln, Lincoln, NE. *Advisor: Dr. Hendrik Viljoen.*

2002 – 2004 **BSc**, Financial and Actuarial Mathematics, University of Pretoria, Pretoria, South Africa

Selected Publications

Pienaar E, Matern W, Linderman JJ, Bader JS, Kirschner DE. Multi-scale model of *Mycobacterium tuberculosis* infection maps metabolite and gene perturbations to granuloma sterilization predictions.

Infection and Immunity. 2016:84(5):1650.

Pienaar E, Dartois V, Linderman JJ, Kirschner DE. *In silico* evaluation and exploration of antibiotic tuberculosis treatment regimens. BMC Systems Biology. 2015;9:79.

Linderman JJ, Cilfone N, **Pienaar E**, Gong C, and Kirschner DE. A Multi-scale Approach to Designing Therapeutics for Tuberculosis. Integrative biology. 2015; 7:591-609.

Pienaar E, Cilfone NA, Lin PL, Dartois V, Mattila JT, Butler JR, Flynn JL, Kirschner DE, Linderman JJ. A computational tool integrating host immunity with antibiotic dynamics to study tuberculosis treatment. J Theor Biol. 2015;367:166-179.

Raffetseder J*, **Pienaar E***, Blomgran R, Eklund D, Patcha Brodin V, Andersson H, Stendahl O, Welin A, Lerm M. Replication Rates of Mycobacterium tuberculosis in Human Macrophages Do Not Correlate with Mycobacterial Antibiotic Susceptibility. PLoS One. 2014;9(11):e112426. *: co-first authors.

Pienaar E, Lerm M. A Mathematical Model of the Initial Interaction between *Mycobacterium tuberculosis* and Macrophages. J Theor Biol. 2014; 342:23-32.

Viljoen S, **Pienaar E**, Viljoen HJ. A state-time epidemiology model of tuberculosis: Importance of re-infection. Computational Biology and Chemistry. 2012;36(0):15-22.

Pienaar E, Fluitt AM, Whitney SE, Freifeld AG, Viljoen HJ. A model of tuberculosis transmission and intervention strategies in an urban residential area. Comput Biol Chem. 2010;34(2):86-96.

Pienaar E, Whitney SE, Viljoen HJ, van Rensburg NF. A model of the complex response of Staphylococcus aureus to methicillin. J Theor Biol. Apr 2009;257(3):438-445.

Pienaar E, Viljoen HJ. The tri-frame model. J Theor Biol. Apr 2008;251(4):616-627.

Pienaar E, Theron M, Nelson M, Viljoen HJ. A quantitative model of error accumulation during PCR amplification. Comput Biol Chem. Apr 2006;30(2):102-111.

Publications in preparation

Pienaar E, Sarathy J, Prideaux B, Dartois V, Kirschner DE and Linderman JJ. Systems Pharmacology Predicts Antibiotic Spatial Distribution And Efficacy In TB Granulomas.

Pienaar E, Kirschner DE and Linderman JJ. Mechanisms driving drug resistance emergence in TB granulomas.

476348 Engineering and Physical Sciences in Oncology: Tumor Cell Adhesion and Treatment in Blood and Bone Marrow

Abstract:

Research Interests: Â

Metastasis through the bloodstream contributes to poor prognosis in many types of cancer. Mounting evidence implicates physical interactions between tumor cells and the local microenvironment as facilitating this process, in a manner similar to leukocyte trafficking during inflammation. In the first part of my talk, I will discuss our work in developing â€œunnatural killer cellsâ€ as an immunotherapy-inspired approach to kill metastatic tumor cells in the bloodstream. Through the use of lipid-based nanomaterials administered intravenously, we were able to coat the surface of circulating white blood cells *in vivo* with the immune cytokine TRAIL and the adhesion receptor E-selectin. This approach is inspired by the cytotoxic activity of natural killer cells, and is effective at killing tumor cells in human blood samples *in vitro* and in the peripheral circulation of mice *in vivo*. This nanoparticle platform holds promise as a means to neutralize circulating tumor cells in the blood and potentially block the progression of metastasis. In the second part of my talk, I will discuss the development of new strategies to deliver RNA therapeutics to the bone marrow microenvironment from the treatment of multiple myeloma, an incurable blood cancer the colonizes within the marrow. Specifically, we have identified novel polymer-lipid hybrid platforms that, for the first time, enable potent gene delivery to the bone marrow microenvironment *in vivo*. These delivery materials were shown to silence key factors within bone marrow that trigger the homing, proliferation, and survival of multiple myeloma *in vivo*. These platforms are now currently being investigated for the prevention of breast and prostate cancer metastasis to bone, the treatment of multiple myeloma, and to increase the therapeutic efficacy of bone marrow stem cell transplantations.

One of the greatest opportunities as a professor is to serve as a mentor, in many different forms, to the next generation of scientists and engineers. Serving as mentor can occur in a variety of settings, ranging from the classroom, the lab, as

well as the next phase of their own independent career. Throughout my academic, I have been committed and actively sought out opportunities to teach future scientists and engineers in various roles including: teaching assistant, guest lecturer, research mentor, and visiting scientist in outreach activities sponsored by federal agencies including the NSF and NIH. I have also had the privilege of mentoring more than 10 trainees in the laboratory, which has shaped myself both as a research and educator, and has been one of the most rewarding aspects of my academic career to date.

Teaching Interests:

I am interested and qualified in teaching both Chemical Engineering and Bioengineering courses (undergraduate and graduate level) in the areas of biomaterials, reaction kinetics, fluid dynamics, and thermodynamics. Additionally, I am excited to create a graduate course in the emerging fields of Molecular Engineering, Drug Delivery, and Nanomedicine.

476598 Novel Bio-Ionic Liquid Functionalized Conductive Hydrogel

Abstract:

Research Interests: Material and Biomaterial synthesis, Intelligent Design, Catalysis, Medicine

Teaching Interests: Transport Phenomena, Thermodynamics, Reaction Engineering, Polymer, and Biomaterial

Conducting polymers (CPs) constitute a new generation of organic materials, which exhibit electrical and optical properties that are similar to those of metals and inorganic semiconductors. They also possess the advantageous properties associated with regular polymers, such as ease of synthesis and processability, when compared to metals [1]. CPs are widely used in microelectronics for the manufacture of batteries, photovoltaic devices, light-emitting diodes and electro-chromic displays, as well as in biomedical applications including neural probes and prostheses, and controlled release systems. CPs have also been shown to modulate cellular processes such as cell growth and migration through electrical stimulation (conductivities from 10-4 to 9 S/cm). These properties make CPs and their derivatives remarkably suitable for tissue engineering applications using excitable cell types, including nerve, muscle, and cardiac tissue. However, there are still several technical challenges associated with the use of CPs for tissue engineering applications. The main drawbacks associated with the existing, CP-based biomaterials are: poor polymer-cell interactions due to the absence of cell binding sites, high hydrophobicity, poor solubility and processability, non-biodegradability, as well as the inability to optimize their mechanical properties. Therefore, the synthesis of biomaterials that are both electroactive and biodegradable is highly desirable. Electroconductive properties have been incorporated into biomaterials through the addition of nanomaterials (e.g. silver nanowires, gold nanoparticles, carbon nanotubes (CNTs), and graphene oxide) or conductive polymers (e.g. polyaniline, polypyrole, polythiophene) to the polymeric matrices. However, these approaches suffer from difficult and prolonged processing procedures, harsh reaction conditions, difficulties in the incorporation of nanomaterials into the polymer networks, limited solubility in aqueous media, and cytotoxicity. Therefore, there is an unmet need for the development of biocompatible, biodegradable, and conductive biomaterials, with tunable conductive and physical properties [1].

Here, I introduce a new class of conductive polymer-based biomaterials through conductive bio-ionic liquid (BIL) functionalization of different biopolymers. BIL are low melting organic salts that exhibit several technical advantages, including low volatility, high ionic conductivity and electrochemical stability, and excellent dissolution capabilities. BILs have been previously used as biocompatible and biodegradable materials for various applications such as cancer therapy, multi-responsive drug delivery systems, sensors, batteries, and biomedical implants [2-5]. Functionalization of biopolymers with BIL can provide conductive properties to the polymer networks, while preserving the biological and physical characteristics of the biopolymers. This property makes them unique candidates for various tissue-engineering applications. I performed chemical conjugation of BIL to various synthetic and natural polymers including gelatin, tropoelastin, elastin like polypeptides (ELP), collagen, poly(ethylene glycol) (PEG), hyaluronic acid (HA), poly(glycerol sebacate) polymer (PGS) and combinations of them, in order to generate conductive biomaterials with high biocompatibility and tunable physical and electrical properties. For example, I engineered a novel 3D gelatin-based hydrogel by using a visible light-activated gelatin prepolymer, functionalized with a BIL for heart tissue regeneration. FTIR and NMR analysis confirmed the successful functionalization of BIL to the biopolymer structure. The mechanical and electrical properties of the engineered hydrogel were then optimized to mimic the properties of the native heart tissue. Our *in vitro*tests confirmed that the engineered conductive scaffolds provide a suitable conductive and mechanical substratum for the culture of cardiac cells. This led to the formation of 3D functional cardiac patches, which can potentially be used for heart tissue regeneration.

References:

[1]- Kaur G , Adhikari R, Cass P , Bown M, Gunatillake P. RSC Adv., 2015, 5, 37553-37567

[2] Brock M, Nickel A-C, Madziar B, Blusztajn JK, & Berse B (2007) Brain Research 1145(0):1-10.

[3] Kumar SSD, Surianarayanan M, Vijayaraghavan R, Mandal AB, & MacFarlane DR (2014) European Journal of Pharmaceutical Sciences 51(0):34-44.

[4] Curto VF, et al. (2014). Physical Chemistry Chemical Physics 16(5):1841-1849.

[5] Dias AMA, et al. (2013) ACS Sustainable Chemistry & Engineering 1(11):1480-1492.

478107 Engineering Surfaces through Sequential Stop-Flow Photopatterning

Abstract:

Poster Abstract

This poster describes the design and application of a novel experimental setup to combine light-mediated and flow chemistry for the fabrication of hierarchical surface-grafted polymer brushes. Using light-mediated, surface initiated controlled radical polymerization and post-functionalization via well-established, and highly efficient chemistries, polymer brush films of previously unimaginable complexity are now shown to be accessible. This methodology allows full flexibility to exchange both lithographic photomasks and chemical environments insitu, readily affording multidimensional thin film architectures, all from uniformly functionalized substrates.

Research Interests:

Many diseases of the human nervous system, e.g., Parkinson's, Alzheimer's, and Huntington's, occur as a result of the progressive loss of structure and function of neurons. Charged proteins – naturally occurring polyelectrolytes and polyampholytes – have been identified as key components to the development and evolution of such neurodegenerative processes. My research interests focus on synthetically mimicking and studying the complex electrostatic interactions inherent to such ampholytic polymer architectures. Beyond these fundamental studies, charged macromolecules and their predicted response to electric fields provide fascinating routes towards advanced materials with tunable and switchable properties.

Teaching Interests:

Education is a continuous and essential component of the academic research environment. I value the opportunity to teach and learn from my students as one of the most important and exciting responsibilities in academia. I want to not only transfer knowledge, but also ignite a passion, a strong internal desire to learn and improve. I wish to emphasize critical thinking & scientific questioning, while conveying the joy of discovery, and helping students develop both professionally & personally.

477586 Biological Applications of Fluctuation Solution Theory

Abstract:

Successful Proposals:

NIH F32 â— NSF GRFP â— NSF GK-12 â— NSF REU â— NSF XSEDE/TeraGrid â— Maryland Advanced Research Computing Center â— Assisted in developing/revising rejected and successful NIH R01 proposals

Research Interests:

Thermodynamics and statistical mechanics â— Solution microstructure â— Molecular Dynamics simulations, including methodology development and force field validation/development â— Biothermodynamics/Biophysics, especially proteins and lipids â— Impacts of microenvironment changes and surfaces on conformational and association equilibria of biomolecules â— Lipid organization in multicomponent, physiologically relevant, lipid bilayers â— Kirkwood-Buff Theory / Fluctuation Solution Theory

Teaching Interests:

Thermodynamics and statistical mechanics â— Computational chemistry (classical) â— Physical chemistry â— General chemistry â— Biochemistry/Biophysics â— Proteins â— Linux â— Fortran/MATLAB/basic scripting

481181 Understanding and Improving Biomanufacturing in Chinese Hamster Ovary Cells through New Gene Expression and Systemic Cell Engineering Platforms

Abstract:

Research Interests:

My research experience has focused on the various subjects of Chinese hamster ovary (CHO) cells, such as -omics studies (genomics, transcriptomics, and proteomics), cytogenetics, genome and metabolic engineering, and bioprocess engineering. CHO cells are the most important mammalian host cells for the biomanufacturing of therapeutic proteins, accounting for more than 50% of global annual biopharmaceutical protein sales. In addition, transformative changes in CHO cell line development are required to meet industrial demands, such as lowering manufacturing cost, producing new medicines, and extending culture duration. I plan to continue this work towards improving CHO biomanufacturing platforms.

While significant productivity improvements have been achieved by extensive efforts over three decades, understanding of CHO biology has been limited. For example, what makes a high productivity cell line, how media components/supplements affect culture longevity and productivity, and how stressful culture conditions (low culture temperature, hyperosmolality, or histone deacetylase inhibitor treatment) can lead to higher recombinant protein production are still largely unknown. In addition, although cell line instability issues, such as unexpected changes in productivity, product quality, and host cell protein (HCP) expression, are increasingly recognized to negatively affect biomanufacturing processes, the causes and underlying mechanisms of cell line instability remain poorly understood. Addressing these important questions and challenges will enable us to develop stable, well characterized CHO cell lines producing proteins and biomolecules with desirable quality attributes, thereby providing more effective, better controlled, and safer therapeutics for human health. Therefore, my research interests include (1) developing new gene expression and amplification methods for the stable and high production of therapeutic proteins and (2) mammalian genome and cell engineering based on genome scale modeling and systems biology.

Postdoctoral Projects:

"Knocking out CHO host cell proteins using CRISPR/Cas9 to improve downstream processes" and "Developing a karyotype-based framework to identify and quantify CHO cells' genomic/chromosomal instability that affects protein productivity" under the supervision of Dr. Kelvin Lee at the University of Delaware.
"Metabolic and bioprocess engineering of CHO cells to produce bioengineered heparin" under the supervision of Dr. Susan Sharfstein at SUNY Polytechnic Institute and Dr. Robert Linhardt at Rensselaer Polytechnic Institute.

Ph.D. Dissertation:

"Engineering of recombinant Chinese hamster ovary cells using a proteomic approach" under the supervision of Dr. Gyun Min Lee at Korea Advanced Institute of Science and Technology (KAIST).

Education:

Ph.D., Biological Sciences, KAIST, 2009
B.S., Biological Sciences, KAIST, 2003

Selected Publications:

1. Baik JY, Lee KH. 2016. Toward improved host cell protein impurity assessment. Biotechnol. J. 11(8):998-999.
2. Baik JY, Dahodwala H, Oduah E, Talman L, Gemmill TR, Gasimli L, Datta P, Yang B, Li G, Zhang F, Li L, Linhardt RJ, Campbell AM, Gorfien SF, Sharfstein ST. 2015. Optimization of bioprocess conditions improves production of a CHO cell-derived, bioengineered heparin. Biotechnol. J. 10(7):1067-1081.
3. Baik JY, Lee KH. 2014. Toward product attribute control: Developments from genome sequencing. Curr. Opin. Biotechnol. 30:40-44.
4. Baik JY, Gasimli L, Yang B, Datta P, Zhang F, Glass CA, Esko JD, Linhardt RJ, Sharfstein ST. 2012. Metabolic engineering of Chinese hamster ovary cells: Towards a bioengineered heparin. Metab. Eng. 14(2):81-90.
5. Baik JY, Ha TK, Kim YH, Lee GM. 2011. Proteomic understanding of intracellular responses of recombinant chinese hamster ovary cells adapted to grow in serum-free suspension culture. Biotechnol. Prog. 27(6):1680-1688.

Teaching Interests:

Learning with understanding: Learning is more than just memorizing and practicing; it is about understanding. Through all my teaching and mentoring experience, I challenged my students' critical thinking skills by asking questions, such as what each procedure meant, why it was necessary, and which steps were more critical than others, and by asking them to explain their thought process. Through classes, I will continually emphasize learning with understanding by asking "how can we understand this concept?".

Applicability: As an engineer, I seek to link knowledge to applications. I will provide a platform for students to translate knowledge gleaned from lectures to problems or applications in various engineering fields.
Communication: As collaboration and interdisciplinary research become increasingly important, good communication skills are essential to successful research. Moreover, strong written and verbal skills are critical to convey the importance and impact of research. To improve communication skills, I will foster discussion-oriented class environments, which will also complement my learning with understanding philosophy. In addition to scientific communication skills, I will also pursue outreach activities to connect research to real life in various ways, such as hands-on activities, demonstrations, or engaging tours of

labs/facilities. In conclusion, I believe that when we are able to understand, apply, and propagate what we learn, knowledge becomes alive and inspires future scientists and engineers.

I would apply my teaching principles to undergraduate and graduate courses, including Molecular Biology, Biochemistry, Genetics, Bioprocess Engineering, and Biotechnology. I am also interested in developing advanced cutting-edge courses such as Genomics, Genome Engineering, and Recent Progress in Biochemical Engineering and Technology.

458653 Human Stem Cell Derived Neutrophils As a Primary Neutrophil Model

Abstract:

As the first responders and the most abundant innate immune cell, neutrophils are a crucial part of the immune system. During an immune response, neutrophils migrate to sites of infection and eliminate bacteria through a series of anti-microbial functions including phagocytosis, production of reactive oxygen species, and NETosis. Improper regulation of neutrophil behavior contributes to many human diseases. Improper neutrophil function leads to chronic infections when neutrophils do not arrive at a site of infection or chronic inflammation when neutrophils persist at a site of infection. Current models for evaluating neutrophil behavior *in vitro* include cell lines and primary human neutrophils. Unfortunately, cell lines do not always accurately model many important neutrophil functions such as NETosis and phagocytosis and primary human neutrophils are hard to genetically manipulate. Recently, stem cell derived neutrophils have emerged as a possible model for primary neutrophils that would allow for genetic manipulation and the modeling of human disease.

In order to properly model primary human neutrophils, hiPSC-derived neutrophils must display the same morphology, migration, and anti-microbial behavior as primary human neutrophils. We have found that the morphology of hiPSC-derived neutrophils is indistinguishable from primary human neutrophils. Furthermore, hiPSC-derived neutrophils have the characteristic tri-lobed nucleus common to primary neutrophils. In order to reach sites of inflammation neutrophils must be capable of migrating to known neutrophil chemoattractants. We studied neutrophil chemotaxis using a previously published microfluidic device that creates a stable chemokine gradient and found that hiPSC-derived neutrophils efficiently chemotax to a source of the bacterial peptide fMLP, a common neutrophil chemoattractant. Finally, hiPSC-derived neutrophils must be able to perform anti-microbial functions. We visualized hiPSC-derived neutrophils in co-culture with *pseudomonas aeruginosa* using time-lapse microscopy. In contrast to popular neutrophil cell lines that lack the ability to phagocytose bacteria, we found that hiPSC-derived neutrophils were capable of efficiently phagocytosing *pseudomonas* over an extended period of time. We have also found that, upon activation, hiPSC-derived neutrophils produce neutrophil extracellular traps, proving these cells are capable of performing several anti-microbial functions.

Overall, we have found that hiPSC-derived neutrophils accurately recapitulate many characteristics and can serve as an excellent model of primary human neutrophils. We have shown that hiPSC-derived neutrophils share the morphology, migration capabilities, and anti-microbial functions of primary human neutrophils. hiPSC-derived neutrophils have the characteristic tri-lobed nuclei, are capable of chemotaxis to fMLP, a physiologically relevant chemokine, and poses the ability to phagocytose bacteria. In the future, hiPSC-derived neutrophils as a model will allow for more in-depth study of primary human neutrophil function.

463229 Mechanistic Model of CD3ζ Immunoreceptor Tyrosine-Based Activation Motif (ITAM) Phosphorylation Sequence

Abstract:

Gaining a detailed understanding of T cell activation has become increasingly important as T cells are used in immunotherapeutic applications. For example, T cells engineered with modified T cell receptors (TCRs) or CD3ζ-bearing chimeric antigen receptors (CARs) have emerged as promising cancer therapies. The CD3ζ chain, a member of the TCR complex, contains three immunoreceptor tyrosine-based activation motifs (ITAMs), which are thought to have different signaling functions: the first N-terminal ITAM is largely stimulatory, while the second and third ITAMs each exhibit some inhibitory signaling characteristics [1]. Lymphocyte-specific protein tyrosine kinase (LCK) is able to phosphorylate two tyrosine residues on each ITAM, resulting in T cell activation. The "kinetic proofreading" hypothesis is the prevailing theory regarding regulation of ITAM phosphorylation. It presumes that the presence of multiple ITAMs on the CD3ζ chain allows T cells to differentiate between foreign and self-peptides; however, the specific mechanisms that control LCK catalytic activity and its phosphorylation of the CD3ζ chain are still not well defined. Therefore, we have developed a mechanistic computational model of CD3ζ phosphorylation by LCK. The model is paired with site-specific mass spectrometry phosphorylation data from both CD3ζ and LCK *in vitro* to better understand the key steps that initiate T cell activation.

We constructed the model using BioNetGen, a rule-based formalism that allows us to account for the many species that arise from the multiple phosphorylation sites on CD3ζ. The model is implemented as a set of ordinary differential equations in MATLAB and fit to experimental data generated using an *in vitro* reconstituted membrane system that mimics the two-dimensional interactions that occur in T cells [2]. We use mass spectrometry to measure the level of phosphorylation at each individual tyrosine residue in the system over time. To determine the order in which each of the six tyrosine residues on CD3ζ are phosphorylated, we first stimulate CD3ζ with a constitutively active form of LCK and fit our mechanistic model to this data. We then combine this model with a previously developed model of LCK autoregulation [3]. The combined model will be validated using experimental data of wild-type LCK autophosphorylation and phosphorylation of CD3ζ. As a result, we generate a model that matches experimental data and can predict data not used in the fitting.

The model predicts the order and rate at which different ITAMs on the CD3ζ chain are phosphorylated, helping to decifer the kinetic proofreading mechanism that is thought to precisely regulate T cell activation. Additionally, we have applied the model to investigate how changes to the CD3ζ chain, such as removing the third ITAM, can affect overall activation. The results predicted by the model can be implemented in CD3ζ-bearing CAR-engineered T cells. Thus, the model generates new hypotheses that can be tested experimentally, allowing us to quantitatively explore T cell signaling and guide the development of immunotherapies. The model is a quantitative framework that can be used to examine the dynamics of CD3ζ chain activation by LCK and its effects on T cell activation.

Citations

[1] Chae, W., et al. *Int. Immunol.*, 2004, 16(9), 1225-36.

[2] Hui, E. and R. Vale. *Nat. Struct. Mol. Biol.*, 2014, 21(2), 133-142.

[3] Rohrs, J. A., Wang, P., and S. D. Finley. *Cell Mol. Bioeng.*, 2016, In press.

463718 Adoptive Transfer of CAR-Engineered T Cells with Surface-Conjugated Synthetic Nanoparticles Containing Small Molecule Inhibitors forÂ Reversing Intratumoral T Cell Hypofunction

Abstract:

Adoptive T cell transfers of genetically modified cytotoxic T cells to express chimeric antigen receptors (CAR) have become a promising immunotherapy method. CARs specifically target tumor-associated antigens or cancer stromal antigens with high binding affinity. Activation of a CAR engineered T cell is independent of the major histocompatibility complex (MHC) and is achieved primarily through the single chain variable fragment (scFv) that is fused with intracellular signaling domains. Multiple researchers have shown that adoptive transfer of CAR T cells is successful in patients with B cell haematological malignancies, but is still in the earlier stages of development for treatment of solid tumors. One limiting factor to adoptive T cell therapy is the suppressive tumor microenvironment that inactivates tumor infiltrated T cell (TIL) function. Moon et al. (2014) showed that hypofunctional TILs express high levels of inhibitors of the intracellular pathways, such as the up-regulation of DGKa and DGKz that phosphorylates diacylglycerol. In addition to intracellular inhibitors, the tumor microenvironment also contains high concentration of TIL suppressor molecules such as adenosine that is up taken by the A2AR receptor expressed on the cell surface of CD4 and CD8 T cells. Adenosine is generated from extracellular ATP through CD39 and CD73 expressed on the surface of tumor cells and regulatory T cells. We hypothesized that *in vivo* delivery of pharmaceutical drugs, namely DGK and A2AR inhibitors, with crosslinked multilayer liposome vesicles (cMLV) would reverse TIL hypofunction. For delivery, cMLVs containing drugs were conjugated with adoptively transferred CAR T cells to target hypofunctional TILs residing in the tumor. We demonstrated that conjugation of cMLVs to CAR T cells does not alter key functions, such as T cell cytokine secretion and cytotoxicity. Furthermore, our *in vivo* study showed that combination therapy of small molecule inhibitors with CAR T cells were able to reverse TIL hypofunction and increase tumor infiltrating CD3+ T cells.

464271 Sequencing the Evolving Functional Antibody Repertoire in Rheumatoid Arthritis

Abstract:

Combining large-scale antibody repertoire sequencing data with a comprehensive analysis of antibody function can provide a more complete understanding of the role of an antibody response in disease pathology. For example, a hallmark of rheumatoid arthritis (RA) is the generation of an adaptive autoimmune B cell response against citrullinated antigens that result from a post-translational modification of arginine. However, the precise specificity and functional properties of the anti-citrullinated protein antibodies (ACPAs) in RA remain poorly understood. To address this, we applied a high-throughput single-cell barcoding methodology to sequence the entire variable region of paired heavy and light chain antibody genes expressed by individual plasmablasts (activated B cells). Bioinformatics analysis of these sequencing results precisely quantifies the degree of clonal expansion and identifies the key affinity matured, clonally expanded antibodies generated in an immune response. Importantly, with the paired heavy and light chain data, we can express these antibodiesÂ *in vitro*, thereby enabling the direct linkage of a patient-derived antibodyâ€™s sequence with its function (e.g. antigen specificity). Furthermore, we evaluated the antibody repertoire over time to characterize the evolving adaptive autoimmune response, elucidate the key autoantibodies, and provide insight into the pathogenic mechanisms underlying RA. We sequenced single plasmablasts from peripheral blood sampled at serial time points from 8 individuals with RA and used a tetrameric citrullinated peptide-based sort reagent to isolate B cells that produce ACPAs. We bioinformatically identified lineages that persisted across the serial time points and characterized the expansion and contraction of clonal lineages, some of which expressed multiple antibody isotypes, in relation to clinical metrics (e.g. swollen and tender joints). We selected antibodies representative of the identified B cell clonal families for recombinant expression and defined their antigen targets with planar and bead-based antigen arrays. By combining detailed sequence information with antibody function, this broadly applicable approach facilitates a comprehensive investigation of the evolution of the antibody response in disease, which can reveal novel diagnostic markers and therapeutic targets.

466794 Elucidating T-Cell Activation Threshold Using Precisely Defined Three-Dimensional Artificial Antigen Presentation

Abstract:

Elucidating T-Cell Activation Threshold Using Precisely Defined Three-Dimensional Artificial Antigen Presentation

Mason Smith, Fei Wen Adaptive immune response is shaped in large part by T cells recognizing and binding cognate antigen on the surface of antigen presenting cells (APCs). While much is known about the behavior of T cells during an immune response, the exact molecular mechanism that translates peptide-major histocompatibility complex (pMHC)- T-cell receptor (TCR) binding to T cell activation remains a controversial topic in immunology. Several studies using antigen presenting fluid lipid bilayers have attempted to shine a light on this mechanism by looking at the minimum number of pMHC required to initiate T cell activation. These studies have found that the early steps of T cell activation (TCR triggering) can be *initiated*by as few as three agonist pMHCs in a membrane, and that there exists a global pMHC density threshold below which T cell activation does not occur. While these studies have provided unique insights into TCR triggering, their predicted T-cell activation thresholds vary in planar intermolecular pMHC spacing from 34 nm to greater than 150 nm. Further, conflicting results regarding the importance of local pMHC density and total pMHC encountered have been reported. Planar fluid lipid bilayers provide robust and scalable experimental systems with incredibly fine control over molecular density; however, it is not immediately clear if their results can be extrapolated to describe T cell response to pMHC on a three-dimensional cell surface. Additionally, planar pMHC display systems assume sustained interfacial contact between a T cell and a flat antigen-presenting surface and are typically characterized by some uncertainty in the number of pMHC molecules immobilized in each defined nanoclusters.

We have engineered a novel artificial antigen presentation system that provides precise control over the local and global arrangement of pMHC molecules on the surface of a living cell using recombinant scaffold protein. Using this artificial antigen presentation system, we are able to display thousands of clusters of up to six individual pMHCs with single-molecule precision on a cell surface. Further, cells presenting unique local pMHC densities can be sorted using fluorescence activated cell sorting (FACS) to isolate cells presenting different global pMHC display levels. By using FACS to isolate cells with the same global pMHC display level but discretely different pMHC cluster size, we can study T cell response to a wide range of local and global pMHC densities in a three-dimensional cell-culture environment. Preliminary T-cell activation results using sorted artificial antigen presenting cells support the existence of a minimum global pMHC display threshold for T cell activation, which is different from that reported using planar lipid bilayers. More importantly, they suggest that local pMHC valency can compensate for a reduction in the global pMHC display level. These results suggest that while two-dimensional systems for *in vitro* T cell activation present an important, scalable research platform, quantitative thresholds for T cell activation derived from these systems may underestimate the activation threshold in physiologically relevant systems.

470621 Development of a Virus-like Particle Based HIV Vaccine Candidate

Abstract:

Worldwide, 35 million people are infected with HIV. Antiretroviral therapies have helped to decrease the mortality rate, but their expensive nature makes them inaccessible to many millions of clinically eligible individuals. Despite over 30 years of intense efforts, an HIV vaccine is still not in sight. A prophylactic HIV vaccine would prevent the infection and do so at a much lowered cost, thus increasing access, particularly in the developing world. Whole-killed vaccines, in which the virus is physically destroyed by chemicals, heat or radiation, are used for many other diseases, but do not lead to a relevant immune response against HIV. The mutation rate of HIV is very high, so variant strains that are resistant to vaccines prepared against individual strains emerge rapidly.

Our approach aims to create an HIV vaccine candidate targeting the process by which the virus infects cells. In order to infect a new cell, the HIV membrane must fuse with the cell membrane. This membrane-fusion event is mediated by a specific protein on the surface of the virus, called gp41, which is highly conserved at both the nucleotide and amino acid levels. Earlier work in the field has elucidated the overall features of this membrane-fusion event, including an intermediate state of gp41 called the pre-hairpin intermediate (PHI). An FDA-approved drug, Fuzeon, binds to the PHI, preventing membrane fusion and blocking HIV infection. Additionally, a monoclonal antibody targeting this PHI (called D5) has been shown to prevent infection of cells by various strains of HIV. Together, these results validate our approach to use the PHI as a vaccine target.

Our proposed vaccine candidate consists of an engineered virus-like particle scaffold displaying both a PHI mimetic and stimulators of the innate immune system. Using the engineered virus-like particle scaffold results in a vaccine candidate with the correct size for efficient trafficking to the lymph nodes. We have produced a PHI mimetic in a scalable cell-free protein synthesis system and have optimized the attachment of the PHI mimetic to the virus-like particle in an ordered, repetitive fashion similar to that of the natural virus. Co-attached innate immune stimulators will act as a non-denaturing adjuvant to further ensure a strong protective response. Consequently, our PHI mimetic will provide a broadly protective immune response to many subtypes of HIV. Promising vaccine candidates will be evaluated in guinea pigs for serum responses which provide viral neutralization in an *in vitro* assay.

449315 Correlations of Antibody Response Phenotype and Genotype Revealed By Immunoglobulin Repertoire Sequencing

Abstract:

It has long been possible to measure the phenotype of antibody responses (antigen-specific titers) through conventional serological assays (e.g., ELISA). In contrast, the ability to measure the genotype of antibody responses has only recently become possible through the advent of high-throughput antibody repertoire sequencing (Ig-seq), which provides quantitative molecular information on clonal expansion, diversity and somatic hypermutation. However, Ig-seq is compromised by the presence of bias and errors introduced during library preparation and sequencing and thus prevent reliable immunological conclusions from being made. By using synthetic antibody spike-in genes, we determined that Ig-seq data overestimated antibody diversity measurements by up to 5000-fold and was less than 60% accurate in clonal frequency measurements. To overcome this, we developed molecular amplification fingerprinting (MAF), which uses unique molecular identifier-tagging before and during multiplex-PCR amplification, enabling tagging of transcripts while accounting for PCR efficiency. Our method allowed us to develop a novel algorithm for bias correction, which improved clonal frequency measurements to an accuracy of up to 99%. Furthermore, we demonstrated nearly absolute error correction (98-100%) of full-length antibody diversity measurements. In order to probe the relationship between antibody genotype and phenotype, we have performed extensive *in vivo* experiments from immunized mice. Specifically, we modulated titer (phenotype) by varying the number of immunizations (boosts) and in parallel performed MAF Ig-seq to assess genotype through a series of quantitative metrics. These metrics describe clonal selection, expansion, and somatic hypermutation. Importantly, we also used different protein antigens, this allowed us to determine how epitope complexity impacted the genotype of antibody repertoires. Finally, by applying multivariate modeling using MAF-corrected data, we were able to predict the immune status of individual antibody clones. This extensive systems-based analysis demonstrates how accurate Ig-seq provides insight on the genotype and phenotype of humoral immunity.

470102 Regulation of Intracellular Delivery through Peptide-Based Nanocarrier Design (Invited Talk)

Abstract:

Nucleic acid cargoes offer unmatched diversity in gene regulatory potential and therapeutics, and understanding of nucleic acid functionality continues to expand rapidly and dramatically through seminal discoveries including RNA interference approaches and gene editing technologies. In nature, the basis for gene regulation is ultimately encoded by the exquisite specificity with which cells are able to control both the location and accessibility of nucleic acid constructs to govern their activation states. My research program seeks to understand and control gene activation using synthetic constructs through nature-inspired approaches to control and quantify cell binding interactions and stability in polymer and peptide nanocarriers. The basis of our approaches is the design of stimuli-responsive polymers and peptides whose interactions with nucleic acids and cells can be controlled dynamically by specific intracellular or external triggers. We exploit our ability to control nucleic acid binding/release and cellular processing to gain new mechanistic insights over nucleic acid delivery, leading to design advances including histone-inspired DNA targeting, light-responsive gene silencing, and collagen turnover-stimulated gene expression. This talk will highlight two ways we have used nature-inspired peptides to control gene transfer in regenerative medicine.

Our approaches are exemplified by our work in histone-targeted nanocarrier design. Histones have received great interest as potential gene carriers for several decades due to their seminal role in chromatin packaging and gene transfer, yet therapeutic efforts with histones have lacked both a well-controlled materials approach and a deeper knowledge of cellular processing mechanisms. Hence, histone-based carriers have failed to reach clinical efficacy. We have capitalized on newly recognized and highly pivotal roles for histone tails in native gene regulatory control to develop a gene transfer method that utilizes native, histone-based processing pathways *via* incorporation of post-translationally modified (PTM) histone tails within controllably-assembled DNA vehicles (polyplexes). Our efforts proved that polyplexes displaying PTM-modified histone tails promote nuclear accumulation, DNA release, transcription, and enhanced transfection. Moreover, our group has combined detailed nanostructure engineering with sophisticated cellular imaging to identify novel aspects in the cell biology framework regulating polyplex transport to the nucleus.

We have also focused on novel mechanisms to exploit natureâ€™s ability to harness extracellular matrix (ECM) proteins such as collagens to sequester and control delivery of bioactive nanostructures. Our specific approaches have capitalized on a class of peptides known as collagen-mimetic peptides, or CMPs, that have been recognized for their unique affinity for native collagen, which can be tailored through alterations in CMP amino acid sequence and molecular weight. CMPs incorporate themselves into the natural collagen triple helical structure via strand invasion, in a reversible process previously that has been used to modify extracted collagen *in vitro* and exclusively target remodeling collagen *in vivo*. In our studies, we employed a proline-rich CMP designed to act not only as an adjustable tether to regulate collagen-polyplex affinity, but also as an adhesive/endocytic ligand for polyplexes. The use of a collagen scaffold afforded our system structural support and innate bioactivity to encourage cellular ingrowth and proliferation, whereas altering the extent of the modification of our vector provided additional tunability to allow tailorable release for prolonged time periods. This CMP-based approach also consistently and fully maintained polyplex activity in the presence of serum for at least a week, whereas most bolus and substrate-mediated gene delivery approaches report rapid reductions within hours or a few days, and the level of transgene expression directly correlated with MMP-concentrations and the extent of collagen remodeling, demonstrating â€œon demandâ€ release. The ability to tailor release over extended periods via physical attachments, combined with the ability to provide cell-trigger release and collagen-mediated uptake, make this approach very attractive for many applications in regenerative medicine.

470495 Role of SNP Characteristics on the Endocytosis and Intracellular Trafficking of siRNA

Abstract:

Short interfering RNA (siRNA) therapeutics, have garnered considerable attention for their potential to treat a wide variety of disease-associated proteins through the use of RNA interference (RNAi). siRNA therapeutics have been developed for cancers, genetic disorders, and infectious diseases, but their progress through clinical pipelines has been limited by delivery vehicles that are toxic, immunogenic, or ineffective. Despite the variety of delivery vehicles that have been developed, the essential chemical and physical characteristics of effective delivery vehicles remain unknown. It is also unclear if RNAi is initiated more potently following endocytosis by particular mechanisms and if this variability is specific to a certain cell type or delivery vehicle chemistry.

Our goal is to define nanoparticle characteristics that are essential for potent siRNA delivery. Using silica nanoparticles (SNPs) for their tunable synthesis, we are investigating delivery criteria in four main categories: sequence specific mRNA degradation (silencing), membrane translocation, endosomal escape, and association with RNAi proteins. Using SNPs, vehicle size, structure, charge, and surface functionalization can be varied to determine optimal vehicle design criteria in a variety of cell types.

The presentation will discuss our novel live cell assay that allows independent intracellular trafficking of delivery vehicle, antisense, and sense siRNA. Using confocal microscopy, we track vehicle-siRNA complexes during endocytosis (whether by Clathrin, Caveolin, ARF6, GRAF1, Flotillin, and Macropinocytosis), endosomal trafficking (Early, Late, Recycling, and Lysosome), trafficking to the endoplasmic reticulum or golgi, and exocytosis. Our results to date demonstrate that multiple vehicle characteristics affect siRNA delivery and demonstrate the critical role of intracellular trafficking in maximizing siRNA potency.

462951 Engineering Periodic shRNA Delivery Systems with High Silencing Efficacy

Abstract:

Challenges in the delivery of small interfering RNAs (siRNAs) to disease sites have hampered clinical translation of RNA interference, particularly in cancer therapeutics. Due to its low valency and high rigidity, siRNA is not readily condensed into stable nanoparticles that can successfully protect it from nuclease degradation and deliver it into target cells. Periodic short hairpin RNAs (p-shRNAs), consisting of repeating RNAi sequences linked together, can potentially address these delivery barriers by improving complexation with cationic delivery materials. Ranging from under 200 to over 5000 nucleotides in length, p-shRNA can be generated in large quantities through rolling circle transcription, in which an RNA polymerase repeatedly traverses a small circular DNA template. We have engineered these p-shRNA molecules via template design and selective enzymatic digestion of their hairpin loops to generate an open-ended p-shRNA (op-shRNA) that induces greater gene silencing than p-shRNA and siRNA in multiple cancer cell lines up to nine days.

The high valency and flexibility of op-shRNA dramatically improves complexation with low molecular weight polycations compared to siRNA. To develop a delivery vehicle optimized for the biophysical properties of op-shRNA, we have synthesized a library of biodegradable poly(beta-amino ester)s (PBAEs) via factorial design. Our designed PBAEs condense op-shRNA into compact nanoparticles below 100 nm in diameter, and demonstrate high transfection efficiencies at polymer and RNA concentrations much lower than typically required for PBAE-mediated gene delivery. By varying the molecular weight and incorporation of alkyl chains in our PBAE library, we were able to develop correlations between these parameters and multiple parameters influencing op-shRNA delivery, including complex assembly, stability, intracellular disassembly, cellular uptake, and cytotoxicity. In particular, PBAEs of low to intermediate molecular weight and increasing alkyl chain content delivered op-shRNA *in vitro* most efficiently, generating levels of gene knockdown comparable to those by commercial transfection reagents. Through nucleic acid engineering and rational carrier design, we have successfully developed a stable, potent RNAi delivery platform that can trigger significant gene silencing at low doses, and potentially enable higher therapeutic efficacy *in vivo*.

460542 Polypeptide/Nucleic Acid Complexes As Delivery Vehicles

Abstract:

Electrostatically driven polymer self-assembly mechanisms, such as polyelectrolyte complexation, are vastly underexplored compared to amphiphilic-based assemblies, and yet offer unique opportunities for both the encapsulation of charged therapeutics and controlled delivery, via the tailoring of intermolecular interactions using pH and salt. Nanoscale polyelectrolyte complexes micelles can be created using block-copolymers that incorporate a neutral, yet hydrophilic block to stabilize the polyelectrolyte complex on the nanoscale. The modular nature of these assemblies enables the addition of a targeting ligand conjugated to the hydrophilic block to increase the efficacy of delivering nucleic acids, creating polyelectrolyte complex micelles with a targeting ligand outside the corona of the micelle. Ultimately, we demonstrate the use of targeted polyelectrolyte complex micelles formed using therapeutic nucleic acids that inhibit microRNAs involved in atherogenesis, with the goal of creating a treatment for atherosclerosis. In doing so, we create trifunctional polymers that contain a targeting peptide (that target fibrin or endothelial cells), a polyethylene glycol domain to facilitate microphase separation, and a polylysine domain used to complex the microRNA inhibitors. We characterize these nanoscale assemblies and proceed to explore the ability of these micelles to inhibit microRNA-92a, which regulates the expression of endothelial transcription factors involved in the progression of atherosclerosis. Also we explore the ability to inhibit microRNA-33a in macrophages, which is involved in the expression of ATP binding cassette transporters involved in the trafficking of cellular cholesterol to the liver. Ultimately, we show the effectiveness and biodistribution of our targeted micelle assemblies using a murine model.

462124 Highly Potent mRNA Delivery In Vivo with Intravenously-Administered Ionizable Lipid Nanoparticles

Abstract:

Introduction: Messenger RNA (mRNA) has tremendous therapeutic applications, including protein replacement therapy, cancer immunotherapy, and genomic engineering. However, successful intracellular delivery of mRNA remains challenging, as nucleic acids cannot readily cross cellular membranes alone due to their relatively large size, negative charge, and hydrophilicity. Over the past decade, researchers have developed lipid nanoparticles (LNPs) to encapsulate and deliver small interfering RNAs (siRNAs) to silence protein expression in the liver, but these LNPs have only recently been reported to encapsulate and deliver mRNA to produce protein *in vivo*. Because of the considerable size, stability, and charge differences between siRNA and mRNA, we hypothesized that LNP formulation parameters of historically-used siRNA-LNPs could be better optimized to more efficiently deliver larger mRNA payloads. To this end, we developed a generalized approach for optimizing LNP formulations for mRNA delivery *in vivo* using Design of Experiment (DOE) methodologies. After using this strategy to significantly improve mRNA-LNP potency *in vivo*, we then applied this mRNA-optimized formulation with new classes of ionizable lipid materials and discovered some of the most efficacious intravenously-administered mRNA delivery vehicles reported to date in the literature.

Materials / Methods: The most common siRNA-LNP formulations contain four components: an amine-containing ionizable lipid, phospholipid, cholesterol, and lipid-anchored polyethylene glycol, the relative ratios of which can have profound effects on the formulation potency. We first encapsulated mRNA encoding for Erythropoietin (EPO), a secreted serum protein, into the original siRNA-based LNP formulation (which employed the ionizable lipid C12-200) to establish a baseline from which to improve. LNPs were synthesized via microfluidic mixing of an ethanolic lipid phase with an aqueous RNA phase to form particles approximately 100 nm in diameter. Then, we employed DOE to modulate several formulation parameters simultaneously (e.g. lipid:mRNA weight ratio, phospholipid structure, molar composition of excipients). DOE allowed for the reduction of the total number of formulations required to deduce statistically significant trends in our large, multi-dimensional design space, which is important for the economical screening of LNP formulations due to poor *in vitro*/*in vivo* translatability and costs associated with screening large libraries *in vivo*. EPO mRNA-loaded C12-200 LNPs were injected intravenously at 15 µg total mRNA per mouse, and blood was collected 6 hours post-injection; an ELISA assay was then used to quantify serum EPO levels. Using Definitive Screening and Fractional Factorial DOE Designs, we screened a total of 38 distinct LNPs *in vivo*and characterized top-performing LNPs via their mRNA encapsulation efficiency, diameter, polydispersity, pKa, zeta potential, and morphology. Then, we applied the formulation parameters from the most efficacious C12-200 LNP formulation to a new class of ionizable lipid materials formed through a ring opening reaction between alkenyl epoxides and a polyamine core; unlike C12-200, which has fully saturated hydrocarbon tails, this new library of alkenyl amino alcohol (AAA) ionizable lipids contained degrees of unsaturation in its tails.

Results: After employing this DOE-based optimization process, we significantly ($p < 0.05$) increased the efficacy of intravenously-delivered liver-targeting C12-200 LNPs seven-fold for EPO mRNA delivery relative to the formulation previously used for siRNA delivery. At a 15 µg (0.75 mg/kg) mRNA dose, we achieved serum EPO protein concentrations in mice of approximately 7,000 ng/mL, which is 5-6 orders of magnitude greater than physiological levels. We found that increased lipid:RNA weight ratios and the use of the phospholipid 1,2-dioleoyl-sn-gylcero-3-phosphoethanolamine (DOPE) were the most crucial factors in determining LNP efficacy. Interestingly, when our mRNA-optimized C12-200 LNP formulation was loaded with siRNA, it silenced protein expression in hepatocytes nearly identically to the original siRNA formulation, suggesting that siRNA-loaded LNPs may be more tolerant than mRNA-loaded LNP of formulation design space differences. Lastly, when C12-200 was substituted with a library of novel alkenyl amino alcohol (AAA) lipids and formulated into LNPs with EPO mRNA, lead lipid OF-02 was discovered to afford EPO serum concentrations of approximately 14,000 ng/mL, the highest efficacy yet reported in the literature. Furthermore, mRNA delivered with OF-02 LNPs was expressed predominately in the liver and had a linear dose-response curve ranging from 0.75 to 2.25 mg/kg RNA dose.

Conclusions: In conclusion, we significantly increased the potency of liver-targeting C12-200 LNPs for mRNA delivery using Design of Experiment methodologies, while showing that the same mRNA-optimized LNPs did not confer additional potency when delivering siRNA. The optimized LNP formulation showed applicability to other types of ionizable lipids, such as the novel lipid OF-02, which made highly potent intravenously-delivered mRNA-LNPs. This work provides evidence that some technologies previously used for siRNA delivery may require further optimization for better mRNA delivery. Additionally, the general strategy presented here could be used to screen multi-dimensional design spaces in vivo for optimizing other biomaterials and payloads.

467122 Folate Receptor-Targeted Aminoglycoside-Derived Polymers for Transgene Expression in Cancer Cells

Abstract:

Targeted delivery of anticancer drugs can overcome the limitations of poor drug bioavailability, drug resistance, and unintended side effects associated with current chemotherapeutic regimens. Folate-receptors are overexpressed in several epithelial malignancies, including ovarian, renal, lung, triple-negative breast and bladder cancers, making these attractive targets for the targeted delivery of small molecule and/or nucleic acid therapeutics to cancer cells. Aminoglycosides are hydrophilic molecules that contain multiple sugars, amines and hydroxyls, and are attractive as monomers for generating diverse libraries of macromolecules. We first cross-linked aminoglycosides with diglycidyl ethers and subsequently derivatized these polymers with folic acid resulting in the formation of folate-targeted polymers. These folic acid-conjugated aminoglycoside-derived polymers were investigated for transgene delivery and expression in triple negative breast and bladder cancer cells at different polymer: plasmid DNA weight ratios. The targeted folic acid-conjugated polymers demonstrated significantly higher levels of luciferase expression in cancer cells compared to their parental polymers. Competitive inhibition of the folate receptor using free folic acid significantly inhibited transgene expression activity, indicating a role for the receptor in transgene expression. The targeted polymers were employed for the delivery of a newly constructed pEF-TRAIL plasmid, which expresses the TNF-alpha Related Apoptosis Inducing Ligand (TRAIL) protein, which is known to iduce death selectively in cancer cells. Cancer cell death following delivery of the pEF-TRAIL plasmid was investigated as a single-agent treatment as well as with other chemotherapeutics resulting in a combination treatment approach. Our results indicate that folate-targeted, aminoglycoside-derived polymers may be promising new materials for delivering nucleic acids to cancer cells in several malignancies.

460866 Sustained Transgene Expression Via Substrate-Mediated Gene Transfer Results from Multiple Transfection Events

Abstract:

Statement of purpose: Sustained delivery of therapeutic genes in vitro and in vivo has a wide range of applications in studying biology and in developing therapies to treat disease. Non-viral vectors such as cationic polymers still present promising approaches; however, bolus transfection methods with polyethyleneimine (PEI)-based DNA polyplexes suffer from considerable levels of cytotoxicity and short-lived transgene expression levels. Here, we designed and characterized a hyaluronic acid (HA)-based porous hydrogel system for non-viral gene delivery by loading with surface-associated DNA polyplexes (Figure 1A). With this, we observed enhanced and sustained transgene expression over 30 days of cell culture, with better cell viability, both marked improvements over comparable bolus transfection techniques. Finally, we investigated mechanisms thought to be responsible for the sustained expression profile.

Methods: HA hydrogels were formed via a Michael-addition reaction between acrylated HA and thiolated peptide crosslinkers with the addition of RGD peptide for cell adhesion. The sphere templating method was used to introduce 60-μm diameter pores. Plasmid DNA was complexed with polyethyleneimine to form DNA polyplexes, which were surface coated by electrostatic association to the pore surfaces of the hydrogel by incubating the formed hydrogel in the polyplex solution. DNA loading was assessed with ^{32}P-labelled DNA and scintillation counting. Transfection profiles were obtained by loading DNA encoding for the Gaussia luciferase gene, seeding D1 mouse mesenchymal stem cells in the hydrogels, and detecting the expressed protein with the associated assay.

Results: ^{32}P analysis of DNA-loaded porous hydrogels showed that up to 4 μg of DNA could be loaded into the scaffold. Although the DNA was electrostatically immobilized to the scaffold, minimal polyplex release was observed with only 5% of the initial loaded DNA released over 7 days. Cell culture within this scaffold resulted in sustained transgene expression over a period of more than 30 days at levels higher than those seen using bolus transfection techniques (Fig. 1B). In addition, adjusting the DNA loading concentration resulted in modulation of transgene expression (Fig. 1B). While repeated bolus transfections in 2D culture result in marked toxicity (Fig. 1C), culture in surface-coated hydrogels was significantly less severe (Fig. 1D). To determine if sustained expression was due to re-transfection events, it was observed that levels of internalized plasmid DNA in cells cultured in surface-coated hydrogels was sustained at higher levels over time than in cells cultured in non-coated hydrogels but administered a bolus polyplex transfection (Fig. 1E).

Conclusions: After DNA loading by surface coating, there was minimal basal DNA release over time, suggesting that this may serve as a robust system for long-term DNA availability and expression. Indeed, sustained expression was seen over thirty days of culture and could be enhanced by increasing DNA loading. Although repeated bolus transfections result in severe toxicity, repeated internalization events are possible in surface-coated scaffolds without the same toxic effects and are responsible for the sustained expression observed. This system is a promising means of sustained gene delivery for various biomedical applications without the need for genomic integration via viral methods, which have innate safety concerns.

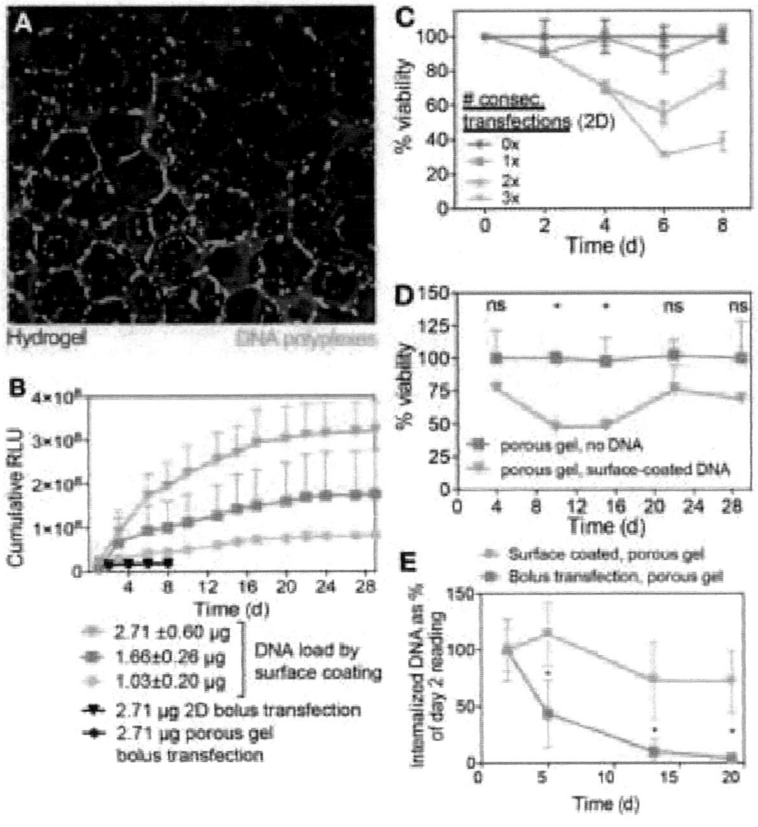

466335 Enzymatic Production of Fructo-Oligosaccharides from Inexpensive and Abundant Substrates Using a Membrane Reactor System

Abstract:

The concept of zero waste in the sugar industry has encouraged the development of technologies that promote utilization of byproducts, including the enzymatic conversion of waste feed stocks into nutraceutical products. Here we investigated a membrane reactor for the production of fructo-oligosaccharides (FOS) with immobilized enzyme (Pectinex Ultra SP-L) system using pretreated molasses as substrate with cleaning efficiency and less fouling as performance tools. The resulting spectra showed a high FOS yield of 63% with 89% conversion from available sucrose. So the EMR system with innovative continuous separation of enzyme was proved to fit for the production of FOS.

Reference: Amad Ur Rehman A, et al, Separation Science and Technology, DOI: 10.1080/01496395.2016.11667740

461271 Consolidated Process for Biobutanol Production from Cellulose By Two Step Fermentation of Different Clostridia Strains

Abstract:

n-Butanol is an important industrial chemical and also has been proposed as an alternative biofuel superior to ethanol. Biological production of *n*-butanol is promising because it can use renewable biomass as feedstock and is environmentally friendly. Conventional biorefinery of lignocellulosic biomass needs separate processes, including feedstock pretreatment, cellulase/hemicellulase production, biomass enzymatic hydrolysis, sugar fermentation, and product recovery, which is not cost-effective due to the complex process and subsequently high equipment capital requirement. Consolidated process only use one tank for enzyme production, biomass hydrolysis, and sugar fermentation, which will greatly simplify the process and reduce the equipment investment. In our research, *Clostridium cellulovorans*, a natural cellulose/hemicelluloses utilizing, but non-butanol producing bacterium, was metabolically engineered for *n*-butanol production from cellulose by overexpressing enzymes in the butanol-producing pathway. The obtained mutants could produce above 2 g/L *n*-butanol from cellulose. However, the *n*-butanol productivity was very low and *n*-butanol titer was hard to be further improved, which could be attributed to the poor specific cellulose hydrolysis activity of the enzyme system from *Clostridium cellulovorans*, as well as limited amount of the secreted enzymes under the circumstances of metabolic burden and butanol stress. To unblock this bottleneck, a novel consolidated process was developed by applying another clostridia strain, *Clostridium thermocellum*, with naturally high cellulose hydrolyzing activity. There two clostridia strains have similar nutrient requirement and could grow in the same cellulose medium. Due to the difference in optimal growth temperature (55°C vs. 37°C), a two step fermentation was necessary. The first step of fermentation under 55°C is a dedicated enzyme production phase by *Clostridium thermocellum*, in which maximal cellulose hydrolyzing activity of the whole culture system is the target of process optimization. The second step under 37°C could be believed as a simultaneous enzymatic hydrolysis and *n*-butanol production phase by *Clostridium cellulovorans* mutant strain. By combining the advantages of these two clostridia strains, this novel consolidated process showed promising potential for cost-effective biobutanol production from lignocellulosic biomass.

Bioprocess Development for Pleuran production by *Pleurotus ostreatus* using submerged cultivation system in Semi-Industrial scale.

El Enshasy H.A[1,2*], *Johari Masri, M.H*[3], *Malek R.A*[1], *Othman, N.Z.*[1], *Leng, O.M*[4], *Aziz, R*[1],

1. Institute of Bioproduct Development, Universiti Teknologi Malaysia, Universiti Teknologi Malaysia, Johor Bahru, Malaysia.
2. City of Scientific Research and Technology Applications, New Burg Al Arab, Alexandria, Egypt.
3. RD Department, All Cosmos Industries Sdn. Bhd., Johor Bahru, Malaysia, Institute of Bioproduct Development, Universiti Teknologi Malaysia, Skudai, Malaysia,
4. Harita Go Green, Johor Bahru, Malaysia and Institute Bioproduct Development, Universiti Teknologi Malaysia, Skudai, Malaysia

(*, corresponding author: Email: henshasy@ibd.utm.my, Phone: +601117739884)

Abstract

Nowadays, the demand of new natural and safe pharmaceutical production of immunomodulatory and anticancer properties is very high. For many years it have been reported that *Pleurotus ostreatus* is characterized by pleuran production, a β-1(1,3/1,6)-D-glucan biopolymer. This polysaccharide has binding affinity to carbohydrate receptors on immune cells and thus have good immunomodulatory activity. When cells cultivated in submerged culture system, they are able to production the pleuran as exopolysaccharide (EPS) in the fermentation medium. In this work, a complete industrial platform was developed to increase the pleuran production using submerged cultivation system in semi-industrial scale. At first, different media formulations were tested for polysaccharide production. The medium of highest EPS production was optimized using statistical approach in shake flask level. After this step, cultivations were conducted in 16-L stirred tank bioreactor for bioprocess industrialization and optimization of different bioprocessing engineering parameters to develop complete industrial platform for this polysaccharide in Semi-industrial scale.

Introduction

Mushrooms have been used since centuries as healthy food and medicine in different cultures. This based on its high nutritional value as rich source of proteins, carbohydrates, lipids, minerals, and vitamins (1-3). Recent research highlighted the medicinal values of macrofungi were highlighted in different research as antibiotic, antioxidant, antitumore/anticancer, immunomodulator, and antiatherosclerosis agents (4). Of different types of medicinal mushrooms known for their high value metabolites, *Pleurotus ostreatus* (oyster mushroom) is world known mushrooms based on its high nutritional value compounds make it one of the mostly used mushroom in food market (5). In addition, this mushroom is characterized by the ability to produce polysaccharides of high medicinal value with many bioactive properties. Pleuran (CAS # 159940-37-1) is one of most important polysaccharides which belong to β-1(1,3/1,6)-D-glucan biopolymer. Like other hexose polymers, it has the general chemical formula of $(C_6H_{10}O_5)_n$ (6). This polysaccharide is characterized by its high immunomodulating and anticancer properties (7-9). Pleuran is traditionally produced by mushroom cultivated in green house under controlled temperature and humidity. However, this method has many disadvantages such as long time cultivation (more than 3 months), low yield of bioactive compound and the ease of contamination by microorganism in this open cultivation system. Therefore, alternative method using submerged cultivation is applied to decrease the production time and increase productivity. In addition, extraction of pleuran from mushrooms in green house require many downstream processes, using a lot of solvent, and many steps of purification. In submerged culture system, the required polysaccharides can be obtained directly through simple purification of the exopolysaccharides from the fermentation broth by direct precipitation followed by drying or ultrafiltration to isolate the required high molecular weight polysaccharides followed by freeze drying. However, the different methods used for pleuran production is presented in figure 1. This work is focused on the development of new cultivation medium which support high yield of pleuran production in both shake flask and in bioreactor culture. Two approaches have been applied for medium optimization: One Factor At Time (OFAT) method and statistical optimization. Furthermore, cultivations were conducted in pilot plat 16-L stirred tank bioreactor to improve the production of pleuran production in semi-industrial scale.

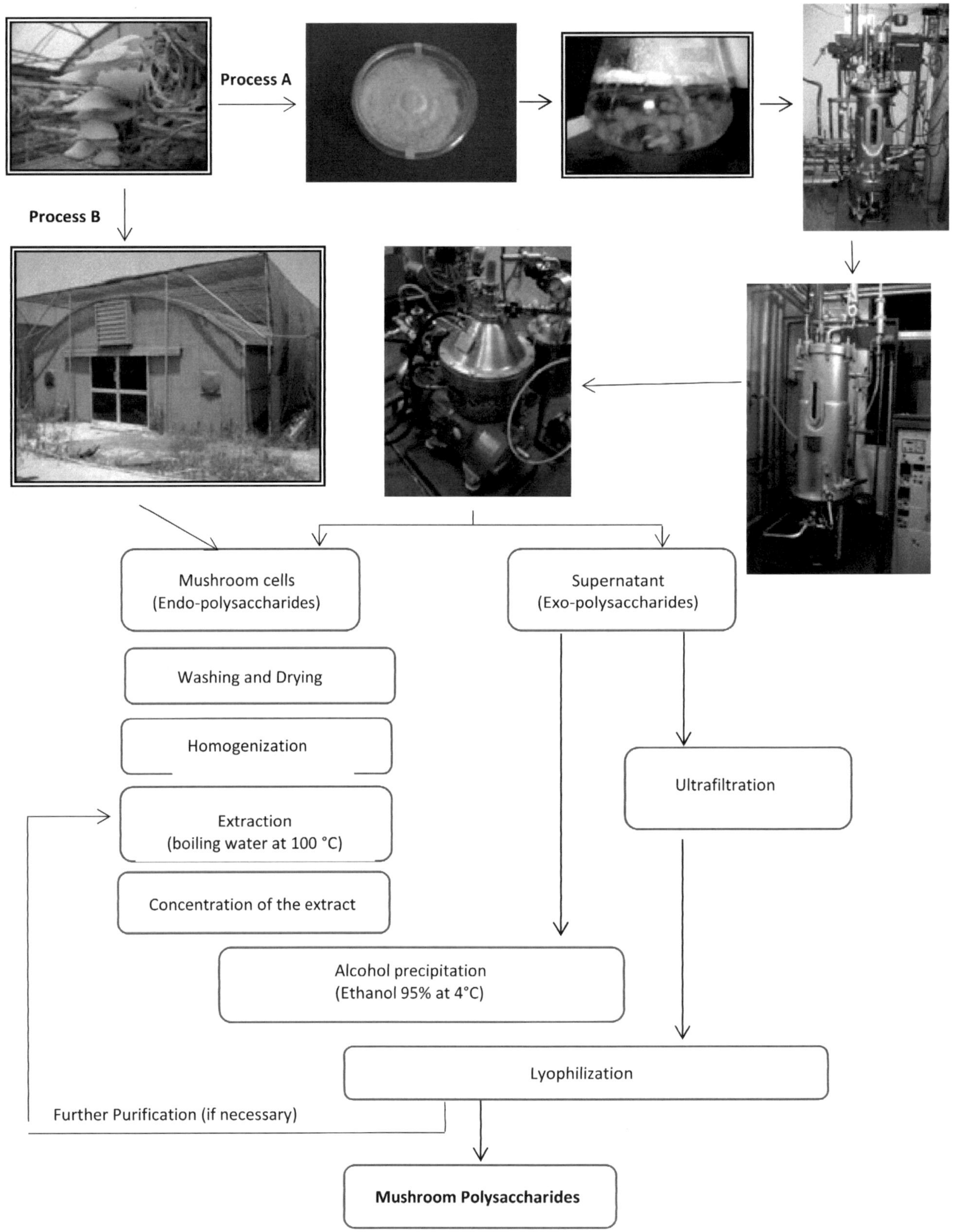

Figure 1. Production process for Pleuran production by *P. ostreatus* using submerged mycelium fermentation (Process A) or solid state fermentation (Process B). Process A started by mushroom

cell propagation on agar plate followed by submerged cultivation of cells in shake flask as intermediate step for inoculum preparation for bioreactor culture. Cultivation in bioreactor carried out by scaling up the process from small to large bioreactor followed by centrifugation to separation cells from the culture broth. If polysaccharides produced extracellular, extraction is usually carried out either through direct precipitation using cold ethanol followed by lyophilization to obtain the crude extract. Other new solvent free method was also developed to separate high molecular weight compounds directly by ultrafiltration followed by direct lyophilization of the polysaccharides (10). For endo-polysaccharides, they will be extracted using the conventional scheme of polysaccharides extraction from mushroom fruiting bodies (Process B). In case of Process using fruiting bodies as starting materials (Process B), cells were first washed and dryed followed by homogenization before many hours extraction using boiling water. The water extract of crude polysaccharide will be concentrated using vacuum extraction method followed by alcohol precipitation using cold ethanol. The precipitate is then lyophilized. In this process, further purification could be also done by repeating the steps between water extraction step and lyophilization. (11).

Materials and Methods

Microorganism

Pleurotus ostreatus NRRL 366 was used in this study. This mushroom was initially obtained from ARS culture collection, Agriculture Research Service, Peoria, USA. Once received from the microbial cell bank, it was subcultured on PDA agar medium (PDA, Oxoid, UK) After inoculation, the inoculated agar plates were incubated at 26 °C for 12 days and stored thereafter in a refrigerator at 5 °C for further use. Cells were activated every 2 months to ensure the cell viability during storage.

Vegetative growth medium and inoculum preparation

Cells were first activated by incubation on PDA agar plate for 12 days, harvested in sterile saline solution, and used to inoculate YM medium composed of (g/L): glucose, 10.0, yeast extract, 3.0, malt extract, 3.0; peptone, 5.0 (pH 5.5). After 7 days cultivation, the grown cells were harvested, homogenized using Omni homogenizer (Omni International, USA) for better inoculum homogenization and prevent the cell aggregation to large pellet.

Production medium

The initial medium used for EPS production before optimization was composed of (g/L): glucose, 60.0; peptone, 2.0; KH_2PO_4, 0.46; K_2HPO_4, 1.0; $MgSO_4.7H_2O$, 0.5 (pH 5.5),. This medium was modified from El Enshasy et al. (11). Glucose was sterilized separately and added to the fermentation medium before inoculation. This medium was further optimized in this study using RSM method in this study.

Shake flask and bioreactor cultivation

In shake flask experiment, 250 mL Erlenmeyer flasks of 50 mL working volume were used, incubated for 14 days at 26 °C on rotary shaker incubator at 200 rpm. Samples in form of two flasks were taken at different time intervals every day for analysis. For bioreactor cultivation, 16-L stirred tank bioreactor (BioEngineering, Wald, Switzerland) with working volume of 8L was used in this study. The stirrer speed with adjusted to 200 RPM and aeration rate of 1 v/v/min.

Determination of glucose and EPS

Glucose concentration in culture broth was determined using glucose lactate analyzer (YSI 2700), which based on enzymatic method. The centrifuged sample was mixed first with cold ethanol in ration (1:3, v/v), kept for 20 hours at 4 °C for polysaccharides precipitation. The precipitated EPS was then centrifuged at 10,000g for 20 min for full separation. The precipitated polysaccharides was then washed consequently with three solvent system (ethanol, acetone, and ethyl ether). The obtained white precipitate was then lyophilized for 3 days, and weighted.

Results and Discussion

Medium optimization for high EPS production by *P. ostreatus*

Set of experiments were conducted to improve the medium components to increase EPS production. Two approaches were applied to achieve this target using one factor at time method (OFAT) and statistical design method using response surface methodology (RSM). The results of medium optimization is summarized in table 1.

Table 1. Medium composition obtained after from different optimization methods to increase EPS production by *P. ostreatus*

Components	**Un-optimized**	**One factor at time (OFAT)**	**Statistical Optimization (RSM)**
Glucose (g/L)	60.0	100.0	105.0
Yeast extract (g/L)	2.0	3.0	5.2
$(NH_4)_2SO_4$ (g/L)	-	2.5	2.6
K_2HPO_4 (g/L)	1.0	1.0	2.2
$MgSO_4.7H_2O$ (g/L)	0.2	0.2	0.2
Peptone (g/L)	1	-	-
EPS (g/L)	**2.13**	**3.33**	**3.65**

As shown optimization of culture medium using statistical optimization increased the volumetric production of pleuran up to 3.65 g/L (almost 71% higher than those value obtained in unoptimized medium. However, the difference between the polysaccharides production in medium optimized using OFAT and RSM was only 9.6% higher in case of RSM culture.

Batch cultivation of *P. ostreatus* in 16-L stirred tank bioreactor

After medium optimization in shake flask level, cultivations were conducted in 16-L stirred tank bioreactor for bioprocess industrialization. Figure 2 shows the growth profile of mushroom cells when cultivated in bioreactor level. As shown, cells grew with low growth rate in the first 48 hours followed by higher growth rate until 240 hours, reaching maximal biomass of 15.22 g L^{-1}. During that time, glucose consumed in culture with rate of about 0.37 g L^{-1} h^{-1}. On the other hand, the maximal volumetric and specific EPS production was 3.75 g L^{-1} and 0.24 g g^{-1}, respectively.

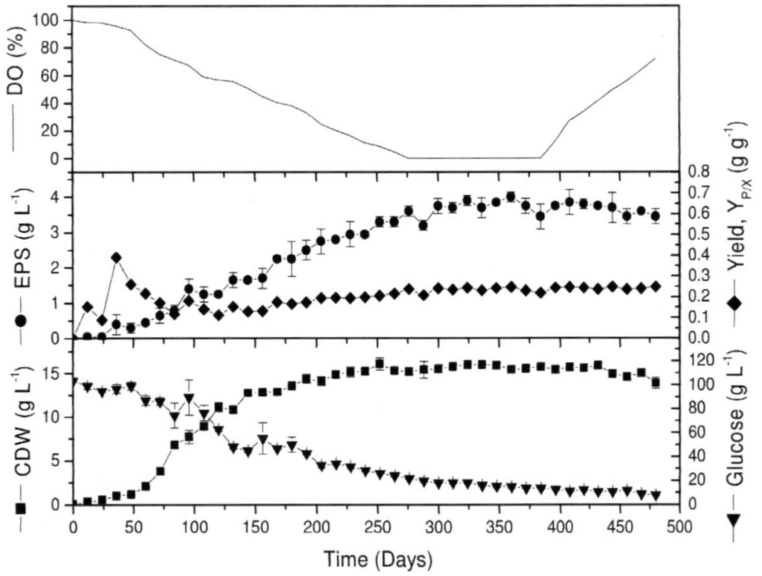

Figure 2. Cell growth, glucose consumption, DO, and EPS production profile during cultivation of *P. ostreatus* in 16-L stirred tank bioreactor.

Fed-batch cultivation of *P. ostreatus* for high EPS production in 16-L stirred tank bioreactor

Based on the results of batch cultivation, fed-batch cultivation was designed using constant feeding approach with either glucose alone or full medium feeding. Figure 3 shows the growth and EPS production profile in cultures fed with two different feeding strategies. Feeding started after 240 hours to prolong cell growth and EPS production up to 384 hours. As shown, the biomass in full medium fed culture reached maximal value of about 23.81 g L^{-1} which was about 16% higher than the value in glucose fed culture. However, the maximal EPS production was also obtained in full medium fed culture and reached 4.5 g L^{-1}, which is about 20% compared to glucose fed culture (which gave the same volumetric production as in batch culture). It is also worthy to note that the specific EPS production in both culture was almost the same. This indicates that the increase of EPS in complete medium fed-culture was due to the increase in biomass not due to higher cell productivity.

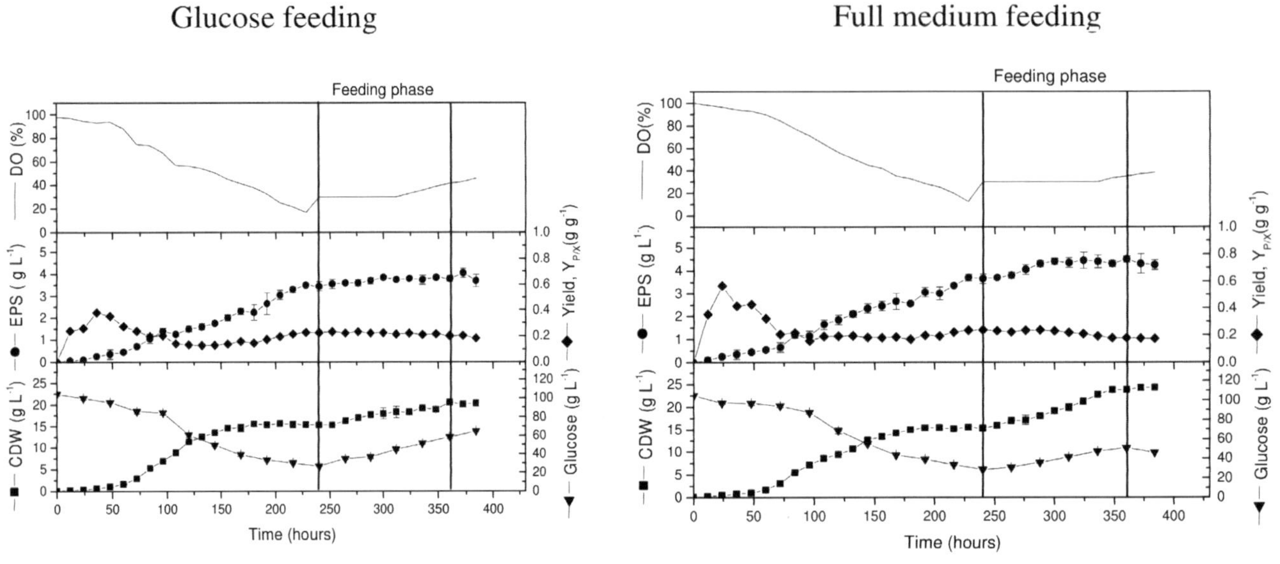

Figure 3. Chang in cell growth, glucose concentration, DO, and EPS production profiles during cultivation of *P. ostreatus* in 16-L stirred tank bioreactor using constant fed-batch cultivation strategy. In this cultivation, two different feeding solutions were applied, either glucose feeding only or complete medium feeding.

Table 2. Comparison between cell growth and EPS production when cultivating *P. ostreatus* in fed batch culture using different feeding solutions.

Parameters for Fed-Batch Cultivation	Exponential time in feeding period (240 hour to 360 hour)	
	Glucose feeding	Full medium feeding
Biomass, X ($g\ L^{-1}$)	20.491	23.81
Specific growth rate, μ (h^{-1})	0.00243	0.0037
EPS production ($g\ L^{-1}$)	3.8	4.5
Yield coefficient product over cell mass $Y_{p/x}$ ($g\ g^{-1}$)	0.185	0.189

Conclusion

This study demonstrates a complete industrial platform for pleuran production in semi-industrial scale. Medium optimization was carried out using two approaches (OFAT) and (RSM), which lead to the development of high volumetric EPS production. The process was scaled up in 16-L stirred tank bioreactor and further optimized by applying fed-batch cultivation strategy. The results clearly demonstrate that fed batch cultivation using complete medium is more suitable than feeding with glucose only. However, the increase of EPS production was due to the higher biomass production and not to the increase in cell productivity.

Acknowledgement
This research was financially supported by Harita Go Green Sdn. Bhd., authors would like to thank also Research Management Centre (RMC), Universiti Technology Malaysia for their support.

References

1. Smith J. E, Rowan, N.J., Sullivan R. (2002). Medicinal mushrooms: Their therapeutic properties and current medical usage with special emphasis on cancer treatment. USA. University of Strathclyde.
2. Pushpa, H.; Purushothama, K.B. Nutritional analysis of wild and cultivated edible medicinal mushrooms. *World Journal of Dairy and Food Science.* 5, 140-144.
3. Chang, S.T., Buswell, J.A. Mushroom nutraceuticals. *World Journal of Microbiology and Biotechnology.* 12, 473-476.
4. Chen, J. and Seviour, R. (2007). Medicinal importance of fungal β-(1→3), (1→6) glucan. *Mycological Research.* 111, 635-652.
5. Maftoun, P., Johari, H., Soltani, M., Malik, R., Othman, N.Z., El Enshasy, H.A. (2015). The edible mushroom *Pleutotus* spp.: I. Biodiversity and nutritional values. *International Journal of Biotechnology for Wellness Industries.* 4, 67-83.
6. El Enshasy, H.A. Maftoun, P., Abd Malek, R. (2012). Pleuran: Immunomodulotor polysaccharide from *Pleurotus ostreatus*, structure, production and application *In.* Mushrooms: Types, Properties and Nutrition (Andres, S. and Baumann, N. Ed.). Nova publisher, NY, USA). pp 153-172.
7. Sun, Y., Liu, J. (2009). Purification, structure and immunological activity of a water soluble polysaccharide from the fruiting body of *Pleurotus ostreatus. Bioresource Technology.* 100, 983-986.
8. Levi, I., Friesem, D., Geresh, S., Hadar, Y., Schwarz, B. (2006). An aqueous polysaccharide extract from the edible mushroom *Pleurotus ostreatus* induces anti-proliferative and pro-apoptotic effect on HT-29 colon cancer cells. *Cancer Letters.* 244, 61-70.
9. El Enshasy, H. (2010). Immunomodulators. In *The Mycota: Industrial Applications,* Esser K., Bennett J.W., Eds.; Springer Verlag: Berlin, Heidelberg, Vol. 10, pp. 165-194.
10. El Enshasy, H. (2006) New method for mushroom cell cultivation and production of polysaccharides/ peptidoglycan in bioreactors. Egyptian Patent No. 24637. (Dated 7/6/2006).
11. El Enshasy, H.A. and Hatti-Kaul, R. (2013). Mushroom immunomodulators: unique molecules with unlimited applications. *Trends in Biotechnology.* 31, 668-677.
12. El Enshasy, H., Daba, A.; El-Demellawy, M.; Ibrahim, A.; Elsayed, S., El Badry, I. (2010). Bioprocess development for large scale production of anti-cancer exopolysaccharide by *Pleurotus ostreatus* in submerged culture. *Journal of Applied Science.* 10(21): 2523-2529.

464571 Process Engineering of Clostridium Cellulovorans for Butanol Production from Biomass

Abstract:

Clostridium cellulovorans can utilize a wild range of low-value lignocellulosic biomass as substrates because of its efficient cellulosome complex. The objective of this study was to achieve a robust lignocellulosic butanol production using metabolically engineered *C. cellulovorans* mutant by the process engineering of consolidated bioprocessing (CBP). The metabolomics was used to create an intracellular metabolite profiling of *C. cellulovorans* and analyze the interaction between cellular metabolism and fermentation process parameters. The bioreactor operation parameters and fed-batch process parameters were specifically analyzed. The changes in core metabolites responsible for carbon, energy and redox balance at different fermentation conditions were identified. The CBP was engineered by optimizing the seeding density, pH, cellulose feeding and trace element. The butanol production was improved to titer of >3 g/L and selectivity of > 0.4 g/g from cellulose. The direct fermentation of corn cob generated butanol titer of > 1.5 g/L.

466640 EthanolÂ Production on Date Syrup Using Yeast: Process Optimization Study for Pilot Scale Production

Abstract:

Yeast is one of a most promising species for ethanol production using various substrates. In the present study bioethanol production from commercially available date syrup was evaluated for yield and process optimization parameters using Central Composite Design (CCD). Date syrup as substrate is quite economic in this part of the world. The ethanol production is specifically chosen to support the industry during falling crude prices globally on one hand, and if successfully produced & economically optimized for large scale production over date syrup; the ethanol production can become a boon to the otherwise struggling date palm agro-economic sector in region; on the other hand. Fermentation temperature, concentration of date syrup sugar, nitrogen source, C/N ratio, feed flow rate, pH, was optimized through CCD in a pilot scale fermentor. The data was compared with production over standard media. Using date syrup, growth kinetics and yields were found to be near double to that of the standard media. With a controlled flow rate; above 97% sugar conversion was achieved. Urea was used as supplemental Nitrogen source for pilot level production and the efficiency of the conversion process was recorded.

Bioprocess Strategies to Optimize Thermostable Phytase Production By *Escherichia coli* B121 (DE3) When Induced with Lactose

Nor Zalina Othman[1], Solleh Ramli[1],Roslinda Abd Malek[1], Mohamad Roji Sarmidi[1], Rajni Hatti-Kaul[3], Hesham A. El Enshasy[1,2]

[1]*Institute of Bioproduct Development, Universiti Teknologi Malaysia, 81310 Skudai Johor, Malaysia;* [2]*Bioprocess Development Department, Mubarak City for Scientific Research and Technology Applications (MuCSAT), New Burg Al Arab, Alexandria, Egypt;* [3]*Department of Biotechnology, Center for Chemistry and Chemical Engineering, Lund University, Box 124, S-221 00 Lund, Sweden.*

Introduction

IPTGis used as a molecular mimic of allolactose, a lactose metabolite that triggers transcription of the *lac* operon. Unlike allolactose, the sulphur atom create a chemical bond which is non-hydrolyzed by the cell, preventing the cell from degradation by the host cells that makes the IPTG concentration remains constant (Fernández-Castané *et al.*, 2012). IPTG is however toxic and comparably expensive (Donovan *et al.,* 1996). Nowadays, lactose has been considered as an alternative inducer in fermentation of recombinant *E.coli.* Since the product yields is compatible with IPTG as inducer however it depends on the strategy of induction to increase heterologus protein expression (Weng *et al.*, 2006; Abdel-Hamid *et al.*, 2014). Induction of promoter gene by lactose during foreign protein expression has a limiting step. It is clearly observed that IPTG can bind directly to the *lac*-repressor molecules and express the phytase protein after one hour of induction. However, the expression of phytase was delayed almost an hour compared induction with IPTG and the amount of lactose needed as inducer should be further optimized for cultivation in different scale (Tran *et al.*, 2010). In case of lactose as inducer, a few steps are needed before induction process. This is due to the time needed by lactose to be transported into the cell by the present of lactose permease and processed to allolactose by β-galactosidase before binding to the repressor (Becker *et al.*, 2012; Santillán *et al.*, 2007). Furthermore, lactose can be metabolized as energy sources. Further catabolic activity of allolactose may produce glucose and galactose that can contribute to increase the metabolic overflow due to excess carbon sources during induction period (Santillán and Mackey, 2008). As a result of metabolite burden, the high concentrations of inducer that are often used in an effort to fully induce the *lac* promoter do not necessarily lead to maximal expression of a target protein. This is due to high residual lactose concentration in the broth led to accumulation of glucose by hydrolysis of excess lactose during post-induction phase (Lim *et al.*, 2004). Therefore the objective of this study to optimize the induction strategy whenlactose as inducer with addition of additive for *E. coli* BL21 (DE3) which growth in glycerol minimal medium to increase productivity of phytase expression and leakage of expression protein from periplasmic space to outside of cell membrane in semi-scale bioreactor.

Materials and Methods

Microorganism

E.coli BL21 (DE3) harboring pE10C2 with the leader peptide that carries thermophile phytase gene from *Bacillus* sp. MD2 under the control of *lac*-O promoter to express thermostable phytase enzyme was used for this research. The strain was originally from of Biotechnology, Center for Chemistry and Chemical Engineering, Lund University and was delivered and stored in Wellness Industries Culture Collection (WICC), Institute of Bioproduct Development, Universiti Teknologi Malaysia. Cells were first propagated on LB agar plate containing 100 μgmL^{-1} amphicillin and were incubated at 30 °C for 24 hours. The arisen colonies were harvested by glycerol solution (50 %) and put in series of 2 mL cryogen vials. The vials were frozen at -20 °C for 24 hours followed by further storage as working cell bank at -80 °C for further use.

Inoculum preparation and cultivations conditions in shake flask

E. coli BL21 (DE3) cells were grown on a LB-agar plate containing 100 μg mL^{-1} of ampicillin at 30 °C for 24 h. A single colony was picked from the plate and cultured for 5h in 50 ml LB broth containing 100 μg mL^{-1} of ampicillin at 30°C. The shake flasks cultivations were carried out in 250 ml Erlenmyer flasks containing 50 ml broth and incubated in a rotary shaker (4230 Innova, New Brunswick, NJ, USA) at 200 rpm. In all shake flask experiments, the media was supplemented with 100 μg mL^{-1} of amphicillin. Inoculation was performed by transferring 5 mL of seed culture to 50 mL medium in a 250 mL Erlenmeyer flask.

Production medium in shake flask cultivation and bioreactor

Medium that used in the shake flask cultivation and the batch culture for bioreactor 16-L was glycerol minimal medium modified from studied of Spexard *et al.*, (2010). For batch cultivation of *E. coli* BL21 (DE3), carbon source was prepared and autoclaved separately in 1000 ml and added aseptically to the bioreactor before inoculation.For the fed-batch pre-induction growth phase, the feeding solution was started using DO-stat feed of a solution containing:yeast extract, 10 g L^{-1}; glycerol, 500 g L^{-1}, trace elements, 10 mL L^{-1}, 1 M $MgSO_4$, $7H_2O$, 50 ml L^{-1}. During the post-induction phase, feeding of glycerol was constantly fed at 15 g L^{-1} h^{-1}. For the induction phase, the induction solution for one-pulse induction strategy and the feeding solution were prepared as batch cultivation. Induction was aseptically carried by using 30 mM of lactose with 10 mM $CaCl_2$ and 0.65 % (w v^{-1}) of glycine which were sterilizing separately, mixed together and added for batch cultivation when the cell mass approximately 6-8 g L^{-1}.

Optimization of induction strategies using RSM

The shake flasks cultivations were carried out in 250 ml Erlenmyer flasks containing 50 ml broth and incubated in a rotary shaker (4230 Innova, New Brunswick, NJ, USA). Induction was aseptically carried when OD_{600nm} of each flask approximately 0.6-0.8. The sample analysis was done after 6 hours of post-induction phase.A Box-Behnken design with a total of 30 experiments was run to further optimize the significant variables. Box-Behnken design was used for optimisation strategy for enhancing leakage of phytase from periplasmic space to the extracellular medium. Concurrently, the correlation of lactose, $CaCl_2$ and glycine concentration to the periplasmic and extracellular phytase activity was studied. Each factor in the design was studied at three different levels that was low, intermediate and high value. Each variable are denoted by the numbers (-1), (0) and (+1), respectively as shown in Table 1; Lactose (**A**), $CaCl_2$ (**B**) and glycine(**C**).

Table 1: Experimental range and levels of factors influencing induction strategy of *E.coli* BL21(DE3) for production of phytase in a Box-Behnken design.

Factors	Low(-1)	Centre (0)	High (+)
Lactose (mM)	15	25	35
$CaCl_2$ (mM)	5	10	15
Glycerine (% w/v)	0.25	0.65	1

Bioreactor operation

Cultivations with growing cells were carried out in a 16-L stirred-tank bioreactor (BioEngineering, Wald, Switzerland). The bioreactor is equipped with pH probe, oxygen probe, foam sensor, and stirrer of two–four bladed Rushton turbines. During the experiments, controlled the pH, temperature, aeration rate and the agitation speed were controlled at 7, 30 ºC, 2 v v^{-1} min^{-1} and 400 rpm, respectively. The dissolved oxygen was adjusted to 100% at the beginning of cultivation and remains controlled at 30 % along the cultivation process. Antifoam reagent (Silicone antifoam, Sigma, USA) was added to suppress the foaming when necessary.

Analysis

Phytase activity assay: Determination of phytase activity

The extracellular phytase activity was determined in the cell-free supernatant obtained after centrifugation of the sample at 8,000 g for 10 min at 4 °C. The cell pellet was washed with physiological saline and the remaining cell pellet was re-suspended in 0.1 M Tris–HCl buffer, pH 7, containing 5 mM $CaCl_2$. The periplasmic content was extracted by using the osmotic-shock method. All the assays were performed in triplicates. The phytase activity was determined from the soluble protein sample using supernatant obtained from extracellular and periplasmic fraction. Base on the method studied by Tran *et al.,*(2010), a mixture of appropriate diluted enzyme sample and sodium phytate (1.5 mM) was dissolved in 100 mM Tris-HCl buffer, pH 7 containing 5 mM $CaCl_2$. Then, the solution was incubated at 70 °C for 20 min. The reaction is stopped by adding an equal volume of 15% trichloroacetic acid (TCA) solution. The precipitate obtained is removed by centrifugation, and 500 μl of the clear supernatant transferred to an Eppendorf tube followed by addition of 500 μl ammonium molybdate reagents (containing 1:4 mixtures of 2.7% $FeSO_4$ and 1.5% ammonium molybdate in 5.5 % H_2SO_4). After 5 min incubation at room temperature, absorbance at 700 nm was read and used to calculate the concentration of P_i from a standard curve made using sodium dihydrogen phosphate solution in the concentration range of 0-600 μmol/l P_i. All the samples were assayed in triplicates. One unit of phytase activity was defined as the amount of enzyme that releases 1 μmole of P_i per minute under the assay conditions.

Acetic acid, glycerol and lactose determination

For analyzing acetic acid, glycerol and lactose in culture supernatant by HPLC. HPLC separation was performed with Waters liquid chromatography (Milford, MA, USA) equipped with Waters 2690 Separation Module and Waters 996 Photodiode Array Detector (DAD). The data was stored and processed by HPLC Empower software. The data was stored and processed by HPLC Empower software. Chromatographic separations were performed on an ROA-Organic Acid H+ (Phenomenex, USA) column (300 x 7.80 mm, 8 micron). As the mobile phase 0.005N H2SO4 was used at a flow rate of 0.6ml/min. The column was kept in column heater at 40 °C. For detection, an optimum wavelength was selected for the acetic acid is 210 nm.

Result and discussion

Optimization of induction strategyfor extracellular and periplasmic phytase activity in the shake flask cultivation.

The aims of this section to optimize excretion of phytase outside the cell membrane towards maximal total phytase activity. The ANOVA for the extracellular and periplasmic phytase activity are shown in Table 2. The calculated value of *F* were compared to the critical value of $F_{(p-1,\ n-p,\alpha)}$. The ANOVA showed that the calculated value of *F* for the extracellular and periplasmic phytase activity that was 3.09 and 6.63, respectively were greater than the critical values for the F distribution table of $F_{(3,26,0.05)}$=2.934. If the calculated value of *F* exceeds the $F_{(3,26,0.05)}$, the null hypothesis is rejected at the level of significance, and it is inferred that the coefficient estimates are not all zero and the variation verified that the model is significantly greater than the unexplained variation (Wonacott and Wonnacott, 1990). Therefore the *p*-value for all of the response is below 0.05 which means that at the 95 % confidence level of the models were significant.The determination coefficient of R^2 which implies that the sample variation of above 95 % for both extracellular and periplasmic activity was attributed to the concentration of lactose, $CaCl_2$ and glycine. This study signifies a good correlation between the experimental data and the predicted values which overall below than 5% was not explained by the model (Varnalis *et al.*, 2004).

Table2: Analysis of variance (ANOVA) of the response variables

Response	Source	DF *[1]	SS*[2]	MS*[3]	F value, F_{cal}	Prob(p)>F	r^2(%)	F (0.05)
Extracellular phytase activity (Y_1)	Regression	3	27.623	9.208	3.09	0.045	97.15	2.934
	Residual	26	77.544	2.982				
	Total	29	105.167					
Periplasmic phytase activity (Y_2)	Regression	3	1.069	0.356	6.63	0.002	98.80	2.934
	Residual	26	0.826	0.032				
	Total	29	1.895					

[1]-Degree of Freedom,[2]- Sum of Square, [3]-Mean Square

Confirmation test of optimization study between non-optimized and optimized induction strategy

Through Box-Behnken design of optimization plot for induction strategy consists of lactose, $CaCl_2$ and glycine of 27.68 mM, 11.32 mM and 0.65 % [w v^{-1}], respectively. The extracellular phytase activity (7.86 U mL^{-1}) and total phytase activity (8.70 U mL^{-1}) obtained in fermentation used the optimal induction strategy that increased after 6 hour induced with lactose by 36.45 % and 29.65 %, respectively. The increment of extracellular phytase activity that excreted outside the cell membrane after statistically optimized induction strategy supplemented with glycine increased up to 90.34 % within 6 hours of the post-induction phase when compared to un-optimization induction strategy only 85.84 % (Table 3).The lost of cell membrane permeability by supplemented with glycine increased the uptake of lactose which reduced the time required for transport of lactose into the cell (Shuman and Silhavy, 2003). In this study, the consumption rate of lactose is not significant but the enzyme production rate was more significant based on glycine alteration on the membrane permeability (Puertas *et al.*, 2010; Ding *et al.*, 2010).This in consequently causes the accumulation of phytase in the cytoplasm and the periplasmic space and consequently secreted outside the cell membrane.

Table 3: The comparison study of the optimized induction strategy with and without addition of ampicillin in the shake flask cultivation.

Parameters	Addition of ampicillin (100 µg mL^{-1})		Without-ampicillin (100 µg mL^{-1})
Induction strategy variables			
Lactose [mM]	30	25.68	25.68
$CaCl_2$ [mM]	10	11.32	11.32
Glycine % [wv^{-1}]	-	0.65	0.65
Growth kinetics and phytase production			
X_{max} [g l^{-1}]	1.79	1.59	1.86
µ [h^{-1}]	0.11	0.12	0.16
$P_{max (Exo)}$ [U mL^{-1}]	5.76(12hr)	7.86 (6 hr)	6.50(6hr)
$P_{max (Total)}$ [U mL^{-1}]	6.71(12hr)	8.70 (6 hr)	7.05(6hr)
$Y_{Totalphytase/cell mass}$ [U g^{-1}]	6,414.91	10,875.00	5875.00
Q_{Exo} [U L^{-1} h^{-1}]	578.00	1,310.00	1,083.33
Q_{Total} [U L^{-1} h^{-1}]	671.00	1,450.00	1,175.00
% excretion outside the cell membrane	83.50	90.34	92.19

X_{max}: maximal cell dry weight, $P_{exo\ or\ total}$: maximal production of extracellular,P_{total}: maximal production of phytase (as sum of periplasmic, cytoplasmic and extracellular activity), $Y_{p/x}$: unit of enzyme produced per unit cell mass (at the time of maximal total phytase activity), Q_p [U L^{-1} h^{-1}]: volumetric phytase production rate

Hence, to study effects of ampicillin for plasmid stability cultivation without and addition of ampicillin were carried out. The cultured grew faster when the cultivation broth not supplemented with ampicillin with the specific growth rate of 0.16 h^{-1}.The maximal cell mass production was increased about 16.98 % (1.86 g L^{-1}) when compared cultured with ampicillin. The growth rate of 0.16 h^{-1} gave high acetic acid production (max=1.55 g L^{-1}). This contributed to rapid declined of plasmid lost after 2 hours of induced with lactose. Finally contributed to the decrease of total phytase activity to 18.96 % with 92.19 % was excreted outside the cell membrane.

Fed-Batch Cultivation to Increase Production of Thermostable Phytase in 16-L Stirred Tank Bioreactor

From the batch cultivation of *E. coli* BL21 (DE3) in the 16-L bioreactor, the addition of lactose as the only carbon source in the fermentation broth in the post-induction phase was not only for cell growth but also for heterologous protein expression by the host cells as also reported by Li *et al.* (2012). Using lactose as the only carbon source in this phase was not efficient for cell growth and protein expression (Lim *et al.*, 2004). Meanwhile, when induced with lactose only, 4.78 U mL^{-1} of total phytase activity was reached after 6 hours of the post-induction phase. Therefore, due to limitations of the carbon source and dissolved oxygen, there was a significant accumulation of acetic acid in the fermentation broth of up to 1.51 g L^{-1}. Therefore, the fed-batch cultivation was carried out by glycerol feeding at the rate of 15 g L^{-1} h^{-1}. From Figure 1, it is shown that this mode of glycerol feeding can maintain the cell growth with the specific growth rate of 0.08 h^{-1}. After 5 hours lactose induction, both of lactose and glycerol was actively consumed in this phase.

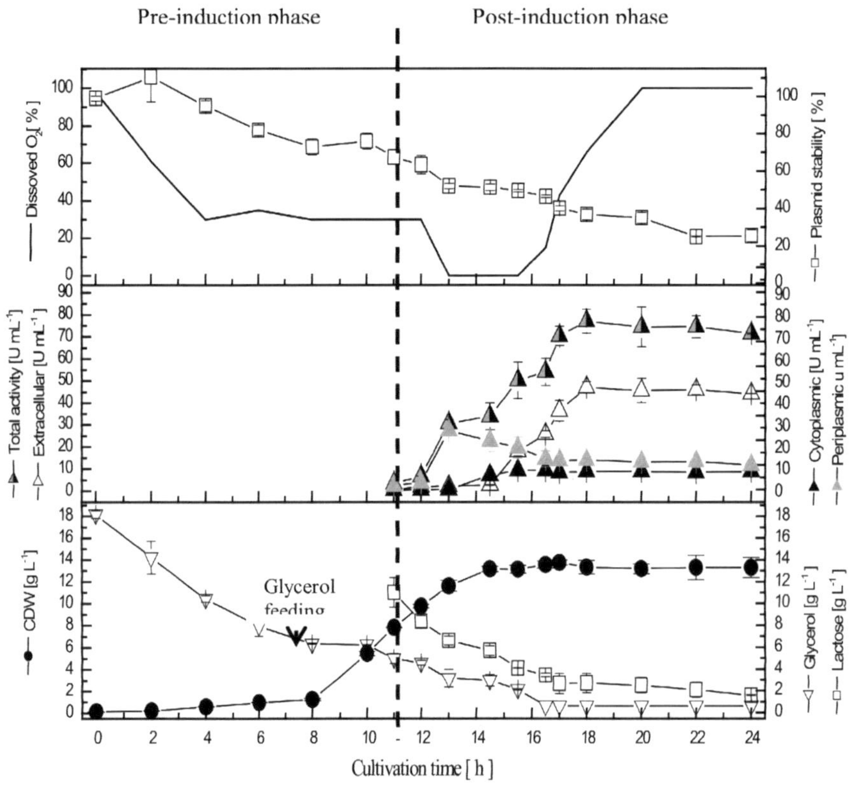

Figure 1: The cell mass production, phytase expression, distribution of phytase activity, residual glycerol/lactose, acetic acid production and plasmid stability in fed-batch cultivation of *E. coli* BL21 (DE3) induced with one-pulse lactose with 15 g L^{-1} hr^{-1} of glycerol feeding during the post-induction phase (----start of induction).

The residual of glycerol was almost 0.1 g L^{-1} and lactose was 3.46 g L^{-1} after 5 hours induced (Figure 1). The presence of glycerol in the post-induction phase contributes, to the carbon stress response of *E. coli* during cultivation (Zhang *et al.,* 2009; Martinez-Gomez *et al*., 2012).

Conclusion

In this study revealed factor that significantly influenced cell mass, extracellular phytase activity and accumulation of acetic acid by *E. coli* BL21 (DE3) in the optimization of induction strategies was lactose. In this study, more than 90 % of the total phytse activity was excreted outside the cell membrane by addition of glycine also contributed to the highest volumetric and productivity induction with lactose. This work has demonstrated the use of the second order polynomial equation to determine conditions leading to the maximum extracellular phytase activity and minimize acetic acid production. The new developed of medium composition and induction strategy can be used for the study in 16 L bioreactor without the addition of ampicillin.

Acknowledgement

The authors would like to thanks the MOHE and Universiti Teknologi Malaysia (UTM) for Contract research grant no R.J130000.7344.andR.J130000.7646.4C101 .

References:

1. Becker,A.N., Peters, P.J., James Maher III, L. (2012) Mechanism of promoter repression by Lac repressor-DNA loops. Nucleic Acids Research. 1-11.
2. Ding, R., Zhaofeng, L. , Chen, S., Wu, D. and Wu, J., Chen, J. (2010).Enhanced Secretion of Recombinant α-Cyclodextringlucosyl Transferase from *E. coli* by Medium Additives. *Process Biochemistry,* 45: 880–886.
3. Donovan, R.S., Robinson, C.W. and Glick, B.R. (1996). Review: Optimizing Inducer and Culture Conditions for Expression of Foreign Proteins under the Control of the *lac* promoter.*Journal Industrial Microbiology*, 16:145-154.
4. Fernandez-Castane, A., Vine, C.E., Caminal, G.and Lopez-Santin, J. (2012). Evidencing the Role of Lactose Permease in IPTG uptake by *Escherichia coli* in Fed-Batch High Cell Density Cultures. *Journal Biotechnology*, 157:391–398.
5. Abdel-Hamid, A. M., Attwood, M.M. and Guest, J. R. (2001). Pyruvate oxidase Contributes to the Aerobic Growth Efficiency of *Escherichia coli.Microbiology,* 147:1483–1498.
6. Li, M., Zhang, X., Agrawal, A. and San, K. Y. (2012). Effect of Acetate Formation Pathway and Long Chain Fatty Acid Coa-Ligase on the Free Fatty Acid Production in E. coli Expressing Acy-ACP Thioesterase From *Ricinus communis. Metabolic Engineering,* 14(4): 380-387.
7. Lim, H.K., Kim, S.G., Jung, K.H. and Seo, J.H. (2004). Production of the Kringle Fragments of Human Apolipoprotein(α) by Continuous Lactose Induction Strategy. *Journal of Biotechnology,*108(3):271-278.
8. Martinez-Gomez, K, Flores, N., Castañeda, H. M, Martínez-Batallar G, Hernandez-Chavez, G., Ramirez, O. T., Gosset, G., Encarnación, S. and Bolivar, F. (2012). New insights into *Escherichia coli* Metabolism: Carbon Scavenging, Acetate Metabolism and Carbon Recycling Responses during Growth on Glycerol. *Microbial Cell Factories,* 11:46.
9. Pei, X.L., Wang, Q.Y., Li, C.L., Qiu, X.F., Xie, K.L., Huang, L.F., Wang, A.M., Zeng, Z.W. andXie, T. (2011). Efficient Production of a Thermophilic 2-Deoxyribose-5-Phosphate Aldolase in Glucose-limited Fed-batch Cultivations of *Escherichia coli* by Continuous Lactose Induction Strategy. *Applied Biochemistry Biotechnology*, 165:416–425.
10. Puertas, J. M., Ruiz, J., de La Vega, M. R., Lorenzo, J., Caminal, G. and Gonzalez, G. (2010). Influence of Specific Growth Rate Over the Secretory Expression of Recombinant Potato Carboxypeptidase Inhibitor in Fed-batch Cultures of *Escherichia coli. Process Biochemistry,* 45(8):1334–1341.

11. Santillán, M., and Mackey, M.C.(2008). Quantitative Approaches to the Study of Bistability in the lac operon of *Escherichia coli. Journal Royal Society Interface,* 5:S29–S39.
12. Shuman, H. A. and Silhavy, T. J. (2003). The Art and Design of Genetic Screens: *Escherichia coli. Nature Reviews Genetics,* 4: 419-431.
13. Santillán, M., Mackey, M. C. and Zeron, E.S. (2007). Origin of Bistability in the *lac* operon. *Biophysiology Journal*, 92:3830–3842.
14. Spexard, M., Beshay, U., Risse, J.M., Miksch, G., Flasche, E. (2010). Screening for conditions of enhanced production of a recombinant *b*-glucanase secreted into the medium by *Escherichia coli.* Biotechnol. Lett .32:243–248.
15. Tran, T.T., Mamo, G., Mattiasson, B. and Hatti-Kaul, R. (2010). A Thermostable Phytase From *Bacillus sp.* MD2: Cloning, Expression and High-Level Production in *Escherichia coli. Journal Industrial Microbiology Biotechnology,*37(3):279-87.
16. Weng, Y.P., Hsu, F.C., Yang, W. S. and Chen, H.P. (2006). Optimization of the Over expression of Glutamate mutase S Component under the Control of T7 System by using Lactose and IPTG as the Inducers. *Enzyme and Microbial Technology,* 38: 465–469.
17. Wonnacott, T.H., and Wonacott, R.J. (1990). Introductory statistics. New York: John Willey and Sons.
18. Varnalis, A., Brennan, J., MacDougall, D., Gilmour, S. (2004). Optimization of high temperature puffing of potato cubes using response surface methodology. Journal Food Engineering, 61: 153-163.
19. Zhang, B. H., Mao, X. O., Wang, Y. J. and Hu, X. Q. (2009). Optimization of Culture Conditions for High-level Expression of Dextransucrase in *Escherichia coli. Journal of Food, Agriculture and Environment,* 7(3&4):75 - 78 .

474193 Laccase: Fermentation Process Intensification and Efficient Purification By Magnetic Separation

Abstract:

A promising farnesol addition strategy was developed to enhance the laccase production from *Trametes versicolor*, and the laccase fermentation was successfully scaled up to a 2 m^3 stirred-tank reactor. The addition of farnesol resulted in an increased expression of the laccase (*lac*) gene and the *RhoA* gene. The *RhoA* gene correlated with hyperbranched mycelia, which facilitated the secretion of the intracellular laccase. An emphasis is highlighted on direct capture of laccase from *T. versicolor* fermentation broth using magnetic mesoporous silica nanoparticles with metal affinity ligands. The Cu^{2+}-chelated magnetic mesoporous silica nanoparticles with an average pore size of 14.5 nm provide 60.6-fold purification of laccase and 114.6% recovery yield of enzyme activity. The simple and efficient bioseparation method has great potential to be used for the robust and inexpensive preparation of purified laccase at the industrial scale.

471297 A Biosensor for Human Milk Oligosaccharide Detection

Abstract:

Human milk oligosaccharides (HMOs) are prebiotic compounds found in breast milk that favor the growth of beneficial bacteria in the infant gut. Often these compounds are derivatives of lactose or n-acetyllactosamine that contain either a fucose or sialic acid moiety. While there are many efforts underway to synthesize these beneficial complex carbohydrates, the current standard for their detection is low-throughput HPLC-MS. We present a high-throughput, fluorescence-based detection system for fucosylated and sialylated human milk oligosaccharides in vivo in *E. coli*. Such high-throughput systems facilitate directed evolution for increased in vivo HMO production. Additionally, the reporter enables the use of HMOs as orthogonal inducers of cellular circuits in synthetic biological applications.

461203 Spatiotemporal Regulation of Calcium Signaling By Morphogens in a Developing Epithelium

Abstract: Organ development and homeostasis rely on patterned morphogenetic pathway activity that is interpreted locally as input into a cellular information-processing network. This network is important for coordinating processes such as proliferation, apoptosis, cell differentiation and metabolism. Calcium ions (Ca^{2+}) are second messengers that are key mediators of signal integration. This regulatory information is encoded by controlling the dynamic concentration of cytoplasmic Ca^{2+}. However, while important for a wide range of medical conditions, control of Ca^{2+} transients by gene-regulatory networks in epithelial tissues is still poorly understood. Here we use an established model system of an epithelial tissue, the *Drosophila* wing imaginal disc, to investigate how morphogenetic signaling impacts Ca^{2+} signaling dynamics in a developing epithelium. Using an optimized genetically encoded Ca^{2+} sensor, GCaMP6, we have elucidated the Ca^{2+} pathway components that are responsible for intercellular calcium transients (here called ICTs) through a combination of genetic and chemical perturbations using an *ex vivo* imaging assay. For genetic perturbations, we created a tester line to systematically perturb a large number of genes in the Bone Morphogenetic Protein (BMP) and Hedgehog (Hh) signaling pathways. Live-imaging of Ca^{2+} signaling in intact organs enabled the extraction of features of the calcium code such as frequency, amplitude, and duty cycle. These summary statistics of Ca^{2+} signaling that affect a wide range of protein functions and enzymatic activities. Principle component analysis of intercellular Ca2+ transients reveal an underlying spatial pattern approximately along the anterior/posterior compartment boundary. Amplitudes of Ca^{2+} oscillations varied with developmental age of wing discs and in different spatial regions. Genetic perturbation experiments reveal that inhibition of specific components of the Bone morphogenetic protein or Hedgehog protein pathway activity increases ICT frequency. Combining a data-driven model with targeted genetic knockdowns reveals the underlying interactions between morphogen-pathway genes and Ca^{2+} signaling. Using a computational model of Ca^{2+} patterning in the wing disc, we find that differential PLC stimulation as the most parsimonious explanation for the observed spatiotemporal patterning of Ca^{2+} waves in the wing disc. A mechanistic understanding of cross-talk and integration between Ca^{2+} and upstream morphogen signaling pathways has important implications in a broad range of basic medical applications, including the development of imaging and computational modeling tools for diagnostics and therapeutics.

- Abstract Text

461595 Competitive Tuning: Competition's Role in Setting the Frequency-Dependence of Calmodulin Target Proteins Important in Learning and Memory

Abstract:

More than 10 million people suffer from learning and memory disorders. Ranging from Alzheimer's and Huntington's disease to autism and schizophrenia, these disorders are caused by disruptions in protein signaling networks that control the strength of neuronal connections. Normally, dynamic changes in the strength of synaptic connections (called synaptic plasticity) are brought about through the exquisite coordination of neurotransmitter release, protein synthesis, protein localization, and cytoskeletal reorganization. The timing, magnitude, and location of these processes are determined by protein binding events within protein signaling networks and, in particular, enzyme activation initiated by Ca2+-flux through NMDA receptors (NMDARs) in the synapse. Therefore, elucidation of the mechanisms that regulate NMDAR-dependent long-term plasticity depends on a complete understanding of the endogenous tuning mechanisms that pair precise patterns of enzyme activation to certain calcium signals.

The calcium sensor calmodulin (CaM) acts as a common activator of the networks responsible for both the strengthening and weakening of synaptic connections. This is possible, in part, because different calmodulin target proteins are 'tuned' to different calcium signals by their unique binding dynamics. Such systems are traditionally studied in vitro or in vivo, but economic and practical limitations have created a demand for computational models capable of predicting molecular phenomena. These predictions can then be used to guide focused experimental studies.

Previous computational studies have demonstrated the role of binding dynamics, feedback loops, and spatial effects in regulating enzyme activation during synaptic calcium signaling, but competition may serve as an additional tuning mechanism. The concentration of CaM binding sites in the cell far exceeds that of CaM itself, and in vitro studies have demonstrated competitive inhibition among neuronal CaM target proteins. But, despite its implicit presence in many computational models, competition among CaM-dependent proteins has not been explicitly investigated as a regulator of enzymatic activation during synaptic plasticity.

In the present study, a mathematical model of eight experimentally-characterized postsynaptic CaM targets is used to investigate competition's potential role as a regulator of synaptic plasticity. Specifically, a pair of mathematical models is developed. A 'competitive' model simulates CaM binding to calcium and eight CaM binding targets– adenylyl cyclase type I (AC1), the adenylyl cyclase type VIII N-terminus (AC8-Nt), the adenylyl cyclase type VIII C_{2b} domain (AC8-Ct), calcineurin (CaN), Ca^{2+}/CaM-dependent protein kinase II (CaMKII), myosin light chain kinase (MLCK), neurogranin (Ng), and nitric oxide synthase (NOS). An 'isolated' model simulates CaM binding to calcium and just one target of interest (TOI) from the eight present in the competitive model. Because our model is devoid of complicating feedback loops and spatial effects, the differences in the CaM-binding of each protein in the competitive and isolated models are the pure competitive effects of this system. Indeed, we find that competition for CaM serves as an additional tuning mechanism; the presence of competitors both shifts and sharpens the Ca^{2+} frequency-dependence of target proteins. Notably, simulated competition is necessary to recreate the in vivo frequency-dependence of several notable CaM target proteins. We conclude that competitive tuning is an important dynamical process underlying learning and memory. In addition, we find that our results explain seemingly contradictory experimental studies wherein genetic knockout of neurogranin leads to increases in CaMKII activation, but has been shown to both decrease and increase synaptic strength. This and future work in characterizing the critical components of Ca^{2+}/CaM protein signaling may lead to the identification of therapeutics that can treat neurological disorders by modulating target protein binding through competition.

472198 Melanoma-Induced Endothelial Barrier Disruption Via VE-Cadherin Disassembly and Cell Contractility

Abstract:

Metastasis is a complex process whereby cancer cells detach from the primary tumor and migrate to remote locations within the body through the vascular and lymphatic systems. During metastasis, breakdown of endothelial adherens junctions is critical for tumor cell extravasation through blood vessel walls and is mediated by a combination of tumor secreted soluble factors and receptor-ligand interactions. Here, we investigate the roles of key intracellular signaling molecules and cytoskeletal contractility in regulating melanoma-induced VE-cadherin disassembly. Through the use of traction force microscopy and pharmacological inhibitors of cytoskeletal contractility we find that endothelial cell contractility is responsive to interactions with metastatic cancer cells and that reducing endothelial cell contractility abrogates melanoma-mediated disassembly of endothelial cell-cell junctions. Furthermore, we find that a combination of tumor secreted soluble factors and receptor-ligand interactions mediate activation of Src within endothelial cells which is necessary for breakdown of the endothelial barrier. Together, these results provide insight into how tumor cell signals act in concert during extravasation and may aid in identification of therapeutic targets to block metastasis.

472631 Using Nucleic Acid Aptamers to Modulate Specificity, Binding Affinity, and Drug Quantity to Create an Individualized Cancer Treatment Approach

Abstract:

Our work uses a gold nanoparticle core coated with DNA strands and aptamers in order to bind cancer cells via an avidity driven process. This results in the creation of a flexible and versatile nanocarrier with detection, drug carrying, and specific recognition capabilities. There are many advantageous properties of gold nanoparticles (AuNps) that can be applied to our DNA based targeting and drug delivery approach. Most notably their detection by x-rays and well studied binding of DNA with a thiol-bond modification. In addition, they are easy to synthesize, modify, characterize, and known to not elicit an immune response. We combine the properties of AuNps with recently developed DNA aptamers in order to target and treat cancer cells. This allows us to construct a nanocarrier with the ability to modulate quantity of drug and specificity towards targeting of distinct cancer cells.

The first step is to functionalize 15-nm gold nanoparticles with single stranded DNA strands (anchor DNAs) modified with a thiol group on the 5â€™ end. Successful attachment of anchor DNAs is validated by surface plasmon resonance and gel electrophoresis. A 15-nm Np is chosen because with the addition of DNA strand attachment to the surface, the particle size is capable of both passive and active targeting. Previous work in our lab has shown that we can bind up to 101 +/- 8 of our anchor DNAs to a 15-nm AuNp using optimized experimental procedures. Since the theoretical maximum is 107 DNA strands, this means we are capable of attaching the entire range of DNA possible. By increasing the size of the AuNps, we can produce nanocarriers that carry more anchor DNAs, which will result in an increased number of aptamers and higher payloads of drug to the surface of the cancer cells. Alternatively, we can produce smaller nanocarriers that can be more readily taken up by cancer cells.

Cancer cell specific aptamers can be equipped with a single-stranded segment that is complementary to the single-stranded DNA anchor. Such design produces a general drug nanocarrier that can be readily programmed with either one or several different cancer-specific aptamers. Additionally, aptamer strands can be engineered to include drug-binding regions in desired locations and quantities. This is done by creating a double-stranded DNA region that will allow for intercalation of chemotherapeutic drugs, such as daunomycin. The length of this region determines how many drugs can bind. Furthermore, the release rate of daunomycin can be regulated using different triplet base pair sequences for varying binding affinity of daunomycin to the intercalating region. This property allows us to control the location, quantity, and release rate of the drug on our nanoparticle platform.

Presently, we have been able to attach a maximum of ~1,100 molecules of drug to a 15-nm gold nanoparticle with our current design, which is much greater than the average drug loading of any 15-nm nanoparticle found in the literature. This was accomplished due to our ability to control the specific amount of anchor DNAs bound to the Np surface, as well as the length of the drug-binding region. Current release studies have indicated that we are able to control the release rate of daunomycin through temperature and by manipulating the triplet base pair binding sequences. *In vitro* viability studies have indicated that our nanocarrier is more efficient at killing cancer cells than free drug at the same concentration. Additionally, our studies show that daunomycin is releasing from the nanocarrier.

Currently, we are investigating regulation of the specificity and binding constant of our nanocarrier by modulating use of the aptamers. By attaching multiple DNA aptamers, the resulting nanoparticle is expected to act as a â€œmolecular octopusâ€ increasing the binding constant of our platform by utilizing avidity, and is able to target different markers on the surface of a heterogeneous cancer tumor. Our preliminary results indicate that by manipulating the size of the nanocarrier, the structure of the DNA aptamers, and its drug-binding segment, we have created an effective yet flexible, versatile, highly tunable platform that is capable of recognizing a specific cancer tumor, and is able to deliver controlled amounts of chemotherapeutics dictated by health status to individual patients.

466970 Membrane-Active PSMA-Targeting Liposomes for Antivascular Chemotherapy

Abstract:

Introduction: Cases of advanced solid cancer as of yet have no cure [1]. A challenge for the medical research community is the development of diagnostic and treatment protocols that significantly extend life expectancy and improve quality of life of these cancer patients. Based on the difficulties encountered in treatment, future effective interventions will likely require a combination of therapies to provide effective tumor control, of which antivascular therapy may be a key element. Antivascular therapy aims to kill tumor cells by cutting off blood flow, depriving tumors of nutrients they need to survive [2].

The objective of this study is to design a new class of liposomes that can target tumor vasculature via Prostate-Specific Membrane Antigen (PSMA) and rapidly and extensively release therapeutic agents intracellularly by using the pH-sensitive peptide GALA conjugated to lipid moieties to destabilize the integrity of intracellular endosomes and achieve better tumor killing outcomes.

GALA is a 30 amino acid peptide that forms an alpha helix at acidic pH [3]. Once formed, this helix can interact with and penetrate lipid bilayers (via pore formation) making it a candidate for inclusion in systems that desire endosomal escape [4, 5]. This study explores the mechanism of GALA anchored to gel-phase lipid vesicles in both clustered (pH and non-pH responsive) and uniform presentations with the goal of optimizing GALA presentation for improving therapeutic efficacy of tumor vasculature targeted liposome compositions.

PSMA is a cell surface glycoprotein receptor present on tumor vascular endothelium of many types of cancer [6, 7]. Although this target is also expressed in the healthy tissues of the prostate, gastrointestinal tract, kidneys, and brain [8-10], normal tissue vasculature is consistently PSMA-negative [11]. Therefore, tumor vasculature is directly and easily accessible to intravenously injected PSMA targeted treatments, whereas healthy vasculature and PSMA-expressing tissues are not expected to exhibit significant localization.

Methods: Liposomes were prepared by hydrating a thin lipid film with the desired aqueous phase buffer, incubating at 55° Celsius for 2 hours, and extruding the resultant suspension through 100 nm -pore polycarbonate membranes to generate unilamellar vesicles. GALA was synthesized by solid phase peptide synthesis (Anaspec) and was conjugated to a DPPE succinyl lipid using click chemistry by reacting the succinyl carbonyl group to the N-terminus amine of the peptide. Calcein at 75 mM was used as a tracking fluorophore for GALA-induced release. Leakage of calcein (EX/EM 495 nm/515 nm) from an endosome analogue EggPC membrane incubated with GALA-bearing liposomes resulting in self-quenching fluorescence release was monitored over time over the pH range from 7.4 to 4.0 (Q_t). pH was adjusted with HCl. The two liposome compositions were incubated at a 1:1 mole ratio (equivalent to 1:100 GALA:lipid mole ratio). Triton-X 100 (5% w/v) was then added to induce the complete release of contents (Q_{max}). The ratio of these two measurements was used to determine the normalized change in self-quenching efficiency of calcein $(Q_t-1)/(Q_{max}-1)$ which is related to the extent of leakage. Calcein containing EggPC liposomes were also exposed to each pH without GALA present as a control.

For targeting experiments, the *in vitro* tumor analogue used consisted of HUVEC induced to express PSMA, as previously described [12]. PSMA induction was achieved by suspending HUVEC in tumor cell conditioned media (CCM; serum-free RPMI1640 incubated with MDA-MB-231 breast cancer cells for 24 hours) and seeding on Matrigel™ (Corning). Liposomes were prepared as described above. Liposomes were incubated with cells on ice for 6 hours at a 1:50 liposome:PSMA-receptor ratio. A peptidomimetic urea-based ligand was used for targeting [13].

Results: Gel-phase, GALA-bearing liposomes induced release of encapsulated contents from endosome analogue vesicles in a pH-dependent manner with clustered GALA inducing faster release at endosomal pH than uniform GALA. The apparent activity of liposome-grafted GALA is affected by the presentation of GALA-grafted peptides (clustered vs. uniformly distributed over the liposome surface).

PSMA-targeted liposomes exhibit increased specific binding to PSMA-positive HUVEC cells. Results showing the effect of targeted, GALA-bearing liposomes on PSMA-positive HUVEC cells under flow will be presented.

Conclusions: A clustered presentation of GALA liposomes has the potential to enhance the delivery of chemotherapeutics directly to the cytosol. It is possible that clustering of GALA allows for each peptide monomer to more readily interact with other peptide units, enhancing induced release in the range of endosomal pH. This is likely the primary variable determining the apparent reactivity of GALA as the comparison between the apparent reactivity of clustered GALA on the surface of non-charged liposomes (where clustering is pH-independent) and the apparent reactivity of uniformly distributed GALA-bearing liposomes showed that the absence of surface electrostatic effects alone does not account for differences in GALA reactivity.

PSMA can be specifically targeted by liposomes using a urea-based ligand as a targeting strategy.

References:

[1] Howlader N NA, Krapcho M, Neyman N, Aminou R, Waldron W, Altekruse SF, Kosary CL, Ruhl J, Tatalovich Z, Cho H, Mariotto A, Eisner MP, Lewis DR, Chen HS, Feuer EJ, Cronin KA, Edwards BK (eds). SEER Cancer Statistics Review, 1975-2008. National Cancer Institute Bethesda, MD. 2011.

[2] Folkman J. Tumor angiogenesis: therapeutic implications. The New England Journal of Medicine. 1971;285:1182-6.

[3] Li WJ, Nicol F, Szoka FC. GALA: a designed synthetic pH-responsive amphipathic peptide with applications in drug and gene delivery. Adv Drug Deliv Rev. 2004;56:967-85.

[4] Shete HK, Prabhu RH, Patravale VB. Endosomal escape: a bottleneck in intracellular delivery. Journal of Nanoscience and Nanotechnology. 2014;14:460-74.

[5] Varkouhi AK, Scholte M, Storm G, Haisma HJ. Endosomal escape pathways for delivery of biologicals. Journal of Controlled Release. 2011;151:220-8.

[6] Chang SS, Reuter VE, Heston WD, Bander NH, Grauer LS, Gaudin PB. Five different anti-prostate-specific membrane antigen (PSMA) antibodies confirm PSMA expression in tumor-associated neovasculature. Cancer Research. 1999;59:3192-8.

[7] Liu H, Moy P, Kim S, Xia Y, Rajasekaran A, Navarro V, et al. Monoclonal antibodies to the extracellular domain of prostate-specific membrane antigen also react with tumor vascular endothelium. Cancer Research. 1997;57:3629-34.

[8] Carter RE, Feldman AR, Coyle JT. Prostate-specific membrane antigen is a hydrolase with substrate and pharmacologic characteristics of a neuropeptidase. Proceedings of the National Academy of Sciences of the United States of America. 1996;93:749-53.

[9] Israeli RS, Powell CT, Corr JG, Fair WR, Heston WD. Expression of the prostate-specific membrane antigen. Cancer Research. 1994;54:1807-11.

[10] Silver DA, Pellicer I, Fair WR, Heston WD, Cordon-Cardo C. Prostate-specific membrane antigen expression in normal and malignant human tissues. Clinical Cancer Research. 1997;3:81-5.

[11] Chang SS, O'Keefe DS, Bacich DJ, Reuter VE, Heston WD, Gaudin PB. Prostate-specific membrane antigen is produced in tumor-associated neovasculature. Clinical Cancer Research. 1999;5:2674-81.

[12] Liu T, Jabbes J, Nedrow-Byers JR, Wu LY, Bryan JN, Berkman CE. Detection of prostate-specific membrane antigen on HUVECs in response to breast tumor-conditioned medium. Int J Oncol. 2011;38:1349-55.

[13] Chen Y, Foss CA, Byun Y, Nimmagadda S, Pullambhatla M, Fox JJ, et al. Radiohalogenated prostate-specific membrane antigen (PSMA)-based ureas as imaging agents for prostate cancer. Journal of Medicinal Chemistry. 2008;51:7933-43.

471130 The Role of Cytoskeleton and Ion Channels in Cell Decision-Making Under Confinement

Abstract:

Introduction: Lymphatic vessels as well as perivascular, perimuscular and perineural spaces provide longitudinal tracks, which facilitate cancer metastasis. The cell decision-making process at intersections of either partially or fully confined microenvironments, has been analyzed in a generic way through paradigms of hydraulic resistance or chemical gradients without any mechanistic explanation. We herein present the first comprehensive analysis on how cross-sectional areas of resulting branches affect tumor cell decision-making, focusing on the role of cytoskeleton and ion channels in the absence of chemical gradients.

Materials and Methods: To examine the effect of channel dimensions on decision-making, we created Ψ-shaped PDMS-based microchannels coated with collagen I. The microchannel array confined MDA-MB-231 breast cancer cells either partially ($\boldsymbol{W}$x$\boldsymbol{H}$=10x10μm^2) or fully ($\boldsymbol{W}$x$\boldsymbol{H}$=3x10μm^2) before reaching a 3-way intersection consisting of channels of prescribed dimensions. Furthermore, we created microchannels with a dead end to account for decisions under infinite hydraulic resistance. To quantify cell speed and persistence we used the MTrackJ plugin of FIJI and a custom made MATLAB code. Statistical significance was assessed with non-paired Student's t-test.

Results and Discussion: Upon decision-making within a bifurcation of an infinite hydraulic resistance (dead end) versus a lower resistance (open end) branch, 20% of MDA-MB-231 cells chose to enter the dead end compared to ~5% of the HL-60 leukocyte-like cells. Hence MDA-MB-231 cells in full confinement don't make decisions solely based on hydraulic resistance. Breast cancer cell decisions in Ψ-shaped asymmetric branches can be predicted in relatively high accuracy by the ratio of the cross-sectional areas that cells encounter within an intersection. Treatment with 125uM colchicine, which disrupts microtubules, lowers significantly the speed and persistence of cells. This, in turn, increases the number of cells which remain undecided (>4hrs to enter a branch). More importantly, in the fully confined regime, the probability of cells continuing on the same base without turning diminishes. We postulate that differences in branch cross-sectional areas reflect differences in cell membrane stretching. Consequently, mechanosensitive and stretch-activated Ca^{2+} channels play a significant role in cell-decision making. When MDA-MB-231 cells are treated with GsMTx4 (pharmacological inhibitor of TRPC1, TRPC6 and PIEZO1 channels), the decision-making differences between the partially and fully confined cases are completely abrogated. Ongoing experiments include transfection of MDA-MB-231 cells with a Ca^{2+} FRET probe to determine Ca^{2+} localization at intersections. Moreover, we are elucidating the contributions of mechanosensitive and stretch-activated Ca^{2+} channels (TRPC1, TRPV6, TRPM7) as well as Ca^{2+} activated K^+ channels (SK3, BK_{Ca}).

Conclusion: Our data provide novel insight into the mechanism of cell decision-making under confinement. Cross-sectional area can be used as a predictor of cell decision, microtubules are key players for maintaining cell directionality, and stretch activated channels help cells sense their microenvironment before making a decision.

470113 Mathematical Modeling and Analysis of BMP-Receptor Complex Formation Elucidates Requirements for Heterodimer Dominant Signaling

Abstract:

Bone Morphogenetic Proteins (BMPs) play a pivotal role in developmental pattern formation and tissue homeostasis throughout the animal kingdom. A consistent feature in many examples of BMP signaling is the requirement for heterodimer ligands to elicit the highest signaling relative to their homodimer ligand counterparts. To investigate the biophysics of heterodimer dominance, we developed a mathematical model of receptor oligomeriztion and found that heterodimer ligands are not kinetically favored over other ligand-receptor combinations. With this as the starting point for analysis, we identified the minimal conditions needed in the model to confer heterodimer-dominant signaling and identified testable hypotheses for heterodimer dominance- 1) homodimer ligands are sequestered in the extracellular environment, 2) heterodimers have a modified kinetic rate that specifically enhances their interactions with receptors, and 3) receptor complexes formed by homodimers have weaker signaling. Testing the alternative explanations in vivo using zebrafish suggests that heterodimer dominance is conferred at the receptor complex formation stage and favors a mechanism where the heterodimers provide a compromise solution that overcomes the kinetic advantages of the stronger binding homodimer ligand.

463137 John Quinn: Master Weaver of a Remarkable Professional Tapestry

Abstract:

Professor John A. Quinn first came to prominence in chemical engineering circles for his fundamental research on mass transfer and interfacial phenomena. His work featured experiments thoughtfully conceived to permit accurate determination and rigorous analysis of interphase mass transfer rates at freshly created interfaces – an achievement made possible by his design of apparatus that brought immiscible fluids together in well-defined geometries at unprecedentedly short phase contact times. These experiments enabled the intrinsic mass transfer resistance of the interface itself to be explored, thus providing John with the opportunity to test the prevailing assumption that solutes diffusing from one immiscible phase into another would always experience equilibrium across the infinitely thin gas/liquid and liquid/liquid interfaces separating them.

While in the process of elucidating the role of the interface in mass transport, John Quinn and his students would initially be plagued by -- and then later intrigued with -- the thin films of surface-active materials that invariably concentrated at phase boundaries. Exploration of these insoluble monolayers revealed that their influence frequently extended well beyond the immediate vicinity of the interface; under some circumstances these interfacial films promoted convective instabilities while at other times they exerted a stabilizing effect on the bulk phases on either side of them.

Given his preoccupation with these thin films or interfacial "membranes" that spontaneously accumulated at phase boundaries, it was inevitable that John Quinn's attention would turn to synthetic membranes, inasmuch as these afforded controllable properties (e.g., porosities and thicknesses) that made them compelling platforms for fundamental transport studies. Track-etched mica membranes provided particularly fertile ground for the investigation of various steric, hydrodynamic, and electrodynamic effects key to the transport of solutes and colloids through submicron pores. Out of this work would come a much-improved theoretical description of sterically hindered diffusion, as well as critical insights on filtration and adsorption in finely porous media. Synthetic membrane composites also informed studies in his group of gas permeation across the skin.

By virtue of his fundamental mass transfer work and his keen observations of transport across thin films, John Quinn was well-positioned to witness and participate in the birth and subsequent growth of a membrane separation industry initially focused on simple polymer films. But John's vision extended far beyond the capabilities of those conventional synthetic membranes to what might be possible if engineers were to mimic elements of the structure and function of vastly more powerful biological membranes. For instance, John noted that cell membranes achieved unparalleled selectivity by exploiting reversible reactions between permeants and membrane-localized carriers, and he was fascinated by the ability of membrane-bound biomolecules to deftly orchestrate key biochemical reactions. He responded by materially advancing the theory of carrier-mediated or facilitated transport and by inspiring the development of enzyme membrane bioreactors capable of chiral separations important to the pharmaceutical industry.

More significantly, John Quinn's enthusiasm for the intellectual interface between biology and engineering – and his influence on the direction that his department at Penn and his chosen field of chemical engineering would take -- amply justify his recognition as a pioneer in the then-emerging field of biomolecular engineering. John and his students would go on to explore several phenomena mediated by biomolecule-rich cell membranes, including, for instance, the attachment of cells to surfaces, the motility of cells on those surfaces, and the directed transport of motile bacteria.

This presentation traces the logic and serendipity that underlie the evolution of John Quinn's varied research interests, identifying the threads that tie together the body of his work into a coherent whole – a truly remarkable "professional tapestry". It documents the considerable impact that his scientific and technical contributions have had on our profession – and it provides an occasion to honor the thoughtful guidance and warm friendship that John so generously offered to so many, his students and colleagues alike.

457447 John Quinn: Advisor, Mentor, and Friend

Abstract:

Much of John Quinn's interest and curiosity centered on surfaces and interfacial phenomena. My thesis on interfacial resistance to mass transfer at liquid/liquid interfaces was based on a most unlikely experiment that somehow John knew had a chance of working. This work inspired much of my career. For example, early in my career I made membranes as thin as 10nm, not much more than two surfaces. Later I helped find a catalytic needle in a haystack that saved a business based on a reaction at a solid surface.

456402 Facilitated Transport Membranes for CO_2 Separation and Capture

Abstract:

This presentation is in honor of Dr. John Quinn for his pioneering work on facilitated transport membranes including those for CO_2 separation, and it covers CO_2 facilitated transport membranes for: (1) hydrogen purification for fuel cells and (2) CO_2 capture from flue gas in power plants and from <1% CO_2 concentration sources. We have synthesized CO_2-selective membranes by incorporating amino groups into polymer networks. The membranes have shown high CO_2 permeability and selectivity vs. hydrogen, carbon monoxide and nitrogen. We have elucidated the effect of amine steric hindrance in the solid membrane, showing significant enhancement for CO_2 transport. The membranes were scaled up to 14 inches in width by using a continuous roll-to-roll machine equipped with a thin-film-coating assembly. For the CO_2 separation from a synthesis gas stream containing CO_2 and H_2S, i.e., hydrogen purification for fuel cells or pre-combustion CO_2 capture in IGCC (integrated gasification combined cycle) technology, thicker amine-containing membranes with around 15 ± 1 microns of selective layer thicknesses were fabricated. For the post-combustion CO_2 capture from flue gas, thinner amine-containing membranes with < 200 ± 20 nm of selective layer thicknesses were scaled up. From gas transport measurements, the flat-sheet samples of the scale-up membranes for both applications exhibited similar performances compared to the membranes synthesized in lab scale. CO_2 permeance of approximately 200 GPU and CO_2/H_2 selectivity > 200 were for the thicker membranes for the purification of hydrogen and synthesis gas along with CO_2 capture (e.g., pre-combustion capture), and CO_2 permeance of about 800 GPU and CO_2/N_2 selectivity > 140 were for the thinner membranes for CO_2 capture from flue gas. The scale-up membrane has been fabricated into spiral-wound membrane modules. Techno-economic analysis has shown that the post-combustion capture process using the thin membrane has approached DOEâ€™s

capture cost target for 2025.

455619 The Force on a Body in Active Matter

Abstract:

The statistical mechanics and microhydrodynamics of active matter systems have seen a dramatic increase in interest in the past several years. Much attention has focused on the fascinating nonequilibrium behaviors of active matter not observed in equilibrium thermodynamic systems, such as spontaneous collective motion and swarming. Even minimal kinetic models of active Brownian particles exhibit self-assembly that resembles a gas-liquid phase separation. Self propulsion allows active systems to generate internal stresses that enable them to control and direct their own behavior and that of their surroundings. Recent work involving the unique 'swim pressure' exerted by active systems offers a perspective on the basic underlying physical mechanism responsible for self-assembly and pattern formation in active matter [1-3].

In this work we take a more microscopic view and present a general theory for determining the force (and torque) exerted on a body (or boundary) in active matter [4]. The theory extends the description of passive Brownian colloids to self-propelled active particles and applies for all ratios of the thermal energy k_BT to the swimmer's activity $k_sT_s = \zeta U^2{}_0\tau R/6$, where ζ is the Stokes drag coefficient, U_0 is the swim speed and τR is the reorientation time of the active particles. The theory, which is valid on all length and time scales, has a natural microscopic length scale over which concentration and orientation distributions are confined near boundaries, but the microscopic length does not appear in the force. The swim pressure emerges naturally and dominates the behavior when the body size is large compared to the swimmer's run length $l = U_0\tau R$. The theory is used to predict the motion of bodies of all sizes immersed in active matter.

References
[1] S.C. Takatori, W. Yan and J.F. Brady, Phys. Rev. Lett. 113, 028103 (2014)
[2] S.C. Takatori and J.F. Brady, Phys. Rev. E 91, 032117 (2015)
[3] S.C. Takatori and J.F. Brady, Current Opinion in Colloid & Interface Science, 21, 24 (2016)
[4] W. Yan and J.F. Brady, J. Fluid Mech. 785, R1 (2015)

459226 A Talk on Bijels, in Honor of John A. Quinn

Abstract:

John A. Quinn was a pioneer who advanced our fundamental understanding of processes that occur at fluid interfaces in the field of chemical engineering with practical implications in separations and reaction processes. John was also an outstanding colleague who formed and nurtured at the University of Pennsylvania our extraordinary department culture of intellectual exchange and interaction. In this talk, I briefly highlight a few examples of his work in the field. Thereafter, I discuss an emerging vehicle for study by chemical engineers in these contexts: Bijel membranes. Bijels, first predicted in simulation (1) and realized by a delicate experimental process, (2) comprise bicontinuous streams of immiscible fluids stabilized by jammed layers of nanoparticles at their interface. We have recently developed a robust and versatile means of fabricating these membranes that exploits a suspension of nanoparticles in a ternary aqueous/oil/co-solvent system. (3) This platform allows a broad range of particles and fluid pairs to be formed into fibers or membranes with tremendous promise in the fields of separations and interphase catalysis. Further, the potential importance of these vehicles in practical application will spur fundamental questions of trapped states and interfacial transport worthy of detailed analysis in the style of Quinn. Finally, this work, performed with colleague Daeyeon Lee at the University of Pennsylvania, embodies Johnâ€™s spirit of collaborative endeavor.

(1) **Colloidal Jamming at Interfaces: A Route to Fluid-Bicontinuous Gels, K. Stratford et al** *ScienceÂ* 2005,**309**, 2198-2201.

Â (2) **Bijels: a new class of soft materials, Michael E. Cates and Paul S. Clegg** *Soft Matter*, 2008, **4**, 2132-2138.

Â (3) **Continuous Fabrication of Hierarchical and Asymmetric Bijel Microparticles, Fibers, and Membranes by Solvent Transfer-Induced Phase Separation (STRIPS)**Â M.F. **Continuous Fabrication of Hierarchical and Asymmetric Bijel Microparticles, Fibers, and Membranes by Solvent Transfer-Induced Phase Separation**

(STRIPS) Haase, D. Lee, K.J. Stebe, Advanced Materials 2015, **27**, 7065-7071.

466714 Reverse Osmosis As a Tool in Water Treatment Strategies

Abstract:

Reverse osmosis is a workhorse tool commonly leveraged in water treatment flowsheets to remove highly soluble salts. This paper will briefly review state of the art reverse osmosis membrane chemistry and the manufacturing process, and then discuss how RO spiral wound modules are employed in conjunction with suitable pre-treatment technologies to create very high quality water for a wide variety of municipal and industrial applications.

463011 Recombinant Oleosin As a Functional Surfactant

Abstract:

Our laboratory is interested in novel surfactants that can be self-assembled into technologically useful suprastructures and can be equipped with novel biofunctionality. Oleosin is a naturally occurring surfactant protein found in plants. The molecule has three domains – two terminal hydrophilic domains, and a central hydrophobic domain with an embedded proline knot, which forces the molecule into a turn. We have obtained the gene for oleosin and expressed it in E.coli, and, using the tools of molecular biology, made hundreds of variants of the molecule, including numerous truncations and additions of functional or targeting domains. Depending on the truncation, oleosin can assemble into vesicles, worm-like micelles, sheets, or spherical micelles – assemblies which have utility in drug delivery and tissue engineering. In one application, we show how spherical micelles equipped with receptor targeting domains can be activated by a protease by protecting receptor-targeting domains with protease cleavable domains. The corresponding spherical micelles bind to target cells, but only when both the protease and the receptor are present, effectively creating an "and" switch for cellular recognition. We also show how oleosin's properties as a surfactant can be exploited to stabilize mondisperse air bubbles for ultrasound imaging. We will also provide a further update on ongoing efforts to incorporate other functional domains into oleosin for a variety of applications.

473326 Driving Change in Pharmaceutical Development & Manufacturing through Innovation and Standards

Abstract:

We are all familiar with the old adage â€œbetter-faster-cheaper, Pick Twoâ€. However in todayâ€™s world of product development and manufacturing, all three are needed and needed now. Organizations are tasked with moving diverse and complex pipelines forward in record time with undivided attention required for product quality, safety and reliability of manufacturing. These challenges do not require incremental change in the operating model but revolutionary change in our model. In this talk I will discuss activities, specifically process and product innovation geared to customer value and standards-based thinking for product lifecycle management and partnership models. I will discuss the implications for the processes, plants and people to ensure successful commercial products.

473183 Biomanufacturing Science and Technology Consortium to Advance U.S. Manufacturing of Biopharmaceuticals

Abstract:

Biomanufacturing has become an emerging industry of strategic importance worldwide and is one of the key technology areas identified by the Advanced Manufacturing Partnership (AMP) as an appropriate target for public-private investment to support advances in manufacturing and U.S. competitiveness. The development of advanced biomanufacturing technologies will foster and support the fast growing biomanufacturing industry and for establishing sustained leadership of the US in the area (PCAST, 2012) (AMP 2, 2014).

The Biomanufacturing Science and Technology Consortium (BSTC) is hereby proposed, which will be led by academic and industry leaders across the nation. The proposed consortium is founded on the operational experiences of the Biopharmaceutical Process and Quality Consortium (BPQC) at University of Massachusetts Lowell (MA), which has been building collaborative efforts in advancing research and practice outcomes in the biopharmaceutical industry for four years. The proposed consortium represents a comprehensive team of actively participating stakeholders covering the entire spectrum of the industryâ€™s value chain to identify challenges that the biomanufacturing industry is currently facing and to develop a dynamic roadmap to accelerate solutions. The consortium will be comprised of biomanufacturers including small and medium size businesses, research universities and colleges, government institutes and non-profits focusing on achieving the sustained U.S. leadership in biomanufacturing. The consortium will encompass multiple geological locations of strategic importance and disciplinary expertise areas, e.g., bio-engineering, regulatory and measurement sciences, education and workforce development. (BPQC, 2012).

The consortiumâ€™s roadmapping exercise will take place at several strategic locations mainly through focused workshops. Each regional workshop will have an open brainstorming session to identify challenges and to encourage innovative ideas. The last concluding workshop will bring an active group of stakeholders and subject matter experts to synthesize and prioritize the outcomes from the multiple national and regional workshops. This consortium will also attempt to leverage the concept of *crowdsourcing* to inform additional ideas for solutions to key questions and to identify research projects. The proposed planning efforts will lead to a clear roadmap for a next generation biomanufacturing paradigm, reliable industry standard systems, and substantive knowledge spanning the industry challenges, as well as a list of potential pre-competitive projects covering the entire spectrum of topics toward the identified new paradigm. Through the roadmapping effort, this proposed consortium aims to lay the foundation for a sustained partnership among the members that will foster the U.S. competitiveness in the global market.

The consortium seeks to make all-encompassing plans to promote the development of advanced biomanufacturing science and technology, and also actively looks for the upgrade and transformation of existing biomanufacturing, as well as sharing resources and promoting innovations. Upon successful completion of this proposed planning effort, the consortium aims to conduct the following tasks:

- Prototyping a system of innovation for advanced biomanufacturing
- Developing breakthroughs in a number of key technologies
- Establishing technological standards and sustained global leadership of the U.S. biomanufacturing industry
- Building a robust and integrated supply chain network of an advanced biomanufacturing

463406 Defining Challenges and Creating Solutions in Biopharmaceutical Manufacturing

Abstract:

In this presentation, we will discuss a number of technology challenges associated with the manufacturing of biopharmaceuticals. The opportunity to address pre-competitive challenges through public-private partnerships is a potential solution to address these challenges. However, careful consideration of the role of relevant federal agencies, health authorities, along with the global nature of the biopharmaceutical industry is required to identify the most effective path forward to advance biopharmaceutical manufacturing.

460344 Technology Advances That Will Enhance Continuous Biomanufacturing

Abstract:

The development of advanced biomanufacturing technologies involves consideration of converting batch manufacturing to continuous operation. Continuous manufacturing has a significant number of advantages that include operation of a manufacturing method that will produce consistent, high quality products at lower costs. This is achieved by using quality by design (QbD) principles which have been historically demonstrated to work in the continuous production of small molecules. It is done by reducing overall variability in the process which enables enhanced control within an optimized operating range. To implement this approach effectively for continuous biopharmaceutical processing, technology is needed in improved sampling, process analytical technology (PAT), and data handling. This will then allow for the determination of the critical quality attributes (CQA) and relate them to the critical process parameters (CPP) and show how these can be used to control process operation within the design space to enable real-time release testing (RTRT).

467812 Energy-Based Medium Design for Mammalian Cell Fed-Batch Cultures

Abstract:

Mammalian cell cultures are the main industrial platform for the production of high-value biologics, such as monoclonal antibodies (mAb) [1]. One way to reduce operational costs is by maximising mAb yield and cell growth whilst minimising cell culture medium expenditure. Current optimization strategies are typically empirical requiring a multitude of expensive, intensive and time consuming experiments. Model-based optimization approaches seek to reduce unnecessary experimentation. Previously, we have applied model-based optimization methods to identify feeding regimes to maximize mAb titter in GS-NS0 cultures limited only on glucose and glutamate supply while neglecting the exhaustion of other essential amino acids [2]. In addition, culture medium design has been based on stoichiometric balances considering only cell composition [3] and metabolite profiling where the medium was designed based on the exhaustion of a limited number of nutrients [4]. Alas, the above-mentioned studies neglect the energy requirements of the cell not only for maintenance and proliferation but also for mAb production.

Herein, a novel predictive model for GS-NS0 cells producing cB72.3 mAb has been developed considering cell and product composition as well as cellular energy requirements for maintenance, proliferation, and mAb production. The development of this model follows the framework described in [5] and describes growth kinetics, metabolism, mAb secretion and the production/consumption of ATP based on glucose and 15 amino acid energy metabolic pathways. The full range of the model parameters was subjected to Global Sensitivity Analysis (GSA) to identify the significant parameters that were then estimated from specifically-designed batch experiments. Subsequently, the validated model was utilised to design a low cost supplemental medium for the implementation of an optimized fed-batch strategy which increased production by 60% and extended viability and culture longevity by 24 hrs. As a result of this feeding strategy, nutrient and ATP concentrations were controlled in a tight range without accumulation. The successful coupling of cell and product composition with energy requirements has enabled the development

of a tailored medium composition reducing operational costs whilst maximising production. This integrated model-based platform, which is experimentally validated, presents a cost-effective and efficient approach for media development and upstream optimisation of a wide range of biopharmaceutical products.

References

1. Zhu, J., Mammalian cell protein expression for biopharmaceutical production. Biotechnology advances. 2012. 30(5): p.1158-1170.
2. Kiparissides,A., Kountinas,M., Kontoradvi,C., Mantalaris, A., Pistikopoulos, E.N. 'Closing the loop' in biological systems modeling — From the in silico to the in vitro. Automatica. 2011. 47(6): p.1147-1155.
3. Xie, L, Wang, D.I.C. High Cell Density and High Monoclonal Antibody Production Through Medium Design and Rational Control in a Bioreactor. Biotechnology and Bioengineering.1996. 51: 725-729.
4. Sellick, C.A., Croxford,A.S., Maqsood,A.R., Stephens, G., Westerhoff,H.V., Goodacre,R., Dickson, A.J. Metabolite Profiling of Recombinant CHO Cells: Designing Tailored Feeding Regimes That Enhance Recombinant Antibody Production. Biotechnology and Bioengineering. 2011. 108 (12): 3025–3031.
5. Kiparissides,A., Pistikopoulos,E.N., Mantalaris,A. On the model-based optimization of secreting mammalian cell (GS-NS0) cultures. Biotechnology and Bioengineering. 2015. 112: 536–548.

461381 Evaluating the Potential of Miniature Scale Chromatography Process Characterization to Support Biomanufacturing Strategies and Regulatory Filings

Abstract:

High throughput downstream chromatography techniques such as resin slurry plates and miniature columns have been well described and implemented for biologics research, discovery, and early process development applications. These techniques have seen relatively little application in late stage development areas, though, mainly due to the necessity of qualified scale-down models and increased precision in this space that is required for regulatory acceptance. However, an automated parallelized scale-down approach could potentially provide much needed reductions in time and resources for activities like formal process characterization where representative data across many conditions is needed to support regulatory filings.

This talk will highlight some of the differences and challenges for utilizing high throughput techniques in late stage development compared to earlier stages of process development and will suggest potential applications in manufacturing space. For miniature columns, theoretical limitations such as transport phenomena and wall effects will be discussed. Practical considerations like pulse testing, pooling strategies, and resin reuse will also be presented. Case study comparisons of formal process characterization strategies for miniature scale and traditional lab scale as well as direct scale-down comparisons across four orders of magnitude will be shown. Finally, recommendations for the ultimate limitations of miniature columns will be provided to suggest future state directions and applications in late stage development spaces.

470644 Ecoprime Twin Continuous Two-Column Chromatography at the Process Scale

Abstract:

The call for implementing continuous processing into the manufacturing of bio-pharmaceuticals requires re-thinking of process chromatography and its equipment designs. Proof-of-concepts of innovative multi-column continuous processes were introduced by the industry, which verified improvements in resin utilization and increased productivity. However, questions remain remains how to scale up the standard chromatographic batch processes to continuous production scale. Our focus is here on the process optimization of the twin-column technology that has a less overall complexity in process and system design and performs comparable if not better than other multi-column setups. The successful scale-up from bench-top to twin-column pilot scale will be supported by performance data. The requirements for the equipment design, process automation and regulatory aspects due to the implementation of the continuous processing are highlighted. Furthermore, future trends how the EcoPrime Twin technology will enable the integrated continuous downstream processing are pointed out.

453796 A Single Amino Acid Change Confers a Radical Change in the Structure of a Virus-like Particle

Abstract:

Single amino acid mutations can alter secondary and tertiary structure in proteins, but mutations that alter quaternary structure are less well documented. Here, we use MS2 bacteriophage as a model for quaternary structure, and we show that a single amino acid mutation can shrink the global virus-like particle (VLP) from 27nm to 18nm in diameter. Crystallizing this unique VLP mutant revealed a uniform, stable shift from a T=3 to a T=1 icosahedral viral capsid. These small VLPs maintain the ability to disassemble and reassemble *in vitro* and can attach to the F pili of *E. coli*â €"which is required for infectivity. Furthermore, these mutant VLPs can be chemically modified to load cargo such as drugs or fluorophores. We are currently designing new strategies to systematically introduce single amino acid mutations into the VLP and screen for altered quaternary structure. Altogether, these structurally-distinct VLPs have implications for *de novo* prediction of quaternary structure, drug delivery, and viral evolution.

467864 A Novel Model for the Osmotic Behaviour of Human Mesenchymal Stem Cells

Abstract:

Stem cells have an enormous therapeutic potential for regeneration and replacement of degenerated tissues. In particular, the ability to readily expand in culture while maintaining a self-renewing phenotype has made human Mesenchymal Stem Cells (hMSCs) a promising candidate for many cell-based therapies. Unlike induced pluripotent stem cells and embryonic stem cells, adult hMSCs do not raise ethical and legislative issues, so their use takes advantage of an increased likelihood of authority approval and public acceptance. Even though bone marrow has been established as the primary source of adult hMSCs, peripheral sources such as Umbilical Cord Blood (UCB) are being investigated because of their non-invasive and painless collection.

Storing of MSCs represents a fundamental step in tissue engineering, gene therapy and regenerative medicine applications. The most common preservation method, to make cells available on demand, consists in cooling to a cryogenic temperature. Cryopreservation allows long shelf lives and genetic stability, reduced risk of microbial contamination, and improved cost effectiveness [1]. Unfortunately, cryopreserved cells are damaged by the process itself [2]. This loss, which amounts to up to 50 %, is unacceptable for the hMSCs from UCB whose collection and isolation is known to be difficult. Even if, in principle, cell expansion/proliferation may solve the problem, an increased number of passages will inexorably lead these cells to lose their peculiar characteristics, and should be avoided.

Cryopreservation consists of different stages: cooling to sub-zero temperatures with or without permeant and non-permeant Cryo-Protectant Agents (CPAs), storage, thawing and return to physiological environment. Any of these steps can potentially lead to cell damage due to the physical and chemical phenomena involved such as intracellular ice formation, excessive solutes concentrations, cell shrinkage or swelling and CPA cytotoxicity. Due to the process complexity, a very high number of trials are required for experimental optimization, which is able to lead to suboptimal solutions only. On the contrary, mathematical modelling may provide considerable time and cost savings as well as mechanistic understanding of the underlying phenomena.

In order to develop a model describing the process of cryopreservation, the osmotic behaviour needs to be investigated first to predict the osmosis-driven change in the amount of intracellular water, which can form lethal ice or glass during the process, as well as to limit excessive cell volume excursions and solute concentrations that might lead to the so-called *solution injury* [3, 4].

As reported in [5] and [6], we first attempted to determine the osmotic properties of hMSC from UCB making use of the typical bi-compartimental two-parameters model used in cryopreservation [7]. This attempt was not successful in simulating the osmotic response of these cells, as they do not behave as perfect osmometers, and show a peculiar osmotic response.

In this work a novel approach is proposed to take this particular osmotic response into account. The model describes the volume variation of intracellular water, dimethyl sulphoxide (DMSO), and sodium chloride (NaCl). The difference of the proposed model from typical models reported in the literature consists in allowing NaCl to permeate the cell membrane. According to this assumption, incomplete volume recovery either upon restoring isotonic conditions after initial contact with hypertonic solutions of non-permeant (sucrose) CPAs or in the swelling phase of the well-known shrink-swell response in presence of permeant (DMSO) CPAs, is attributed to a net efflux of NaCl (low-permeant). The low-permeant solute efflux reduces its intracellular concentration, thus reducing the influx of water and the final equilibrium cell volume [8].

The NaCl flux can be rationalised by invoking the notion of mechanosensitive (MS) channels. There are two main theories purporting to explain mechanosensitive channel gating: (i) the bilayer model hypothesising that the tension of the lipid bilayer alone is sufficient to gate the MS channels, and (ii) the tethered model involving relative displacement of the MS channel with respect to the cytoskeleton or extracellular matrix proteins [9]. Whatever the exact microscopic mechanism involved, the overall effect is that the membrane tension is reduced in the shrinking phase leading to the closure of MS channels, while its increase during swelling results in their opening with the overall effect of changing NaCl permeability.

The proposed model was found to be able to describe the osmotic behaviour of hMSC from UCB in presence of permeant and not permeant CPAs. Using this model, the permeabilities of water, NaCl and DMSO have been estimated from experimental data, while the inactive cell volume was fixed at the value previously obtained in [5]. According to the MS channels hypothesis, NaCl permeability has been found to depend on whether the cell is shrinking or swelling with higher values of permeability associated with the latter.

It should be noted that different parameter values were estimated from experimental runs in presence of sucrose and DMSO. More precisely, lower values for NaCl and water permeability were found in presence of DMSO. Regarding water permeability, this behaviour is well known in literature. A decrease in the hydraulic conductivity has been found in presence of different CPAs (glycerol, ethylene glycol and propylene glycol, DMSO) and at increasing CPA concentrations [10]. It has been hypothesised that CPAs block water channels changing bilayer permeability [11]. Also molecular dynamics simulations have indicated that, depending on concentration, DMSO is able to fluidise the membrane [12] leading to the relaxation of membrane tension that could alter the permeability of MS channels. It is therefore not unlikely that NaCl permeability is affected by permeant CPAs in a similar manner, in accordance with the estimated parameter values.

Further validation of the model was achieved by predicting experimental cycles of sucrose addition and restoring isotonic conditions, as well as runs with concentrations of DMSO different from those used for the determination of model parameters.

1. Karlsson, J.O.M. and M. Toner, *Cryopreservation*, in *Principles of Tissue Engineering, 2ndEd.*2000, Academic press: San Diego. p. 293-307.

2. Mazur, P., *Life in Frozen State*. 2004, London, UK: CRC Press.

3. Fadda, S., A. Cincotti, and G. Cao, *The Effect of Cell Size Distribution During the Cooling Stage of Cryopreservation without CPA.* Aiche Journal, 2010. **56**(8): p. 2173-2185.

4. Fadda, S., A. Cincotti, and G. Cao, *Rationalizing the Equilibration and Cooling Stages of Cryopreservation: The Effect of Cell Size Distribution.* Aiche Journal, 2011. **57**(4): p. 1075-1095.

5. Casula, E., et al., *hMSCs from UCB: Isolation, Characterization and Determination of Osmotic Properties for Optimal Cryopreservation.* Chemical Engineering Transactions, 2015. **43**: p. 265-270.

6. Casula, E., S. Fadda, and A. Cincotti. *Isolation, Characterization and Analysis of the Osmotic Behaviour of hMSCs from UCB for Optimal Cryopreservation.* in *AIChE Annual Meeting*. 2015. Salt Lake City, UT.

7. Kleinhans, F.W., *Membrane permeability modeling: Kedem-Katchalsky vs a two-parameter formalism.* Cryobiology, 1998. **37**(4): p. 271-289.

8. Katkov, I.I., *A two-parameter model of cell membrane permeability for multisolute systems.* Cryobiology, 2000. **40**(1): p. 64-83.

9. Martinac, B., *Mechanosensitive ion channels: molecules of mechanotransduction.* Journal of Cell Science, 2004. **117**(12): p. 2449-2460.

10. Wang, J.Y., et al., *Dual Dependence of Cryobiogical Properties of Sf21 Cell Membrane on the Temperature and the Concentration of the Cryoprotectant.* Plos One, 2013. **8**(9).

11. Gilmore, J.A., et al., *Effect of Cryoprotectant Solutes on Water Permeability of Human Spermatozoa.* Biology of Reproduction, 1995. **53**(5): p. 985-995.

12. de Menorval, M.A., et al., *Effects of Dimethyl Sulfoxide in Cholesterol-Containing Lipid Membranes: A Comparative Study of Experiments In Silico and with Cells.* Plos One, 2012. **7**(7).

463310 Water-Restructuring Mutations Can Reverse the Thermodynamic Signature of Protein-Ligand Association

Abstract:

The organization of water surrounding interacting biomolecules can influence—even dictate—the strength of their interaction; the mechanisms by which that influence is exerted, however, remain incompletely understood. This talk dicusses a study in which mutants of human carbonic anhydrase are used to examine how changes in the organization of water filling a binding pocket can alter the thermodynamics of protein-ligand association. Results from empirical (thermodynamic and crystallographic) and theoretical analyses suggest that most mutations strengthen networks of water-mediated hydrogen bonds, and reduce binding affinity by increasing both the enthalpic cost and, in an incompletely compensating fashion, the entropic benefit of rearranging those networks during binding. The organization of water filling a binding pocket can thus determine whether hydrophobic interactions involving that pocket are enthalpy-driven or entropy-driven. Findings highlight a possible asymmetry in protein-ligand association by suggesting that binding events associated with enthalpically favorable rearrangements of water—rearrangements that can be enhanced by mutations that destabilize water networks over nonpolar regions—are stronger than those associated with entropically favorable rearrangements of water.

472833 Mitochondrial Metabolism in the Regulation of Human Mesenchymal Stem Cell Immunomodulatory Property

Abstract:

Introduction

Human mesenchymal stem cells (hMSCs) isolated from various adult tissues are primary candidates in cell therapy and tissue regeneration. The recovery and regenerative effect of hMSCs is largely attributed to their release of immunosuppressive factors and growth factors that coordinately modulate the progression of inflammation, and enhance the propagation and differentiation of endogenous progenitor cells. Similar to macrophage polarization into the proinflammatory M1 or anti-inflammatory M2 phenotype, hMSCs are sensors and switchers of inflammation via their ability to adopt a proinflammatory (MSC1) or anti-inflammatory (MSC2) phenotype in order to either promote or inhibit an immune response, depending on the severity of the inflammation, and the types of inflammatory cytokines present [1]. While much effort has been put on the interplay of hMSCs and immune cells via paracrine signaling, the regulatory mechanisms that mediate the phenotypic switch from MSC1 to MSC2 remain to be fully explored. Recent studies suggest that extensive metabolic reconfiguration of hMSCs occurs and plays central role in hMSCs fate decision and adaptation to environmental stimuli, such as during hypoxia, ischemic stress, and chemokine or cytokines stimulus. Identification of the metabolic profile and key metabolic components that control the immunophenotype of hMSCs and their response to inflammatory environment could potentially be used to enhance therapeutic applications of hMSCs. The major metabolic consequences of environmental stress involve the sequential inhibition of mitochondrial respiration and activation of ROS mediated redox signaling. Indeed, inflammatory cytokines IFN-γ and TNF-α have been reported to insert synergetic effect on the production of nitric oxide. Whether these metabolic events and in particular mitochondrial ROS are connected to IFN-γ-induced immunosuppressive cytokines production and act in concert to help establish or enhance the MSC2 phenotype is unknown. Our preliminary data demonstrated the important role of mitochondrial metabolism in hMSC 'stemness' decision that mitochondrial electron transport chain (ETC) and adaptive metabolic changes are not only correlated but also instrumental for MSCs pluripotent gene expression. The objective of present study is to explore the role of mitochondrial metabolism in hMSCs immunomodulatory property with aim to develop approachable strategy to enhanced hMSCs therapeutic potential.

Materials and Methods

To obtain MSC2 phenotype, hMSCs were incubated with IFN-γ (40 ng/ml) containing complete culture medium for 24 to 48 hours. The key immune suppressive factor released by hMSCs, indoleamine 2,3-dioxygenase (IDO), was measured for mRNA level and protein level. Kynurenine level in culture conditioned medium was quantified to assess the activity of IDO, which catalyzes the conversion of tryptophan to kynurenine. A metabolic profile for glucose metabolism was generated for unstimulated MSCs and MSC2 using ^{13}C-labeling based GCMS. Several metabolic modulators were used to probe the impact of specific metabolic pathway in the regulation of hMSCs IDO activity.

Results and Conclusion

We observed spontaneous induction of mRNA for IDO and increased kynurenine level in MSCs conditioned medium after IFN-γ (40ng/ml) treatment. In addition, we observed metabolic changes of hMSCs associated with the raise of MSC2 that both glucose consumption and lactate release were increased. MSC2 were associated with changes in mitochondrial respiration that ETC complex II succinate dehydrogenase complex, subunit D (SDHD) was lower expressed in MSC2, with occurrence of mild decrease in mitochondrial membrane potential and accumulation in intracellular succinate level, indicating reduced ETC activity. As a result of interruption of mitochondrial respiration, both mitochondrial reactive oxygen species (ROS) and total cellular ROS level were increased in MSC2. Importantly, enhancement of mitochondrial function by PDK inhibitor DCA or reducing ROS level by mitochondrial targeted antioxidant Mito-Q both potently reduced IDO activity in hMSCs treated with IFN-γ. In contrast, restriction of cell respiration by hypoxia or ETC complex I and III inhibitor antimycin A and rotenone shifted MSCs into a MSC2-like phenotype in that it strongly potentiated IFN-γ mediated IDO activation. Together, the results revealed the mechanistic connection between metabolic regulation and MSC2 phenotype, and demonstrate the regulation of metabolism as a strategy in potentiating MSCs properties for cell therapy.

References:

[1] Wang Y, Chen XD, Cao W et al. Plasticity of mesenchymal stem cells in immunomodulation: pathological and therapeutic implications. Nat Immunol. 2014;15:1009-1016.

462247 Engineering Artificial Metabolons for Substrate Channeling

Abstract:

Enzymes are valuable catalysts in many applications due to their high catalytic activity and substrate specificity. For multi-enzyme systems, limiting conditions may arise including unfavorable equilibrium and unstable intermediates. In order to better control these reactions, we can look to how enzymes evolved. In the cell, most enzymes belong to cascades, where they exist as parts of sequential multi-enzyme pathways. Over the past few decades, highly organized multi-enzyme complexes within these pathways, termed metabolons, have been the focus of intense research in the field of enzymology. In a metabolon, intermediates can be channeled directly from one enzyme to the next without diffusing into the bulk phase through mechanisms including physical tunnels and electrostatic channels connecting active sites. Metabolon formation and substrate channeling phenomena have several advantages over free enzymes including high local metabolite concentrations, promoting the progression of reactions as well as minimizing the cross-talk between competitive pathways. Overall, better mass transport within the pathway can be achieved, resulting in enhanced kinetics of the metabolic process, allowing the cell to operate at high efficiency.

Mitochondrial malate dehydrogenase (mMDH) and citrate synthase (CS) from the TCA cycle have been well-studied for metabolon formation and substrate channeling of the intermediate oxaloacetate (OAA). We examined the structure of a recombinant mMDH/CS metabolon and located a patch of positively charged amino acids connecting the enzyme active sites along which the negatively charged OAA could travel by bounded diffusion. By mutating a single amino acid along this channel, we inhibited metabolon formation and substrate channeling, decreasing the probability of channeling from 0.99 to 0.23.[1] Next, we use several strategies in order to assemble multi-enzyme complexes to mimic nature. These strategies include protein scaffolds to which enzymes are attached by covalent bonds, DNA scaffolds in combination with DNA-binding proteins, and peptide staples genetically fused to one enzyme with binding affinity for the next enzyme. In this work, we compare the substrate channeling in these engineered metabolons to free enzymes and natural recombinant metabolon complexes.

1. Bulutoglu, B.; Garcia, K. E.; Wu, F.; Minteer, S. D.; Banta, S., Direct evidence for metabolon formation and subatrate channeling in recombinant TCA cycle enzymes, submitted.

472412 Design and Evaluation of Engineered Protein Biosensors for Live Cell Imaging of Epidermal Growth Factor Receptor Phosphorylation

Abstract:

Live-cell imaging studies can provide mechanistic insight into dynamics of receptor-mediated cell signaling; however the availability of intracellular biosensors is presently limited. The tandem SH2 domain from PLCÎ³1Â (tSH2-WT) has been used as a marker of phosphorylated epidermal growth factor receptor (EGFR). Consistent with the promiscuity of SH2 domains, we show that tSH2-WT lacks specificity for phosphorylated EGFR. Further, EGF-stimulated recruitment of tSH2-WT differs qualitatively from the expected kinetics of EGFR phosphorylation. Analysis of a mathematical model shows that the high avidity of tSH2â€‘WT, because of sequential binding of both SH2 domains, explains the discrepancy between the actual and expected biosensor readout. To address the limitations of the tSH2-WT biosensor, we constructed a combinatorial library through random mutagenesis of the C-terminal SH2 domain (cSH2) of PLCÎ³1. The library was screened using yeast surface display to isolate a mutant protein (mSH2) with enhanced specificity for the Y992 phosphorylation site in EGFR (pY992). Accordingly, a biosensor based on mSH2 faithfully reports kinetics of EGFR phosphorylation in live-cell imaging experiments. This approach, integrating theoretical considerations with a protein engineering strategy, can be generalized to design and evaluate suitable biosensors for specific intracellular targets.

470105 Exploiting Structural Effects of Post-Transcriptional RNA Modifications to Engineer Orthogonality of Amber Suppressor Transfer RNAs for Genetic Code Expansion

Abstract:

The ability to introduce non-canonical amino acids through expansion of the genetic code is also of great interest for producing proteins with new functionalities[1]. To this end, suppressor tRNAs (tRNAs that suppress stop codons by decoding them for a 21st amino acid), are often taken from other organisms with their corresponding synthetases and transplanted into a more conventional production organism such as *E. coli*. These suppressor tRNAs often reduce cellular fitness due to mischarging by host synthetases or inaccuracies in decoding such as frameshifting[2]. As such, approaches for improving the orthogonality (not being charged by host aminoacyl-tRNA synthetases) of suppressor tRNAs are critical for effective expansion of the genetic code. One striking feature of tRNAs is their abundance of post-transcriptionally modifications (PTMs), which are generally believed to play pivotal roles in fine tuning the structural flexibility of RNAs to maintain robust functions even under non-ideal conditions[3,4]. Furthermore, many positions known to be modified in *E. coli* tRNAs are tightly linked to determinants of identity for various tRNA isotypes[5,6], highlighting the potential for taking advantage of PTMs for improving orthogonality of engineered suppressor tRNAs.

The main premise of this work is that strategic engineering locations to rationally introduce tRNA orthogonality can be identified an exploited by a combination of conventional genetics and local intracellular structural accessibility probing methods. This strategy is of impact as inferences about PTMs are difficult to make in vivo. Specifically, we have used a modified version of a tool developed in our lab for studying RNA structural accessibility in living cells[7] to characterize the structural flexibility for various domains in native *E. coli* tRNAs as well as in several engineered *Methanocaldococcus jannaschii* suppressor tRNAs. The changes

in flexibility associated with various PTMs were then characterized using genetic strain that were presumed to lack the PTMs of interest (given deletions of the corresponding enzyme (s) responsible for these modifications). After correlating the orthogonality and mischarging frequency with changes in PTM-dependent structural flexibility, this work highlights the importance of PTMs in modulating molecular flexibility as a key characteristic for engineering strong suppressor tRNAs. As such, this project demonstrates the value of considering RNA post-transcriptional modifications in the design of orthogonal systems for applications in synthetic biology.

1. Young, T. S. & Schultz, P. G. Beyond the canonical 20 amino acids: expanding the genetic lexicon. *J. Biol. Chem.* **285,** 11039–44 (2010).
2. Kurland, C. G. Translational accuracy and the fitness of bacteria. *Annu. Rev. Genet.* **26,** 29–50 (1992).
3. Jackman, J. E. & Alfonzo, J. D. Transfer RNA modifications: nature's combinatorial chemistry playground. *Wiley Interdiscip. Rev. RNA* **4,** 35–48 (2013).
4. Baldridge, K. C. & Contreras, L. M. Functional implications of ribosomal RNA methylation in response to environmental stress. *Crit. Rev. Biochem. Mol. Biol.* **49,** 69–89 (2014).
5. Giege, R., Sissler, M. & Florentz, C. Universal rules and idiosyncratic features in tRNA identity. *Nucleic Acids Res.* **26,** 5017–5035 (1998).
6. Freyhult, E., Moulton, V. & Ardell, D. H. Visualizing bacterial tRNA identity determinants and antideterminants using function logos and inverse function logos. *Nucleic Acids Res.* **34,** 905–16 (2006).
7. Sowa, S. W. *et al.* Exploiting post-transcriptional regulation to probe RNA structures in vivo via fluorescence. *Nucleic Acids Res.* **43,** e13 (2015).

467421 Hybrid Soluble/Cellular Target Selection Schemes Improve Discovery of Translatable Ligands

Abstract:

Precision medicine necessitates molecular targeted agents to match the growing diversity of clinical biomarkers. The current standard, high-throughput screening against soluble domain fragments of target molecules, can result in failure when selected ligands are tested for binding to genuine, complete, membrane-bound target. Multiple factors may contribute to this lack of translatability including: 1) improper folding of the soluble domains due to lack of transmembrane domain, 2) differential post-translational modification by the production host, 3) binding to a non-natural epitope resulting from chemical or biological tags, or 4) lack of accessibility of the bound epitope in the cellular context. This work compares five selection strategies to incorporate translatability as a selective pressure in ligand discovery campaigns against CD276 and Thy1, two biomarkers that are expressed in tumor vasculature.

Combinatorial libraries of fibronectin domains and affibodies were displayed on the yeast surface using an extended linker. Binders to CD276 and Thy1 were selected by (a-c) magnetic bead sorting and (a) FACS with soluble extracellular domains, (b) FACS with detergent solubilized cell lysates, or (c) panning on mammalian cell surfaces; or (d,e) direct panning on mammalian cell surfaces with or without depletion of non-specific protein binders. Enriched subpopulations were characterized by clonally assessing target-specific binding in a cell-based phage ELISA or yeast panning assay.

Enrichment of binders to recombinant target followed by direct cell panning on mammalian cell lines yielded the best outcomes as three of four campaigns (all but affibodies against Thy1) exhibited strong selectivity for target-positive cells (ranging from 3:1 to 86:1 selectivity). 51% to 85% of lead molecules were successful hits in a clonal cell assay. Enrichment to recombinant target followed by cell lysate FACS yielded modest selectivity (and a 23% lead hit rate) for CD276-targeted affibodies and Thy1-targeted fibronectins but not the other two scaffold/target combinations.

Enrichment only with recombinant target has yielded a 35% lead hit rate in the affibody/CD276 effort. None of the campaigns directly using cell panning yielded ligands with a strong preference for target-expressing cells. Sequence diversity, affinity, stability, and structural characterization of lead molecules will be presented.

In conclusion, selections utilizing soluble target methods followed by direct cell panning yield optimal selectivity for the membrane-bound targets of interest. Several small protein ligands for CD276 and Thy1 have been identified.

452951 Enzymatic Synthesis of Complex Natural Products

Abstract:

On account of their structural and biosynthetic complexity, most natural products have traditionally been discovered and produced from fermentable microorganisms. Within the past decade however, there has been an explosive growth of "orphan" biosynthetic pathways whose products are unknown. Many of these compounds cannot be accessed from their natural sources. In vitro reconstitution of these multi-enzyme systems has enabled efficient synthesis of relatively small quantities of some of these products for discovery applications. The benefits of scaling up these preparative methods are enormous, but significant challenges remain to be addressed. Using glycolipids and polyketides as examples, I will illustrate some of these opportunities and challenges.

466988 Quality-by-Control (QbC) Technologies for Pharmaceutical Process Development

Abstract:

The advent of process analytical technologies (PAT) more than a decade ago has brought the applications of advanced control of pharmaceutical manufacturing processes in the realm of possibility. The role of advanced feedback control concepts in improving the performance of batch processes, as well as enabling technologies in the paradigm shift from batch to continuous manufacturing will be corroborated. Examples how advanced control can be used for the improved production of pharmaceutical particulates will be provided, with application to both continuous drug substance and drug product manufacturing. Novel control approaches for integrated continuous crystallization and wet milling systems will be presented that can produce crystalline materials with tailored maid properties with excellent consistency. The role of rigorous mathematical modelling in the systematic optimal design, start-up and control of continuous crystallization processes to achieve and maintain the desired controlled stage of operation will be illustrated through examples for both cascade of mixed suspensions mixed product removal (MSMPR) as well as plug flow crystallization systems. Anti-fouling control (AFC) will be presented as an illustrative example demonstrating how active feedback control can be the key enabling technology for the implementation and smooth operation of continuous crystallization processes. The application of a systematic framework for the design of hierarchical control systems for continuous tablet production will also be described. A novel view for implementing continuous drug manufacturing processes that are based on spatially distributed control of various particle formation processes within a single continuous plug flow system, rather than the integration of individual unit operations, will be presented. The system uses spatially distributed antisolvent addition, heating/cooling cycles and/or binder/excipients addition to control the individual mechanisms of nucleation, growth and spherical agglomeration to produce particles with controlled internal and external size and micromeritic properties. The talk intends to provide motivating examples of the next stage of innovation in pharmaceutical manufacturing, illustrating the potential benefits of the new quality-by-control (QbC) framework in improving product quality and process efficiency while reducing costs and time-to-market.

470571 Drug Product Continuous Manufacturing Control Strategy Considerations

Abstract:

This presentation will cover the key components of a fully automated drug product continuous manufacturing control strategy. In-process control, real time decision making, sampling plans and sampling plan justification, and real time release testing will be discussed. Experience with filings and inspections along with a few learnings and considerations for continuous manufacturing will be shared.

470871 Regulatory Approach for Supporting Emerging Pharmaceutical Technology

Abstract:

Office of Pharmaceutical Quality (OPQ) supports modernization of pharmaceutical development and manufacturing. Therefore, OPQ encourages the adoption emerging pharmaceutical technology that has potential to enhance the process capability and therefore product quality. Emerging technology may include novel process and product designs, as well as innovative analytical technology. Examples of emerging technology are continuous manufacturing and 3-D printing technologies. Because emerging technologies, by definition, will generally be unfamiliar, in both industrial and regulatory contexts, OPQ established the Emerging Technology Team (ETT) to facilitate the identification and implementation of emerging technology in at least three ways. 1) by enhancing communication among sponsors, manufacturers, and regulators, both early in the engagement of emerging technology design and development and at appropriate later time points (including the course of regulatory review); 2) by leading or otherwise supporting the quality assessment team, in partnership with relevant OPQ offices, in regulatory reviews; and 3) by documenting ETT/OPQ resolutions to scientific and policy issues that may inform FDA approaches to the review and implementation of emerging technologies. Furthermore, OPQ have initiated a regulatory science program to enhance the agencyâ€™s knowledge in the emerging technology area. For example, OPQ initiated several research projects in continuous pharmaceutical manufacturing, including the development of integrated process models similar to the approaches utilized by the chemical industry; the development of a database of excipient properties to capture the behavior of common raw materials within different processing environments, the standardization of processing equipment integration components to increase flexibility; and the advancement of automatic process control systems.

470944 The Integrated Quality Assessment: Â Office of Process and Facilities'Â Perspective on the Holistic Evaluation of Pharmaceutical Quality

Abstract:

The Office of Pharmaceutical Quality (OPQ) was formed for the purpose of ensuring a consistent and unified approach to FDAâ€™s assessment of pharmaceutical quality for pharmaceutical products. This restructuring involved a dramatic modification in the structure of the organization, changes in roles and responsibilities for review and inspection functions, and an elevated stature for the quality assessment function within CDER. The reorganization and resulting formation of OPQ has been in effect since January 2015. Since that time there has been continued implementation and improvement upon the integrated quality assessment within OPQ. This assessment now includes a greater level of communication and interdependency within the assessment team, simultaneous with the specialization of the assessment disciplines for products, processes and facilities. In particular, there has been an evolution of the facility assessment: from the initial risk-based for the assessment of the facility; to the composition of the inspection team and focus of the pre-approval inspection; concluding with the final integration of the inspection results into the quality assessment. The changes have positioned OPQ to provide most appropriate level of engagement with sponsors and the development of expertise necessary to perform the technical assessment of increasingly complex technologies, such as continuous manufacturing of pharmaceutical products. The basis for the changes with OPQ will be provided, along with a description of important initiatives within the office, concluding with a description of the programmatic successes resulting from the formation of OPQ.

470987 Engineering a Bacterial System for the Production of Biologics Leads to Discovery of Two New Functions of a Pathogenesis-Related Protein

Abstract:

Secretion is emerging as a useful strategy for exporting proteins from bacteria for the production of biologics. The type III secretion system (T3SS) in ***Salmonella enterica*** is an ideal path to protein export because it is non-essential for bacterial metabolism and allows for target proteins to cross both bacterial membranes in one step, via characteristic needle-like protein structures. We engineered a super-secreting strain of ***Salmonella*** for the high-titer production of a variety of biochemically challenging heterologous proteins, including enzymes, antibody fragments, and antimicrobial peptides. We achieve titers of over 100 mg/ml for a variety of proteins – a 100-fold improvement on wild type levels – at relative purity. To design the super-secreting strain, we investigated SipD, a structural T3SS protein. SipD is one of the first three proteins to be secreted through the assembled T3SS. These "translocon" proteins form a complex at the tip of the T3SS needle that interacts with a mammalian host cell. We find that exogenous addition of SipD results in high secretion titers even in the absence of a host cell. Moreover, our results indicate that SipD also acts as an intracellular regulator in a tightly controlled secretion hierarchy, inhibiting protein secretion. Interestingly, these two roles have opposite effects on protein secretion, and occur on different sides of the cell envelope. Furthermore, evidence suggests that different domains of the protein are involved in each function. We propose a physical and chemical mechanism for the dual role of SipD. This study serves as an example of how a complicated multi-protein apparatus can be rationally engineered for biotechnological protein production while also revealing how it functions in its native role of bacterial pathogenesis.

470985 Developing a Modular Virus-like Particle Vaccine Platform for Viral Diseases

Abstract:

Recent Zika outbreaks in French Polynesia and Brazil have brought the association between the Zika virus and multiple serious conditions including birth defects such as microcephaly and the neurological condition known as Guillain-BarrÃ© Syndrome to the worldâ€™s attention. The Zika virus is spread to people primarily through the bite of an *Aedes ageypti* mosquito (the same mosquito that spreads dengue and yellow fever) and has symptoms similar to many other diseases including fever, rash, joint pain, and red eyes. Because the symptoms of Zika virus are mild and resolve within less than a week, people do not generally require hospitalization and cases are not reported, making Zika infection difficult to track. Currently the only way to prevent Zika infection is to avoid mosquito bites and relies strongly on regular patient compliance. A prophylactic vaccine is especially necessary in the case of a virus like Zika where the consequences are potentially severe but symptoms are typically mild and overlap with many other common illnesses.

Here we discuss the development of a modular, virus-like particle based vaccine platform in the context of our work to address the need for a prophylactic Zika vaccine. Through cell-free protein synthesis, we can produce and stockpile an engineered virus-like particle. This virus-like particle can be conjugated to a relevant antigen (Protein E Domain III for our Zika vaccine candidate) using click chemistry. The virus-like particle scaffold allows for multivalent display of the relevant antigen in a manner that mimics presentation on the natural virus. Multiple orientations of the antigen on the virus-like particle scaffold are easily evaluated by creating antigen variants with a non-natural amino acid incorporated in different positions within the protein. Additionally, innate immune stimulators will be co-attached to the virus-like particle to boost the immune response to the antigen by acting as a non-denaturing adjuvant. This platform is easily translated to other diseases, including the influenza and HIV efforts currently underway in our lab.

462511 Bio-Click Chemistry for Modular Modification of Hepatitis B Viral-like-Particles As a Biosensor Platform for Cancer Detection

Abstract:

In an effort to diagnose and treat disease, we desire to create smart nanodevices which recognize molecular signals, transduce, and amplify desired therapeutic and imaging outputs. The Hepatitis B Virus-Like-Particle (HBV VLP) has been extensively studied and engineered into such a device for a variety of applications including vaccine delivery, drug delivery, biosensing, and imaging. In order to customize HBV VLPs for these applications, it is necessary to covalently attach multiple components. This has previously been done using chemical conjugation, unnatural amino acids, and genetic modification; however, chemical based strategies suffer from harming the integrity of more delicate protein attachments, while unnatural amino acid and genetic fusion strategies can create challenges in protein expression. We have developed an alternative strategy for the decoration of HBV VLPs using â€œbio-clickâ€ chemistry, allowing a wide range of peptides and proteins to snap directly onto the modified HBV capsids in a controlled, biocompatible manner. Our technology allows us to modularly modify the surface of HBV VLPs with multiple components. We have demonstrated this by creating a biosensing platform that incorporates an array of sensing and reporting domains which can be used to detect cancer biomarkers.

472425 Improving the Gut Residence Time of Probiotics through Functional Metagenomics

Abstract:

Engineering human-associated microbes holds great promise for treatment of chronic conditions, but the resulting health-promoting strains may not be well-suited for persistence in their target environment. As one solution, human-associated microbial communities may contain genes or pathways which improve the colonization potential of synthetic, therapeutic strains. To test this hypothesis, we focused on the trillions of bacteria which inhabit the human gut. This community performs many functions critical to the maintenance of health, including nutrient extraction, vitamin production, and defense from pathogens. Disruptions to the structure of this bacterial community can have significant negative impacts to the human host, including obesity, malnutrition, and cancer. As one method for correcting these pathologies, probiotics are orally ingested microorganisms which improve health, potentially through provision of additional metabolic functions, modulation of the host immune response, or competition for niche space with pathogens. In particular, *E. coli* Nissle (EcN) is an approved probiotic with demonstrated efficacy against inflammatory bowel disease. Unfortunately, the residence time of EcN in the gut is quite low, necessitating continual administration for treatment of this chronic condition. In order to discover factors limiting EcN colonization and improve the gut residence time of this important probiotic, we extracted metagenomic DNA from both established and developing gut microbial communities and shotgun-cloned tens of millions of 2-5kb fragments of this DNA into an expression vector. We next selected colonization-enhancing gene fragments via passage of EcN containing this metagenomic library in cell culture, germ-free mice, and mice containing a human-like microbiota consuming one of several human-relevant diets. We observed a reproducible, million-fold reduction in diversity of cloned fragments as our selections proceeded, indicating robust enrichment for factors which improve the persistence of EcN in these environments. Bioinformatic and biochemical analyses of these fragments indicated mechanisms of improved colonization, and specificity of recovered mechanisms to particular dietary contexts was observed. Furthermore, we found that EcN containing these colonization factors exhibited improved persistence over wild-type in a mouse model of *Salmonella typhimurium* infection, indicating the potential for exclusion of this pathogen via competition for niche space. Taken together, this work has the potential to significantly improve our understanding and use of probiotics while providing a framework for developing robust chassis strains for future synthetic biology efforts in the human gut microbiota.

468251 Engineering Probiotic Bacteria for Pathogen Reduction in Poultry Production

Abstract:

Foodborne illnesses are a costly public health concern affecting nearly 1 in 6 Americans each year. The effective treatment of these infections is being greatly compromised by the continuing emergence of antibiotic-resistant strains. Healthcare professionals are running out of viable treatment options for resistant infections which results in prolonged illnesses, increased mortality rates, and higher medical expenses. Estimates predict that, in the absence of a solution, drug-resistant infections will surpass cancer in total annual deaths by 2050.

With a major driving force for resistance development being the overuse of antibiotics, technologies are being sought to limit their injudicious use, particularly in food-producing animals. Over 50% of the antibiotics produced in the U.S. are consumed by livestock to promote animal growth and disease prevention. Often, this sub-therapeutic administration breeds bacterial resistance which can be transmitted to humans through food sources.

We have developed an antibiotic-alternative to decrease *Salmonella* carriage in poultry by engineering safe-to-consume probiotic bacteria with significant anti-*Salmonella* activity.Â *Salmonella* enteritidis (SE) is a leading bacterial agent of disease outbreaks and is primarily linked to poultry products. We modified the probiotic, *E.coli* Nissle 1917 (EcN), to produce high-titers of the antimicrobial peptide (AMP) , Microcin J25, which exhibits remarkable antagonistic activity against SE. AMPs have been researched for years for their antibiotic properties but have failed in translational success due to their high synthesis costs and rapid degradation rates in the body. Our novel use of a probiotic delivery vector has overcome these hurdles, enabling localized production among the hostâ€™s natural microflora.Â By developing a strong synthetic-hybrid promoter system we were able to re-engineer the native microcin operon to enable controlled, robust, and well-characterized peptide production and secretion.Â Our *in vivo* studies have repeatedly demonstrated that a single treatment of our modified EcN is capable of dramatically reducing

SE carriage in the GI tract of turkey poults, with better clearance than the traditional antibiotic, enrofloxacin. Treatment with the modified EcN has no adverse effects on bird growth. In performing an in-depth microbiome analysis, we have shown that there is no discernible impact on the health or diversity of the native microflora.

We will present the results of experiments that have involved over 600 turkey poults. We will discuss the strengths and limitations of synthetic biology techniques when used to develop a new antibiotic technology.

465124 An Integrated Synthetic Biology Pipeline for Novel Plant Natural Product Discovery in Yeast

Abstract:

Natural products from plants and their derivatives are an abundant source of valuable compounds such as drugs. However, the discovery of plant natural products and their biosynthetic pathways remains challenging, which may be partially due to the large size and complexity of plant genomes, and the difficulties in plant gene isolation and pathway validation. Recent research progress in plant genome sequencing and transcriptomics analysis allows better understanding of plant genomics, which offers the potential to expedite the pathway discovery process. However, there are no approaches to efficiently decode the complex plant genomes and to precisely elucidate these pathways in plants. Here we report an innovative pipeline for systematic and accelerated discovery of novel plant natural products and associated biosynthetic pathways. This versatile pipeline combines leading techniques in bioinformatics, synthetic biology, and analytical chemistry, and enables fast-paced, high-throughput discovery of plant natural products through predicted gene cluster reconstruction and metabolite analysis in heterologous hosts such as *Saccharomyces cerevisiae*. This approach comprises three main steps: 1) new plant metabolic pathway prediction by bioinformatic tools for gene cluster identification and prioritization, which is inspired by recent bioinformatics analysis revealing large numbers of metabolic pathways clustered in plant genomes; 2) cluster reconstruction in yeast using established synthetic biology toolkits and gene assembly pipeline for heterologous production of novel compounds and functional validation of predicted biosynthetic pathways; 3) novel compound identification by optimized mass spectrometry-based untargeted metabolomics analysis. We demonstrated this approach by discovering novel phenylpropanoid derivatives and elucidating their biosynthetic pathways in yeast, which not only sheds light on multiple branched phenylpropanoid biosynthetic pathways, but also generates engineered yeast strains providing a large amount of pathway intermediates, thereby establishing a platform for advancing drug discovery and production. This integrated approach will support the future development of scalable drug manufacturing platforms, ultimately advancing drug discovery, development, and production in yeast.

458910 Tumor Microenvironment and Cancer Cell Dormancy

Abstract:

Breast cancer affects 1 in 8 women in the United States, and has a high risk of recurrence that can develop over the course of decades due to dissemination of cancer cells from the primary tumor to secondary sites. The time from the formation of the first micrometastasis to that of a clinically detectable macrometastasis is the rate-limiting step in metastasis and is called cancer dormancy. The mechanisms by which the host niche regulates tumor dormancy, or the characteristics of a dormant niche, are unknown. However, cells respond to the mechanical properties of their surrounding microenvironment. Common chemotherapeutic agents tamoxifen and 5-fluorouracil that are used in breast cancer treatment induce dormancy in breast cancer cells, but the underlying mechanisms and regulation by the microenvironment remain unclear. Here we investigated whether the physical microenvironment, specifically host tissue stiffness, affects the propensity of disseminated tumor cells to become dormant at secondary sites. We used a two-dimensional polyacrylamide (PA) gel-based culture model to investigate the effects of host tissue elastic modulus on drug-induced cancer cell dormancy. Breast cancer cells were cultured on PA gels of different stiffness to mimic the microenvironments of the normal mouse mammary gland and breast tumors respectively. We found that both tamoxifen and 5-fluorouracil induce dormancy in estrogen receptor-positive breast cancer cells as determined by immunofluorescence of Ki67 and flow cytometry analysis of propidium iodide staining. Drug-treated groups had a larger G0/G1 population, as well as reduced cyclin D1 and cylin A2 transcript levels compared to untreated controls. Surprisingly, cyclin-dependent kinase inhibitors p21 and p27 were also downregulated in the drug-treated groups. Our data suggest that tumor cells become dormant through regulation of cell cycle control proteins in a manner that differs from that of normal cells. Defining the characteristics of the microenvironment that control these mechanisms, and targeting the host niches that harbor dormant tumor cells before they form clinically detectable macrometastases, would pave the way for new treatments that will keep this incurable disease under control.

456460 *Human Serum-like Medium Promotes Extensive Metabolic Alterations in Cancer Cell Lines and Affects Cellular Sensitivity to Chemotherapeutic Drugs.*

Abstract:

Altered cancer metabolism arises through a combination of genetic lesions and conditions subjected by the surrounding microenvironment. Indeed, the interplay between genetic and non-genetic determinants can dynamically affect metabolic regulation, and moreover, influence the heterogeneous metabolic dependencies across different cancers. Efforts to decipher this interplay and, more generally, the context-dependent heterogeneities of cancer cell metabolism may enable the identification of novel targets for anticancer therapeutic intervention. The interrogation of human cancer cell lines in culture is one of the most common approaches exploited to improve our understanding of altered cancer metabolism. However, existing cell culture medium formulations were designed primarily with the objective to promote cell survival and proliferation, and consequently, they poorly reflect *in vivo* nutrient conditions. Therefore, in an effort to better model the cancer cell microenvironment *in vitro*, we developed a novel formulation that contains metabolite and salt ion components at concentrations described for human serum. We then utilized a panel of blood cancer cell lines as a model system to determine the metabolic consequences of culturing cells in human serum-like medium (HSLM) relative to RPMI – the most commonly used medium for culturing this class of cell lines otherwise. We first validated that HSLM supported cell survival and proliferation across our cell line panel despite having a nutrient composition that was vastly different than that of RPMI and even of serum isolated from mice. We then quantified the net exchange of metabolites between cells and exogenous medium, and found that the rates and – in some cases – the directionality of metabolite exchange were substantially different in HSLM-cultured cells relative to their RPMI-cultured counterparts. We next quantified the intracellular concentrations of nearly 200 polar metabolites upon culturing in both formulations, and observed extensive alterations across several metabolic pathways in a medium-dependent fashion. In particular, we observed that intermediates along the *de novo* pyrimidine biosynthesis pathway were among the most drastically affected in both the net nutrient exchange and intracellular quantification data sets. As several widely used chemotherapeutics are known to target different facts of nucleotide metabolism, we then evaluated whether cellular sensitivity to these drugs would be changed upon culturing in HSLM. Indeed, we found that sensitivity to the drug 5-fluorouracil, which is commonly used in the treatment regimen of several types of cancer, was significantly decreased relative to analogous evaluation in RPMI. Our findings demonstrate the importance of efforts to better model *in vivo* nutrient conditions for interrogation of cancer cell lines *in vitro* towards improving our understanding of cancer cell metabolism. Moreover, such modeling may also enable a more accurate reflection of cellular responses to anticancer drugs relative to those described in existing medium formulations or, perhaps in some cases, those described in mouse models.

467410 Glycosylation Dynamically Tunes the Biophysical Properties of the Cancer Cell Glycocalyx

Abstract:

Muc1 is a large, heavily O-glycosylated membrane protein that is overexpressed in the majority of cancers and constitutes the major structural element of the cancer cell glycocalyx. While the biophysical properties of the glycocalyx directly regulate cell-ECM adhesion, cell-cell interactions, and receptor spatial organization, how these properties change dynamically during acquisition of a metastatic phenotype is largely unknown. To dissect the relationship among glycan biosynthesis, mucin rigidity, and the bulk properties of the cancer-specific glycocalyx, we isolated sub-populations of Muc1-expressing breast epithelial cells that present with distinct mucin glycophenotypes. Remarkably, alterations in glycosylation were associated with profound differences in glycocalyx structure, even in isogenic cell lines. In one stable glycophenotype, cellular mucins were highly rigid and mechanically disrupted integrin adhesion, resulting in complete dissociation of cells from the substrate. The cells exhibit striking and rapid 3D motility compared to wild-type cells without Muc1. On the other extreme, we isolated a stable population of cells that expresses high levels of Muc1, but these mucins were apparently flexible and had no observable effect on cell adhesion. In motility assays, the phenotype of the wild-type cells is recovered and, despite expressing similar protein levels of Muc1, cells show limited 3D motility. Based on lectin profiling, we identified specific patterns of Muc1 O-glycosylation associated with the two extremes in glycocalyx structures. Of notable significance, we discovered that metabolic reprogramming by the HRas oncogene could induce a glycophenotype consistent with mechanically rigid mucins. Future studies will identify the specific glycosylation machinery associated with the different glycophenotypes, with the hope of identifying new avenues of therapeutic intervention based on normalizing the mechano-phenotype of the cancer cell glycocalyx.

471926 A 3D Biomaterial Model for Patient-Derived Ovarian Cancer Drug Screening

Abstract:

Ovarian Cancer (OvCa) is poorly understood and has many challenges, including poor screening techniques, late diagnosis and lack of effective second line therapy. Recurrence rate after first line therapy (surgery + platinum-based chemotherapy) is more than 70%. There is a need for a second line therapy, but there have only been two drug approvals by the FDA since 2006 (Olaparib & Bevacizumab). 3D culture systems that mimic the peritoneal microenvironment of progressed OvCa could provide important insights into OvCa drug response, and potentially help the field find new, more efficacious treatments; however, there are very few studies describing 3D models for OvCa.

A significant number of drugs work well during in vitro preclinical trials, but fail in later stages (animal models or Phase I-II clinical trials). We hypothesize this is because 2D models used in most preclinical studies are inaccurate representations of OvCa. We hypothesize that 3D models can more accurately predict the usefulness of a drug, leading to fewer false leads, and fewer animal models or clinical studies which test drug candidates that will ultimately fail. Thus, 3D models of OvCa for drug screening can provide significant savings in time and resources to researchers in the industry and academia. In this study, we evaluate response to three chemotherapeutic drugs (cisplatin, carboplatin, and paclitaxel) and one targeted drug (sorafenib) of two OvCa cells lines (SKOV-3 and OVCAR-3), both on plastic surfaces and 2D gel surfaces coated with collagen. 2D gels were made by co-polymerization of poly(ethylene glycol) dimethacrylate (PEG) and the zwitterion 2-Methacryloyloxyethyl phosphorylcholine (PC) [1, 2] . We also evaluate the response to the chemotherapeutic drugs of patient ascites-derived cells on plastic, and patient ascites- derived spheroids on 3D gels obtained from UMass Medical School. For the 3D gels, spheroids were encapsulated in poly(ethylene glycol)-Maleimide (PEG-Mal) gels. PEG-Mal prepolymer solution, which contains spheroids, was crosslinked by PEG-dithiol. We measured drug response by using the CellTiter-Glo ATP assay and the required concentration of drug to kill 50% of the cells (IC_{50}) was calculated.

Thus far, our results show that SKOV-3 and OVCAR-3 are more resistant in 2D gels than plastic for all the drugs and combination of cisplatin+paclitaxel, and carboplatin+sorafenib. We also showed that patient ascites are more resistant than SKOV-3 and OVCAR-3 to cisplatin, paclitaxel, cisplatin+paclitaxel on plastic. Most importantly, spheroids encapsulated in 3D gels are more resistant than patient ascites-derived cells on plastic (where cells cannot be seeded as spheroids). This is a possible confirmation that 2D in vitro models show inaccurately high efficacies of cancer drugs, while more realistic 3D in vitro models are able to better represent drug resistance mechanisms found in vivo. The 3D culture system will be used to test drug resistance of multiple patient samples and we will validate our results with an in vivo model.

References

1. Herrick, W.G., et al., *PEG-phosphorylcholine hydrogels as tunable and versatile platforms for mechanobiology.* Biomacromolecules, 2013. **14**(7): p. 2294-2304.

2. Nguyen, T.V., et al., *Sorafenib resistance and JNK signaling in carcinoma during extracellular matrix stiffening.* Biomaterials, 2014. **35**(22): p. 5749-5759.

469193 A 3D Air-Grown Lung Cancer Tumor Spheroid Assay As a Preclinical Model for the Evaluation for Anticancer Therapeutics

Abstract:

Cancer is the second most common cause of death in United States, with lung cancer resulting in the highest mortality rate and being ranked as the top malignant cancer killer in the world [1, 2]. Despite significant advancements in the cancer therapeutics, lung cancer remains one of the most challenging malignancies to treat. In 2016, 224,390 cases of lung cancer are anticipated, with 158,080 cases likely resulting in death [3]. The relative survival rate for lung cancer is lower (18%) than other cancers partly because most of the cases are diagnosed at advanced stage [3, 4]. Therefore, there is a need for improvement in early diagnosis and development of more effective anticancer therapeutics.

The development of lung cancer therapeutics currently relies on two-dimensional (2D) cell culture assays, which does not recapitulate *in vivo* conditions. Due to this unnatural microenvironment, 2D cell assays often provide unpredictable and contrasting results in comparison to animal studies and clinical trials, which contributes to the low success rates and high costs associated with drug development. Three-dimensional (3D) spheroids provide improved representation of cancer tissues that tend to form solid tumors *in vivo*. By exhibiting intrinsic physiological and morphological characteristics similar to tumor tissue, 3D spheroids have been reported to closely mimic *in vivo* biology [5] and are characterized for having hypoxic cells and a necrotic core, which often gives them high resistance to the treatments [6].

The objective of the present study was to develop air-grown *in vitro* 3D multicellular spheroids and evaluate their potential to act as preclinical models for the evaluation of anticancer therapeutics. Growing the spheroids at an air-liquid interface (air-grown) aids in mimicking the environment where lung tumors typically present themselves, e.g. in the airways. Alginate gels were used as scaffold matrix and 3D Petri Dishes® were used to imprint uniform-sized micro-wells in the alginate scaffolds. A549 (lung adenocarcinoma) cells were seeded in the alginate micro-wells after

they were transferred to Transwells in a multi-well plate, which allowed for the cells to grow in air on the apical side of the platform and have nutrient access through media on the basolateral side. The air-grown spheroids were assessed for their growth and viability to ensure that they are optimal in the evaluation of aerosol anticancer therapeutics. A chemotherapeutic drug, paclitaxel (PTX) was used against A549 spheroids to study drug penetration behavior and cell apoptosis in the spheroids.

Keywords: lung cancer; spheroids; alginate; paclitaxel, A549

Reference:

[1] A. Jemal, F. Bray, et al., *Global cancer statistics*, CA Cancer J Clin, 2011. 61 69-90.

[2] J. Wang, Z.H. Li, et al., *Lung cancer stem cells and implications for future therapeutics*, Cell Biochem Biophys, 2014. 69 389-398.

[3] R.L. Siegel, K.D. Miller, et al., *Cancer statistics, 2016*, CA Cancer J Clin, 2016. 66 7-30.

[4] M. Reck, D.F. Heigener, et al., *Management of non-small-cell lung cancer: recent developments*, Lancet, 2013. 382 709-719.

[5] D. Loessner, K.S. Stok, et al., *Bioengineered 3D platform to explore cell-ECM interactions and drug resistance of epithelial ovarian cancer cells*, Biomaterials, 2010. 31 8494-8506.

[6] A.I. Minchinton, I.F. Tannock, *Drug penetration in solid tumours*, Nature Reviews Cancer, 2006. 6 583-592.

470688 Implantable Bioengineered Microenvironments to Study Human Tumor-Immune Interaction

Abstract:

Introduction: Recent studies have revealed the important role of local stromal and immune cells in recruiting and regulating disseminated circulating tumor cells while direct investigation of natural metastatic niches represent multiple difficulties due to the vast stromal diversity and potential inaccessibility of organs like the bone marrow. Implantable â €œhumanizedâ€ microenvironments in immunodeficient mice have emerged as a tunable surrogate for human tissues for studying functional interaction between local tissue cells and human tumor cells.Â Yet, an *in vivo* system that encompasses the human tumor-immune cell interaction has not been fully developed. In this work, we introduce a tissue engineering approach to generate an ectopic microenvironment and its application for studying the interaction of human immune cells and human cancer cells in an immunodeficient murine host.

Materials and Methods: Scaffolds fabricated as previously describedÂ were seeded with 0.5 x 10^6 human bone marrow stromal cells. After 3 days in culture 4 scaffolds were subcutaneously implanted into an immunodeficient, NSG mouse. Two weeks following implantation 2 x 10^6 PC-3 cells expressing luciferase and GFP were injected into the mouse prostate to develop an orthotopic prostate tumor xenograft that subsequently released circulating tumor cells. Four weeks after tumor injection, 10 x 10^6 human peripheral blood mononuclear cells (PBMCs) were injected intravenously. One week later the scaffolds were retrieved from primary mice and re-implanted into new NSG mice at the same anatomical location. Bioluminescent images were taken weekly to non-invasively monitor tumor growth *in vivo*for up to 12 weeks. Finally mice were sacrificed and inter-scaffold vasculature development and recruitment of human tumor and immune cells was characterized via immunohistostaining. Tissue and scaffold samples from a small group of mice were prepared for the intact tissue clearing technique.

Results and Discussion: Immunohistostaining analysis revealed that both cancer and immune cells were successfully recruited to the scaffolds before transplantation and remained twelve weeks post-transplantation. Immune recruitment was also confirmed via FACS. Interestingly our implantable scaffolds displayed similar compositions to other immune related organs, the bone marrow and spleen. In the primary mice, none of the scaffolds displayed visible metastases under bioluminescent imaging. Following serial transplantation, six of the seventy-two scaffolds (8.3%) had visible metastases. Bioluminescence imaging revealed a steady growth in signal in control mice, while mice that received an injection of PBMCs had restricted growth.

Conclusions: Our results demonstrate a tissue engineering approach to study human cancer metastasis and immune cell interaction. We envision that implantable humanized microenvironments in NSG mice will be an invaluable tool for studying the role of tumor microenvironments and human disease.

470529 Systematic Modulation of the Physical Microenvironment for Characterization of Cancer Cell Lines and Primary Tissue

Abstract:

INTRODUCTION

The tumor microenvironment includes a range of physical, chemical, and cellular components such as pressure, hypoxia, acidosis, the extracellular matrix (ECM), stromal, and immune cells. These factors can influence the gene and protein expression of the various types of cells that make up the tumor as well as influence the selection of cells that can thrive in that environment; however, typical cancer cell culture rarely uses hypoxia and pressure nor utilizes substrates similar to the native ECM. We designed a system study the influence of hypoxia, pressure, and native ECM conditions on cancer cell lines and primary cells, with the goal of creating a more relevant culture environment for translational studies.

METHODS

Here, we comprehensively studied how hypoxia, pressure, and ECM composition and organization influences cell biology and gene expression, using transcriptome profiling across a range of physiologically-relevant culturing conditions to mimic various tumor microenvironments found within the body. To do this, we utilized Xcell's primary cell culture platform which includes a custom bioreactor that allows us to control both oxygen concentration (0.1%-20% O2) and hydrostatic pressure levels (26 to 260 mmHg / 0.5 psig to 5 psig). Additionally, we studied the influence of biomimetic substrates by controlling ECM composition and organization (aligned or unaligned collagen at concentrations from 1-2.5 mg/ml +/- fibronectin 0.1-10 microgram/ml) as both 2D ECM coatings and 3D hydrogels. Cell lines studied included models for brain cancer (U-87, A172), pancreatic cancer (PANC10.05), and prostate cancer (DU-145, PC-3, 22Rv1, LNCaP). We performed high-resolution immunofluorescence imaging and western blot protein expression analysis of

key targets, including immunotherapy targets CTLA-4, PD-1, and PD-L1.

Additionally, we applied our ability to tune culturing conditions to represent physiological conditions in the body for the successful culture of fresh primary tissues from normal volunteers as well as cancer patients for downstream analysis studies focused on biomarker discovery and as a clinical tool for disease monitoring and treatment decision-making.

CONCLUSIONS

We identified both common and unique gene signatures across these different cells lines, with the hypoxic conditions activating HIF1 pathway activation having common across the cells studied whereas pressure results in more restricted signatures. Some cancer cell lines and PBMCs differentially express immunotherapeutic targets, at low oxygen and high pressure culturing conditions, resulting in reduced expression of key targets. Moreover, biomimetic ECM studies revealed modulation of ECM to represent a "tumor"-like ECM dramatically altered cell growth, morphology, and focal adhesion organization. In a 3D hydrogel context, a "tumor" like ECM and a hypoxic environment dramatically enhanced tumor cell invasion. Consistently, analysis of mRNA-seq data revealed alterations in gene expression profiles of immunotherapeutic and drug-target pathways involving CTLA-4 and AR signaling as well as EMT targets. In contrast, we observed increased CD47 and CD44 expression at low oxygen and high pressure culturing conditions in cancer cell lines and PBMCs. Thus, supporting the notion that these are possible drug targets for tumors that are characterized by low oxygen levels and high interstitial fluid pressure.

469365 Engineering Novel Modular Biosensors to Confer Metabolite-Responsive Regulation of Transcription

Abstract:

Efforts to engineer microbial factories have benefitted from increasingly facile technologies for synthesizing novel enzymatic ensembles and mutating entire organisms, yet screening and optimizing metabolic pathways remain rate-limiting steps. Metabolite-responsive biosensors may help to address these persistent challenges by enabling the monitoring of metabolite levels in individual cells and the implementation of metabolite-responsive feedback control. To date, however, we are currently limited to the use of naturally-evolved biosensors, which are insufficient for monitoring many metabolites of interest.

To help address this need for novel biosensors, we have developed a modular and general approach for engineering novel transcriptional regulators that are actuated by the presence or absence of metabolites of interest â€“ the Small molecule Responsive Transcriptional Regulator (SmaRTR) biosensor platform. The central goal of this approach is to develop a systematic strategy for converting proteins that bind small molecules into proteins that regulate gene expression in a manner that is regulated by the presence of those small molecule analytes. In our initial evaluation of this strategy, we utilized the BCR-ABL1 zinc finger protein as a model DNA-binding domain. We built and characterized a library of promoters including BCR-ABL1 binding sites at various positions to elucidate which promoter design features are most important for achieving effective zinc finger mediated-transcriptional repression. Using this rich data set, computational feature identification analysis was applied to quantitatively assess the contribution of each design feature to promoter repressibility, ultimately generating new insights into promoter output tuning.

We next evaluated several strategies for converting the well-characterized maltose binding protein (MBP), a model ligand-binding protein, into a novel maltose-actuated transcription factor that regulates gene expression in a ligand-dependent manner. Through this approach, several robust and functional novel biosensors were identified, establishing the fundamental feasibility of the SmaRTR approach. Moreover, we evaluated a series of strategies for design-driven tuning of biosensor and promoter performance and ultimately identified biosensor systems that substantially outperformed our initial designs. Finally, we developed methods for evaluating the extent to which our biosensor design strategy is generalizable to the conversion of ligand-binding proteins into biosensors. Ultimately, this technology may provide new tools for high throughput screening and improved biomanufacturing through implementation of metabolic feedback control.

472170 Engineering Fatty Acid Responsive Elements for Metabolic Engineering in the Oleaginous Yeast, *Yarrowia Lipolytica*

Abstract:

In the past decade, *Y.lipolytica* has emerged as an excellent example of a microorganism with several biotechnology applications such as the production of citric acid, lipases and an assortment of lipids. One of the most interesting facets of *Y.lipolytica* is that its native metabolism is wired towards superior de-novo lipogenesis and an ability to accumulate considerable amounts of lipids. This native oleaginous and lipolytic potential has been further explored to engineer a strain capable of accumulating upwards of 90% lipid content. Most of these feats, however, have been accomplished with a handful of genetic tools to direct metabolic pathway engineering. To tap into the full potential of lipid engineering in *Y.lipolytica* for efficient and cost effective ways of producing value added lipid products, there is a need for the development of more lipid responsive genetic tools such as promoter systems that are inducible and have varying degrees of fatty acid responsiveness. To date, the only fatty acid inducible promoter system used is the acyl CoA oxidase, POX2, promoter. This native promoter spans 2000 bps and confers weak fatty acid responsiveness relative to engineered constitutive hybrid promoters. In this study, we have identified the fatty acid responsive motifs in the POX2 promoter. Initially, a series of conventional 5â€™ truncations of the long promoter was performed. Oleic acid induced transcriptional strength of the truncated promoters was characterized using qPCR at a transcriptional level and using a GFP reporter at the translational level. DNA fragments in between regions of promoter truncations that demonstrated considerable decrease in strength were further screened for transcriptional binding motifs using EMSA coupled with DNase I foot printing. A library of â€œprotectedâ€ sequences was created and alignments were performed to identify testable fatty acid responsive motifs. Here, we have demonstrated an effective methodology to identify substrate inducible elements that can be used to engineer tunable substrate responsive hybrid promoter switches to advance the toolkit for metabolic engineering applications in *Y.lipolytica.*

464709 Orthogonal Regulation of Human Genes Using Chemically Induced Crispr/Cas9 Activators

Abstract:

Synthetic transcription factors (sTFs) represent a powerful tool for regulating endogenous gene expressions in human cells. With the rapid advancement of CRISPR technologies, catalytically dead Cas9 (dCas9) has become the go-to choice as DNA binding domain (DBD) for the creation of sTFs. dCas9 can be readily targeted to virtually any DNA sequences within the human genome, by simply changing the guide sequence of dCas9 guide RNA (gRNA). This feature makes dCas9 an ideal DBD for simultaneously regulating multiple endogenous gene targets, as is required in programming embryonic stem cell or induced pluripotent stem cell differentiation. To achieve efficient, fast and reliable programming, it is desirable to impart spatial and temporal control over endogenous genes and gene networks. Current studies directly fuse dCas9 with effector domains such as activators, repressors and epigenetic modifiers. A lack of control over these constitutive sTFs makes it difficult to reset, fine tune and orchestrate expression levels of key lineage-associated endogenous TFs, which is crucial for directed programming of cell fate. To achieve this level of regulation, we have engineered orthogonal transcription activators based on two commonly used CRISPR/Cas9 systems from *Streptococcus pyogenes* and *Neisseria meningitidis*. We have shown that these two systems are orthogonal in terms of gRNA recognition. When fusing these dCas9s and a synthetic transactivation domain to chemically induced dimerization (CID) proteins, we were able to activate both transgenes and endogenous human genes in a ligand-dependent manner. We further showed that multiple endogenous human genes can be simultaneously activated with addition of ligands. Finally, we used these two orthogonal systems to regulate the expression pattern of two different pairs of endogenous human genes. In each pair of genes, each gene was only activated in the presence of the corresponding ligand. With both ligands present, both genes could be activated. To our knowledge, this is the first example showing coordinated regulation of multiple endogenous human genes. Given the existence of various orthogonal CRISPR/Cas9 and CID systems, this study serves as the first step toward achieving precise regulation of complex gene networks and may elicit new therapeutic applications.

464313 Library and Model Derived Synthetic Terminators for Yeast

Abstract:

Control of transcription and translation are critical parts of yeast synthetic biology and metabolic engineering. While significant efforts have historically focused on the engineering of promoter elements, more recent attention has shifted toward terminator elements as a means to influence mRNA half-life and resulting protein expression. Our group has recently demonstrated that short, synthetic terminator sequences (ranging from 35 to 70 bp in length) can result in superior synthetic parts for the yeast *Saccharomyces cerevisiae*. We show that these more minimal elements (nearly an 80% reduction in sequence space) increase transcript levels nearly 4-fold over commonly used terminators such as CYC1. We then show that these short, minimal terminators can function across different fungal host systems. In an effort to move toward the model-based design of these elements, we have extracted a functional model effective at *de novo* design of these terminators. We demonstrate the utility of this model by establishing a library of synthetic terminator elements with a range of activity. In doing so, we create some of the first model-specified terminator elements for yeast. Moreover, we highlight that this model-based approach can be used to re-design both *de novo* and native terminator elements for yeast. This work both expands the set of synthetic parts available for fungal systems as well as provides insights into the sequence properties and underlying mechanisms of yeast transcriptional termination. Furthermore, these new elements provide an added level of control for gene expression in fungal systems.

470314 Properties of Intragenic microRNA-Based Gene Regulation in Single Cells and Population Measurements

Abstract:

Intragenic microRNAs can regulate their host genes through targets in the gene's 3' untranslated region (UTR). This regulation is an incoherent feed-forward loop (IFFL), with a transcription factor producing both the host gene and the negative regulator of the output, the microRNA. To characterize the properties of a microRNA based IFFL, we used CRISPR/ Cas9 to stably integrate an engineered genetic circuit into the AAVS locus of human kidney cells. The genetic circuit consists of two fluorescent proteins produced by a bi-directional TRE promoter in response to the transcription factor rtTA and the small molecule doxycycline. One protein, DsRed, is produced as two exons with an intronic synthetic microRNA; the other protein, AmCyan, is used as a control. The DsRed gene has microRNA targets in its 3'UTR. We performed flow cytometry measurements on our circuit and control monoclonal cell lines under a range of doxycycline concentrations and at multiple time points. We also performed time-lapse fluorescence microscopy measurements on our cell lines under full doxycycline induction. We analyzed the data to study the properties of the intragenic microRNA based regulation, in particular how the IFFL regulation affects the transient behavior of the output in single cells, the type of distribution of the output in the population, and finally the impact on the noise levels.

471426 Rational Modular Engineering of a Global RNA Regulator for Tunable Metabolic Control

Abstract:

Global regulators enable organisms to respond to diverse and rapidly changing environmental stimuli by affecting vast networks of targets at, frequently, multiple biological levels. For this reason, there is ample interest in tuning their regulation as a strategy for strain engineering. Using the Carbon storage regulator (Csr) as a model system, we present a novel and modular redesign approach that focuses on the rational recombination and shuffling of substructures within a regulatory molecule of interest to engineer new schemes of metabolic control. We specifically feature this approach in the context of the non-coding RNA global regulator *csrB* (~400nt) to tune expression of a number of mRNAs that encode for key enzymes in many essential metabolic pathways. The *csrB* sRNA contains repeating stem loop structures that frequently present a known consensus CsrA binding motif, GGA. In this way, *csrB*can bind a CsrA molecule, the primary Csr protein regulator, at each of these structured binding sites, blocking CsrA-mRNA interactions and thus CsrA-mediated regulation of targets.

Given that *csrB* acts as a sponge to titrate away the amount of CsrA that directly interacts to (up or down-) regulate an mRNA, a major premise of this work is that various engineered interactions between the *csrB* RNA and its cognate protein CsrA lead to a regulation gradient of all mRNAs affected in this pathway. Moreover, given the modular composition of the *csrB* regulator (18 GGA stem loops that are not identical and exist in different contexts within the *csrB* molecule), this system was ideal to test the notion of combinatorial shuffling of structural motifs to engineer many regulatory variants. In this work, we determine the regional accessibility of *csrB* in *Escherichia coli* as a measure of its stem loop-specific binding preferences for CsrA by means of an *in vivo* RNA structural sensing system (previously developed by the group). We then engineer 25 variants of the molecule based on rational recombination of *crsB* stem loop structures and demonstrate the effect of this approach in engineering regulation on a number of mRNA targets. Lastly, we demonstrate process-level effects of this regulation gradient on yields of industrially-relevant

metabolites as proof of concept of tunable phenotype engineering.

475001 Synthetic Biology Platforms for Natural Product Biosynthesis and Discovery (Featured Presentation)

Abstract:

Plants are a rich source of unique scaffolds, including 25% of natural-product-derived drugs. However, the discovery, synthesis, and overall material supply chains for sourcing plant natural products and their derivatives remain ad hoc, biased, and tedious. While microbial biosynthesis presents compelling alternatives to traditional approaches based on extraction from natural plant hosts, many challenges exist in the reconstruction of plant specialized metabolic pathways in microbial hosts. My laboratory has developed approaches to address the challenges that arise in the reconstruction of complex biosynthesis schemes, including spatial engineering strategies to direct the activities and specificities of pathway enzymes and recoding strategies to address folding, processing, and stability issues that may arise with the expression of plant enzymes in heterologous microbial hosts. We have utilized these strategies to develop yeast-based production platforms for an important class of plant alkaloids, the benzylisoquinoline alkaloids, including the medicinal opioids. These synthetic biology platforms will lead to transformative advances in natural product discovery, drug development, and production.

474476 Regenerative Engineering: The Convergence Approach

Abstract:

Regenerative Engineering represents the Convergence of Advanced Materials Science, Stem Cell Science, Physical Forces, Developmental Biology and Clinical Translation.

The field works to capitalize on advances in these an other newly emerging fields and is especially suited to address grand challenges, and to challenge conventional norms and thinking on science.

The development of new technologies, new ways of thinking and new science through Convergence offers outstanding opportunities for creating new solutions to important problems.

474477 Regulation of Growth Factor Signaling and Regeneration by Modifications of Heparan Sulfate Proteoglycans in the Extracellular Matrix

Abstract:

Axolotls (and other Urodele amphibians) are unique among adult vertebrates in their ability to regenerate complex body structures after traumatic injury. We used an *in vivo* gain-of-function assay in the axolotl to demonstrate both qualitative and quantitate differences in extracellular matrix (ECM)-associated heparan sulfate proteoglycans (HSPG) regulate growth factor signaling that in turn regulates formation of limb skeletal structures. Most importantly, this activity is present in mammalian HSPG present in mouse skin in vivo as well as in commercially available, purified preparations of HSPG. This activity can both inhibit regeneration or induce *de novo* limb structure, thus regulation of HSPG is a mechanism for the regulation of regeneration, and establishes a crucial functional link between regeneration in salamanders and mammals. In light of recent findings that not all preparations of HSPG have activity in our regeneration assay, we have begun to screen for pro-regenerative HSPG activity between difference HSPG sources. Understanding the signals and responses of cells, whether they are pro- or anti-regenerative will be critical to achieving the goal of regenerative engineering to reform complex structure and restore function in humans.

474480 Cell, Peptide and Protein Delivery Problems in Regenerative Engineering

Abstract:

Convergence in the biomedical field rehires synergism of different factors, parameters, ideas and disciplines. This convergence allow for translation of bench-top research to bedside methods. We will discuss a range of recent efforts to provide intelligent cell, protein and peptide delivery systems with high output and exceptional precision for the treatment of diseases in the hospital or at home. We show how the basics of chemical engineering-based controlled release have led to a number of major improvements in the fields of cardiovascular diseases, diabetes, neurological diseases and autoimmune diseases.

475505 Nano- and Microfabricated Hydrogels for Regenerative Engineering

Abstract:

Engineered materials that integrate advances in polymer chemistry, nanotechnology, and biological sciences have the potential to create powerful medical therapies. Our group aims to engineer tissue regenerative therapies using water-containing polymer networks, called hydrogels, that can regulate cell behavior. Specifically, we have developed photocrosslinkable hybrid hydrogels that combine natural biomolecules with nanoparticles to regulate the chemical, biological, mechanical and electrical properties of gels. These functional scaffolds induce the differentiation of stem cells to desired cell types and direct the formation of vascularized heart or bone tissues. Since tissue function is highly dependent on architecture, we have also used microfabrication methods, such as microfluidics, photolithography, bioprinting, and molding, to regulate the architecture of these materials. We have employed these strategies to generate miniaturized tissues. To create tissue complexity, we have also developed directed assembly techniques to compile small tissue modules into larger constructs. It is anticipated that such approaches will lead to the development of next-generation regenerative therapeutics and biomedical devices.

476338 Review and Perspective on Nexus System Analysis

Abstract:

As the global population grows and increasingly lives in cities with lifestyles based on greater material consumption, more attention is being given to the integrated system that supplies our food, energy and water, the Nexus. The interaction between these three essential commodities can be seen in many ways. For example, production of sufficient food requires use of man-made fertilizer, to an extent that food and energy prices have become linked. Agriculture now accounts for 70% of water withdrawn from aquifers, streams and lakes. In many places fresh water is critically constrained, which affects our ability simultaneously to produce more food, grow crops for biofuels, and maintain natural ecosystems. And there is mounting concern about the effects on the Nexus of climate change, itself largely the result of fossil fuel consumption and land-use change (for agriculture).

Evidence for this growing interest in the Nexus is illustrated by a review of published literature (1). Over the 5-year period 2011-2016 the rate of publication of studies on the Food-Water-Energy Nexus in the scientific literature grew by a factor about ten. These publications underline the importance of Nexus concerns to policy-makers, transnational and regional institutions, and business/commerce. The study purposes of this body of literature can be summarized as

- reviews and Nexus issue discussion papers;
- quality-of-life studies to understand Nexus trade-offs or develop responses to achieve sustainability objectives (e.g., human well-being, healthy ecosystems);
- product studies to understand Nexus implications or develop responses related to providing products (e.g., a renewable energy technology, a service such as clean water); and
- studies to develop integrated system modelling tools.

Case studies tend to be of two types: those related to regional development, and those related to specific sectors. So Hoff's (2) background paper for the Bonn2011 conference outlines ten Nexus case studies, including Jordan and Ethiopia, the arid Ningxia region of China, and Gujarat. The Brazilian Cerrado savanna area offers a study in the competition between biofuel and food crops for land and water resources; the Nile Basin offers the same trade-off, with hydropower also playing a role (2). Responses by several companies to Nexus challenges in their own business areas are outlined in a report by Conservation International (3). These responses cover a wide range of policies and activities, but there are several common themes, including

- developing greater resource efficiency
- conducting resilience planning for price volatility and climate change
- systems thinking to design synergistic solutions (e.g., business decisions that also yield multiple environmental and social benefits)
- gathering new data to understand current and future trends
- stakeholder engagement to better understand the Nexus
- ecosystem valuation to help decision-making and balance trade-offs.

Particularly important in view of the trend of more people to live in cities are the (relatively few) studies of the "Urban Nexus", dealing with what may be termed the "metabolism" of the city. It is concluded that this metabolism, a hybrid of natural and artificial structures presents an opportunity for greater efficiency through better design and regulation. At present it is poorly understood, and studies are hampered by lack of data.

Case studies show the vital role played by stakeholder involvement in clarifying issues, priorities, and values. They also demonstrate the importance of an integrated systems view of the complex interrelationships of the Nexus when planning effective remedies. Assessments conclude that transformative social and political change is needed to create new structures, markets, and governance to deal with the Nexus if we are to meet agreed-upon sustainable development goals.

Needs for the future of Nexus studies include

- the publication of more case studies, particularly including stakeholder involvement;
- the creation of new frameworks and methodologies of analysis, including some that will integrate societal, environmental, and technical dimensions;
- the development of modelling tools accessible to a wider range of investigators (at present these tend to be highly specialized);
- the promotion of best practice in Nexus evaluation and effective use of study results;
- improved accessibility of Nexus data;
- metabolism modelling and engineering for the urban Nexus; and
- promotion and support from funding agencies and institutions for multidisciplinary and multinational studies of Nexus analysis, issues, and applications.

(1) Keairns DL, Darton RC, Irabien A 2016 ***The Energy-Water-Food Nexus*** Annu. Rev. Chem. Biomol. Eng. 7:239–62

(2) Hoff H. 2011. ***Understanding the Nexus. Background Paper for the Bonn2011 Conference: The Water, Energy and Food Security Nexus.*** Stockholm: Stockholm Environ. Inst.

(3) Gerholdt J, Pandya S. ***Resources: The Energy, Water, Food Nexus***. Arlington, VA: Conserv. Int. Bus. Sustain. Counc.

454095 Construction of Programmable Genetic Circuits through Generalizable RNA Regulators and Crispr Interference

Abstract:

The past decades have witnessed the power of RNA regulators for precise and orthogonal gene expression control. Although gene regulation studies have focused more on protein regulators than RNA regulators, RNA regulation has many advantages over protein regulation. RNA regulators are easier to design than protein regulators, due to the simplicity of base-pairing rules and our increased understanding of RNA folding and functions. Additionally, RNA regulation is faster and has a lower metabolic burden than protein regulation because regulatory RNAs are shorter than regulatory proteins, and RNA regulators do not undergo the cellular resource-intensive process of translation. Still, synthetic RNA regulators for precise and tunable gene expression control are in great demand, and we aim to provide generalizable RNA regulators that allow for rational reprogramming of cellular functions. To this end, we have employed the CRISPR interference system (CRISPRi) from *Streptococcus pyogenes* and synthetic antisense RNAs (asRNAs) to repress or derepress target genes in a programmable manner. Specifically, we demonstrated that the repressed target output by the CRISPRi system can be derepressed by activating transcription of asRNAs that sequester their target small guide RNAs (sgRNAs). Our regulation system is unique in that it combines dCas9-sgRNAs with asRNAs, which can control the dynamics of gene expression. Furthermore, these RNA-based genetic circuits have been optimized by varying design parameters (e.g., the hybridization free energy of the sgRNA-asRNA complex), and tunable levels of derepression have been achieved (up to 95%). These RNA regulators can also target multiple genes orthogonally, allowing for multiplex gene regulation. To our knowledge, this is the first demonstration of a tunable RNA circuit that integrates multiple types of RNA regulators in bacteria. This new system will allow for rational reprogramming of cellular processes, facilitating diverse synthetic biology applications. We will present progress towards development of the design principles for generalizable RNA regulators.

466934 Crispr-Based Antimicrobials for Treating Multidrug-Resistant Infections

Abstract:

Once considered a revolution in modern medicine, antibiotics are becoming obsolete. Multidrug-resistant infections are increasingly being reported, where many resistance mechanisms can be readily passed between unrelated bacteria. Furthermore, traditional antibiotics are broad-spectrum, decimating the multitude of beneficial bacteria in the human body and leaving us susceptible to opportunistic infections and chronic illnesses. These challenges demand novel approaches to treat multidrug-resistant infections while discriminating between pathogenic and symbiotic bacteria.

Here, we report a targeted strategy for treating ranging bacterial infections that relies on CRISPR (Clustered Regularly Interspaced Short Palindromic Repeats)-Cas (CRISPR associated) systems. These RNA-directed defense systems are naturally employed by bacteria and archaea to recognize and cleave the DNA of foreign invaders. Small CRISPR RNAs bind their complementary sequences in the foreign DNA directing cleavage or degradation by Cas proteins. By designing and expressing synthetic genome-targeting CRISPR RNAs, we can reprogram CRISPR-Cas systems to induce bacterial cell death. We found that this strategy could achieve potent and sequence-specific killing of multidrug-resistant bacteria regardless of the type or location of the targeted region and using either imported or native CRISPR-Cas systems. Additionally, we were able to distinguish between even highly related strains as well as between commensal and pathogenic bacteria.

For this strategy to further develop as a therapeutic, a delivery method for the CRISPR-Cas DNA is needed. Toward this goal, we are engineering the broad-host range P1 bacteriophage as a delivery platform that can be used for multiple enteric pathogens. We identified and validated multiple landing sites to introduce foreign DNA while maintaining the bacteriophage function. The engineered P1 bacteriophage showed significant success in delivering the DNA of interest to the target bacteria compared to the state-of-art P1 phagemid systems. Finally, we utilized the same approach to deliver genome-targeting CRISPR-Cas systems to a number of gram-negative pathogens and eradicate them. We envision using this strategy to develop â€œsmartâ€ non-antibiotic treatments capable of circumventing common modes of resistance while readily distinguishing between pathogens and symbiotic bacteria.

461946 Improving RNA Regulator Dynamic Range with a Dual Transcription-Translation Control Mechanism

Abstract:

RNA repressors are emerging as a highly versatile component in genetic circuit construction.Â However, the dynamic range of RNA transcriptional repressors still doesnâ€™t match that of their protein counterparts. This incomplete repression can cause circuit leak, which impedes the construction of large predictable synthetic regulatory networks. Here we demonstrate how naturally-derived antisense RNA-mediated transcriptional regulators can regulate both transcription and translation, increasing repression from 85% to 98% and activation from 10 fold to over 900 fold, using transcriptional termination and RBS sequestration in a single compact RNA molecule that functions in *Escherichia coli*. We also show that orthogonal versions of this mechanism can be created through engineering minimal versions of the antisense RNAs. To demonstrate the utility of this dual control mechanism, we use it to show that they greatly reduce circuit leak when used in RNA-only transcriptional cascades that activate gene expression as a function of a small molecule input. We anticipate these regulators will find broad use as synthetic biology moves beyond parts engineering to larger and more sophisticated circuits.

463470 Acoustic Reporter Genes for High-Resolution Ultrasound Imaging of Engineered Bacteria In Vivo

Abstract:

More than 90% of the cells in a typical human body are microbes, and there is a rapidly growing appreciation of the crucial roles these organisms play in human health and disease. Additionally, advances in synthetic biology have enabled the development of genetically engineered bacterial therapies and diagnostics that are starting to advance into the clinic. These diverse activities of natural and engineered microbes depend critically on their anatomical location within host organisms, such as the various regions of the mammalian gastrointestinal tract. However, no effective methods are currently available to image microbes inside mammalian hosts with high spatial and temporal resolution. The development of such methods would transform the study of microbial-host interactions and facilitate the development of an entirely new class of bacterial diagnostics.

We address this challenge by developing acoustic reporter genes (ARGs) to enable the imaging of microbes with ultrasound. Unlike optical techniques, which have poor spatial resolution in vivo due to the scattering of light by tissue and are not commonly used in the clinic, ultrasound is capable of very high spatial resolution in vivo (tens of microns in mice) and is used widely in medicine. We recently discovered a unique class of protein nanostructures encoded in the genomes of certain photosynthetic microbes that can serve as imaging agents for ultrasound[1]. These nanostructures, called gas vesicles (GVs), comprise all-protein shells that are ~ 250 nm in size and are hollow on the inside. We showed that the hollow interior and unique mechanical properties of GVs allow them to scatter sound waves and thereby serve as sensitive in vivo imaging agents for ultrasound.

Now, we have succeeded in adapting the gene clusters encoding GV formation for heterologous expression as ARGs in Escherichia coli and Salmonella typhimurium, two important model organisms and chasses for synthetic biology. ARG expression allows us to image these bacteria with ultrasound in vitro and in vivo. Cells can be detected at

concentrations below 10^8 cells/ml (representing < 0.1% of voxel volume). In addition, engineering of the GV gene cluster has yielded variants that can be â€œerasedâ€ with acoustic pressures above specific thresholds, thus enabling more sensitive detection within biological tissues. Furthermore, the engineering of ARGs with different thresholds for acoustic collapse enables multiplexed imaging of more than one cellular population.

ARGs will allow microbes to be imaged dynamically in vivo with unprecedented precision, unlocking a wide range of new biological research and the development of spatially informative bacterial diagnostics.

1. Shapiro, M.G. et al. Biogenic gas nanostructures as ultrasonic molecular reporters. Nature nanotechnology 9, 311-316 (2014).

458211 Mix and Match Sweets: Modular, Coordinated Cell-Free Transcription, Translation, and Glycosylation of Proteins Using Selectively Enriched *Escherichia coli* Cell Lysates

Abstract:

Glycosylation, or the attachment of glycans (sugars) to proteins, is one of the most abundant post-translational modifications in nature and plays a pivotal role in protein folding, sorting, and activity. In molecular medicine, the compositions and structures of glycans on recombinant therapeutic glycoproteins are known to impact pharmacokinetics and drug activity. However, unlike DNA, RNA, and protein synthesis, protein glycosylation is not a template-driven process. Instead, glycans are assembled by coordinated expression of glycosyltransferase enzymes and attached to proteins by oligosaccharyltransferase (OST) enzymes. The complexity of this biosynthetic process results in inherent heterogeneity of glycan compositions and structures in living organisms. The inability to precisely control protein glycosylation in eukaryotic protein expression systems (*e.g.*, yeast and CHO cells) represents a key challenge in the fields of glycoprotein synthesis and glycoprotein therapeutics. Bacterial glycoengineering, or the expression of orthogonal glycosylation machinery in *Escherichia coli*, is an alternative strategy for production of proteins with controllable glycosylation. However, *in vivo* bacterial glycoengineering is limited by i) the need to maintain cell viability during glycoprotein synthesis while simultaneously overexpressing multiple non-native, membrane-bound glycosylation enzymes and ii) lengthy design-build-test cycles for novel glycosylation pathway engineering. To address these limitations, we have developed a bacterial cell-free glycoprotein synthesis (CFGpS) system with the potential to enable controllable, user-specified protein glycosylation. Uniquely, in our CFGpS system, i) all the biosynthetic machinery for protein synthesis and glycosylation is supplied by *E. coli* lysates and ii) transcription, translation, and glycosylation occur in an all-in-one *in vitro* reaction. We engineered glycosylation chassis strains that are optimized for glycosylation and produce ~1.5 g/L protein in cell-free protein synthesis, which represents a 50% increase in potential glycoprotein yields compared to state-of-the-art *E. coli* glycosylation chassis and recombinant protein expression strains. We then demonstrated that *E. coli* cell lysates can be selectively enriched with active glycan precursors and/or the OST from *Campylobacter jejuni via* overexpression of orthogonal genes in the chassis strain. Next, we used these lysates individually and as a mixture to carry out one-pot synthesis of glycoproteins. By mixing and matching lysates, we also used the CFGpS system to characterize the *in vitro* activity of four additional OSTs with varying sequence homology to the archetypal *C. jejuni* OST. The CFGpS platform is modular, flexible, and has promising applications as a high-throughput prototyping platform for glycosylation pathway design and synthesis of glycoproteins of biotechnological interest. This system promises to advance our understanding of bacterial glycosylation and aid development of new glycoengineering tools.

463037 Engineering Intracellular Trafficking for Pathway Co-Localization

Abstract:

Many essential and industrially useful metabolic pathways are assembled on intracellular membranes or are sequestered in organelles. Connections and organization between enzymes in these structures produce optimized kinetics through controlled local concentrations of enzymes and pathway intermediates. When engineering new pathways such organizations and intracellular localizations are often lost, thus eliminating any kinetic advantages that may have resulted. We have discovered this to be true with ester biosynthesis pathways engineered to produce fragrance and flavor compounds, solvents, and biofuels. The key reaction step, acetyl transfer from acetyl-CoA to an alcohol, localizes to the endoplasmic reticulum and lipid droplets in yeast and disruption of this localization significantly limits enzymatic activity. By first understanding how alcohol-O-acetyltransferases (AATases) from yeast and fruit species express and localize, we have developed engineering solutions that overcome the losses in activity due to disrupted intracellular localization. When highly active AATases are heterologously expressed in *E. coli* activity is decreased by nearly three orders of magnitude from levels observed in native hosts. Controlling expression level and preventing aggregate formation partially recovers the lost activity. An alternative solution is to introduce new multienzyme structures that create spatially organized metabolic pathways around the native intracellular localization of highly active AATases. Co-localization of upstream enzymes to lipid droplets locally synthesizes pathway intermediates, producing kinetic enhancements in the AATase pathway and increased fermentation titers. The solutions developed for ester biosynthesis were designed to be broadly applicable and promise to help solve challenges in the disruption of native enzyme localization and pathway structure that limit the engineering of other potentially useful pathways.

474999 Invited Presentation: Metabolite Valves: Dynamic Control of Metabolic Flux for Pathway Engineering

Abstract:

Microbial strains have been successfully engineered to produce a wide variety of chemical compounds, several of which have been commercialized. As new products are targeted for biological synthesis, yield is frequently considered a primary driver towards determining feasibility. Theoretical yields can be calculated, establishing an upper limit on the potential conversion of starting substrates to target compounds. Such yields typically ignore loss of substrate to byproducts, with the assumption that competing reactions can be eliminated, usually by deleting the genes encoding the corresponding enzymes. However, when an enzyme encodes an essential gene, especially one involved in primary metabolism, deletion is not a viable option. Reducing gene expression in a static fashion is possible, but this solution ignores the metabolic demand needed for synthesis of the enzymes required for the desired pathway.

We have developed "metabolite valves" to address this challenge. The valves are designed to allow high flux through the essential enzyme during an initial period where growth is favored. Following an external perturbation, enzyme activity is then reduced, enabling a higher precursor pool to be diverted towards the pathway of interest. We have designed valves with control at both the transcriptional and post-translational levels. In both cases, key enzymes in glucose metabolism are regulated, and two different compounds are targeted for heterologous production. We have measured increased concentrations of intracellular metabolites once the valve is closed, and have demonstrated that these increased pools lead to increased product yields. We have also demonstrated valve construction and operation in two different organisms. These metabolite valves should prove broadly useful for dynamic control of metabolic flux, resulting in improvements in product yields.

461558 Rapid Uptake and Ubiquitination of Fluorescent Peptides into Intact Cancer Cells Using a β-Hairpin Sequence Motif

Abstract:

Cell penetrating peptides (CPPs) have emerged as powerful tools for delivering bioactive cargoes, such as biosensors and chemotherapeutics, to cells. CPPs possess several advantages over competing techniques like microinjection, including their ease of synthesis, compatibility with high-throughput analysis techniques, and decreased effect on cellular viability; however, the application of CPPs may be limited as a result of their degradation by intracellular proteases. In this research, a library of short peptide sequences, termed 'protectides', were synthesized using solid phase peptide synthesis that exhibit β-hairpin motif. These short peptides exhibited a prominent secondary structure which resulted in increased stability under cytosolic conditions. The effect of incubation time, temperature, and concentration was correlated to cellular uptake efficiency and long-term stability in cell lysates. Interestingly, the ornithine-rich members of the peptide library were also found to act as ubiquitination sensors capable of *in vitro* ubiquitination in intact cells. All members of the β-hairpin peptide library demonstrated the ability to rapidly penetrate the cell membrane of intact HeLa cells, using both fluorometry and microscopy techniques, although to various extents. Peptide uptake was found to be both temperature and concentration dependent. The arginine-rich peptides named RWRWR [Ac-RWVRVpGO(FAM)WIRQNH_2] and OWRWR [Ac-OWVRVpGO(FAM)WIRQ-NH_2], were discovered to exhibit the highest permeability efficiency. The stability of the members of the peptide library under conditions similar to the cytosolic environment was verified using a degradation assay and reverse phase high performance liquid chromatography (RP-HPLC). The β-turn was confirmed using circular dichroism (CD). While commercial peptides TAT and Arg(9) demonstrated superior permeability efficiency, they also exhibited decreased intracellular stability compared to members of the protectide library. The scrambled versions of the peptides revealed that the secondary structure is not necessary for peptide internalization; however the unstructured variants showed increased cellular uptake in intact cells, but decreased stability in

cell lysates. Finally, as a proof-of-concept, the protectide was incorporated into the N-terminus of an enzyme substrate to generate a long-lived, cell permeable fluorescent probe for detecting intracellular deubiquitinating enzyme activity in intact cancer cells.

470027 The Role of Glycocalyx and Shear Stress on Endothelial-Cancer Cell Attachment

Abstract:

Introduction: Cancer and endothelial cell (EC) interactions [Giavazzi, R., et al., Journal of Clinical Investigation, 1993] are likely mediated by glycocalyx (GCX), a complex of sugar molecules that envelop both cell types and can regulate the interactions of their intercellular adhesion molecules. It was recently shown that metastatic cancer cells up-regulate the expression of their surface sugar molecules, which mechanically regulate integrin-mediated intercellular interactions and amplify cancer cell growth and survival [Paszek, M.J., et al., Nature, 2014]. Compared to cancer GCX, endothelial GCX composition, structure and function during intercellular interactions are understudied but of great clinical importance. EC GCX consists of membrane bound glycoproteins, proteoglycans, and glycosaminoglycans side chains [Pries, A.R., et al., Pflugers Arch, 2000; Boddohi, S., et al., Adv Mater, 2010]. EC GCX structure is dependent on active EC shedding and synthesis of GCX, in response to the microenvironment [Arisaka, T., et al., Ann N Y Acad Sci, 1995; Elhadj, S., et al., J Cell Biochem, 2002; Koo, A., et al., Am J Physiol Cell Physiol, 2013].The glycocalyx structure is particularly responsive to shear stress and has been shown to remodel when exposed to flow [Tarbell, J., et al., PloS one, 2014]. Typically, degraded GCX coincides with diseased states, whereas intact EC GCX coincides with normal or healthy conditions. For example, endothelial cell glycocalyx is often degraded in areas prone to atherosclerosis, coinciding with abnormal levels of shear stress in these regions. It is therefore believed that EC GCX is substantially degraded in settings of secondary tumor formation due to irregularities in the tumor vessels; which are mostly disorderly organized [Jain, R., J Clin Oncol, 2013]. The resultant structure and composition of the GCX is presumed to pave the way for cancer cells to invade the endothelium for transendothelial migration. The goal of this present study is to test the hypothesis that GCX controls the attachment of cancer cells onto the endothelium.Â **Materials and Methods**: To test this hypothesis, we conducted both static enzyme degradation experiments and shear stress induced flow experiments using rat fat pad endothelial cells (RFPECs), which naturally express a robust glycocalyx. We verified the condition of the RFPECsâ€™ GCX by immunostaining the sialic acid

glycosaminoglycans of the GCX. Sialic acid was used to characterize GCX due to its implication in the attachment of cancer cells to the endothelium. For the static experiments, degradation of glycosaminoglycans *in vivo* was mimicked by using neuraminidase to degrade sialic acid, respectively. Immunofluorescence confocal interrogation was used to examine the effects of degrading enzymes on sialic acid thickness and coverage on RFPECs. RFPEC monolayers with intact or enzyme degraded glycocalyx were co-incubated with metastatic mouse breast cancer cells (4T1) labeled with CellTracker Red. The fluorescently-labeled 4T1 cells were easily identified during co-incubation with glycocalyx-rich or glycocalyx-deficient RFPEC monolayers, enabling quantification of the effect of glycosaminoglycan removal on cancer cell attachment to ECs. For the shear stress induced flow experiments, a custom parallel plate-step flow chamber was developed to expose RFPECs to shear stresses at 10 dynes/cm^2. Hypothetically, areas of recirculation, characterized by low shear stresses, will produce RFPECs with degraded glycocalyx, representing degraded EC GCX *in vivo*. Attachment of cancer cells to RFPEC monolayers from shear stress induced flow experiments will be tested for cancer cell attachment using the same methods as previously described. **Results and Discussion:** In control conditions, the thickness of sialic acid was 1.6 ± 0.01 µm. Enzyme doses of 15, 135, 1215, and 3645 mU/ml reduced thickness to 1.41 ± 0.04 µm, 0.97 ± 0.03 µm, 0.88 ± 0.02 µm and 0.80 ± 0.01 µm respectively. These results indicate that an enzyme concentration of 3645 mU/ml was required to degrade ~50% of the thickness of sialic acid in RFPEC GCX. To see the impact of sialic acid thickness reduction on cancer cell attachment, we co-incubated CellTracker Red stained 4T1 breast cancer cells with untreated RFPEC versus neuraminidase-treated RFPEC. On control RFPEC the number of attached cancer cells was 310.77 ± 30.40 for every 1 cm^2 (or for every 2000 adherent RFPEC at confluence). RFPEC treated with 15, 135, 1215 and 3645 mU/ml of enzyme showed cancer cell attachment levels of 305.43 ± 43.07, 244.92 ± 29.16, 432.71 ± 39.23 and 740.66 ± 92.20, respectively. In flow conditions, it is expected that areas close to the disturbed flow step will experiment irregular flow patterns which will result in the disorganization and subsequent degradation of GCX. This will then lead to the increased attachment of cancer cells in the disturbed flow areas in comparison with laminar and static conditions. The enzyme degradation results point to the fact that degrading GCX to ~50% of its thickness could be very instrumental in the attachment of circulating cancer cells at secondary tumor formation sites. We hope to see a disrupted GCX at disturbed flow areas which should result in an increase in cancer cell attachment. **Conclusion:** We have shown that GCX degradation may initiate secondary tumor formation. As a future direction we hope to dive deeper into the subject of whether other glycosaminoglycans are also partly responsible for enhancing circulating cancer cell attachment to the endothelium. We also look forward to examining the role of enzymatic GCX degradation in cancer metastasis *in vivo*. **Acknowledgements:** Northeastern University provided start-up and pilot funding to support this work.

474057 Mitogen-Activated Protein Kinase-Driven Sprouty2 Expression Mediates Resistance to Receptor Tyrosine Kinase-Targeted Therapeutics in Glioblastoma Cells

Abstract:

The therapeutic efficacy of inhibitors of oncogenic kinases in glioblastoma multiforme (GBM) has been disappointing to date. In glioblastoma and other malignancies, resistance to targeted therapeutics can arise through dynamic rewiring of cell signaling pathways responsible for tumorigenesis and survival. Here, we identify one such signaling rewiring process in glioblastoma cells that promotes expression of Sprouty2 (SPRY2). SPRY2 expression is of interest because we recently identified SPRY2 as a driver of glioblastoma cell and tumor proliferation and resistance to kinase inhibitors. Specifically, we found that SPRY2 depletion reduced the ability of glioblastoma cells to form colonies in soft agar, to form subcutaneous tumors in mice, and to resist co-inhibition of the EGFR and MET receptor tyrosine kinases. In the present study we found that, in a panel of glioblastoma cell lines, SPRY2 expression was initially reduced in response to inhibition of EGFR and MET. However, at later time points (24-48 hr after receptor inhibition) resurgent SPRY2 expression was observed even as EGFR and MET phosphorylation remained suppressed. Further inspection of the experimental data revealed that phosphorylated ERK, which was initially suppressed in response to EGFR and MET inhibition, also displayed a resurgence that paralleled the trend in SPRY2 expression. This correlation between ERK phosphorylation and SPRY2 expression is consistent with the documented ability of ERK to regulate SPRY2 expression transcriptionally. The importance of resurgent ERK phosphorylation was confirmed in experiments showing that its suppression through the use of a MEK inhibitor prevented resurgent SPRY2 expression and augmented cell death in response to EGFR and MET inhibitors. We hypothesized that resurgent ERK phosphorylation and resultant SPRY2 expression occurred due to the ability of glioblastoma cells to activate receptor tyrosine kinases other than EGFR and MET in response to inhibition of those receptors. Antibody microarray experiments confirmed this hypothesis, and ongoing experiments demonstrate the efficacy of co-targeting the receptors that rescue ERK phosphorylation in augmenting response to EGFR and MET inhibition.

470759 A Model for the Evolution of Tumor Populations and It s Application to Human Hepatocellular Carcinoma and to Zebrafish Melanoma

Abstract:

We recently proposed a new mathematical model for tumor growth, shrinkage. Unlike previous mathematical models that try to describe the dynamics of a typical or mean tumor and generally do not include immunity or therapy, our model describes an entire tumor population's evolution, where each tumor is subject to mitosis, immunity, chemo/immunotherapy and metastasis. The individual terms in our model naturally (and surprisingly) combine to generate a qualitatively new type of behavior – a diffusive process in tumor-size space. This diffusion turns a collection of tumors of similar sizes into one whose sizes are more spread apart, without changing their mean size. When coupled with terms that either grow or shrink all tumors, and noting evidence that patients with small tumors respond to therapy better than patients with larger ones, it can predict the emergence of a tumor that suddenly grows rapidly after a long period of stable tumor load and of relapse after surgery. The former is reminiscent of certain patient cohorts, whose members persist with apparently stationary tumors for long periods of time until suddenly one or more tumors begin to grow rapidly. We use fit our model to literature data on the progression of human hepatocellular carcinoma and make predictions as to the evolution of this disease with and without therapy. We also present new data taken on zebrafish whose genome has been altered to remove its stripes, which leaves the fish nearly clear, that have been inoculated with a virulent green fluorescent protein-labeled melanoma cell line. We follow the evolution of both the primary tumor and the formation and growth of its metastases and compare these data to our model. Our long-term goal is to try to explain confounding patient cohorts, potentially propose new treatment strategies and, eventually, to use patient-specific information to guide treatment and to potentially predict likely time-to-recurrence.

460502 Fluid Shear Stress Activates Epithelial-to-Mesenchymal Transition Genes in Luminal Breast Cancer Subtype

Abstract:

Metastasis is the leading cause of cancer deaths due to the spread of cancer cells through the blood vessels leading to the formation of secondary tumors. Metastasizing cancer cells in the human vasculature are called circulating tumor cells (CTCs) and are characterized to express epithelial cell adhesion molecule (EpCAM). They are further known to survive the harsh physiological fluid shear stress (FSS) conditions. However, the effect of FSS on CTC molecular phenotype, such as the epithelial-to-mesenchymal transition (EMT) and cancer stem cell (CSC) expression, has not been extensively studied. Here, we created an in vitro model system using COMSOL Multiphysics and a syringe pump for experimental FSS to evaluate CTCs in blood flow. Breast epithelial cell line 184A1 and breast cancer cell lines MCF7, SKBR3, and MDA-MB-231 were tested for cell viability. Gene expression and CSC expression were tested on the MCF7 and MDA-MB-231 cell lines. Results show that cell viability under FSS is variable in different breast cancer cell types with breast cancer cells having a higher viability under fluid shear stress compared to the 184A1 cell line. The difference in breast cancer cell type also correlated to differences in epithelial and mesenchymal gene expression with the more invasive MDA-MB-231 having higher mesenchymal gene expression. Upon application of FSS, there was an upregulation of genes in MCF7 cells and MDA-MB-231 cells. However, cancer stem cell expression which is seen in metastatic cancer cells was more resistant to expression changes under fluid shear stress. These results show that the luminal breast cancer subtype is more susceptible to the effects of FSS with greater changes in their genetic and protein expression.

470619 Interference-Based Label-Free Probe of Cancer Cells

Abstract:

Cancer diagnostics and treatment requires identification of tumor cancer cells and determination of their metastatic potential. Current standard methods make cancer cell identification time-consuming and costly, since they involve the use of specific antibodies and biomarkers for immuno-histochemical staining. Here, we describe work aimed at overcoming these barriers. Interference-based techniques have demonstrated great spatial and temporal resolution capabilities in the study of microscopic objects near surfaces, with reflection interference contrast microscopy (RICM) offering a unique non-invasive â€œview from belowâ€ perspective. Therefore, to identify cancer cells and determine their metastatic potential, cells are studied using RICM, where images captured in just microseconds contain interferograms that embed detailed topographical information of the cells, down to the nanometer-scale. The computational analysis of RICM interferograms, using fast custom-developed algorithms, facilitates label-free quantification of cell adhesion, deformability, and membrane fluctuations. Direct observation of cell adhesion by RICM enables both static and dynamic determination of the contact boundary fractal dimension, measurement of the size and distribution of adhesion patches within the contact area, and evaluation of adhesion strength. In addition, interference fringes can be conveniently analyzed for fast reconstruction of the cellâ€™s surface contour with unprecedented nanometer-scale resolution, while fringe visibility analysis allows quantification of nanometer-scale cell membrane undulations or roughness. These static and dynamic measurements using RICM produce valuable information that could reveal characteristic cell signatures and open new possibilities for identification of tumor cancer cells and determination of their metastatic potential.

470450 Transport and Reaction Modeling of Liposomal Nanocarriers for Chemotherapeutics in Solid Tumor Analogues

Abstract:

Liposomal carriers have been successfully used for many years in chemotherapy of solid tumors in order to improve delivery of different chemotherapeutic drugs. Currently, ligand-labeled liposomes for targeted chemotherapy are evaluated on clinical trials which suggest promising potential. Liposomes' usage has both increased the killing efficacy of the drug within the tumor, and reduced its uptake by the healthy tissues. The most advanced of these liposomes can be equipped with masking/unmasking mechanisms of specific ligands and/or triggered release of their contents - affected by the pH gradients within these solid tumors - which can further improve their performance. Despite the advanced state of experimental research in this field, there is still the need to develop a general and consistent mathematical model, which can accurately predict the drug delivery efficiency of liposomal carriers (and any other type of nanoparticle) in diverse situations, thus allowing the optimization of their features. In this communication we present such a theoretical model, we discuss its origin and range of validity, and we show results from numerical simulations.

455933 Simulation and Experimental Analysis of an Automated System for Magnetically Labeling and Collecting Rare Blood Cells

Abstract:

To test the ability of a point-of-care cell detector to automatically label and trap cells, simulations and laboratory experiments on cell labeling and separation were performed. A static mixer's cell labeling efficiency and a magnetic trap's capture ability were simulated using computational fluid dynamics and tested in cell labeling and trapping experiments in the presence and absence of whole blood to determine the effect of red blood cells on labeled tumor cells detection. After each cell labeling and separation test a particle tracking microscope was used to detect particle motion, size and magnetophoretic mobility distribution thereby identifying and counting magnetically labeled cells. A statistical analysis of data was obtained during the test allowing a better understanding of cell labeling. The results of cell counts indicated the immunomagnetic particles were able to perform tumor cell labeling and separation tasks with good capture efficiency. The static mixer proved to have a similar cell labeling efficiency to that of the traditional multistep methods using shake-bed labeling. The velocimeter plots showed cells were successfully labeled by magnetic beads in the static mixer, and the magnetic trap was capable of capturing labeled particles. Whole blood was found to hinder cell detection by the velocimeter, and a simple post- processing step was introduced to ensure unlabeled cells are washed away properly.

Keywords: cell labeling, magnetic particles, magnetic separation, Hyperflux^{tm} analysis, velocimetry, computational fluid dynamics

462662 Genome-Scale Engineering for the Construction and Optimization of Yeast Cell Factories

Abstract:

Biological conversion of renewable feedstock into fuels and chemicals has been intensively investigated due to increasing concerns on sustainability and global climate change. Compared with its counterparts, *Saccharomyces cerevisiae*, the bakerâ€™s yeast, is more industrially relevant thanks to its well-studied genetic and physiological background, the availability of a large collection of genetic tools, the compatibility of high-density and large-scale fermentation, the resistance to phage infection, and the high tolerance against toxic inhibitors and products. Therefore, we aim to develop platform yeast cell factories for efficient and cost-effective production of biofuels and chemicals. To achieve this, the host genome should be engineered such as the disruption of the competing pathways to redirect the metabolic fluxes to biosynthesis of the desired products. Although a wide variety of cell factories have been constructed, we are still facing the challenges of constructing efficient cell factories in a short period of time. Therefore, we took advantage of the recently developed CRISPR/Cas9 system to facilitate the development of cell factories. The CRISPR/Cas9 system was adopted to achieve both gain of function (knock-in and gene activation) and loss of function (knock out and gene repression) in yeast. More importantly, we combined gene deletion, gene repression, and gene activation into a single system. By doing this, we could engineer any gene of interest with a full spectrum of gene expression profiles (zero expression, down-regulation, and up-regulation). Finally, genome-scale studies were performed to develop a rapid and efficient strategy for the construction of cell factories and unveil the synergistic effects between gene deletion, gene repression, and gene activation in yeast.

Keywords: Yeast cell factory; Genome engineering; CRISPR/Cas9; Gain-of-function; Loss-of-function

472667 Building a Full Set of Genetic Toolkit to Engineer *Scheffersomyces Stipitis*

Abstract:

Many of the 1800 other known yeast species have highly unusual metabolic, biosynthetic, physiological and fermentative capacities that are not possessed by model yeasts such as *Saccharomyces cerevisiae*. These nonconventional yeasts remain largely unexplored due to the lack of efficient genetic engineering tools. In general, stable episomal expression platforms and precise genome-editing tools are the two foundational technologies used for engineering nonconventional species. We recently succeeded in establishing the first stable expression platform for engineering *Scheffersomyces stipitis*, one of the species with the highest native capability for xylose fermentation. The current expression plasmid is extremely unstable, being lost within 2 days of cultivation. Through the perfect integration of *in silico* centromere prediction and library-based screening, we were able to rapidly isolate all eight centromere sequences from *S. stipitis* genome. The identified centromeres significantly enhanced the stability of the episomal vector, resulting in homogenous expression of target protein and significant increase in production of valuable biochemicals. In addition, many nonconventional species rely on non-homologous end joining (NHEJ) mechanism for DNA repairing, which makes targeted gene knockout extremely challenging. We addressed this limitation by adapting CRISPR-Cas9 system for markerless gene knockouts and insertions in *S. stipitis*. A functional plasmid-based system was created using a codon-optimized Cas9 from *Streptococcus pyogenes*Â and single guide RNA (SgRNA). An efficiency of â‰¥80% indel was achieved. *Ku70* and *ku80* genes, which were required for NHEJ, were subsequently disrupted to suppress NHEJ and make homologous recombination (HR) as the primary double strand break repair mechanism.Â Using *Î”ku* mutant strain and the CRISPR-Cas9 system, we successfully constructed *ura3*, *trp1*, *his3*, and *leu2* auxotroph mutants, and demonstrated that *S. stipitis* is well suited for producing shikimate pathway derivatives such as shikimate and muconic acid. Our work represents the new explorations in expanding the current collection of microbial factories.

457853 Engineered Signal Transduction Coupled with Gene Circuits Enables Efficient Assimilation of Native and Foreign Nutrients in *Saccharomyces Cerevisiae*

Abstract:

Metabolic engineering has enabled production of bio-based chemicals in organisms, usually by overexpressing the genes (heterologous or native genes) involved in the biochemical pathway to debottleneck rate limiting steps. Recent studies have shown that engineered regulatory systems, such as removing feedback inhibition, can further improve the performance of engineered strains. We hypothesize that engineering a global regulatory system (regulon) could provide a new paradigm in engineering biological systems and complement current tools available for metabolic engineering. To demonstrate this, we use the assimilation of xylose by *S. cerevisiae* as a test case. Xylose is a non-native sugar to this yeast, but an abundant natural sugar.

Currently, engineering xylose assimilation for biomass or ethanol production in *S. cerevisiae* has been limited towards overexpression of initial genes in the pathway to convert xylose to xylulose-5-phosphate followed by expression of non-oxidative Pentose Phosphate Pathway genes to increase the flux towards glycolysis. But the phenomenon of growth or biomass production involves coordinated control of multiple pathways involving carbon metabolism, cofactor regeneration, amino acid synthesis, nucleotide synthesis, cell cycle maintenance etc., and debottlenecking rate limiting reactions in all of the necessary pathways required for growth in xylose would involve extensive pathway engineering. To get around this, we hypothesized that further enhancement in xylose utilization can be made by addressing the issue from a regulatory perspective rather than metabolic. To that end, we decided on a regulon engineering strategy whereby a sugar sensing regulon can be engineered to trigger transcriptional machinery when xylose is encountered thereby enhancing the growth and biocatalytic fitness in this non-native sugar.

Previous studies have shown that presence of xylose weakly upregulates galactose catabolic genes (GAL). This suggested that xylose can mimic galactose as an agonist of the GAL regulon, and that this system could serve as a platform to develop a xylose-dependent regulatory system. We first engineered the signal transduction step that increases the sensitivity and response kinetics of the GAL regulon for xylose as well as its native ligand, galactose. We further show that by switching ON the regulon using the positive feedback loop of the galactose regulatory circuit, growth rates comparable to already existing evolutionary engineered strains can be achieved with minimal metabolic engineering. In this process, we also enhanced the galactose sensing capabilities of the sensor, thereby achieving higher growth rates on galactose than the wild type strains. Finally, we also show that under non-inducing conditions, strains carrying the xylose regulon show better growth fitness than strains with constitutive expression of xylose metabolic genes. Further increase in growth rates, xylose uptake and specific chemical (including bioethanol) production can be achieved by expanding the genes under the synthetic xylose regulon.

472365 Optimization of a Tunable RNAi Screening Approach for Systems Engineering of Complex Phenotypes in Yeast

Abstract:

Complex phenotypes are notoriously difficult to rationally engineer as they result from the combined activities of multiple gene products, the identities and roles of which are often unknown. However, most current methods for non-rational strain improvement suffer from a variety of limitations, including laborious transfer of beneficial mutations to alternative strains, restricted (often binary) exploration of expression space, requirement for detailed knowledge of a strainâ€™s genetics, or inefficiency in eukaryotic hosts. To address these issues, we synthetically optimized RNA interference (RNAi) in yeast to enable genome-wide screening of downregulation space for effectors of complex phenotypes, and demonstrated this approach by identifying novel targets for improved 1-butanol and isobutanol tolerance. First, we optimized the efficiency and tunability of RNAi library screening in yeast, ultimately enabling downregulation efficiencies from 0% to 94% to be attained through tuning dsRNA cassette design and expression level. Using this system, we generated a genome-wide RNAi library from yeast cDNA which collectively spanned a wide range of downregulation capacities. By screening this library in inhibitory concentrations of isobutanol, we identified key regulators of isobutanol tolerance in a single round, with downregulation of recovered genes conferring up to 64% increased growth rate in 12 g/L isobutanol. We further found, through two iterative rounds of screening this RNAi library, that the combined downregulation of two genes improves growth rate up to 3100% in 10 g/L 1-butanol. For both compounds, the best-performing hits downregulate multiple homologous genes and elicit phenotypes which are fundamentally different from that of the corresponding gene knockout, thus validating the utility of the RNAi screening approach for yeast strain optimization. Collectively, this work improves the tunability of RNAi in yeast, uncovers novel effectors of butanol tolerance, and demonstrates a powerful approach for improving complex phenotypes in yeast.

461391 Crispr-Cas9 Based Transcription Inactivation and Elevation System for Multiplex and Tunable Control of Metabolic Flux in One Step

Abstract:

The success of the metabolic engineering requires efficient regulation of native and heterologous genetic elements in multi-enzyme pathways. Therefore, it is crucial to achieve the multiplex and tunable control of pathway activities. Previous approaches to achieve such goal often rely on engineering DNA-binding proteins (e.g., transcription factors), which is a multistep biomolecular processes for construction and optimization of function, and could only be applied to a small amount of genes. Recently, CRISPR-Cas9 system of *Streptococcus pyogenes* was proved to be capable of either activating or repressing multiple genes in prokaryotic and eukaryotic systems. In this study, we designed a CRISpr-cas9 based Transcription Inactivation aNd Elevation System (CRISTINES) to achieve multiplex and tunable control of metabolic flux in one step. In general, we found that by selectively designing and positioning guide RNAs with base-pairing complementarity to target DNA sites, a synthetic dCas9 system fused with VP64 (a commonly used eukaryotic transcription activator domain) could activate and repress the expressions of multiple genes at the same time. In order to uncover the designing principle for the guide RNAs to achieve the precise transcriptional regulation, we next systemically screened a library of guide RNAs that targeted on over 400 different sites on multiple fluorescent protein-encoding genes, followed by measuring the gene expression (i.e., fluorescent signal) and developing a regression model to accurately simulate the quantitative relation between the gene expression and the guide RNA design. Finally, the functionalities of CRISTINES were confirmed in the violacein pathway by simultaneously tuning the expression level of multiple genes to control synthetic flux towards the production of various chemicals.

469672 Genome-Wide Genetic Screening in Mammalian Cells Using TALE Libraries

Abstract:

Genome-wide screens to identify genes connected to a phenotype of interest have been performed using RNAi libraries and recently CRISPR-Cas9 libraries. However, both approaches are limited by inherent disadvantages including off-target effects and inefficient gain-of-function screenings. Recently, we introduced an assembly method to efficiently construct TALE (transcription activator-like effectors) libraries which cover all possible combinations of the nucleotides, and demonstrated genome-wide screening experiments in yeast cells. In this study, we constructed biased TALE libraries to target desired motifs and subsequently performed genome-wide screens in human cells to probe for desired change in phenotype.

458491 Theoretical Analysis of Enhanced Mass Transfer in Modules with Novel Twisted Hollow Fiber Membrane

Abstract:

A number of approaches have been used to develop membrane modules with enhanced mass transfer. Rotating membrane systems generate Taylor vortices that can significantly increase flux and reduce fouling, but these modules have very low membrane packing densities (membrane area per unit volume) and are very difficult to scale-up. Helically wrapped hollow fibers generate Dean vortices in a much more scalable geometry, although these modules have not yet been commercialized for large-scale applications. The objective of this work was to examine the behavior of twisted hollow fiber modules in which Dean vortices are generated by the coiled flow that occurs within a twisted elliptical flow channel. Computational fluid dynamic simulations are used to examine incompressible flow through twisted elliptical and helically-coiled cylindrical fibers for typical flow Reynolds numbers employed in membrane filtration processes ($Re < 200$). The resulting flow fields are compared in terms of the distributions of vorticity and wall shear rate, the pressure drop, and the azimuthally-averaged normal velocity in the vicinity of the wall. The secondary flow in twisted elliptical fibers generates significant recirculation flow leading to near-wall velocities directed normal to the surface of the membrane (below figure). The azimuthally-averaged normal velocity near the wall of a twisted elliptical fiber is found to be two to four orders of magnitude larger than typical ultrafiltration velocities. For $Re < 75$, the average wall shear rate and normal velocity near the wall are found to be significantly higher in twisted elliptical fibers compared to their helical counterparts. Modules produced using twisted elliptical hollow fibers would be linearly scalable by changing the number of parallel fibers, without any alteration in the device hydrodynamics. These findings highlight the potential for mass transfer and flux enhancement using modules with novel twisted elliptical hollow fiber membranes.

473805 Advances and Challenges in Membrane Desalination and Osmotic Power Generation

Abstract:

Membrane based water treatment and desalination technologies have been gaining momentum in potable water production from seawater and brackish groundwater, in municipal and industrial water reuse, as well as for management and reuse of agricultural drainage water. Considerable progress has been made since the early days of Reverse Osmosis (RO) and more recent development in Forward Osmosis (FO) with respect to water development of high performance fouling and chlorine resistant membranes, improvements of in feed pretreatment and reduction in RO energy consumption. Given the increased level of water scarcity in various regions of the world there has been growing efforts to expand the use of otherwise unutilized inland groundwater resources, lower the cost of seawater desalination and even produce energy from salinity gradients (i.e., pressure retarded osmosis (PRO). Here a fundamental description and evaluation of RO, FO and PRO will be provided with a focus on assessment of the merits of various process alternatives to common RO process configurations, the viability of PRO and challenges in FO. In particular, recent results from CFD analysis of FO and PRO and fundamental theoretical analysis of RO energy consumption (for different configurations) will be presented to clarify overoptimistic expectations and address the remaining major challenges.

462750 Cactus Inspired Self-Humidifying Membranes for Energy Generation

Abstract:

Cactus inspired self-humidifying membranes for energy generation

Young Moo Lee

Department of Energy Engineering, College of Engineering, Hanyang University, Seoul 133-791, Korea

The regulation of water content in polymeric membranes is important in a number of applications, such as in reverse electrodialysis and in proton-exchange fuel-cell membranes. External thermal and water management systems add both mass and size to systems, and so intrinsic mechanisms of retaining water and maintaining ionic transport in such membranes are particularly important for applications where small system size and density is important. For example, in proton-exchange membrane fuel cells, where water retention in the membrane is crucial for efficient transport of hydrated ions, by operating the cells at higher temperatures without external humidification, the membrane is self-humidified with water generated by electrochemical reactions.

Here we report an alternative solution that does not rely on external regulation of water supply or high temperatures. Water content in hydrocarbon polymer membranes is regulated through nanometre-scale cracks in a hydrophobic surface coating. These cracks work as nanovalves to retard water desorption and to maintain ion conductivity in the membrane upon dehumidification. Hydrocarbon fuel-cell membranes with surface nanocrack coatings operated at intermediate temperatures show improved electrochemical performance, and coated reverse electrodialysis membranes show enhanced ionic selectivity with low bulk resistance.

472029 Organic and Biofouling of Polymeric Membranes: Mitigation Using Surface Graft Polymerization and Assessment with Fouling Analytics

Abstract:

Membrane processes require both mitigation and management of fouling to ensure viable process economics, and performance that meets required standards in terms of selectivity or separation effectiveness.

In this talk we will discuss mitigation approaches that focus on surface modification of poly(ethersulfone) membranes using plasma or photo-initiated graft polymerization. We will show that selection of graft moieties is aided by employing a high throughput screening process, which is able to identify surfaces having antifouling properties for a range of feeds. Examples will be provided for feeds of interest in water purification, including humic substances, extracellular polymeric substances modeled by sodium alginate, and microbial fuel cell effluent. In addition, we will show that biofouling can be controlled using both amines and zwitterionic surfaces (betaines).

Membrane surface modification can be employed to fulfill multiple objectives of reducing flux decline and improving rejection. An example will be presented to show how surface charge can be manipulated to improve the rejection of oxyanions found in contaminated groundwater while preserving antifouling properties toward humic substances.

Fouling analytics â€“ including models that describe the rate of flux decline â€“ can be used to help manage the impact of fouling on membrane processes. In this talk we will show how such analysis can be used to interpret data from screening studies and aid in membrane selection, reduce energy consumption, and optimize cleaning schedules to control the temporal progression of fouling, especially as different fouling mechanisms evolve.

463629 Surface Modification of Ultrafiltration Membranes : Where Are We Now ?

Abstract:

The surface modification of ultrafiltration membranes to increase their resistance to fouling started at the beginning of the 90'. Lots of works have been published, with some of them turned into successful commercial products. If commercial membranes are nowadays more fouling resistant than ever, there is still room for improvement and we can try to understand how this could be achieved. Models borrowed from polymer physics can help understanding the behaviour of hydrophilic moieties attached by one end to a membrane surface. As for an example, a model known as the Alexander's model, further developed by de Gennes and co-workers, suggests that when the grafting density is is high enough, the polymer chains stretch towards the bulk solution and form a so called "brush". Then, the foulant molecules/particles can:
Diffuse through the brush, if their dimension is small compared to the average distance between the grafted polymers on a membrane surface
Compress the brush if they are too large to penetrate it.
This helps to define some criteria for the membrane modification to be efficient. In particular, the trade-off between length of the polymer chains and grafting density can be discussed. A comparison of the model predictions to data of adsorption of proteins on solid surfaces taken from the literature shows the relevance of this model and its limitation. The experimental minimum grafting density required to avoid the adsorption of proteins seems to be larger than the threshold density for which a "brush" of polymer" would form at the surface of the membrane. We analyse, by FTIR microspectrometry, the surface of modified membranes to reveal
a) the heterogeneity of the surface modification at the scale of the millimeter, showing that parts of the surface of the membranes are not covered with the copolymer, even after several hours of contacting the membrane to the solution,
b) that the dynamics of the adsorption of the polymer to the surface is slower than expected.
In cases where the surface modification is unevenly distributed over the surface of the membrane, the existence of a protective brush at the surface of the membranes is also probably unevenly distributed. This may explain this discrepancy between the experimental threshold density and the one predicted by the theory.
In the presence of a filtration flux, this picture can be significantly changed. When comparing adsorption (no flux through) and fouling (in the presence of flux) experiments run with a membrane modified with a given co-polymer, we find that the optimum grafting density in terms of resistance to fouling is larger than the grafting density which gives a high resistance to adsorption.
This can be analysed by accounting for the contribution of the flux to the compression of the brush. By combining the transport equations at the surface of the membrane to the Alexander model, one can show that this contribution can be significant, especially for short polymer length. A higher flux would then require a higher grafting density.

Figure 1: Depending on the fouling molecule/particle radius and on the density of hydrophilic polymers grafted at the surface of a membrane, a molecule/particle can approach the surface either by insertion into the brush formed by the polymers or by compressing the brush. The "resistance" to the compression (V/kT) requires higher grafting densities for small molecules/particles. If the grafting density is lower than a threshold value (around 0.5% here) the grafted polymers are not expected to form a brush but may mask part of the surface of the membrane and reduce adsorption.

Some references:

[1] Surface modification of ultrafiltration membranes by low temperature plasma, II. Graft polymerization onto polyacrylonitrile and polysulfone, M. Ulbricht, G. Belfort, J Membr Sci, 111 (1996) 193-215

[2] S. Alexander, Adsorption of chain colloids with a polar head : a scaling description, J Physique 38(1977) 983-987

[3] S. I. Jeon, J. H. Lee, J. D. Andrade, P. G. de Gennes, Protein-Surface Interactions in the Presence of Polyethylene Oxide, J Coll Interface Sci 142(1991)149-158.

452700 Experimental Validation of Membrane-Vapor-Extraction Separation Process

Abstract:

Our group recently proposed a novel separation process called membrane vapor extraction (MVE), to separate dilute (bio)solutes from an aqueous fermentation broths [1]. In MVE, semi-volatile aqueous solutes vaporize at the upstream side of a membrane, diffuse in through the membrane pores, and subsequently dissolve into a nonpolar solvent highly favorable to the solutes but not to water. Since the two liquids are not in contact, MVE avoids formation of difficult to separate microemulsions.

Design analysis of a 1.5-m long, 30-m^2 membrane-area countercurrent MVE unit for processing 2-wt. % aqueous butanol by a prototype solvent (dodecane) at 40 °C demonstrated over 90 % recovery of the feel butanol with essentially no water loss [1]. The separation factor is well over 1000. Our initial design study, however, gave no experimental evidence for the calculated MVE-unit performance [1].

The present study experimentally validates the MVE process. A 6-cm wide by 10-cm long Teflon channel-flow cell with 0.1 mm gap thicknesses permits countercurrent (or concurrent) extraction studies. The flow cell is operated in the transient mode since axial concentration changes are minimal. 0.2-µm pore-diameter VersaporeR membranes (Pall Corporation, New York) are both hydrophobic (water contact angle = 128°) and oleophobic (dodecane contact angle = 105°). Accordingly, VersaporeR membranes are nonwetting to both feed and solvent, as required for MVE.

Figure 1 shows a typical result for extraction of 2-wt. % aqueous butanol into dodecane at 40 °C. Filled triangles illustrate loss of butanol from the aqueous feed stream, whereas filled circles demonstrate absorption of butanol into the countercurrent dodecane solvent stream. Solid lines represent *a priori* predictions using mass transfer coefficients from the Graetz-Lévêque analysis for laminar channel flow. The excellent agreemen confirms that vapor-phase diffusion resistance is negligible in the MVE-membrane pores. We present similar results for different operating conditions in MVE including feed concentration, temperature, and flow rate. In consort with the initial design analysis [1], our new experimental results confirm the viability of MVE to remove and recover dilute aqueous (bio)solutes with minimal energy expenditure.

1. D.E. Liu, C. Cerretani, R. Telles, A.P. Scheer, S. Sciamana, P.F. Bryan, C.J. Radke, and J.M. Prausnitz, "Analysis of Countercurrent Membrane Vapor Extraction of a Dilute Aqueous Biosolute," AIChE Journal, 61(9), 2795-2809 (2015).

Figure 1. Transient extraction of 2-wt % aqueous butanol (filled triangles) into dodecane (filled circles) using countercurrent MVE at 40 °C. Solid lines are predicted *a priori* with Graetz-Lévêque mass transfer resistances and a measured partition coefficient.

459986 How to Describe and Predict Plasticization of Glassy Polymeric Membranes in Gas Separations

Abstract:

<span">How to describe and predict plasticization of glassy polymeric membranes in gas separations

<span">

<span">Francesco M. Benedetti, Matteo Minelli and Giulio C. Sarti*<span">*

<span">Department of Civil, Chemical, Environmental and Materials Engineering (DICAM)

<span">Alma Mater Studiorum - University of Bologna, via Terracini 28, 40131, Bologna, Italy.

<span">

giulio.sarti@unibo.it

<span">

<span">Plasticization of glassy membranes due to gases or vapors is typically recognized through an increase in gas permeability as the upstream pressure increases. Qualitatively that is associated to an enhancement of polymer chain mobility with consequent increase in penetrant diffusivity, starting from a certain relatively high upstream pressure below which such effects are not present. The phenomenon is normally associated to the sorption of large quantities of low molecular weight species into the polymer matrix, leading to important volume dilation, reduced mechanical rigidity, depression of the glass transition. The increased mobility affects the transport properties of the plasticizing agent as well as of the other gases possibly present in the feed, thus reducing the membrane mixed gas selectivity. The quantitative description of gas permeability and, possibly, the prediction of its dependence on operating pressure is thus relevant for a better understanding of the phenomenon and for the selection of suitable operating conditions.

By considering several different glassy membranes, we have observed that the various different behaviors shown by gas permeability, including plasticization, can be described by considering only a solution-diffusion model in which penetrant molecular mobility varies with its concentration through an exponential law, with two adjustable parameters only. Diffusivity is taken as the product of molecular mobility and a thermodynamic factor, calculated by using the NELF model for thermodynamic properties of the glassy phase.

In particular, the fitting of the only two adjustable parameters of the model to the initially declining branch of the permeability-pressure curve allows the prediction of the plasticization pressure, if existing, by using only the solubility diffusivity model without invoking the onset of any additional physical phenomenon.

Interestingly, we have also inspected the permeability dependence on upstream pressure for CO_2in a commercial polyimide (Matrimid), at different downstream pressure values. The resulting behaviour is indeed unexpected and not fully consistent with the qualitative description recalled above. The model analysis has been thus focused on the experimental data obtained at different downstream pressures, offering a deeper insight on the so-called plasticization phenomenon.

The simultaneous consideration of permeability and solubility isotherms, together with the concentration and swelling profiles in the membrane, reveals that in some cases the plasticization pressure takes place after part of the membrane has turned into rubbery phase; while in other cases, e.g. for Matrimid, plasticization pressure takes place when the entire membrane is still glassy but with a polymer swelling sufficient for a permeability increase.

Finally, it is observed that all parameters used have a defined physical meaning which lead to good general correlations with properties of both polymer and penetrant, based on which permeability predictions can be obtained.

469279 A Phenomenological Model for the Prediction of Granule Drying Times in a Fluid Bed Dryer

Abstract:

A computational model designed to predict the drying times of pharmaceutical granules in a fluid bed dryer is described. The model is comprised of two major components, the first of which utilizes a well-established two-stage phenomenological drying model, and assumes that granule drying consists of evaporation of liquid 'domains' from the granules surfaces followed by liquid evaporation from within the intra-granular porous network. The transient evolution of the granule moisture content is solved by numerically integrating the one-dimensional heat equation. The moving liquid/air interface is solved in a stationary computational domain via a Landau-style mapping. The second component numerically simulates the stochastic nature of the fluidization behavior of granules inside the fluid bed dryer. The combination of this stochastic simulation with the existing phenomenological approach allows for the study of mechanistic correlations between the granule physical characteristics (size, porosity, permeability, and initial moisture content) and the process conditions (e.g., drying air flow rate and temperature, and fluidization "efficiency"). We demonstrate that center-point conditions from an extensive series of process development runs can be used to derive the empirical inputs to the model, and that these inputs can be scaled according to the aforementioned process conditions—ultimately allowing for the prediction of drying times across a broad range of processing conditions.

458414 Practical Application of a Mechanistic Model to Scale up Fluid Bed Drying in Drug Product Development

Abstract:

Fluid bed drying is one of the core processes in pharmaceutical manufacturing of oral dosage forms. For several decades models based on mechanistic understanding of the fluid bed drying process have been used. In practice however, in the pharmaceutical industry extensive experimentation is still carried out during drug product development, to investigate fluid bed drying of wet granules at different operating conditions and equipment scales. Typically experimentation is guided by Design of Experiments and results are studied by multivariate analysis. This procedure does not take advantage of the existing fluidized bed drying theory that is already available, and that would allow to reduce the number of experiments performed, at the same time providing better process understanding and more reliable scaling to different operating conditions [1].

In this work a case study is presented, the goal of which was to show that it is possible to use a mechanistic model to scale up fluid bed drying of granules containing active pharmaceutical ingredients (API) from the laboratory to the pilot scale. To this end, two different formulations containing two model APIs were chosen and granulated via high shear wet granulation. The obtained wet granules were then dried in a fluid bed dryer at the laboratory scale. The fluid bed drying model available in the software gSolids v.4.1 (PSE, London, UK) [2, 3], was calibrated for the two formulations with the laboratory scale experimental results. The model was then used to predict the fluid bed drying process of the two formulations at pilot scale. Experiments were run at the pilot scale and model predictions could be confirmed without further adjustments.

It was so demonstrated that it is possible to scale up the fluid bed drying process of API containing granules from laboratory to pilot scale by using a mechanistic model. A workflow was established based on the experience gathered in the case study that details experimental data needed for model set up and calibration and possible solutions to overcome challenges in using the model to predict the process at pilot scale.

It can be concluded that using a mechanistic model constitutes a viable alternative to empirical modelling allowing the saving of API during drug product development by reducing experimentation at large scale.

References

[1] Kemp, I. (2011) Drying models, myths and misconceptions. *Chem. Eng. Technol.,* **34** (7), 1057-1066.

[2] Burgschweiger, J., Groenwold, H., Hirschmann, C. And Tsotsas, E. (1999) From hygroscopic single particle to batch fluidized bed drying kinetics. *The Canadian Journal of Chemical Engineering* **77**, 333-341.

[3] Burgschweiger J, Tsotsas E (2002) Experimental investigation and modelling of continuous fluidized bed drying under steady-state and dynamic conditions. *Chem. Eng. Sci.*, **57**, 5021-5038.

461578 Modelling of Co-Rotating Twin-Screw Extruders in the Pharmaceutical Industry I: Single Component Model

Abstract:

Modelling of co-rotating twin-screw extruders in the pharmaceutical industry I: single component model

Pavol Rajniak[1], Maria Fuentes-Gari[1], Niall Mitchell[1], Tariq Islam[2], John Robertson[2] and Sean K. Bermingham[1], (1)Process Systems Enterprise, London, United Kingdom, (2)CMAC, Strathclyde University, Glasgow, United Kingdom

Twin-screw extruders are being increasingly applied in the pharmaceutical industry for the manufacture of solid dispersions. In particular, Hot Melt Extrusion (HME) is a viable manufacturing alternative for poorly soluble drugs that are difficult to process. This is due to the high shear stress applied in the process, which enhances mixing of the base polymer with the dispersed API. Delivery systems that can be obtained using this technology include pellets, granules, sustained release tablets and implants. Experiments for HME are typically very labour-intensive, involving the use of highly viscous polymers, high pressures, and require proper cleaning between runs. Identifying the critical experiments to perform based on model simulations would thus be highly desirable. Single-component modelling of twin-screw extrusion processes could be beneficial in the following areas: identifying optimal screw configurations, tracking the degree of melting and when identifying the area within the extruder where the polymer becomes completely melted.

An axially distributed mechanistic model of HME was previously developed by Eitzlmayr et al.[i], wherein the extruder elements were described using physically-meaningful parameters estimated from experimental data and any given extruder could be built into the model in a modular way. Their model was implemented into gPROMS

FormulatedProducts®, and the simulation results obtained for different HME configurations using Soluplus® were reproduced. Important process variables such as filling ratio or element pressure, which are often difficult to measure, were given as simulation outputs; measurable variables, including the temperature of the screws and the melt and residence time distribution, were validated against experimental results obtained in a 40:1 ThermoFisher Eurolab co-rotating twin-screw extruder at CMAC. In addition, experiments with a chosen extruder configuration operating under fixed conditions were carried out with Soluplus®, Plasdone™ and Affinisol™. The effect of polymer nature on screw element parameters (A1, A2 model parameters) was investigated by comparing the results obtained across the three polymers.

Overall, the existing model was proven to perform well for the single component polymer case. Further work focuses on extending this model to account for multi-component mixtures and capturing API solubility.

i. Eitzlmayr, A., Koscher, G., Reynolds, G., Huang, Z., Booth, J., Shering, P. and Khinast, J. (2014) "Mechanistic modeling of modular co-rotating twin-screw extruders", *International Journal of Pharmaceutics,* 474(1), 157-176.

**Corresponding Author: m.fuentesgari@psenterprise.com*

464894 Modeling and Optimization of Agglomeration in Suspension through a Coupled Population Balance Framework

Abstract:

Since its introduction by Randolph and Larson[1] (1971) and the solution methods described by Ramkrishna[2] (2000), the population balance model (PBM) has been widely used and accepted as model formulation method for simulation and prediction of the size distribution and other properties of particulate systems. PBMs allow for systems that include any or all of the following mechanisms: nucleation, growth, breakage and agglomeration. Following the initial work by Smoluchowski[3] (1917) on the rate of aggregation for spherical particles, there have many contributions for systems that exhibit agglomeration including dispersion (bubble) coalescence[4,5], granulation[6,7] and particle aggregation during crystallization[8,9,10]. The commonality in the limitations of many of the previous studies were physically irrelevant and/or empirically based agglomeration kernels, difficulty in assessing the influence of process conditions (*e.g.* hydrodynamics, particulate physical properties), solution method efficiency for optimization and control applications, and loss of information of constituent particles. These limitations present obstacles for the simulation, optimization, and control of the increasingly popular agglomeration technique of spherical crystallization.

Peña and Nagy[11] studied and showed the benefits of spherical crystallization as a process intensification technique whereby both internal (primary crystals) and external (agglomerates) properties can be controlled experimentally through a decoupled continuous spherical crystallization (CSC) approach; providing the means by which both biopharmaceutical (bioavailability, dissolution) and manufacturing (flowability, filtration, drying) properties can be simultaneously adapted to meet desired quality specifications. This technique opens the door for combined experimental and modeling approaches for the optimization and control of both the primary crystal and agglomerate properties in spherical crystallization processes. However, many of the PBMs currently in literature would fail to accomplish this because of the aforementioned limitations and loss of information.

To overcome the issues of loss information a coupled PBM formulation is required. A coupled PBM formulation could simultaneously track the evolution in the primary crystals and the evolution of the agglomerates. The relationship between primary crystal properties and their effect on final agglomerate properties would thereby be more evident and more efficient than traditional approaches. To the best of our knowledge, the only previous work that presented this approach is that of Ochsenbein et al.[12]. In their study, Ochsenbein et al.[12] focused on the agglomeration of needle-like crystals in suspension. Through a coupled PBM framework Ochsenbein et al.[12] were able to develop a population balance equation (PBE) to describe the evolution of the primary crystals by a two-dimensional growth rate to represent the needle like structure of the crystal. They then used another PBE to describe the evolution of the agglomerates as a function of the primary crystals. For the agglomeration kernel, they derived their own modified kernel that include both characteristic lengths of the primary crystals participating in the agglomeration. The new PBM formulation also allowed for the development of new parameters that add value to the simulations due to their experimental relevance. However, the work of Ochsenbein et al.[12] neglected nucleation something common to previously developed agglomeration models. The coupled population balance model framework will be extended herein.

The contribution of this work is the extension of the coupled PBM framework for application in the simulation and optimization of a spherical crystallization system. A coupled PBM framework has been developed for a semi-batch, reverse addition, anti-solvent crystallization system with agglomeration. The system includes nucleation and growth the primary crystals and subsequent agglomeration. The purpose of the work is to exploit the advantages presented by a coupled PBM framework; for example, the ability to optimize for specific primary and agglomerate sizes. This presents an opportunity to find optimal operating conditions that meet both bioavailability and manufacturability demands. It also allows for the ability to develop first principles based parameters for agglomeration efficiency and porosity. Additionally, through the retention of the information of the primary particles the interplay between the effects of operating conditions on the properties of the primary crystals versus the agglomerates will be clear.

1. Randolph, A.; Larson, M. Theory of particulate processes; analysis and techniques of continuous crystallization. 1971. New York: Academic Press.
2. Ramkrishna, D. Population balances theory and applications to particulate systems in engineering. 2000. San Diego, CA: Academic Press.
3. Smoluchowski, M. V. Z. *Physical Chemistry*, 1917, 19, 129–168.
4. C. a. Coulaloglou and L. L. Tavlarides. Description of interaction processes in agitated liquid-liquid dispersions. *Chemical Engineering Science*, 1977, 32(11), 1289–1297.
5. M. J. Prince and H. Blanch. W. Bubble coalescence and break-up in air-sparged bubble columns. *AIChE Journal*. 1990, 36(10), 1485–1499.
6. Simon M. Iveson. Limitations of one-dimensional population balance models of wet granulation processes. *Powder Technology*, 2002, 124(3), 219-229.
7. L. X. Liu and J. D. Litster. Population balance modelling of granulation with a physically based coalescence kernel. *Chemical Engineering Science*, 2002, 57(12), 2183-2191.
8. P. Marchal, R. David, J.P. Klein, and J. Villermaux. Crystallization and precipitation engineering—I. An efficient method for solving population balance in crystallization with agglomeration. *Chemical Engineering Science*, 1988, 43, 59–67.
9. R. David, P. Marchal, J.P. Klein, and J.V. Klein. Crystallization and Precipitation Engineering III. A Discrete Formulation of the Agglomeration Rate of Crystals in a Crystallization Process. *Chemical Engineering Science*, 1990, 46(1), 205–213.
10. Sanjeev Kumar, D. Ramkrishna, On the solution of population balance equations by discretization—III. Nucleation, growth and aggregation of particles. *Chemical Engineering Science*, 1997, 52(24), 4659-4679.
11. R. Peña and Z. K. Nagy. Process Intensification through Continuous Spherical Crystallization Using a Two-Stage Mixed Suspension Mixed Product Removal (MSMPR) System. *Crystal Growth & Design*, 2015, 15(9), 4225-4236.
12. D.R. Ochsenbein, T. Vetter, S. Schorsch, M. Morari, and M. Mazzotti. Agglomeration of Needle-like Crystals in Suspension: I. Measurements. *Crystal Growth & Design*, 2015, 15(4), 1923–1933.

457293 Modelling of Water Absorption By a Pharmaceutical Tablet

Abstract:

Aqueous film coating is a crucial step in the manufacture of solid-dosage drugs in pharmaceutical industries. It is well understood that the shelf life of pharmaceutical tablets depends on the amount of humidity to which they are exposed during the coating process, the handling of the intermediate coated product and the packaging. Understanding, and being able to predict the mechanisms of water absorption onto and into tablets, is therefore important, to avoid accelerating the degradation mechanisms caused by high water content.

Mathematical models that efficiently simulate the overall coating process have already been developed. Recent models permit calculating the size of the spray droplets before impact, simulating the tablet movement inside the coater and predicting coating uniformity. However, there is little, if any, fundamental understanding of the sorption phenomena at the point of contact between the coating droplet and the porous tablet. This information is essential to perform a full-system global optimisation and determine the optimal design of the tablet.

This work aims to define and describe the driving forces that play a role in water-based coating sorption after impingement on a porous substrate. We aim to demonstrate two mathematical models. The first takes into account the spreading and sorption phenomena, whereas the second predicts the evaporation rate of the water absorbed in the porous matrix of the tablet. The ambient conditions; such as temperature, pressure and humidity, are set to be similar to the conditions present inside a pan-coater.

The spreading sorption model is divided into two separate phases: the kinematic phase and the capillary phase. During the first, inertial forces govern the spreading, while during the second, capillary forces become dominant. To acquire results for the kinematic phase, we developed a 1D model that solves the energy balance equation for the first milliseconds after impact. The results for the capillary phase come from the solution of the Navier Stokes equation, obtained using the lubrication approximation theory.

For the second model, of the evaporation process, we adopted a model for the second drying stage of a slurry droplet inside a spray dryer. During this stage, we can no longer describe the droplet as a liquid system containing solids, having to regard it as a wet particle with a dry crust and a wet core. The dry core represents the surface of the tablet, while the wet core represents the wetted areas inside the tablet.

The information provided by these mathematical models, is important to understand the spreading, absorption and evaporation phenomena and predict the moisture content inside the tablet after the impingement of a liquid coating droplet. We implemented the models in gPROMS, employing the Modelbuilder modelling platform. The results (droplet height, spreading and wetting front profile) are in good agreement with recent experimental data and CFD simulations.

455995 Design Problem of Pharmaceutical Manufacturing Processes

Abstract:

The modern pharmaceutical industry is requested to realize superior manufacturing processes regarding multiple aspects such as quality, economy, supply reliability, and many others. To this end, the setup of design problem is critical because the choice of design variables, constraints and objective functions determines the capability of the process to be realized. The aim of this presentation is to explore the characteristics of design problem of pharmaceutical manufacturing processes based on the two research examples from the presenterâ€™s group.

The first example is on the design of pharmaceutical tablet manufacturing processes considering the choice between batch and continuous technologies. While pharmaceutical tablets are usually manufactured in batch processes, continuous manufacturing technology is in intensive development in both industry and academia. In this work, a decision-support method was proposed for the choice between two technologies with four steps of: (i) process modeling, (ii) evaluation, (iii) sensitivity analysis, and (iv) interpretation. The method was applied to the following design situation, i.e., â€œa company launches a new product which is supposed to be marketed for 30 years. The tablet product can be manufactured by either technology which needs to be specified based on the economic evaluation.â€ After performing the four steps of method, a better process regarding net present cost could be suggested depending on the API price as well as the maximum demand quantity over the production period of 30 years.

The second example is on the design of solvent recovery processes in API manufacturing. Generally in API manufacturing, various types and large amount of organic solvents are used, which raises concerns regarding environment, health and safety (EHS) risks. In this work, an actual API manufacturing was investigated, and a process was designed for recovering tetrahydrofuran (THF) from the azeotropic mixture with methanol and water. The objective functions were set as the net present value as well as EHS risk evaluation scores. As the process, four alternatives were generated regarding the core separation technology, i.e., (a) zeolite membrane (b) pressure-swing distillation (c) azeotropic distillation, and (d) entrainer. The target THF concentration was set as 99.95 wt% in the initial evaluation, which was in the secondary evaluation relaxed to 95 wt%, a sufficient concentration in the actual API manufacturing. The initial multiobjective evaluation identified the alternatives (a), (b) and (d) as Pareto optima, among which, the alternative (b) was the best in all aspects in the secondary evaluation.

The outcomes in these two research examples were sensitive to the setup of the design problem, i.e., design variable, constraint and objective function. In the first example, treating the choice of batch and continuous as a design variable was meaningful because the superior technology found was depending on the API price and the product demand. In the second example, different results were obtained by regarding the target concentration changeable, which would be considered fix in reality due to the required effort of change management in GMP. The design problem of pharmaceutical manufacturing process has large degree of freedom in nature, however in practice, many variables are considered as hard constraints or determined on an empirical basis. In this regard, MSO can be regarded as the supportive technology for investigating wide range of items in a speedy manner in the design phases of pharmaceutical manufacturing processes.

467775 Multi-Scale Modeling in Pharmaceutical Drug Substance Manufacturing

Abstract:

The role of modeling in pharmaceutical drug substance manufacturing is becoming increasingly more important as the regulators push to the use of more sophisticated methods and tools to address process development and manufacturing. Those who adopt the best modeling approaches benefit from such decision in terms of right-first-time and confidence resulting from enhanced process understanding.

This work presents a multi-scale modeling paradigm applied to drug substance manufacturing. Starting at the molecular level, it is shown how Ab initio modeling (Density Functional Theory) can be useful for determining the IR spectra of the molecules that in specific cases can facilitate analytical work. As the modeling scale is increased, CFD simulations are shown to be key to study in detail the flow patterns inside a reactor. At the unit operation level, Dynochem is used to model the reaction kinetics and crystallization steps, as well as the work-up part of the process. Process trains are then addressed resorting to SuperPro Designer flowsheet software that can be useful for determining bottleneck and COGs. In the next modeling scale scheduling of a multi-process train facility is simulated with SchedulePro, that includes details also on operators and stocks. Finally, company-wide capacity (multi-site) is assessed by making use of Monte-Carlo stochastic simulations.

468026 A SimultaneousÂ Approach for Kinetic Parameter Estimation and Curve Resolution Based on Spectroscopic Measurements

Abstract:

Identification of reaction kinetics plays an important role in the development of the commercial product and process development for chemical and bio-chemical systems. Precise knowledge of kinetic parameters is critical in the reactor scale-up from laboratory to industrial levels and the design of robust, controllable and safe processes. Spectroscopic techniques are commonly used for chemical reaction monitoring especially within ultraviolet-visible, infrared and near-infrared wavelength regions; these measurements provide vast amounts of data related to the concentration of the species in the reaction medium. Ample research has been conducted in developing techniques to resolve the information in the spectral data, and identify the kinetic parameters of the underlying chemical reaction (MCR-ALS).

This work describes an efficient, unified framework for kinetic parameter estimation and curve resolution that is based on maximum likelihood principles and a simultaneous collocation approach that leads to improved estimates of parameters.

This study considers the dynamic equation model for the reaction kinetics as a stochastic differential-algebraic equation system,Â $dc = f(c, y, q) + k\, dW(t)$, $g(c, y, t) = 0$, where $c(t)$ is the concentration vector at time t, $y(t)$ are additional algebraic states, q is the kinetic parameter vector and $W(t)$ is the reaction system noise described by standard Brownian motion or Wiener processes. The spectroscopic measurements $d(t, l)$ are governed by Beer-Lambertâ€™s law, $d(t, l) = S_k\, c_k(t)\, s_k(l) + z(t, l)$, with absorbance $s_k(l)$ for component k and wavelength l, and measurement noise $z(t, l)$.

Determination of q by parameter estimation requires consideration of a dynamic optimization problem derived from maximum likelihood principles, and this leads to two important challenges. First, an efficient and reliable optimization strategy is required, even for kinetic systems that exhibit open loop instability and dependent kinetic equations. We successfully address this task by applying a direct transcription nonlinear programming (NLP) approach that determines the optimal kinetic parameters and the concentration profiles simultaneously. Second, since only spectroscopic measurements are available, covariances and other distributional information for $W(t)$ and $z(t, l)$ cannot be determined directly. Instead, we apply maximum likelihood principles in order to derive the covariances needed for parameter estimation. Coupled solution of these two tasks leads to an iterative optimization-based approach that provides covariance estimates and initializes the kinetic parameters. These results allow the subsequent solution of the parameter estimation problem with fixed covariance terms. Confidence regions for the kinetic parameter estimates are then obtained through the application of NLP sensitivity, a byproduct of the direct transcription approach.

We demonstrate our proposed approach on six case studies, including a comprehensive treatment of aspirin kinetics. These cases include simulated as well as actual experimental datasets. Detailed numerical results for these cases are presented and compared with the MCR-ALS GUI 2.0 toolbox. These show significant improvements in performance and computational cost over state-of-the-art approaches.

465303 Application of Computational Chemistry in Pharmaceutical Process Development

Abstract:

Increasing speed of drug discovery and chemical development often requires complementing experimental data with a variety of modeling tools to provide fundamental knowledge at a molecular level. With advances in computational power and theoretical techniques (e.g. density functional theory), implementation of quantum mechanical (QM) calculations for systems relevant to small molecule pharmaceutical process development has become more readily available. Process design and understanding in several areas of chemical development are significantly enhanced by utilization of QM modeling. In this work, several industrially relevant case studies will be presented to provide an overview of QM modeling and its applications in both early and late phase development. The case studies will discuss QM calculations in three key areas related to reaction and crystallization development: a) evaluation of relative stabilities and activation energies (pKa estimation, conformational analysis, transition structures); b) investigation of parallel reaction pathways (chemo- and stereo-selectivities); c) spectral predictions for reactive intermediates (NMR, FTIR, Raman). Impact to process understanding and design gained from computational chemistry modeling will be highlighted.

466350 Systematic Optimisation of Continuous Separation for Artemisinin Recovery in Continuous Pharmaceutical Manufacturing

Abstract:

An environment of growing R&D expenditure and increasing competition from generic manufacturers has led to a sustained research interest in Continuous Pharmaceutical Manufacturing (CPM), which has emerged as a promising alternative to the current paradigm of batch production. The latter, while possessing benefits including equipment flexibility and the extensive industry know-how of a mature technology, is hampered by disadvantages including low efficiency (of both material and energy use), problematic process scale-up and poor heat and mass transfer. Moreover, its status as a mature technology means improvements are incremental at best. Novel CPM methodologies provide enticing options toward fostering the concept of Quality by Design (QbD) which is enthusiastically pursued by the US pharmaceutical industry, and the closely related development of Process Analytical Technology (PAT), which focuses on efficient, online pharmaceutical process monitoring and control. The concepts of CPM, QbD and PAT have accordingly received attention and encouragement by influential, world-leading regulatory authorities (Lee et al., 2015).

Artemisinin is one of the most important antimalarial substances available today. First identified and isolated from the plant *Artemisia annua* in the late 1970s (Tu, 2011), work for which the *2015 Nobel Prize in Physiology/Medicine* was awarded, artemisinin is currently produced via batch extraction from the cultivated plant. However, long product lead times, and fluctuating demand due to unpredictable burden of malaria, lead to highly variable prices and production levels (Jolliffe and Gerogiorgis, 2015a). Recent research has demonstrated the continuous synthesis of artemisinin (Kopetzki et al., 2013; Seeberger et al., 2014). This novel synthesis uses a waste product from the current batch production process as a feedstock, and produces artemisinin via two sequential plug flow reactors, the first of which employs photo-oxidation. Remarkable product yields can be achieved, and the process has been further adapted to produce several artemisinin derivatives (Gilmore et al., 2014).

Critical to harnessing the full potential of CPM is the design and employment of continuous product separation systems. A recent published study of ours has explored the potential performance of eight solvents for continuous crystallisation of artemisinin via both computationally predicted (UNIFAC) and experimentally reported solubilities; results indicate promise for the use of ethanol or ethyl acetate (Jolliffe and Gerogiorgis, 2015a, 2016). We have also previously employed nonlinear optimization toward finding ideal continuous separation scenarios for another API: promising temperatures, solvent and liquid-liquid extraction (LLE) configurations have been determined toward minimising the total CPM process cost, by means of comprehensively considering mass transfer, thermodynamics and detailed economic cost modelling (Jolliffe and Gerogiorgis, 2015b).

The present paper expands upon our previous contributions in the field of CPM: we investigate further the continuous separation of artemisinin by formulating a nonlinear optimization problem comprehensively considering mass transfer modelling via empirical correlations, solubility prediction via group contribution methods (UNIFAC, NRTL), and environmental and sustainability metrics (E-factor, Mass Intensity), for several potential separation unit operations and solvents. The CPM process studied here, based on the chemistry developed by Kopetzki et al. (2013), has been shown to be compatible with a range of separation methods (Horváth et al., 2015). The design space illustrates clear nonconvexities for several key process variables, and potential local minima in the final objective surface arise from regions of applicability of empirical relations. The solved problem and our novel results illustrate how the crucial consideration of less easily quantifiable characteristics (e.g. environmental impact, safety, toxicity) can qualify optimised solutions of superior technical performance.

LITERATURE REFERENCES

1. Gilmore, K., Kopetzki, D., Lee, J.W., Horváth, Z., McQuade, D.T., Seidel-Morgenstern, A., Seeberger, P.H., 2014. Continuous synthesis of artemisinin-derived medicines. *Chem. Commun.* **50**: 12652–12655.

2. Horváth, Z., Horosanskaia, E., Lee, J.W., Lorenz, H., Gilmore, K., Seeberger, P.H., Seidel-Morgenstern, A., 2015. Recovery of artemisinin from a complex reaction mixture using continuous chromatography and crystallization. *Org. Process Res. Dev.* **19**: 624–634.

3. Jolliffe, H.G., Gerogiorgis, D.I., 2015a. Process modelling and simulation for continuous pharmaceutical manufacturing of artemisinin. *Chem. Eng. Res. Des.* (in press, available online).

4. Jolliffe, H.G., Gerogiorgis, D.I., 2015b. Technoeconomic optmisation for green Continuous Pharmaceutical Manufacturing (CPM) of analgesics, *Proceedings of the AIChE Annual Meeting*, Salt Lake City, UT, USA.

5. Jolliffe, H.G., Gerogiorgis, D.I., 2016. Systematic solvent evaluation for artemisinin recovery in continuous pharmaceutical manufacturing, *Comput.-Aided Chem. Eng.*(in press, available online).

6. Kopetzki, D., Lévesque, F., Seeberger, P.H., 2013. A continuous-flow process for the synthesis of artemisinin. *Chem.–Eur. J.* **19**: 5450–5456.

7. Lee, S.L., O'Connor, T.F., Yang, X., Cruz, C.N., Chatterjee, S., Madurawe, R.D., Moore, C.M.V., Yu, L.X., Woodcock, J., 2015. Modernizing pharmaceutical manufacturing: from batch to continuous production. *J. Pharm. Innov.* **10**(3): 1–9.

8. Seeberger, P.H., Kopetzki, D., Levesque, F., 2014. *Method and device for the synthesis of artemisinin.* US20140364630 A1.

9. Tu, Y., 2011. The discovery of artemisinin (qinghaosu) and gifts from Chinese medicine. *Nat. Med.* **17**: 1217–1220.

468494 Prediction of Liquid-Liquid Partitioning and Solubility of APIs with the SAFT-γ Mie Approach

Abstract:

Lipophilicity is a fundamental physicochemical property in drug discovery and is usually quantified from the partitioning equilibrium of solute molecules between water and an immiscible organic solvent. The most popular system in this context is the 1-octanol + water mixture. The partitioning of solutes in aqueous alkane systems (*e.g.*, hexane + water and hexadecane + water) is also of interest for drug discovery and the difference between the partition coefficients serves as a measure of the hydrogen bonding potential of a molecule. The solubility of APIs in both aqueous and organic solvents is an essential property for drug development and manufacturing. It serves as essential information in the design of the downstream processes of APIs.

A general approach to predict the solubility and partition coefficients in octanol+water (K_{OW}) and alkane+water (K_{alk}) systems is developed. The core of our predictive framework lies in the use of the recently developed group-contribution SAFT-γ Mie approach [1-4]. SAFT-γ Mie is a predictive equation of state which allows one to determine the thermo-physical properties of molecules in terms of the constituent functional groups that represent their unique molecular identity. These functional groups are then modelled as fused spherical segments that interact via Mie (generalized Lennard-Jones) potentials of variable repulsive and attractive ranges. The parameters for each functional group are developed from fluid-phase equilibrium data for simple compounds and, once estimated, they are applied to the study of more complex multifunctional molecules in a predictive manner. This key feature of the SAFT-γ Mie approach enables the prediction of the properties of organic molecules in various solvents and solvent mixtures over wide thermodynamic conditions without the need for experimental solubility data for a specific molecule.

The necessary alkane-water and octanol-water group interactions are established providing an excellent description of the partition coefficients of several families of organic compounds which are used to test the

reliability of the models. The SAFT-γ Mie models are then applied to predict the K_{OW} and solubility of various APIs in different organic solvents; the results are found to be in excellent agreement with the experimental data. The APIs employed in this study range from a relatively simple molecule, azelaic acid, to multifunctional molecules, such as ibuprofen, ketoprofen, lovastatin, and simvastatin. No experimental data related to the API are used in the K_{OW} prediction. For the solubility calculation, in addition to the molecular structure, a few properties of the pure API are required, namely the melting temperature, heat of fusion, and the heat capacity difference between the liquid and the solid state of API. This work highlights the extension of the SAFT-γ Mie equation of state in predicting important thermodynamic properties of complex multifunctional molecules in different solvents and demonstrates that the appraoch can be employed as a novel predictive platform for the prediction of important API properties from molecular structure alone.

References

[1] T. Lafitte, A. Apostolakou, C. Avendaño, A. Galindo, C. S. Adjiman, E.A. Müller, and G. Jackson "Accurate statistical associating fluid theory for chain molecules formed from Mie segments" The Journal of Chemical Physics, 2013, 139: 154504

[2] V. Papaioannou, T. Lafitte, C. Avendaño, C. S. Adjiman, G. Jackson, E. A. Müller, and A. Galindo "Group contribution methodology based on the statistical associating fluid theory for heteronuclear molecules formed from Mie segments" The Journal of Chemical Physics, 2014, 140: 054107

[3] S. Dufal, V. Papaioannou, M. Sadeqzadeh, T. Pogiatzis, A. Chremos, C. S. Adjiman, G. Jackson, and A. Galindo "Prediction of thermodynamic properties and phase behavior of fluids and mixtures with the SAFT-γ Mie group-contribution equation of state" Journal of Chemical & Engineering Data, 2014, 59: 3272-3288

[4] S. Dufal, T. Lafitte, A. J. Haslam, A. Galindo, G. N. Clark, C. Vega, and G. Jackson, "The A in SAFT: Developing the contribution of association to the Helmholtz free energy within a Wertheim TPT1 treatment of generic Mie fluids" Molecular Physics, 2015, 113: 948-984

Acknowledgements

We gratefully acknowledge financial support from Pfizer Inc.

461660 Use of a First Principles Model during QbD Studies to Assess Impurity Formation for a Coupling Reaction

Abstract:

The overall strategy that was used in QbD studies for the evaluation of reactions in the synthesis of an API will be assessed by reviewing the specific case for the coupling reaction of an amine with an acid chloride to form the corresponding amide. For each reaction in the experimental design set, a model was developed based on kinetic data that was acquired. Kinetic parameters were then obtained for each reaction and were assessed as a set to determine if the studied variables impacted impurity formation over their desired manufacturing range. In the reaction presented, two main impurities are tracked: 1) an imide impurity is formed from the product and excess acid chloride and 2) a carboxylic acid impurity is formed by hydrolysis of the acid chloride. The formation of impurities was investigated varying the amount of acid chloride, reagent level, catalyst loading, solvent amount and temperature. The resulting model was used to calculate impurity levels at reaction completion for a variety of conditions. The development and use of the kinetic model during QbD studies allowed for a deeper understanding of the reaction over its entire time course and enhanced confidence in how to control the formation of impurities.

461275 Verification and Use of CFD Modeling in Stirred Tank Reactors for Pharmaceutical Applications

Abstract:

Mixing often determines the success or failure of pharmaceutical process scale-up. Many unit operations, including heterogeneous reactions, fed-batch processes, liquid-liquid extractions, and crystallizations are susceptible to changes in performance due to variable mixing. To predict mixing performance between scales, we have adopted a computational fluid dynamics (CFD) modeling approach using MixIT, a CFD package specialized for mixing in stirred tanks. In this talk, we present a comparison of CFD modeling with experimental measurements.Â To assess the model's accuracy, we measured agitator power consumption, a representative mixing attribute in stirred tanks, up to kilo-scale and compared the measurements with the model's predictions. The measurements were taken across a variety of common agitator types and baffle configurations, with a wide range of power per unit volume values. The results demonstrated good agreement between experiments and simulations, showing the modelâ€™s accuracy in predicting mixing attributes. We then demonstrated the use of CFD to predict performance of a mixing sensitive liquid-liquid reaction as it was scaled up to 20L. In this example, reaction monitoring with process analytical technology (PAT) was combined with CFD predictions to develop correlations between mixing attributes and the mass transfer coefficient.

469808 Product-Agnostic Computational Fluid Dynamic (CFD) Model for Single-Use Mixing System

Abstract:

As the interest in single-use systems continues to grow in the biopharmaceutical community, mixing is a phenomenon that needs to be understood and characterized routinely, taking effort and time. Insufficient mixing can lead to content inhomogeneity, while excessive mixing can lead to protein denaturation as well as an increase in cycle time. To help reduce experimentation for future process transfer and scale-up, a product-agnostic multiphase computational fluid dynamic (CFD) model was developed. The CFD model establishes a process design space for 10L and 50L single-use mixing bags. The design space can be used by product teams to define impeller speeds in order to achieve the desired blend times based on working volumes of fluid, and fluid properties. These studies will accelerate the design and use of single-use stirred bags to formulate drug solutions.

455964 Fast, Efficient and Gentle Transfection of Human Adherent Cells in Suspension

Abstract:

The current strategy for delivering genetic materials to adherent cells using gene delivery vehicles (GDVs), involves incubating the GDV solution over a cultured cell monolayer, and its performance is limited by slow mass transport of GDVs to the cell surface. We demonstrate a novel suspension-based transfection method that provides efficient delivery of genetic materials to clinically relevant human cell types, such as induced pluripotent stem cells and fibroblasts, using lipid- and polymeric-based commercial transfection reagents (FugeneHD and Glycofect). Our method reduces the time required for transfection by one full day. The reason behind the success of our method is that suspension transfection leads to faster transport of GDVs to the cell surface. This resulted in maximum GDVs attachment to cell surface within 15 minute of transfection time, with atleast 2-fold higher binding efficiency of GDVs as compared to the plating method. Our method is also compatible with common techniques to improve gene delivery, such as cell cycle synchronization (6-fold increase in gene expression compared to control), facilitating easy adaptability in existing setups. Furthermore, to preserve cell physiology during transfection in suspension, we designed a microfluidic version of our method where the transfection takes place in a flow. In this device, the GDV and cell contact time is below one minute, and we achieved high binding of GDVs to cells (>90% GDV-bound live cells) and significant gene expression in several human cell types with minimal effect due to the device flow rates. The fast, efficient and generally applicable transfection methodologies that we will present can be used for rapid screening of different delivery systems and have significant potential for scale-up and high-throughput cell therapy applications.

Keywords: pluripotent stem cells, fibroblasts, suspension transfection, flow cytometry, microfluidics

462134 Poly(β-amino ester) Terpolymer Nanoparticles As Delivery Vehicles for mRNA and DNA

Abstract:

Introduction: A host of human diseases are caused by missing or aberrant proteins. Effective treatment of the root cause of these maladies therefore requires replacement of such proteins. The most common means of protein replacement therapy is simply to administer the protein itself, but protein stability concerns often limit the effectiveness of this strategy. Alternatively, one can deliver nucleic acids coding for the aberrant protein. Nucleic acids are more easily stabilized than proteins, and can be condensed and encapsulated within delivery vehicles which can facilitate their effective *in vivo* transport. To date, DNA has been the most-studied nucleic acid for protein replacement therapy due to its relatively high stability and low immunogenicity when compared to mRNA. However, mRNA does not pose a risk of insertional mutagenesis, and it need only reach the cytoplasm to be effective. Additionally, recent innovations in mRNA synthesis have yielded more stable and potentially less immunogenic nucleic acids, further increasing the interest in delivering mRNA for protein replacement therapy. Thus, there is now a significant interest in designing materials to deliver mRNA *in vivo*. Previously, our lab reported the synthesis of poly(Î²-amino ester) terpolymers (PBAE terpolymers), copolymeric materials capable of electrostatically condensing DNA into nanoparticles that can additionally be formulated with other hydrophobic excipients. In this work, we utilize PBAEs to deliver mRNA and DNA *in vitro* and *in vivo.Â*

Materials and Methods: PBAE terpolymers which had previously been synthesized for DNA delivery were synthesized via step polymerization, with synthesis conditions optimized for purification in diethyl ether to remove excess monomer. PBAE terpolymer nanoparticles were formed by nanoprecipitation via charge-based interactions with nucleic acids, as well as hydrophobic interactions between the PBAE terpolymer and a lipid-anchored PEG (PEG-lipid). Particle serum stability was assessed by doping nanoparticles with fetal bovine serum and measuring optical absorbance of the particle/serum solution over a 2-hour span at 37°C. *In vitro* nanoparticle efficacy was assessed by

formulation PBAE nanoparticles with luciferase-encoding mRNA or DNA and transfecting HeLa cells, with cellular luminescence measured as an indicator of luciferase protein production. *In vivo* experiments were performed by delivering nanoparticles intravenously to C57/BL6. In a typical example, a mouse was dosed with 0.5 mg/kg of luciferase-coding mRNA loaded into a PEGylated PBAE terpolymer, and luminescence was quantified after 24 hours using an IVIS imaging apparatus.

Results and Discussion: We demonstrate that PBAE terpolymers can also be used to encapsulate and deliver mRNA both *in vitro* and *in vivo*. Specifically, we show that mRNA-loaded terpolymer nanoparticles can be co-formulated with (PEG-lipid) to improve the serum stability of terpolymers, as well as to increase *in vivo* potency. Nanoparticles that were stabilized *in vitro* were also successful in delivering nucleic acid to the lungs following intravenous delivery to mice, although biodistribution experiments revealed that nucleic acid localization was spread throughout major organs. Additionally, we utilize PBAEs as a platform to compare the protein production of mRNA and DNA. Our data shows that certain aspects of PBAE structure, especially the diacrylate used to form the polymer backbone, influence whether the material will be more successful for mRNA or DNA delivery. We further compare the role of other factors, such as polymer molecular weight and formulation parameters, for PBAE-mediated mRNA and DNA delivery. Furthermore, we plan on taking advantage of the *in vivo* potency of PEGylated PBAE nanoparticles to compare the potency of DNA and mRNA-based protein production.

Conclusions: In total, our research has shown that PBAE terpolymers are a therapeutically-relevant material for nucleic acid delivery. PBAEs are also useful given that they can deliver both mRNA and DNA. Their ability to deliver both nucleic acids successfully, along with their facile synthesis, can be leveraged to investigate chemical and formulation parameters that favor delivery of one nucleic acid over the other, as well as protein production afforded by both nucleic acid types *in vivo*.

462851 Enhancement of Gene Therapy Via Inhibition of the Innate Immune Response

Abstract:

While the adaptive immune system detects extracellular pathogens, most cells also have an innate immune system that recognizes intracellular pathogens. For example, when viruses inject their DNA into cells, it is quickly detected by cytosolic â€œDNA sensorsâ€ that trigger a signaling cascade that leads to the expression of inflammatory cytokines, inhibition of translation, or even apoptosis to prevent the spread of the virus. Unfortunately, this defense mechanism can also significantly inhibit gene therapy, since most methods also introduce therapeutic DNA into the cytoplasm. In this work, we describe several different strategies to improve non-viral gene therapy by inhibiting the innate immune response. Specifically, we show that addition of small molecule inhibitors for several different enzymes (e.g. TLR9, TBK-1, Î²-catenin, and p38 MAPK) involved in this response can significantly improve transgene (luciferase) expression in cancer cells (e.g. prostate cancer, PC3 cell line). Alternatively, we have also enhanced transgene expression by constructing a plasmid that contains several repeats of a specific DNA sequence that is known to inhibit Toll-Like Receptor 9 (TLR9, a DNA sensor that binds DNA inside endosomes). ELISA assays show that the inhibitors also decrease the expression of inflammatory cytokines (e.g. IL-6), thereby confirming that the enhancement of transgene expression is due to inhibition of the innate immune response. Overall, these results clearly show that the innate immune response is a significant obstacle that must be considered in the development of gene therapy techniques and our work provides multiple alternative strategies to address this issue.

465582 Engineering Megakaryocyte-Derived Microparticles for Gene Delivery

Abstract:

Cell-derived microparticles (MPs) or microvesicles are small particles (100-1000 nm) shed from the plasma membrane from various types of cells, such as hematopoietic cells, endothelial cells, and embryonic stem cells (Cocucci et al., 2009). MPs play an important role in cell-to-cell communication by carrying and transferring various signaling molecules to target cells, such as RNAs, proteins or lipids. MPs have several key characteristics, and notably target specificity, biocompatibility and immunological inertness, that make them a vehicle for potential cargo delivery to desirable cell targets for therapeutic applications.

Megakaryocytic MPs (MkMPs) are the most abundant MPs in blood circulation, and were shown by our group to induce and promote the megakaryocytic (Mk) differentiation of human hematopoietic stem and progenitor cells (HSPC) (Jiang et at., 2014). We have recently shown that MkMPs have exquisite specificity in targeting only HSPCs, and preferably the more primitive HSCs. In this study, we aim to engineer MkMPs as vectors for delivering nucleic acids to HSPCs. We largely employed CHRF cell-derived MPs (CMPs) as a model MP in lieu of MkMPs, since CMPs can be produced readily and in large quantities, and are able to induce and promote Mk differentiation of HSPCs as well as MkMPs can. CHRF cells, a human megakaryoblastic cell line, have been validated as an excellent model of Mk differentiation of human cells. However, we also used primary human megakaryocytes and MkMPs when appropriate.

First, we desired to reduce, if not completely eliminate, the endogenous nucleic-acids (notably RNAs as most MPs contain no DNA) in CMPs so that we can load them with desirable cargo. We developed an RNase-treatment protocol so that the amount to RNA in CMPs decreased by 40 to 60 percent, and this was sufficient to abolish the function of CMPs in inducing and promoting Mk differentiation of HSPCs. This indicated that our RNase treatment protocol was

effective in eliminating most of the endogenous RNA load. Then, we aimed to develop a reliable protocol for loading of desired cargo (and notably plasmid DNA (pDNA)) into CMPs. We examined two different methods for loading: endogenous (cargo loaded into parent cells before CMP generation) and exogenous (cargos loaded directly into CMPs). In endogenous loading, pDNA cargo was loaded into parent cells (both CHRF cells but also primary megakaryocytic cells were examined) by transfection, and MPs were collected after 48 hrs. In exogenous loading, pDNA cargo was directly loaded into MPs by electroporation at various conditions. The loading efficiency was quantitated by either measuring the fluorescent positive MPs if fluorescent-labeled pDNA was loaded, or quantitating pDNA amount from MPs and confirmed with DNA gel electrophoresis. We are detail our studies to optimize the loading efficacy and quantitate the efficiency of cargo delivery by CMPs or MkMPs to HSPCs.

Cocucci, E., G. Racchetti, and J. Meldolesi. 2009. Shedding microvesicles: artefacts no more. *Trends Cell Biol.* 19:43-51.

Jiang, J., D.S. Woulfe, and E.T. Papoutsakis. 2014. Shear enhances thrombopoiesis and production of megakaryocytic (Mk) microparticles that induce Mk differentiation of stem cells. *Blood*.

Acknowledgements: This work was supported by a State of Delaware Bioscience Center for Advanced Technology (CAT) grant (# 15A01570).

467718 Microfluidic Production of Ternary Polyplexes for Non-Viral Gene Delivery

Abstract:

Microfluidic Production of Ternary Polyplexes for Non-Viral Gene Delivery

Gene therapy has the potential to treat a variety of diseases by delivering DNA that codes for a therapeutic protein into human cells. Recombinant viruses are the primary vector choice in the majority of gene therapy clinical trials due to their high gene delivery efficiency. However, viral vectors suffer from serious disadvantages including poor transduction of many human cell types, immunogenicity and lack of targetability. Non-viral polymeric gene delivery vectors (polyplexes) provide an alternative solution but are limited by poor efficiency and cytotoxicity.

Improvements in polyplex-mediated gene delivery can be achieved through control of the polyplex structure and composition. Microfluidic nano-precipitation is an emerging field in which researchers seek to tune the physicochemical properties of nanoparticles by controlling the mixing of particle components during fabrication. Using this approach, several groups have demonstrated the successful production of enhanced polymeric gene delivery vectors. Controlled mixing of polymer and DNA in the laminar flow environment resulted in higher transfection efficiency and lower cytotoxicity. Other groups have demonstrated that incorporating a polyanion to form ternary polyplexes improves transfection efficiency and serum stability.

To advance non-viral gene delivery towards clinical relevance, we are developing a microfluidic platform that provides more controlled and reproducible layer-by-layer assembly of ternary complexes. Our approach provides diffusion-controlled mixing of plasmid DNA and polycations to form polyplexes, separation of excess free polycation, and diffusion-controlled introduction of polyanions to form ternary polyplexes in a sequential manner on a single device. We will describe fabrication and characterization of polyglutamate (PGA)/polyethylenimine (PEI)/DNA ternary complexes. We will also compare these ternary polyplexes to conventional PEI/DNA and PGA/PEI/DNA polyplexes formed by conventional methods in terms of cytotoxicity and transfection in several human cell lines.

468252 Combinatorial Treatment Using Lipopolymer-Mediated TRAIL Gene Delivery and Kinase Inhibitors for Bladder Cancer

Abstract:

The estimated number of new cancer cases within the US for 2016 is 1,658,270 with 595,690 deaths projected during the same year. Urinary bladder cancer is the 6th most common type of cancer in men with 77,000 new cases and 16,400 deaths expected for both men and women in 2016. Treatments for bladder cancer depend on the type and stage of cancer and may involve surgical removal of the tumor, immunotherapy, chemotherapy (intravenous or intravesical), radiation, and/or surgery for bladder removal. Limitations of current chemotherapy regimens include poor drug bioavailability, drug resistance and toxicity, and additional side-effects on healthy cells. Therapies that can specifically target tumor cells can improve patient treatment and quality of life. Tumor Necrosis Factor-alpha Related Apoptosis-Inducing Ligand (TRAIL) induces programed cell death in several cancer types with minimal effect on normal cells, making it a desirable target for cancer treatment. However, resistance to TRAIL (either intrinsic or acquired) has been observed in cancers possibly due to overexpression of anti-apoptotic proteins or low expression of TRAIL receptors. Combination therapy using TRAIL may be a key strategy to sensitize cancer cells to TRAIL-induced apoptosis.

The current work investigates the treatment efficacy of using kinase inhibitors in combination with TRAIL gene delivery in bladder cancer cells. Protein kinases are involved in metabolism, transcription, cell cycle progression, and apoptosis and are linked to TRAIL resistance. The two kinase inhibitors chosen, TG101209 (a Janus kinase II inhibitor) and HMN-214 (a polo-like kinase inhibitor), were based on a previous kinase screen that showed enhanced transgene expression using these inhibitors. Novel aminoglycoside-derived lipopolymers were used for the delivery of a plasmid expressing TRAIL (pEF-TRAIL) in UMUC3 bladder cancer cells. These aminoglycoside-derived lipopolymers displayed enhanced transgene expression compared to parental polymers as well as PEI. A dose-response study suggested that the combined treatment of pEF-TRAIL and TG101209 or HMN-214 demonstrated significantly enhanced cell death compared to either treatment alone. In vivo studies confirmed a significant decrease in tumor volume in mice treated with the TRAIL plasmid alone. Experiments with the combination treatments in vivo are currently underway. This study demonstrates the effectiveness of aminoglycoside-derived lipopolymers for combinatorial treatments using pEF-TRAIL and kinase inhibitors, and suggests potential applications of this treatment in the clinic.

462095 Localized and Sustained Delivery of siRNA from Hydrogels Expedites Fracture Healing

Abstract:

Introduction: Impaired fracture healing, which stems from reduced mesenchymal stem cell (MSC) osteogenesis, is a major clinical challenge [1]. To augment MSC-mediated fracture healing, known inhibitors of bone formation can be downregulated using siRNA-mediated gene knockdown. For example, mouse knockouts of WW domain containing E3 ubiquitin protein ligase 1 (WWP1) exhibit robust fracture healing [2]. However, siRNA delivery suffers from inefficiencies that preclude fracture-targeted and sustained delivery of protected siRNA [3]. Thus, we developed a hybrid nanoparticle (NP)/hydrogel system where NPs protect and increase siRNA delivery efficiency and hydrogels provide localized and sustained delivery by controlling release of embedded siRNA/NPs.

Materials and Methods: Diblock copolymers composed of cationic dimethylaminoethyl methacrylate (DMAEMA) blocks (m~140) and endosomolytic blocks of DMAEMA, propylacrylic acid (PAA), and butyl methacrylate (BMA) (n~80) were synthesized by reversible addition-fragmentation chain-transfer (RAFT) polymerization [3]. Self-assembled siRNA/NPs embedded in hydrolytically degradable hydrogels prepared using 10 wt% poly(ethylene glycol)-b-poly(lactide)-b-dimethacrylate, 1 Î¼M siRNA/NPs, and 0.05 wt% lithium arylphosphonate (LAP) photoinitiator, following exposure to ~5 mW/cm^2 365 nm light for 10 min. Hydrogels with embedded siRNA/NPs were characterized via scanned electron microscope (SEM). Characterization of localization of both Cy5-siRNA and Cy7-PEG hydrogels at Mid-diaphyseal femur fractures (Balb/c mice) was accomplished using XENOGEN/IVIS imaging. Fractures were treated with WWP1 siRNA/NPs hydrogels, followed by RT-PCR, histology (Alcian blue & Orange G) and microcomputed tomography (ÂµCT) to assess fracture healing.

Results and Discussion: siRNA/NPs were loaded into degradable hydrogels for treatment of femur fractures. SEM

imaging showed uniform distribution of NPs within hydrogels. According to IVIS imaging, siRNA and hydrogels were co-localized at fractures for ~21 days, whereas free siRNA/NP dispersed from injection sites rapidly. 14 days after WWP1 siRNA treatment, we observed significantly decreased WWP1 expression compared with untreated controls. Histological analysis showed enhanced bridging bone formation in siRNA/NPs hydrogel treated fractures, whereas extensive unmineralized cartilage and undifferentiated mesenchyme tissue were still present in non-treated and hydrogel only controls. µCT showed 1.3-fold increase in bone callus volume at week 3 in siRNA/NPs hydrogel treated fractures compared with untreated fractures, indicating accelerated bone formation and mineralization.

Conclusions: PEG hydrogels were designed for siRNA/NPs delivery to fracture sites in a sustained and localized manner. Target knockdown of Wwp1 using siRNA/NPs hydrogels showed significantly increased bone formation and accelerated healing. In sum, the development of the hydrogel system has outstanding therapeutic promise.

Acknowledgements: Funding: NSF-DMR1206219; NSF-CBET1450987 NYSTEM N11G-035; NIH-R01AR056696; NIH-R01 AR064200; NIHS10 RR027340.

References: **[1]** Hofmann, A., et al., Bone, 2008. 42(5): p. 894-906 **[2]** Shu, L., et al., J Bone Miner Res, 2013. 28(9): p. 1925-35 **[3]** Benoit, D.S., et al., Biomacromolecules, 2012. 13(11): p. 3841-9

468817 Development and Impacts of Intein-Based Bioseparations

Abstract:

Discovered in the early 1990â€™s, inteins are self-slicing proteins with the ability to excise themselves from a variety of host proteins. This self-splicing activity allows inteins to be moved into many contexts for a range of applications, and mutated inteins have now been demonstrated in a wide variety of protein engineering, labeling and purification approaches. One of the most impactful of the intein applications has been the development of self-cleaving affinity tags for recombinant protein purification. In this strategy, a conventional affinity or purification tag is appended to a self-cleaving mutant intein. The intein is further appended to a target protein, which can be easily purified via the tag. Once purified, the intein is induced to self-cleave, releasing the purified native target with minimal expense or additional processing. Early intein applications were plagued by issues with premature cleaving, required modifications to the target protein, or the need for high concentrations of thiol compounds to induce cleaving. These issues limited the applications where inteins could be used successfully, and prevented inteins from gaining widespread adoption in scientific research and industrial protein production. Recent work has solved many of these problems, however, most notably allowing inteins to be used for disulfide-bonded recombinant glycoproteins expressed in mammalian and other eukaryotic hosts. Here we will report a split intein system developed in our laboratory, where a small intein-segment tag on the target protein is captured by novel affinity resin. Controlled self-cleavage of the smaller segment releases the native protein, without the need for modification of its sequence or additional processing. This approach has been highly successful in solving the remaining problems with inteins, and has potential as a platform for the purification of virtually any recombinant protein expressed in any host system.

467599 Trends in Preparative Chromatography

Abstract:

Besides the widespread application of chromatography as an analytical technique, there is a large and growing interest in the pharmaceutical industry and in biotechnology to isolate and purify value added products using preparative liquid chromatography. This led to remarkable progress in the areas of a) developing and commercializing highly selective and efficient stationary phases, b) understanding and quantifying the consequences of overloading chromatographic columns and c) exploiting more efficiently the driving forces for the separation using advanced operating regimes [1]. In the lecture selected new process concepts of preparative chromatography will be explained. Hereby, in particular multi-column configurations and various ways of exploiting gradients will be illustrated [2-5]. These concepts offer potential for significant improvements of essential performance parameters as productivity, yield and solvent consumption.

[1] Guiochon G., Felinger A., Shirazi D.G., Katti A.M., Fundamentals of Preparative and Nonlinear Chromatography, 2nd Edition, Academic Press, New York, 2006 [2] Seidel-Morgenstern A., Keßler L. C., Kaspereit M., New Developments in Simulated Moving Bed Chromatography, Chem. Eng. & Techn. 31, 2008, 826 [3] Seidel-Morgenstern A., Preparative Gradient Chromatography, Chem. Eng. & Techn. 28, 2005, 1265 [4] Schmidt-Traub H., Schulte M., Seidel-Morgenstern A. (Eds.), Preparative Chromatography, 2nd Edition, Wiley-VCH Verlag, Weinheim, 2012Â [5] Kaspereit M., Seidel-Morgenstern A., Process Concepts in Preparative Chromatography, in: â€œLiquid Chromatographyâ€ (Eds. Fanali S., Haddad P., Poole C., Schoenmakers P., Lloyd D.), Elsevier, Oxford, 2013, Ch. 19, 427

466847 Molecular Insights into Protein Separations from Scattering Methods

Abstract:

Obtaining molecular-level insights into protein separations requires spatial resolution on the order of nanometers, which can be especially challenging for systems that include high concentrations, multiple phases and separation agents such as chromatographic media. Drawing on methods that are widely used in the soft matter community, we have used small-angle neutron scattering (SANS) and small-angle x-ray scattering (SAXS) to characterize the structure of systems arising in protein separations. Because both neutrons and x-rays have wavelengths on the order of Ångströms, these methods are able to access the size range of proteins, albeit results are obtained in reciprocal space and generally require model fitting to interpret. Examples that will be presented include protein precipitates and gels as well as ion-exchange adsorbents with and without adsorbed protein.

464362 Understanding and Enhancing Selective Fab Separations Using Multimodal Chromatography

Abstract:

This research seeks to determine what conditions are required to achieve selective separations of similar protein variants and to provide fundamental insight into the mechanisms underlying these separations. The retention of protein libraries on several multimodal cation-exchange systems demonstrated that the retention of many proteins proved to be sensitive to subtle changes in the ligand chemistry and geometrical presentation. All-atom explicit Molecular Dynamics (MD) simulations were then carried out to shed light on the multiple weak interactions that resulted in the unique selectivities achieved in these multimodal chromatographic systems. A range of biophysics techniques including NMR, ITC and AFM were also employed to study the energetics, kinetics and thermodynamics of protein binding to self-assembled monolayers (SAMs) of MM ligands. Finally, quantitative structure activity relationship models were developed using unique molecular descriptors inspired by the biophysics and simulation studies for predicting the chromatographic behavior of proteins on a many multimodal resin systems and to provide insight into the relative importance of various regions on Fabs for interacting with these ligands. This work provides fundamental understanding of the nature of these interactions at the molecular level and insight into the design of MM ligands with important implications for addressing challenging problems in downstream bioprocessing.

466693 Insights on Ligand-Protein Interactions in Affinity Membranes for Antibody Purification

Abstract:

Recently, we designed and synthesized a series of small molecules acting as synthetic ligands for IgG. These ligands are the results of several years of study of the nature of interaction of monoclonal antibodies, and in particular IgGs, with their natural ligand, namely protein A and with the available synthetic ligands. An integrated approach that combines molecular dynamics with experimental activity and crystallographic studies was used to find of a new binding site of IgG, which lies in proximity of the main binding site of protein A, known as the consensus binding site, but is not overlapped with it. These small ligands were thus specifically computationally designed to bind to the new binding site of IgG that is, according to our simulations, more easily accessible than the consensus binding site.

Preliminary experiments with commercial cellulosic membranes decorated with these ligands, in particular with one of them called HPTA, showed promising properties in terms of binding capacity, selectivity and overall recovery of IgG from complex mixtures. Indeed, there are several issues related to residual non-specific binding and to the antibody recovery with the elution step, that appear to be associated to the membrane surface and to the spacer arm and not to the ligand itself. This was demonstrated in fundamental studies in which the influence of the endcapping protocol and of the spacer arm on the new ligand selectivity was studied, using pure solutions of IgG, BSA and lysozyme, as well as mixtures of IgG and BSA and lysozyme.

Current work is focused on the modification of the properties of the membrane surface in proximity of the affinity ligand by functionalizing part of the surface with histidines or similar residues containing amines, that able to change their protonation state as a function of pH. A modification of the charge of the surface is indeed expected to change drastically the energy of interaction of the adsorbed antibody with the surface, thus improving selectivity.

459137 Polyelectrolyte Functionalized Membranes for Tunable Separations and Enzyme Reactions

Abstract:

The development of separation and greener chemical synthesis processes with reduced energy consumption and minimal environmental impact is critical for sustainable operation. Membrane processes are finding wide applications ranging from water treatment to reactors to advanced bio-separations. The development of bioinspired, tunable, nanocomposite membranes provide added opportunities in permeate flux and separation selectivity control. Depending on the types of functionalized groups (such as, chain length, charge of groups, biomolecule, etc.), these membranes could be used in applications ranging from metal (or oxyanions) separation to enzymatic catalysis. These membranes (with functionalized polypeptides or other polyelectrolytes) are also used for obtaining nanofiltration type separations stimuli responsive (pH and temperature). For example, we have demonstrated 90 % separations of arsenic oxyanions with high water flux at pressure of about 1 bar. By taking advantage of helix-coil transitions one can get tunable separations by altering pH. In addition, electrostatic self assembly in pores (layer-by-layer, LbL) can also be achieved through alternate adsorption of cationic and anionic poly-aminoacids or polyelectrolytes under convective flow conditions. Non-stoichiometric immobilization of charged multilayers within a confined pore geometry leads to an enhanced volume density of ionizable groups in the membrane phase leading highly tunable separations. Multilayer assemblies of polyelectrolytes provide excellent platform for protein/enzyme immobilization by providing reusability and high reactivity. Our results with Glucose oxidase enzyme with glucose as substrate showed very high (70 to 80% of homogeneous activity) activity, and continuous synthesis of hydrogen peroxide and gluconic acid. The presentation will include the role of nano-domain interactions for selective separations, polypeptide/polyelectrolyte assembly in membrane pores, and enzyme reactions for chemical synthesis and water applications. This research is supported by NSF KY EPSCoR program and by NIH-NIEHS-SRC.

470642 Acoustic Wave Separation: An Alternative Single-Use Technology for Harvest Operations

Abstract:

As mammalian cell culture has become more productive, there has been a drive to scale down manufacturing processes into a smaller footprint and shift to single-use technologies. The high cell densities that drive this productivity create significant harvest challenges that must be overcome in order to enable a scaled-down, disposable process. Beyond direct depth filtration (DF) and early prototype single-use centrifugation systems, there are few options for single-use harvest technologies that are commercially available. Acoustic wave separation (AWS) is a novel technology that clarifies cell culture in a continuous manner utilizing 3D standing acoustic waves. Acoustic forces generated by the standing waves trap cells and cell debris, which agglomerate to form large cell particles that settle out of solution. As the primary harvest step, this technology has the potential to reduce high cell density cultures to levels that can be managed by the secondary DF step using significantly less filter area.

In this presentation, the results of an evaluation of AWS technology for a range of cell culture processes will be reviewed. Parameters affecting AWS performance were explored, including: feed packed cell mass, feed and solids stream flow rate, voltage, and number of stages. Optimized process conditions delivered greater than 95% reduction in normalized turbidity, 92% in total cell density and 80% in packed cell mass, while maintaining a step yield of greater than 85%. In comparison to a platform direct DF harvest, pre-processing with AWS was able to reduce the DF area requirements by approximately 6x, providing an estimated 6.7x cost savings. Based on these results, AWS provides a promising alternative to current single-use harvest technologies.

450680 Efficient Screening of Rabbit Scfv Antibodies with High Affinity By Use of Antigen-Coupled Multilamellar Vesicles (MLVs)

Abstract:

In this study, we developed efficient screening method for isolation of single-chain Fv antibodies with high affinity by use of antigen-coupled multi-lamellar vesicles (Ag-MLVs) as a solid support. MLVs are a kinds of liposomes that are composed of phospholipid bilayer. Antigen-coupled liposomes as a solid support have several advantages, compared with the conventional antigen-adsorbed plastic tube and even antigen-coupled magnetic particles. Non-specific interaction between phages and MLVs almost eliminated without blocking operation because the surfaces of MLVs with optimized phospholipid composition are rigid, hydrophilic and negatively-charged. Much high accessibility of antigen increases a chance to interact with specific phages in a library due to lateral diffusion of phospholipid bilayer. It is possible to display various proteins on surfaces of MLVs with high biological activity because the surface hydrophilicity of biological membrane stabilizes the proteins immobilized. We first demonstrated a model biopaning procedure using Ag-MLVs and Ag-Tube as solid surfaces, and positive phages showing specific affinity towards the antigen were tried to be isolated. Consequently, Ag-MLVs successfully isolated specific phages with 10% probability after 2nd round of biopaninng and the probability reached to 100% after 3th round of biopaning. Under the same condition, positive phage was not obtained by use of Ag-Tube. Furthermore, we also investigated to isolate the specific phages displayed with rabbit scFvs possessing high antigen-binding affinity toward a target antigen. Biopaning procedure was performed by use of antigen-coupled MLVs and scFv-displayed phage library. Consequently, 60 % of all phages were idensitfied as positive after the 1st biopaning procedure, and probability of positive clones were increased up tp 100% after the 3rd round of biopanning. According to the analyses of dissociation rate constants (k_{off}), dissociation rates of scFvs became gradually smaller, indicating that binding affinity of scFv moelcules isolated by Ag-MLVs must be increased by cycling a biopaning operation. Thus, the screening method developed in this study will be significantly useful for isolation of specific antibodies with high affinity, and the

isolated antibodies would be valuable for various industorial use such as diagnostic agent and affinity ligand of chromatography.

451934 Immunogenomic Engineering of a Plug-and-(dis)Play Hybridoma Platform

Abstract:

Hybridomas, a fusion of primary mouse B cells and myelomas, are stable, rapidly proliferating cell lines widely utilized for monoclonal antibody screening and production. The antibody specificity of a hybridoma clone is determined by the immunoglobulin sequence of the primary B cell partner. Here we report a platform for rapid reprogramming of hybridoma antibody specificity by immunogenomic engineering. We used CRISPR-Cas9 to generate targeted double-stranded breaks in immunoglobulin loci. Multiplexed-targeting enabled deletion of the native variable light chain and homology directed repair allowed replacement of the endogenous variable heavy chain with a fluorescent reporter protein, mRuby. New antibody genes were then introduced by using Cas9 to target mRuby and promote replacement with a donor construct encoding a light chain and a variable heavy chain, resulting in full-length antibody expression from the immunoglobulin heavy chain locus. Since hybridomas surface express and secrete antibodies, reprogrammed cells were isolated using flow cytometry and cell culture supernatant was used for antibody production. The simplicity of our approach was exemplified in that we generated a homogenous population of antibody-producing hybridoma cells with just a single transfection and screening step. Our hybridoma plug-and-(dis)play platform represents a rapid method for generating antibody-producing stable cell lines.

452036 Prediction of Bacterial Fluxomics Using Machine Learning and Constraint Programming

Abstract:

Metabolic information is important for disease treatment, bioprocess optimization, environmental remediation, biogeochemical cycle regulation, and our understanding of life's origin and evolution. ^{13}C-MFA can quantify microbial physiology at the level of metabolic reaction rates. Mining the relationship between environmental and genetic factors and metabolic fluxes hidden in existing fluxomic data will lead to predictive models that can significantly accelerate flux quantification. In this paper, we present a web-based platform (MFlux: http://mflux.org/) that predicts the bacterial central metabolism via machine learning, leveraging data from ~100 ^{13}C-MFA papers on heterotrophic bacterial metabolisms. Three machine learning methods, namely Support Vector Machine (SVM), k-Nearest Neighbors, and Decision Tree, were employed to study the sophisticated relationship between influential factors and metabolic fluxes. We performed a grid search of the best parameter set for each tested algorithm and verified their performance through 10-fold cross validation. SVM yielded the highest accuracy of all three algorithms. Further, we employed quadratic programming to adjust flux profiles to satisfy stoichiometric constraints. Multiple tests have shown that MFlux can predict fluxomes as a function of bacterial species, substrate types, growth rates, oxygen conditions, and cultivation methods.

In summary, aided by constraint programming and quadratic optimization, our machine learning platform can predict meaningful metabolic information about bacterial species in their environments. Further, it can offer constraints to improve the accuracy of flux balance analysis. This study infers that the bacterial metabolic network has a certain degree of rigidity in allocating carbon fluxes, and different microbial species may share common regulatory strategies for balancing carbon and energy metabolisms. As a proof of concept, we demonstrate that the use of data driven models (e.g., artificial intelligence) may assist mechanistic based models to elucidate the topology of microbial

fluxomes.

452531 Advanced Analytical Methods to Assess the Impact of Intercellular Reactive Oxygen Species (ROS) on Mammalian Cells Productivity and Product Quality

Abstract:

Reactive oxygen species (ROS), a phrase used to describe a number of reactive molecules and free radicals derived from molecular oxygen, can cause negative impact to cell growth, productivity and product quality. We developed an easy, quick, reliable, near real time, and inexpensive method to detect ROS level in production mammalian cell lines. MitoSOXâ„¢ Red Molecular ProbesÂ®, a fluorescent tool, was used to establish an analytical method to measure ROS level in mammalian cells. The implementation of this method and other florescent tools could increase the speed and quality of bioprocess development, improve cell productivity and product quality.

455168 Engineering *Saccharomyces Cerevisiae* for Monoterpene Bioconversion By Modulating α,β-Unsaturated Aldehyde Metabolism.

Abstract:

Monoterpene indole alkaloids (MIAs) represent a structurally diverse, medicinally essential class of plant derived natural products. *Catharanthus roseus* is a prolific producer of MIAs, as well as the source of the indispensable chemotherapeutic MIAs vinblastine and vincristine. A key enzymatic step in MIA biosynthesis involves the non-canonical, reductive cyclization of the monoterpene 8-oxogeranial by iridoid synthase to form nepetalactol. Generation of a bicyclic molecule with four chiral centers, from a precursor that is both acyclic and achiral demonstrates that Mother Nature is truly the best synthetic chemist. In the process of leveraging a bioconversion production strategy in *Saccharomyces cerevisiae*, we have identified 8-oxogeranial as the primary pathway intermediate from which a number of shunt products are derived. By modulating yeast's metabolism of 8-oxogeranial, which contains two α,β-unsaturated aldehydes, we illustrate the roles played by several key enzymes in disrupting the heterologous production of nepetalactol. Furthermore, we clarify confusion in the field regarding reduction of pathway intermediates in yeast by endogenous and exogenous enzymes. We anticipate that our engineered platform will play an important role in development of *S. cerevisiae* for production of iridoids and MIAs.

456395 Microbiome-Bots for the Treatment of Inborn Errors of Metabolism

Abstract:

Probiotic bacteria have been commercially sold as dietary supplements to treat or prevent various gastrointestinal disorders. Their inherent safety, abundance, and high-level of interaction with the hosts digestive tract coupled with recent advances to the genetic tools for manipulation have opened opportunities to exploit these organisms for unique therapeutic applications. We hypothesize that these bacteria can be genetically modified into â€œmicrobiome-botsâ€ capable of substituting catalytic function to the human digestive system to treat various inborn errors of metabolism in a safe and effective manner. This talk will primarily focus on our efforts in developing a lactic acid bacteria-based microbiome-bot that provides non-native metabolic function through delivery of engineered enzymes for treating galactosemia. Galactosemia is the general description for a group of rare autosomal recessive metabolic disorders that results from the improper folding of one of three enzymes involved in galactose metabolism. In all cases incomplete metabolism of galactose generates an accumulation of a toxic metabolite. Pathologies related to this disorder include permanent damage to major organs such as the liver, kidneys, brain, and ovaries. There are no current therapies approved for the treatment of galactosemia to date. The microbiome-bot platform provides a targeted solution for disorders arising within the human digestive tract. We also describe our efforts in expanding this technology beyond galactosemia to other inborn errors of metabolism for which there are currently few treatment options.

457932 Elucidating the Neuroprotective Mechanism of Curcumin Against Amyloid Beta-Induced Toxicity in Alzheimer's Disease

Abstract:

Alzheimer's disease is the most prevalent form of dementia with 5.2 million people affected in the U.S. This neurodegenerative disease is characterized by the aggregation of the amyloid-beta (Aβ) protein into extracellular senile plaques. The aggregation mechanism and toxicity pathway of this protein is not well understood yet. However, its capability to bind directly to cell membrane could mediate both processes. Aβ binding to lipid membrane has been shown to seed aggregation and disrupt the lipid membrane.

We have previously shown that the polyphenolic compound, curcumin, reduces the Aβ toxicity by attenuating Aβ-lipid membrane interactions. Additionally, curcumin can directly reduce the toxicity of Aβ oligomers and modify the Aβ aggregation pathway. However, the detailed mechanisms by which curcumin affects Aβ toxicity and aggregation are still unknown.

In this study, we investigate the effect of curcumin on the aggregation of Aβ by monitoring the kinetics and morphology of Aβ oligomers and fibrils formed in presence of curcumin using a combination of characterization methods, including Thioflavin T (ThT) assay, size exclusion HPLC, Transmission Electron Microscopy (TEM) imaging, and dynamic light scattering. In particular, curcumin binding to oligomers and fibrils will be assessed. Finally, the effect of curcumin on Aβ aggregate toxicity will be tested. Results so far show that the presence of curcumin does not reduce protein aggregation and instead leads to increased levels of oligomeric and fibrillary forms of the protein. Additionally, a significant reduction of the Aβ toxicity was observed when the protein was incubated with curcumin.

Investigations are currently on going to determine whether a direct binding between curcumin and Aβ oligomers is leading to the reduction of oligomer toxicity. We propose to verify this interaction by monitoring the fluorescence of curcumin. We will also test the effect of curcumin on Aβ aggregate interaction with the lipid membrane by performing insertion assays of the Aβ aggregation-curcumin complex into lipid monolayers using a Langmuir Trough. This investigation will lead to a better understanding of the molecular mechanism(s) of curcumin's inhibition of Aβ toxicity, which will contribute towards the development of therapeutic approaches to prevent or treat Alzheimer's disease.

Challenges in predicting protein-protein interactions from measurements of molecular diffusivity

Lea L. Sorret, Madison A. DeWinter, Daniel K. Schwartz, Theodore W. Randolph
Department of Chemical and Biological Engineering, University of Colorado - Boulder

Objectives

This study aims to address some of the limitations of using the interaction parameter, k_D, to measure protein-protein interactions through the following goals:

Aim 1: To show that protein diffusion is dependent on electrolyte concentrations and local electric potential gradients, not reflective of protein-protein interactions.
Aim 2: To demonstrate that changes in the infinite dilution diffusion coefficient, $D_{s,2}$, obtained from DLS need not imply changes in protein conformation.
Aim 3: To show that at high protein concentrations, viscous effects arising from excluded volume may largely determine the value of k_D.

Motivation

Weak protein-protein interactions are a major determinant of protein stability:

Attractive interactions — Result from forces such as van der Waals, hydrophobic, dipole-dipole interactions etc...

Repulsive interactions — Result from electrostatic forces

Problems associated with attractive interactions: protein aggregation, high solution viscosity, off-target binding, and fast antibody clearance.[1]

The osmotic second virial coefficient B_{22}, normalized by its hard-sphere (HS) contribution, B_{HS}, can be used to measure protein-protein interactions:[2,3]

$B_{22}/B_{HS} - 1 \geq 0$ dominance of repulsive forces
$B_{22}/B_{HS} - 1 < 0$ dominance of attractive forces

Methods: static light scattering (SLS) & membrane osmometry
Disadvantages: low throughput, high amount of protein required

High-throughput alternative, the interaction parameter k_D:[4]

$$k_D = 2MB_{22} - (\xi + \bar{v}_2) \quad (1)$$

$$D_{m,2} = D_{s,2}(1 + k_D c_2) \quad (2)$$

ξ: first order friction coefficient
$\bar{v}_2$: protein excluded volume
M: protein molecular weight
$D_{m,2}$: protein mutual diffusion
$D_{s,2}$: protein intrinsic diffusion
c_2: protein concentration

Methods: dynamic light scattering (DLS) & taylor dispersion analysis (TDA)[5]
Advantage: high throughput, low amount of protein required

Protein diffusion in ionic solutions

The diffusion of the three species (K^+, Cl^-, protein) in Figure 1 can be represented mathematically using Nernst-Planck equation, where N_i is flux, F is Faraday constant, z_i the net charge of species i, and $D_{s,i}$ single-particle diffusion coefficients of species i.[6,7]

$$-N_i = D_{s,i}\left(\nabla c_i + \frac{z_i c_i F}{RT}\nabla\varphi\right) \quad (3)$$

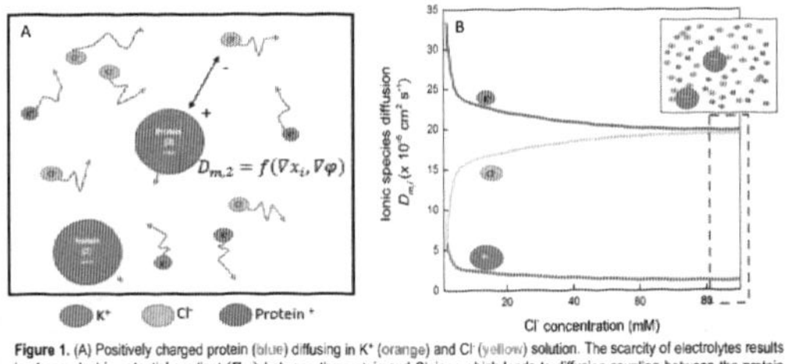

Figure 1. (A) Positively charged protein (blue) diffusing in K^+ (orange) and Cl^- (yellow) solution. The scarcity of electrolytes results in strong electric potential gradient ($\nabla\varphi$) between the protein and Cl^- ions which leads to diffusion coupling between the protein and its counter-ions. (B) Figure adapted from Cadman & Fleming.[8] Protein diffusion is sped up and Cl^- is slowed down at low counter ion (Cl^-) concentrations due to local electric potential gradients arising from the separation of oppositely charged species in solution.

Methods

Model proteins:
1. Lysozyme: z = +4, 14.3 kDa
2. Monoclonal antibody (mAb): z = +10, 145.5 kDa

Electrolyte solution: 2 to 90 mM KCl pH 6.0

Method for aim 1:
Dilute dialyzed protein stock to 2-20 mg mL^{-1}.
Use CONTIN algorithm to extract $D_{m,2}$ from DLS results.

Method for aim 2:
Record Far-UV CD spectra from 190-250nm in triplicates of ten scans each.
Blank solution spectrum is subtracted and spectra normalized by converting to mean residue molar ellipticity.[9]

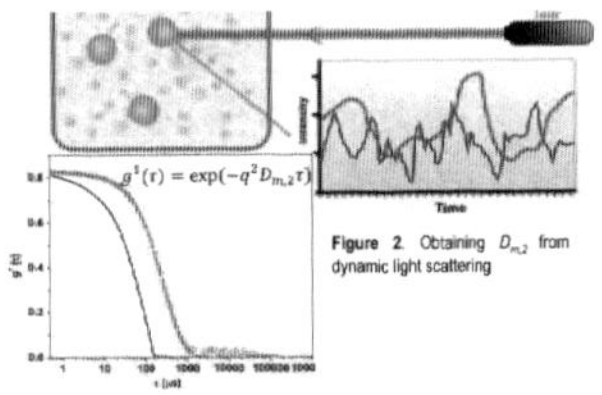

Figure 2. Obtaining $D_{m,2}$ from dynamic light scattering

Results

Comparison between experimental and theoretical protein diffusivities

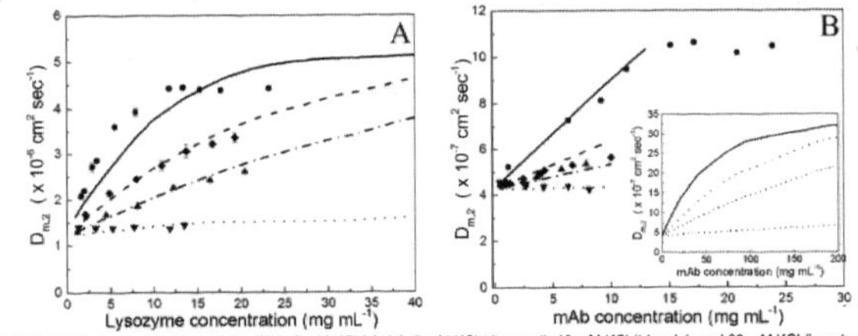

Figure 3. DLS data for lysozyme (A) and mAb (B) in 2 mM KCl (circle), 5 mM KCl (diamond), 10 mM KCl (triangle), and 90 mM KCl (inverted triangle), at pH 6.0. Error bars represent the standard deviation of triplicate measurements. Most error bars are smaller than symbols. Lines show theoretical $D_{m,2}$ versus c, calculated from the Nernst-Planck equation (Eq. 3) in 2 mM KCl (solid), 5 mM KCl (dash), 10 mM KCl (dash-dot), and 90 mM KCl (dot), at pH 6.0. Insert in (B) shows a zoomed-out version of the model to demonstrate curvature. Both experimental and theoretical results demonstrate that at low ionic strength and low protein concentration, protein diffusivity increases with increasing protein concentration. At higher ionic strength however, this trend becomes weaker. The agreement between the model – which does not incorporate protein-protein interactions – and DLS data for both proteins demonstrate that local electric field gradients at low ionic strength account for the majority of protein mutual diffusion behavior.

Conformational changes from diffusivity data

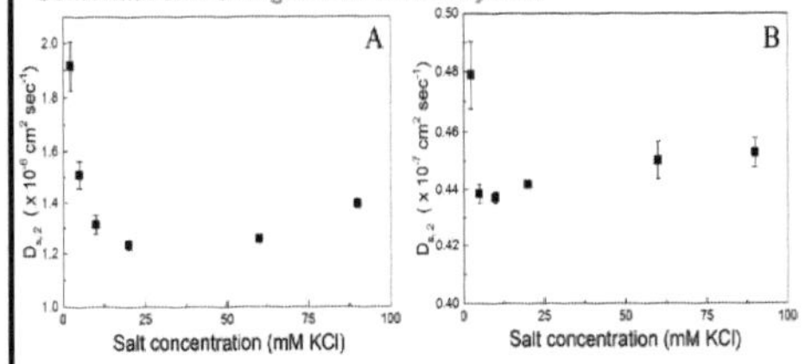

Figure 4. Self-diffusion coefficient $D_{s,2}$ as a function of ionic strength calculated by linear extrapolation of DLS data for lysozyme (A) and mAb (B) using Eq. 2. Results are obtained by finding the intercept of the linear portion of DLS data from Fig. 3. Error bars represent extrapolation to zero protein concentration. The increase in diffusivity at low ionic strength need not imply a conformational change as mutual diffusivity of a protein depends not only on its concentration but also on the ratio of its concentration to that of other ionic species.

Secondary structure determination

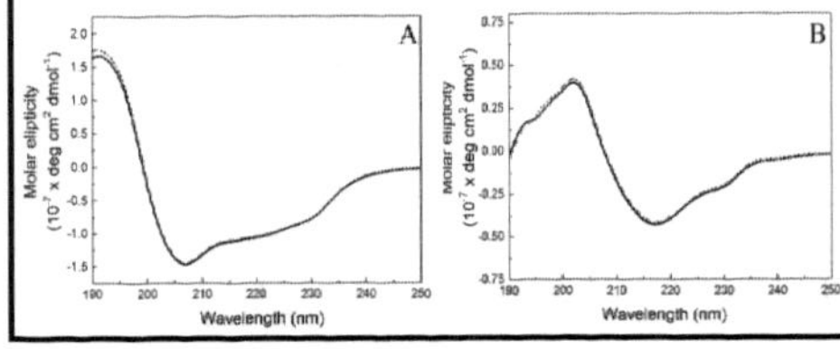

Figure 5. Normalized far-UV CD spectra in KCl at pH 6.0 for lysozyme (A) and the mAb (B) at 2 mM KCl (solid black), 10 mM KCl (dashed gray), and 90 mM KCl (dashed black), at pH 6.0. Each curve represents the average of triplicates of ten scans each. Error bars not shown for clarity. Data show that no conformational changes occur for both lysozyme and the mAb as a function of ionic strength.

Results

Excluded volume contribution to k_D

B_{22} is often normalized by its B_{HS} component so that comparisons between proteins of different sizes and molecular weights are possible. A similar step should be undertaken for k_D as viscous effects arising from excluded volume will have an impact on the hydrodynamic portion of k_D.

To calculate the effect of excluded volume on k_D, Mooney's equation of viscosity is used and plugged into Stokes-Einstein's equation of diffusivity, resulting in Eq. 4, where S is a structure factor and k is a packing factor, ϕ is protein excluded volume and η_0 is the viscosity of solvent:[10-12]

$$D = \frac{k_B T}{6\pi\eta_0 \exp\left(\frac{S\phi}{1 - k\phi}\right)} \quad (4)$$

For a hard sphere, S = 2.5 and k = 1.35, for lysozyme (thick line) S = 3.01 and k = 2.91, and for the mAb (dashed line) S = 2.6 and k = 2.3.

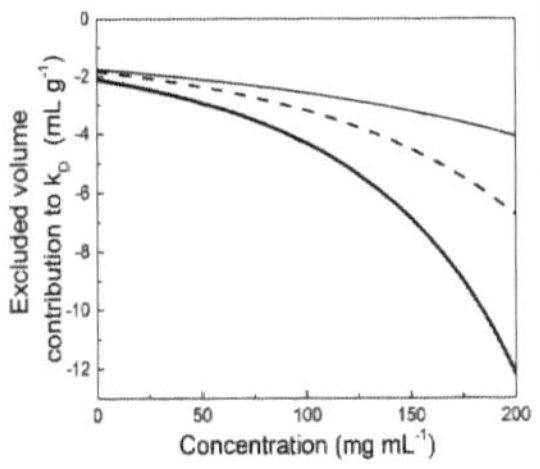

Figure 6. Excluded volume contribution to k_D for a hard sphere (thin black line) with. The excluded volume contribution to k_D is always negative and becomes increasingly large at high protein concentration although no thermodynamic forces are represented beside pure steric effects.

Conclusions and recommendations

- k_D reflects both hydrodynamic and thermodynamic effect on protein solutions, so careful interpretation is needed if it is to be used to measure protein-protein interactions and colloidal stability of protein formulations.
- Local electric potential gradients induce coupled diffusion between the protein and its counter-ions. This can drive protein diffusion, not representative of true protein-protein interactions.
- Although DLS is a promising and versatile technique, orthogonal techniques should be used to quantify protein stability.
- Caution should be taken when using DLS to measure protein conformational changes as the mutual diffusivity of protein even at infinite protein dilution depends on the ratio of its concentration to that of other ionic species in solution.
- At high protein concentrations, viscous effects arising from excluded volume may largely determine the value of k_D.
- We recommend comparing experimentally-determined k_D values against calculated contributions to k_D excluded volume effects.

References

1. Geng, S.B., J.K. Cheung, C. Narasimhan, M. Shameem, and P.M. Tessier. 2014. Improving monoclonal antibody selection and engineering using measurements of colloidal protein interactions. J Pharm Sci. 103: 3356–63.
2. George, A., and W.W. Wilson. 1994. Predicting protein crystallization from a dilute solution property. Acta Crystallogr D Biol Crystallogr. 50: 361–5.
3. Mcmillan, W. G., and J. E. Mayer. 1945. The statistical thermodynamics of multicomponent systems. J. Chem. Phys. 13:276–305.
4. Harding, S.E., and P. Johnson. 1985. The concentration-dependence of macromolecular parameters. Biochem J. 231: 543–547.
5. Hassan, P. a, S. Rana, and G. Verma. 2015. Making sense of brownian motion: colloid characterization by dynamic light scattering. Langmuir. 31: 3–12.
6. Krishna, R., and J.A. Wesselingh. 1997. The Maxwell-Stefan approach to mass transfer. Chem Eng Sci. 52: 861–911.
7. Lightfoot, E. 1962. Applicability of the Stefan-Maxwell equations to multicomponent diffusion in liquids. AIChE J. 8: 708 – 710.
8. Cadman, A. D., R. Fleming, and R. H. Guy. 1981. Diffusion of lysozyme chloride in water and aqueous potassium chloride solutions. Biophys J. 37:569–574.
9. Kelly, S.M., T.J. Jess, and N.C. Price. 2005. How to study proteins by circular dichroism. Biochim Biophys Acta. 1751: 119–39.
10. Mooney, M. 1951. The viscosity of a concentrated suspension of spherical particles. J Colloid Sci. 6: 162–170.
11. Einstein, A. 1905. On the Motion of Small Particles Suspended in a Stationary Liquid, as Required by the Molecular Kinetic Theory of Heat. Ann Phys. 322: 549–560.
12. Monkos, K. 1997. Concentration and temperature dependence of viscosity in lysozyme aqueous solutions. Biochim Biophys Acta. 1339: 304–310

Acknowledgements

The authors gratefully acknowledge financial support from the National Institute of Health (RO1 EB0060006). Lea Sorret acknowledges support from the National Science Foundation fellowship.
The authors also would like to thank Dr. Annette Erbse for letting them use the dynamic light scattering instrument as well as the circular dichroism spectrometer from the Biochemistry shared instrument facility.

458005 Gene Regulation and Signal Integration Across Organelles: CX_9c Isoform Proteins As Bi-Organellar Transducers Between Mitochondria and the Nucleus

Abstract:

The regulation of gene expression is a crucial component in responding adaptively to stressors that disrupt cellular homeostasis. The regulation of energy metabolism is particularly urgent when a cell is faced with a life-threatening emergency and calls into play multiple gene regulatory systems. A special and interesting case is integration of gene regulatory systems across organelles in a process by which mitochondria signal to the nucleus, called retrograde regulation. We have uncovered two such regulators, members of the CX_9C family of proteins that are typically found in both the nucleus and the mitochondrial intermembrane space. These are MNRR1 (Mitochondrial Nuclear Retrograde Regulator 1, also called CHCHD2) and CHCHD10. The importance of these regulators is emphasized by their being found associated with a number of peripheral neuropathies such as ALS, Parkinsons Disease, and frontotemporal dementia. In the mitochondria both are physically bound to, and co-purify with, cytochrome *c* oxidase (COX), the terminal protein of the electron transport chain, as well as bound to each other. MNRR1 requires Tyr-99 phosphorylation, carried out *in vivo* by Abl2 kinase (ARG), for efficient binding to COX. Lack of MNRR1 produces mitochondrial dysfunction, including reduced oxygen consumption, increased free radical production, and mitochondrial fragmentation. However, both proteins are also present in the nucleus, where they serve as modulators of transcription. MNRR1 is maximally produced at 4% oxygen and activates a novel oxygen responsive element (ORE) in the promoter for a COX subunit gene, *COX4I2*. Similar OREs are present in the promoters of a number of genes that are stress responsive such as SOD, MFN1, PGC-1Î±, Sirt1, Atg7, and Hsp60; all respond to a change in MNRR1 levels. Fundamental mitochondrial processes such as oxygen consumption, ROS production, growth rate, and mitochondrial shape are regulated by MNRR1 levels. An MNRR1 isoform, CHCHD10 is also present in both the mitochondria and the nucleus. It is maximally induced at 8% oxygen and in the mitochondria it also binds to COX and also causes increased oxygen consumption. However, in the nucleus it functions oppositely at the ORE by repressing transcription. We propose that these proteins provide a gradual cellular systems response to decreasing oxygen concentrations prior to the very low concentrations that trigger a response by the well-known hypoxia inducible factor (HIF).

458747 Characterizing Complex Gene-Drug Interactions: Gene Perturbation and Antibiotic Treatment As Combination Therapy

Abstract:

Bacteria possess the ability to survive in and gain resistance to diverse antibiotic treatments, necessitating new approaches by which to combat bacterial infection. One useful strategy has been to utilize antibiotics in combination. Combination therapies can lead to increased antibiotic effect especially in the case of synergistic interactions, or could potentially be used to select against resistance in the case of antagonistic interactions. As new multi-drug resistant strains continue to emerge, the need to locate and pursue alternative antibiotic targets increases in priority. In the face of new technologies allowing for precise interference with the expression of specific genes, next-generation therapeutics do not need to be compounds that specifically bind to an enzyme or cellular components; any region of the genome has the potential to be utilized as a therapeutic target. We hypothesized that combining antibiotic treatment with gene perturbation would reveal complex interactions, advantageous in the design of new therapeutics. We characterized the impact of single gene knockout as well as single gene perturbation with CRISPR-dCas9 in combination with antibiotic agents targeting various pathways, including cell wall synthesis, translation (impacting 50S, 30S, and RNA), and nucleic acid synthesis. The thirty target genes included many global regulators as well as other genes involved with the general stress response, the SOS response, the mar-regulon, efflux, motility, and several of unknown function that have been previously connected to adaptive resistance. Using a growth assay, we demonstrate that the gene-drug combinations result in synergistic, antagonistic, and additive relationships. We describe the landscape of select interactions by titrating the degree of repression in conjunction with varying concentrations of antibiotic. Finally, network analysis reveals how specific functional classes of genes generally interact with various drug classes. This work provides evidence for the applicability of drug-gene combination therapies in promoting complex synergistic and antagonistic behaviors, an area which has great impact on the design of future treatments.

457538 Mechanically Fingerprinting Blood to Predict Pathological and Physiological Conditions

Abstract:

American Institute of Chemical Engineering

Tyler T. Helton and Matthew J. Armstrong

Department of Chemistry and Life Science, United States Military Academy, West Point, NY 10996

Complex, viscoelastic materials exhibit a change in viscosity depending on the extent of shearing forces placed upon them. The study of these materials is called Rheology. Among these types of materials is blood, a well documented viscoelastic material. An underdeveloped area of study in this field analyzing these mechanical measurements of blood for use in clinical diagnoses. Rheologists have long studied a unique condition of shear flow called large amplitude oscillatory shear (LAOS). The challenge of this condition is its nonlinearity, stemming from the difficulty of modeling the unique internal structure of each material. There is a possibility that blood could be classified using its own unique LAOS signature, and with the addition of large quantities of data for comparison, could be used as a method of diagnoses.

Blood is a great example of a non-Newtonian fluid. It is a suspension red blood cells (RBCs), white blood cells, proteins, hormones, platelets, and other ionic solutes in the blood plasma. This internal microstructure affects the dynamic viscosity of blood when forced into flow conditions. The interactions of the microstructure play a major role in determining the viscosity of blood. RBC aggregation, deformability, and membrane rigidity all contribute to the blood's viscosity [1]. Plasma proteins, most notably fibrinogen, contribute to the interactions among RBCs as well as platelets [2].

Recent work with blood using LAOS has shown that its unique rheological signature can be used as a

psuedo- "fingerprint" [3-7], able to clearly show and delineate elastic and viscous regions of the LAOS cycle. Applying large amplitude oscillatory shear (LAOS) to complex fluids induces nonlinear rheological responses, that, with proper LAOS analysis technique, can be used to sensitively probe the underlying microstructure and its dynamics. Using the previously published blood LAOS data of Sousa *et al.* [10] we compare LAOS using three published LAOS analysis frameworks, including the aforementioned by Bharadwaj and Ewoldt as well as a novel Series of Physical Processes (SPP) framework [8,9] by Rogers. The techniques will be compared and contrasted and respective strengths and weakness discussed.

For many years, the standard means of analyzing LAOS has been the use of discrete Fourier transforms in order to break down the sinusoidal stress signal, $\sigma(t)$, into elastic and viscous components respectively [14,15]. These components can be used to analyze the extent of the elastic and viscous contributions to the stress signal. A new method of analyzing LAOS data, (SPP), uses all the components of LAOS data used by the rheometer and treats the now three dimensional data as a space curve. SPP also allows for a comparison between the G'and G" analog at every given point. With this advantage, the materials movement between the elastic and viscous regimes can be analyzed.

References

[1] O.K. Baskurt, H.J. Meiselman, Seminars in Thrombosis and Hemostasis. 29(5) (2003) 435-450.

[2] G.D.O. Lowe. Clinical Science. 71 (1986) 137-146.

[3] C.J. Dimitriou, R.H. Ewoldt and G.H. McKinley. J. Rheol. 571(1), 27-70.

[4] B.C. Blackwell and R.H. Ewoldt. JNNFM 227, (2016) 80-89.

[5] B.C. Blackwell and R.H. Ewoldt. JNNFM 208-209, (2014) 27-41.

[6] R.H. Ewoldt and N.A. Bharadwaj. Rheol. Acta. (2013) 52:201-219.

[7] N.A. Bharadwaj and R.H. Ewoldt. J.Rheol. 59(2), (2015) 557-592.

[8] S.A. Rogers. J. Rheol. 56(5), (2012) 1129-1151.

[9] S.A. Rogers and M.P. Lettinga. J. Rheol. 56(1), (2012) 1-25.

[10] P. C. Sousa. Biorheology 50 (2013), 269 – 282.

[11] Moreno et al., Korea-Australia Rheology Journal Vol. 27, No.1 (2015) 1 – 10.

[12] J. Mewis and N. J. Wagner, Colloidal Suspension Rheology, Cambridge Univ. Press, 2012.

[13] O.K. Baskurt, H.J. Meiselman, Seminars in Thrombosis and Hemostasis. 29(5) (2003) 435-450.

[13] G.D.O. Lowe. Clinical Science. 71 (1986) 137-146.

[14] R.H. Ewoldt, A.E. Hasoi, G.H. McKinley. J. Rheol. 52(6). (2008) 1427-1458.

[15] M.J. Armstrong. PhD Thesis, University of Delaware (2015).

458573 *S.Epidermidis* Alters a Fibrin Clot Insights into Septic Thromboembolism

Abstract:

Pathological blood clots, for example, venous thromboembolism (VTE), are a leading cause of death and disability worldwide. The 3-fold increased risk of VTE by the presence of bacterial infection motivates work to establish the basis for the high correlation between infection and thrombosis; the specific impacts of bacteria on clot formation, remodeling, and embolization are unclear. Here we demonstrate that the common pathogen and biofilm former *Staphylococcus epidermidis* RP62A has significant effects on the formation as well as the structural and mechanical properties of an *in vitro* model of a blood clot (i.e., fibrin clot). By mechanical rheometry, we find that the presence of *S.epidermidis* cells retards the formation of a fibrin clot. The resultant steady state infectious clot has a bulk elasticity that is about four times larger than a pure fibrin clot. From confocal microscopy images, we observe that the infectious clot has a heterogeneous structure with two characteristic mesh sizes rather than a single one as in a pure fibrin network. We correlate the higher bulk elasticity to this heterogeneous microstructure via fiber motility using multiple-particle tracking microrheology. Finally, we observe that the fibrin network continues to remodel in the presence of *S. epidermidis* and eventually ruptures. The higher elasticity of the infectious clot may be associated with embolic diseases that have been related to high fibrin rigidity. The rupture of the infectious clots provides insights into the correlation between bacterial infection and embolism.

458848 Investigating Cancer Cell Migration Guided By Stiffness Gradients

Abstract:

Metastasis is the primary cause of death in cancer patients; yet, there is a shortage of highly potent anti-metastatic drugs. Traditional cancer cell migration studies and screening for anti-metastatic compounds have been primarily performed using two-dimensional platforms that only allow cell migration along a single plane. However, spatial arrangement of in vivo biological signals are much more complicated, especially in scenarios where cell migration is guided by matrix stiffness gradients. In this work, we sought to develop a 2.5D culture platform where cells are placed at the interface between stiff TCPS and soft alginate to study the role of transitions in stiffness on cell migration. The 2.5D platform was used to study the effects of both matrix stiffness and cell-matrix adhesions on cell migration. We confirmed the roles of mechanotransductive signaling on cell migration using small molecule inhibitors of the RhoA-ROCK and Rac1 pathways, as well as the ability of cells to switch between adhesion dependent and independent modes using the integrin inhibitor, Cilengitide. The results reported in this study will facilitate the development of optimal in vitro platforms to discover therapeutic strategies against tumor cell motility and invasion.

460225 Designer Bicelles: Peptoid Incorporation into the Edges and the Planar Region

Abstract:

The study of membrane-affiliated interactions is significant to the understanding of cell function, detecting biomarkers to diagnose disease, and testing the efficiency of new therapeutic targets. The amphiphilic character of membrane proteins makes them difficult to work with due to low stability. Model membrane systems can be used to stabilize membrane protein structure and maintain native activity. The model membrane system of focus in this study are bicelles, disc-shaped lipid bilayers created by combining long- and short-chain phospholipid pairs. We have modified the chemistry at the edges and at the planar region of bicelles by incorporating peptoids, which are non-natural protein mimics. Peptoids are ideal for this application since a large diversity of side chains can be easily incorporated in a sequence-specific manner. In these preliminary studies two sequence of peptoids were designed to contain a lipophilic side chain at the C-terminus for interactions with the short and long-chain lipid tail groups to for the edge and planar modification, respectively as well as thiol groups at the N-terminus for characterization. Peptoid-bicelles were imaged using transmission electron microscopy and size analysis was performed by dynamic light scattering. The presence of peptoids in the edges and planar regions of the bicelles was confirmed by attachment of gold nanoparticles to the thiol groups within the peptoid sequence. Our results show that the designed peptoids incorporate preferentially into the edges and planar regions of the bicelles, respectively as expected. Further studies will be performed to alter the diameter of the bicelles to allow for study of membrane proteins and complexes of various sizes.

460560 Developing a Sterile and Scalable Polishing Step for a Large Live Virus Purification

Abstract:

Live attenuated virus (LAV) vaccines offer the benefit of stimulating humoral and cellular immune responses, which is recognized as an advantage over many subunit vaccines. However, process development of LAV vaccines presents many unique challenges due to their size, molecular complexity, and structural heterogeneity. Downstream purification is further limited by a number of factors, including low binding capacity for many viruses on traditional chromatography resins, constraints on operating conditions (e.g. pH, use of detergents), and the requirement of closed-system aseptic processes for those viruses that are too large for terminal sterile filtration. This work examines several purification methods as a polishing step for a candidate LAV vaccine requiring significant purification as well as aseptic processing. A variety of purification approaches were evaluated including membrane supports, newly designed bead-based media for viruses and large biomolecules, and non-chromatographic methods such as selective precipitation and centrifugation. Polishing steps were evaluated based not only on usual performance criteria such as purification potential and yield but also on their ease to be performed aseptically and their potential for scale up.

460619 Metal-Catalyzed Fragmentation of a Mab: a Case Study

Abstract:

Metal-Catalyzed Fragmentation of a mAb: a Case Study

Michael Grooms, Jennifer Fetter, T. Bruce Vickroy, Rianna Gallo Collier, Myrna Monck, George Alkire, Matthew Henry, Jennifer Dally, Robert G. Luo, Kent E. Göklen

GlaxoSmithKline

King of Prussia, PA, USA

Poster Abstract

The level of metals in pharmaceuticals has come under scrutiny lately due to recent regulatory guidance requiring a clearer understanding and better controls of these impurities. This is motivated primarily by concerns of metal toxicity, but of equal concern for the producers of biopharmaceuticals is metal-catalyzed fragmentation of their product. In this case study, a mAb product is found to unexpectedly suffer fragmentation during accelerated stability experiments, where none had been seen earlier in the program. The investigation that led to an understanding of the phenomenon and the change that caused it will be presented. In addition, spiking studies to understand the impurity levels and their clearance during the downstream process, along with approaches to control the problem will also be presented.

460690 Mimicking Mussel Adhesives with Peptide Glue

Abstract:

The development of glues that adhere to wet surfaces is important for medical, dental, and marine applications. However, glues that are effective in dry conditions often fail in aqueous conditions, especially in salt solutions or when the surfaces are contaminated, for example by a biofilm. Nevertheless, many marine organisms adhere to surfaces under water. The marine mussel attaches itself to rocks in the ocean by secreting proteins which adhere to fouled mineral surfaces via chemical and physical bonds. To study these interactions, a peptide glue was synthesized to mimic a common type of mussel adhesive protein (MFP-3). The adhesion energies between the peptide and hydrophobic (CH_3-SAM) and hydrophilic (OH-SAM, mica) surfaces were measured experimentally using a surface forces apparatus (SFA) and computed from molecular dynamics simulations. Changing surface hydrophobicity resulted in different peptide-surface molecular geometries, and suggested a synergistic and adaptive adhesion mechanism based on adjacent hydrophobic and hydrophilic residues. Understanding the dependence of peptide adhesion on peptide primary structure and interaction geometry will allow the design of adhesives for use in aqueous environments.

461226 Production and Kinetics of Metallic Nanoparticles from Phototrophic Cell Culture

Abstract:

Biosynthesis of nanomaterials, such as noble metal nanoparticles, has experienced a tremendous expansion during the last two decades. Among the proposed biological production platforms, algae have attracted increasing attention from research scientists worldwide because they represent a trend towards greener, more sustainable processes which reduce or eliminate potentially toxic solvents and reagents, minimize energy requirements and use renewable natural resources. These "green" synthesized nanobiomaterials can be used to fabricate biosensors, therapy and diagnostics tools, medical imaging agents, as well as tissue engineering scaffolds and biomaterials. Indeed, microalgae are produced in photobioreactors in aqueous solutions at ambient temperatures and pressures and at a near neutral pH. Additionally, under the appropriate bioprocess conditions phototrophic algal cell cultures can catalyze the conversion of soluble metal cations, such as silver cation (Ag^{+}
literature that the reducing power required to drive this reaction is derived from the electron flux produced in the algae's photosynthetic apparatus, or specifically, in the thylakoid membrane. Furthermore, the as-produced NPs show a narrow size distribution because the thylakoid membrane provides a shape-directing template. This research carries out controlled photobioreactor cultivations for the biological production of Ag^{0} NPs. Photon, mass and energy balances are used to determine rate NP production kinetics and identify potential rate-limiting steps. This study represents a path forward towards the scalable, sustainable, economical production of metallic NPs from photosynthetic cell culture.

461539 Engineered Sigma Factors Increase Full-Length Antibody Expression in *Escherichia coli*

Abstract:

Escherichia coli **is one of the organisms of choice for production of antibody fragments such as Fabs and half-antibodies, however, obtaining high yields of full-length antibodies in *E. coli* remains a challenge. Global transcription machinery engineering (gTME), a powerful approach to improve cellular phenotypes, was used to improve expression of a full-length antibody (mAb1) in *E. coli*. A mutant library of one of the primary sigma factors in *E. coli*, RpoD (Ïf^{70}), was generated and cells expressing RpoD variants were screened for high levels of mAb1 expression using fluorescence assisted cell sorting (FACS). Three RpoD mutants were isolated via FACS and were shown to increase mAb1 expression by ~2â€“fold compared to the wild type RpoD. Transcriptional analysis using RNAseq of the cells expressing the sigma factor mutants revealed several interesting insights into the mechanistic role of these mutants in improving expression of mAb1.**

461706 Molecular Basis of 5-Hydroxytryptophan Synthesis in *Saccharomyces Cerevisiae*

Abstract:

5-Hydroxytryptophan (5-HTP) is an intermediate in the biosynthesis of the neurotransmitter serotonin1 and has been used in the clinic for over 30 years. 5-HTP is mainly extracted from the seeds of the African plant *Griffonia simplicifolia*, which, however, is largely limited by the season- and region-dependent supply of the raw materials. Recently, microbial production of 5-HTP has been achieved successfully in *Escherichia coli*[1,2], which offers an alternative route for producing this small molecule drug. Encouraged by the pioneering studies on *de novo* synthesis of 5-HTP in *E. coli*, a model prokaryotic microorganism, in this study we aim to explore the possibility of using eukaryotic organisms such as *Saccharomyces cerevisiae* to synthesize 5-HTP and uncover the molecular basis of 5-HTP synthesis in eukaryotic cells. We report that 5-hydroxytryptophan can be synthesized in *S. cerevisiae*by heterologously expressing prokaryotic phenylalanine 4-hydroxylase or eukaryotic tryptophan 3/5-hydroxylase, together with enhanced synthesis of MH4 or BH4 cofactors. The results showed that the activities of OcT5H and SmT5H were 50-75% higher than that of XcP4H and BpP4H when using BH4 as the cofactor. We also find that DFR1, a native enzyme that catalyze reactions for folate acid synthesis, can also catalyze the reaction for MH4 synthesis. It is the first time, to the best of our knowledge, that the molecular basis of 5-HTP synthesis was revealed in a eukaryotic organism, which paves the ways for future metabolic engineering (e.g., increasing the supply of tryptophan precursor and cofactors) to promote yeast-based production of 5-HTP.

1 Lin, Y., Sun, X., Yuan, Q. & Yan, Y. Engineering bacterial phenylalanine 4-hydroxylase for microbial synthesis of human neurotransmitter precursor 5-hydroxytryptophan. *ACS Synth Biol* **3**, 497-505, doi:10.1021/sb5002505 (2014).

2 Sun, X., Lin, Y., Yuan, Q. & Yan, Y. Precursor-Directed Biosynthesis of 5-Hydroxytryptophan Using Metabolically Engineered *E. coli*. *ACS Synth Biol* **4**, 554-558, doi:10.1021/sb500303q (2015).

461731 Pathway Compartmentalization in Peroxisome of *Saccharomyces Cerevisiae* to Produce Versatile Medium Chain Fatty Alcohols

Abstract:

Fatty alcohols are value-added chemicals and important components of a variety of industries, which have a >3 billion-dollar global market annually. Long chain fatty alcohols (>C12) are mainly used in surfactants, lubricants, detergents, pharmaceuticals and cosmetics while medium chain fatty alcohols (C6-C12) could be used as diesel-like biofuels. Microbial production of fatty alcohols from renewable feedstock stands as a promising strategy to enable sustainable supply of this important chemical. In this study, we report, for the first time, that medium chain fatty alcohols could be produced in *Saccharomyces cerevisiae* via targeted expression of a fatty acyl-CoA reductase (TaFAR or AdFAR) in the peroxisome. By tagging FAR enzymes with peroxisomal targeting signal peptides, the FAR could be compartmentalized into the matrix of the peroxisome to hijack the medium chain fatty acyl-CoA generated from the beta-oxidation pathway and convert them to versatile medium chain fatty alcohols (C10&C12). The overexpression of genes encoding PEX7 and acetyl-CoA carboxylase further improved fatty alcohol production by 1.4-fold. Fatty alcohols were produced at 1.3 g/L in fed-batch fermentation under nitrogen limited condition, including 6.9% 1-decanol, 27.5% 1-dodecanol, 2.9% 1-tetradecanol and 62.7% 1-hexadecanol. This work revealed that peroxisome could be engineered as a compartmentalized cell factory for producing versatile fatty acid-derived chemicals in *S. cerevisiae.*

462116 Metabolic Modeling and Pathway Engineering of an Industrially Relevant *Pseudomonas Putida* KT2440 Strain to Produce Muconate from Glucose

Abstract:

The biological production of muconic acid has received significant interest because of the metabolite's facile conversion to adipic acid and terephthalic acid, commodity chemicals used in large quantities as plastic precursors. The latter's traditional production route from petrochemicals involves the generation of toxic byproducts, further motivating microbial production strategies. A key limitation in the production of muconate from biomass has been the relatively low yields achieved from industrially relevant strains. In this talk, we describe a core-carbon metabolic model to guide strain engineering efforts to improve flux to muconate in a genome-integrated *Pseudomonas putida* KT2440 strain. Using elementary flux modes and constrained minimum cut sets, we determine an optimum set of knockout targets to force carbon flow into the engineered muconate pathway. After further pathway de-bottlenecking, we achieve muconate yields nearly equal to those expected from the stoichiometric model. This work highlights the metabolic flexibility of P. putida, as well as the advantages of the Entner-Doudoroff pathway in coupling growth to product formation.

462269 Encapsulating Yeast in Agarose Microparticles for High Throughput Microcolony Analysis

Abstract:

Yeasts and fungi are powerful hosts to produce numerous value-added chemicals. Metabolic engineering efforts in yeasts can be coupled with high-throughput analysis platform to effectively analyze and select desired phenotypes. However, choices of analyzing yeast cells using high-throughput are still limited. Traditional plate based colony screening can be non-quantitative and time-consuming while FACS based single cell screening can be associated with high false positive rate. To provide an alternative method for high-throughput analysis in yeasts, we performed experiments in a droplet microfluidic platform. We encapsulated single yeast cells into pico to nano-liter sized agarose droplets and allowed them to culture and to form isolated microcolonies in agarose microparticles. These microparticles can then be analyzed by high-throughput analysis platform such as flow cytometry and next generation sequencing. This workflow enables a quantitative and high-throughput analysis of yeast colonies.

462624 A Plug and Play Pathway Refactoring Strategy for Natural Product Discovery in *Escherichia coli* and *Saccharomyces Cerevisiae*

Abstract:

Natural products remain one of the major sources for drug discovery. However, expression of the corresponding biosynthetic gene clusters in heterologous hosts requires a pathway refactoring strategy with high efficiency, fidelity and modularity. Ideally, such a strategy should be automation-compatible in order to increase the throughput by robotic systems. Here, we report a plug and play pathway refactoring strategy for our automation platform, Illinois Biological Foundry for Advanced Biomanufcturing (iBioFAB), based on Golden gate reaction. Briefly, a total of 9 destination sites were predefined in a receiver plasmid and the genes to be refactored can be inserted into any of these sites by using helper plasmids. For destination sites without genes, we also computationally designed 9 linkers, which makes the assembly condition consistent and genes interchangeable. Two sets of helper plasmid were built for pathway expression in *Escherichia coli* and *Saccharomyces cerevisiae*, and high fidelity of assembly was observed in both cases. With the usage of counter-selection marker, the whole workflow can be finished in 2 days. As a proof of concept, 96 combinatorial carotenoid pathways were built automatically and a library of products with varying colors were generated successfully.

462501 Diminished Effectiveness of Neutralizing Antibodies for Respiratory Illnesses Due to Mass Transfer Limitations

Abstract:

Respiratory illnesses, such as whopping cough, RSV, and influenza, present a two-faced challenge for the design of bio-therapeutics. Immune recognition is easily overcome by common techniques such as library screening, phage display, and single-cell phenotyping. Yet, effectively neutralizing antibodies are often ineffective *in vivo*. We identify this discontinuity as a result of a mass transfer limitation. The common routes of antibody introduction to the body lack the lung localization that is seen in respiratory illnesses. We show here that an anti-*Bordetella pertussis* antibody is over 90% effective at neutralizing immune cell invasion *in vitro*, but exhibits nominal protection in an *in vivo* murine model when injected intraperitoneally. Rather, when directly administered to the lungs by a nanoparticle-assisted aerosol route, the *in vitro* predicted protection is observed. While further work is investigating a more practical route for larger proteins, we propose here an essential consideration in respiratory antibody engineering design.

462957 Development of a Surface Plasmon Resonance and Plasmon-Waveguide Resonance Combined Chip for Studying the Transport Behaviors of Cell Membrane Transport Proteins

Abstract:

Surface plasmon resonance (SPR) is a powerful label-free and contact-free technique for chemical and biological sensing experiments. However, the traditional use of SPR instruments is for molecular interactions on the surface of metallic film. In order to extend the application of SPR, we integrated the concept of plasmon-waveguide resonance (PWR) and proposed the idea of PWR/SPR combined chip in order to measure the transport behaviors of cell membrane transport proteins. The PWR/SPR combined chip is composed of a silica layer with sub-micron sized pores on a thin gold film. The geometry allows us to use SPR to detect the refractive index change in the pore region, which is correlated to the target species concentration inside the pore, and PWR to simultaneously monitor the change of refractive index at the top silica surface. We deposited the giant plasma membrane vesicles (GPMV) derived from cells onto the PWR/SPR combined chip to construct the lipid membrane with the membrane proteins suspending over the sub-micron sized pores. Consequently, the detection area can be divided into two regions including the space inside the pores and the region above the grating structure. By using COMSOL simulation, we confirmed that this system allows us to simultaneously measure the change of refractive indices in the two regions across the lipid membrane. We experimentally demonstrated how the platform can be used to study the glucose transport through the corresponding membrane transport proteins (Glut1, Glut2). In the future, we plan to use these platforms to monitor how various inhibitors or ligands could influence the transport dynamics of interested membrane transport proteins.

462625 Mass Spectrometry-Based Screening for Protein Engineering

Abstract:

Mass spectrometry imaging (MSI) has been instrumental to characterize spatial chemical heterogeneity of biological systems due to genetic diversity, cell differentiation, and intra-species interactions. Here we applied MSI in screening microbial strain libraries for directed protein evolution. First, we sought to produce designer rhamnolipids with defined molecular compositions by modifying the substrate specificity of the biosynthetic enzymes. In particular, rhamnolipids are a class of glycolipids with varying lipid chain lengths. We targeted the RhlA protein, which is responsible for the lipid moiety synthesis, and performed error-prone PCR for variant creation. The abundance ratios of different rhamnolipids were directly measured using MSI from individual colonies of the strain libraries. As a result, RhlA mutants were isolated with modified preferences towards the beta-hydroxylacyl ACP substrates, enabling production of rhamnolipids with specific lipid chain lengths. Second, we tried to activate cryptic pathways for natural product (NP) production in high throughput. We isolated strains with spontaneous mutations in ribosomal proteins, as ribosomal mutations may activate secondary metabolism (ribosome engineering). Using MSI, the isolated strains were rapidly screened to discover new peptidic NPs, and the results were used to guide the fermentation and purification of new compounds for structural elucidation. Overall, we established a metabolomics-based screening platform that may greatly expand the range of phenotypes that can be improved using protein engineering.

463195 Profiling the Humoral Antibody Response in Type 1 Diabetes Via Peptide Display

Abstract: Type 1 Diabetes Mellitus (T1DM) is characterized by an aberrant immune response that results in the production of autoantibodies that target islet cell molecules. While their exact role in the pathology is unclear, the number and type of autoantibodies present correlates with risk of T1DM, which is valuable in screening genetically at-risk populations. Since antibodies bind with high specificity, they can be screened using peptides that mimic their epitopes, providing information about their target molecules, or antigens. Bacterial cells displaying peptides can be used to represent a large, unbiased library of epitope mimics. The physical and chemical properties of ligands can be used to develop markers for disease and search for antigens that may play a role in disease genesis.

Using random peptide libraries coupled with fluorescent and magnetic-based cell sorting techniques (Figure 1), we performed an initial screening of 10 T1DM and 10 non-T1DM (control) plasma samples to uncover disease-specific peptide patterns, or motifs. 60% of the T1DM samples contained several motifs that were not prevalent in the control samples. However, these discovered motifs, when screened against a validation set of 78 additional T1DM and control samples, reacted with only a few additional disease samples (Figure 2).

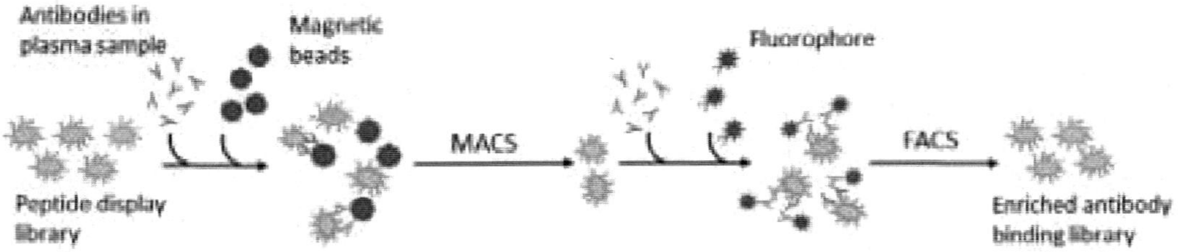

Figure 1: Screening bacterial display libraries with patient sera using magnetic-activated (MACS) and fluorescence-activated (FACS) cell sorting. Antibodies in plasma bind to peptides expressed by bacterial cells, allowing preferential collection using either magnetic beads or flow cytometry. Multiple rounds of sorting result in a library enriched in cells displaying peptides that bind to antibodies in the patient plasma.

Based on these initial findings and the knowledge that T1DM has a highly heterogeneous pathology, we expanded the screening strategy, organizing the samples into well-characterized subgroups, such as those patients positive for a specific T1DM-associated autoantibody. Peptides isolated from subgroups may have properties that confer T1DM specificity. For example, we screened an additional 11 T1DM samples predominately positive for the insulin autoantibody (IAA) to facilitate comparison with sequences from patients testing negative for IAA, with the aim of identifying candidate antigens with similarities to peptides isolated from these distinct T1DM patient subgroups.

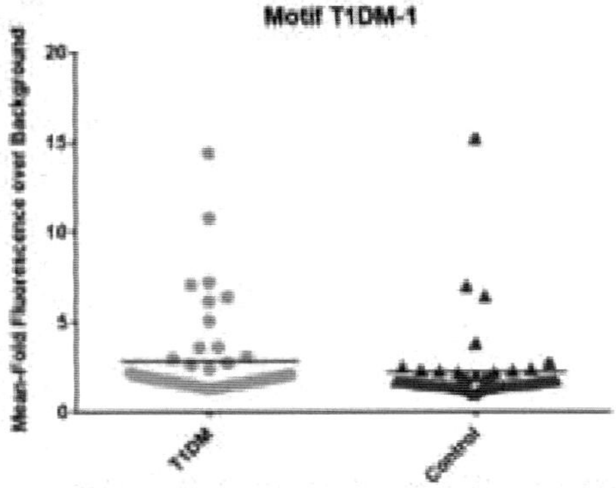

Figure 2: Flow cytometry screening of a motif. The discovered motifs were not cross-reactive with most of the validation samples, demonstrated by fluorescence similar to the background signal.

- Abstract Text

463495 Structural Basis of Macrocyclization in a Fungal Nonribosomal Peptide through a Condensation-like Termination Domain

Abstract:

Nonribosomal peptide synthetases (NRPSs) in fungi biosynthesize some of the most important pharmaceutical compounds, including penicillin, cyclosporine and echinocandin. The key difference between fungal and bacterial NRPSs is the use of a condensation like (C_T) domain instead of the canonical thioesterase (TE) domain in catalyzing the macrocyclization reaction. In contrast to the TE domain, the C_T domain does not form covalent intermediate during catalysis and requires specific protein-protein interactions with the upstream T domain. To understand the fungal strategy of forging the macrocyclic peptide linkage, we determined the crystal structures of the C_T domain and the holo T-C_T complex of *P. aethiopicum* TqaA, a trimodular NRPS involved in fumiquinazoline F biosynthesis. Our work revealed unique structural elements in C_T that differentiates from the canonical C domains, including structural elements that seal off the acceptor site and prevent solvent entry or further peptidyl chain elongation. Solution small angle scattering (SAXS) analysis revealed phosphopantheteine prosthetic attachment in the T domain modulates the dynamic interactions between the T and C_T domains. The structure of the TqaA C_T domain represents the first C or C-like domain determined for a fungal NRPS, and provides a molecule blueprint for engineering the sizes and structures of this large family of natural products.

463530 Structural Insights into Catalytic Function of Non-P450 Hydroxylase (HpaB) from *Escherichia coli*

Abstract:

Hydroxylation plays a pivotal role in the synthesis of various natural products. Regio-specific hydroxylation is usually difficult to be achieved via chemical synthesis approaches due to its low selectivity and requirement for excitation, protection and de-protection of other groups. In biological systems, a number of cytochrome P450 hydroxylases can catalyze hydroxylation of inert carbon. However, it is hardly used in metabolic engineering owing to its low stability, lower activity and the requirement for a reductase partner. 4-hydroxyphenylacetate 3-hydroxylase (HpaB) is a flavin adenine dinucleotide (FAD) dependent hydroxylase existing in *Escherichia. coli*. It has broad substrate spectrum and can efficiently catalyze *ortho*-hydroxylation towards a series of phenylpropanoids to produce the hydroxyl derivatives, while the latter are widely used in clinical research and pharmacological industry. In this work, a structure-based study is conducted to expand the catalytic function of HpaB. Firstly, we elucidated the crystal structure of HpaB by X-ray crystallography. Based on the structure, we predicted that a loop covering the entrance of the catalytic site may play a significant role in substrate selection and catalytic activity. In order to extend the substrate spectrum and improve the catalytic efficiency to large phenylpropanoid molecules, seven variants have been generated to change the rigidity of the loop structure. Four different substrates with different molecular sizes, including *p*-coumaric acid, 7-hydroxycoumarin, resveratrol and naringenin, were used to test the activity of those variants. We found one mutant enzyme with the most flexible spatial conformation in the loop structure has shown the highest catalytic activity towards all four substrates. *In vivo* bioconversion and *in vitro* enzyme assay of the mutant indicated that the conversion rate was increased by 80% and the *Km* value was decreased dramatically towards the largest substrate naringenin compared with wild type HpaB. Especially, this mutant exhibited the highest *ortho*-hydroxylation activity towards naringenin reported so far. These results suggest that the flexible nature of the loop is an important determinant for the enzyme to accommodate larger molecules. This research increases the substrate accommodations and improves the activity of hydroxylase HpaB via structure-based rational design, which implied potential application in metabolic engineering of large phenylpropanoids and provided new thoughts in protein engineering of spatial conformation of the enzyme.

463719 UsingÂ Plasmon Waveguide Resonance Spectroscopy to Study the Conformational Change of Membrane Proteins

Abstract:

Membrane proteins are unique and essential for carrying out cellular operation. Studying conformation change of membrane proteins upon stimulation is important for investigating their functions. However, measuring the real-time dynamic change at nanometer scale is still challenging. Here, we developed a method to estimate how the structures of membrane proteins change in supported lipid bilayers by plasmon waveguide resonance (PWR) spectroscopy, which is a birefringent optical label-free sensing technique. The method uses both p- and s-polarized lights to detect the membrane protein mass density change and structure change in the directions perpendicular to and parallel to the lipid membrane plane. We built a PWR sensor platform with a silicon dioxide layer on a thin gold layer, and used 2D COMSOL Multiphysics material model and Maxwell equation to simulate the influence of protein configuration alternation on the PWR sensor chip. We reported the correlations between the PWR resonance angle shifts and the protein mass density and conformational changes. The overall mass density change is found to be majorly correlated to the resonance angle shift of s-polarized light, and the conformational change can be derived from the resonance angle shift of p-polarized light. We used bacteria rhodopsins with known structural responses to light to verify that the sensor chip and the methodology we developed can be used to estimate the conformational change of membrane proteins in supported lipid bilayers.

463828 Artificial Cellulosome Designed from Cellobiohydrolase Module Library on Streptavidin

Abstract:

Celluloses is naturally occurring plant cell wall components and most abundant polysaccharides on earth. An economical conversion from cellulosic biomass materials to water-soluble sugar units is expected to produce food, feed, chemicals and fuels. Microbial cellulolytic enzymes, called cellulases, can convert cellulose to water-soluble sugar units with low energetic and environmental loads. However, highly crystallized structure of cellulose in nature makes efficient hydrolysis hard, even using cellulases from cellulolytic fungus *Trichoderma reesei*, so that drastic improvement of enzyme activity is one of serious issues for efficient cellulose conversion. Many cellulases are consist of modular units of independent catalytic domains (CDs) and carbohydrate-binding domeins (CBDs), and on some rumen bacteria, CDs with different functions are clustered on a giant scaffold protein containing a CBD, via cohesin-dockerin interaction, called cellulosome, to efficiently degrade cellulose. To degrade cellulose effectively, two kinds of cellulase are essential, such as endoglucanases (EG), which randomly cleave internal bonds at amorphous cellulose, and cellobiohydrolases (CBH), which move processively along the cellulose chain and liberate cellobiose unit from the end of cellulose chain. Especially, CBH is a key enzyme capable of degrading crystalline cellulose. Recently, we proposed a new design of artificial cellulosome to drastically enhance cellulase activity.[1-3] The clustering of CDs of EG with CBD on streptavidin via biotin-avidin interaction the reducing sugar produced by free CD. In this study, we clustered various CDs of CBH and CBD on a streptavidin to enhance the degradation activity. To identify the clustered enzyme with optimal CBH and CBD combination, 5 kinds of CBH and 8 kinds of CBD were clustered on streptavidin and analyzed the reducing sugar produced from phosphoric acid swollen cellulose. Consequently, 19 kinds of clustering enzymes with CBH showed the increase of produced reducing sugar, while 21 kinds of clustered enzymes decrease of the amount of sugar. The results showed that the clustered enzymes with CBH have a different behavior from those with EG, and indicates that module structure in CBH might be repuired to express the function in cellulose degradation.

[1] Kim *et al.*, *Small.*, 7 (5), 656-664(2011)

[2] Kim *et al.*, *Catalysis Science and Technology.*, **2** (3), 499-503(2012)

[3] Nakazawa *et al.*, *ACS Catalysis.*, **3,** 1342-1348, (2013)

463531 *De Novo* Biosynthesis of 3,4- Dihydroxybutyric Acid in Engineered *Escherichia coli.*

Abstract:

3-Hydroxy-Î³-butyrolactone (3HBL) is a building block widely used in pharmaceutical industry. Many important pharmaceuticals are derived from 3HBL such as anti-cholesterol medicines Lipitor and Crestor, lipid-lowering drug Zetia, nutrition additive L-carnitine and HIV inhibitors. Its market price is about $450 per kilogram. Owing to its great commercial value, US Department of Energy listed 3HBL as one of the top value-added compound. Currently, 3HBL is mainly produced by chemical synthesis, which involves hazardous materials and processing conditions. Only one biosynthetic pathway has been reported that use glucose and glycolate as the substrates and coenzyme A as the activator, leading to 221 mg/L of 3HBL and 555 mg/L of 3,4- dihydroxybutyric acid (3,4-DHBA, hydrolyzed form of 3HBL). However, this pathway involves multiple steps (>10 steps) and needs exogenous supplement of glycolate that limit the production and yield. Here we established a novel and efficient biosynthetic pathway for 3,4-DHBA that can be easily converted to 3HBL by acid treatment using xylose as the substrate in *Escherichia coli.* The new pathway contains only 4 steps. We first screened efficient enzymes for each step through *in vitro* enzyme assay. Then we introduced the optimized pathway into *E. coli*, resulting in production of 3,4-DHBA *in vivo*. Additional host strain engineering enables the titer of increased up to 2.1 g/L 3,4-DHBA in shake flask experiments, which is the highest titer reported so far. This study demonstrates a novel biosynthetic pathway for the production of 3,4-DHBA from xylose in *E. coli*, and this engineered pathway indicates the potential for industrial manufacturing of 3,4-DHBA and 3HBL.

463924 Domain Library Approach for Designing Therapeutic Small Antibody with High Cytotoxicity for Tumor Cells

Abstract:

Antibody has a hierarchical structure constructed from structurally-independent domains, and various recombinant antibodies with artificial structural formats are designed from the domains. Bispecific Diabody (BsDbs) is a small antibody designed only from two kinds of the fragments of variable region (Fvs), and the diabody composed of the Fvs from anti-cancer and anti-lymphocyte antibodies can form a bridge between cancer and lymphocyte cells to make an effective damage on cancer cells. The cytotoxicity of diabody is dependent on the characters of Fv used, that is, targets on cancer cell / T-lymphocyte and the epitopes of the targets. Here, we propose a methodology for rapid screening of the most effective bispecific antibody of more than 100 diabodies designed from several anti-cancer and anti-lymphocyte IgG type antibodies.

We applied each medium fraction to immobilized metal affinity chromatography (IMAC) to purify expressed diabodies. We step-wisely measured the cytotoxicity of IMAC-refined diabodies at the various diabody concentrations by MTS assay. In this assay, all samples were prepared to less than 50 mM imidazole concentration, because imidazole showed the cytotoxicity against cancer cells at more than 75 mM. In the result of step-wise screening, we got 6 kinds high cytotoxic diabodies which damaged at cancer cells at 100 fM concentration

By means of the screening, we found some rules of high cytotoxicity, specific domain order, target molecules, and specific antibodies for highly active diabody. These results imply that high cytotoxic diabodies have universal rules of structure and using antibodies for diabody.

463829 Selection of CDR-Mimic Peptide By Using Phage Display Method

Abstract:

Biological medicines, such as peptides and antibodies, have been expected to substitute for chemical medicines, because biomolecules have higher specificity for target molecules than chemical compounds. Peptides as drugs with target specificity are more easily selected by in vitro evolution technology than antibody, as well as low production costs that can be synthesized with a general chemical synthesis. However they tend to be readily degraded in vivo and have a limited specificity for target molecules. Although antibody recognizes target molecules with high binding affinities and specificities and shows long biological half-life, development of antibodies for therapy needs for complicated process with a long period. Here, we designed peptide format which can be used not only for peptide medicines, but also for the grafting into antibody: the grafting of the peptide into complementarity determining region (CDR) of camel-type single variable antibody fragment (VHH) to express the function of grafted peptide in antibody.

Â To design the peptide format for the grafting into CDR of VHH, we employed the crystal structure of VHH of camel anti-BcII Î²-lactamase antibody cAbBCII-10. In previous research, the grafting of foreign peptide into CDR1 and CDR3 induced no denaturation of framework, but some trials could not functionalize the grafted VHH. We therefore measured the length of Î±-carbon atoms between N- and C-terminus of CDR loop based on the crystal structure to make the peptide format that form the same steric structure in CDR as free format has. The measurement gave that the length of Î±-carbon atoms between N- and C-terminus of CDR3 is 8.70 Ã…. This result showed that CDR3 loop could be mimicked with introduction of disulfide bond.

Â Next, we reproduced the CDR mimicked peptide loop on M13 phage surface. To easily construct a large scale library, we applied inverse PCR method to prepare M13 phage libraries. Our method could be expected to reduce loss of library DNA for following electroporation because of 4 steps manipulation procedures. As a result, we constructed M13 phage library with a complexity of $\sim 10^7$ clones.

464604 Rationally Metabolic Engineering of *Clostridium Cellulovorans* for Butanol Production

Abstract:

Clostridium cellulovorans can produce butanol from biomass through consolidated bioprocessing. In this study, we rationally metabolically engineered the *C. cellulovorans* to produce lignocellulosic butanol. Our proteomics analysis showed that the high expression of thiolase (*thl*), which catalyzes the reaction from acetyl-CoA to acetoacetyl-CoA, had strong correlation with butanol production, so the heterologous *thl* gene from *C. tyrobutyricum* was overexpressed in *C. cellulovorans*. To selectively produce butanol, the aldehyde dehydrogenase (*ald*) and butanol dehydrogenase (*bdh*) were synthesized to covert butyryldehyde to butanol. The butanol production is conventionally limited by the byproduct formation during the fermentation, such as acetate. ClosTron-mediated engineering and CRISPR interference technology were applied to decrease the acetate kinase and phosphotransacetylase activities, which are directly related to acetate formation. Finally, the constructed engineered mutants of *C. cellulovorans* were evaluated in consolidated bioprocessing. The acetate was decrease by > 50% and the selectivity of butanol was improved to >0.5 g/g. The lignocellulosic fermentation produced butanol with titer of >2 g/L with the yield at 0.2g/g. The changes in carbon, energy and redox balance in response to metabolic engineering were also identified to assist further engineering of butanol production process.

464091 Engineering of Microbial Cell Factories for Biobased Chemicals Production

Abstract:

The production of chemicals through biological systems offer promising solutions to global challenges presented by a carbon constrained future and an increasing global population. In the past decade, rapid advances in DNA sequencing and synthesis and genome editing technologies have accelerated our ability to program biological systems for the production bio-based chemicals from renewable feedstocks. However, challenges such as feedstock utilization, microbial cell factory engineering and stabilization of the production strainâ€™s genetics, still remain.

The Metabolic Engineering Research Laboratory in Singapore is focused on addressing these challenges through the design and engineering of recombinant microorganisms. In this presentation, I will highlight our accomplishments including: (a) deconstruction of cellulose through engineering a *Saccharomyces cerevisiae* pentafunctional minicellulosome system with enhanced cellulose degradation capabilities, which can also be applied as a cellulase engineering platform, (b) metabolic engineering of microbial cell factories for the production of biochemicals, such as n-butanol and adipic acid, and (c) development of an efficient technological platform for the integration of large genetic pathways and circuits (up to 24 kb) into the yeast genomes to create genetically stable production hosts.

464711 Characterization of Molecularly-Complete Planar Supported Cell Plasma Membranes for Biomedical Applications

Abstract:

Emerging technologies to study membrane proteins and protein-lipid interactions use cell plasma membrane blebs as an intermediate to form planar bilayers that are compatible with a variety of characterization tools and microscopy methods. This technology allows membrane proteins to be incorporated into supported lipid bilayers without using detergents and reconstitution, preserving the native lipids and other species within the plasma membrane. However, the impact of chemicals used to induce cell blebbing on protein properties is still unknown. Our research focuses on characterization of the cell blebs under various blebbing conditions and the result on protein behavior (mobility, activity, orientation, *etc.*) in this emerging platform. Bilayers formed from cell blebs expressing GPI-linked yellow fluorescent protein (GPI-YFP) provide a simple way for us to study the mobility of proteins through the single particle tracking (SPT) and how properties change, e.g., confinement effects, immobility, *etc.*, with the conditions used for bleb generation. The orientation of GPI-YFP in the cell blebs and bleb bilayer is revealed using a protease cleavage assay. We will discuss these characterizations of supported cell plasma membranes and the application of this platform to create adult stem cell bilayers for the study of oncogenic microvesicle interactions.

465061 Electric Field Enabled Continuous Oxidation Catalyzed By Laccase

Abstract:

Electric Field Enabled Continuous Oxidation Catalyzed by Laccase

Liwei Ren; Diannan Lu*; Zheng Liu

Key lab of Industrial Biocatalysis, Ministry of Education

Department of Chemical Engineering, Tsinghua University, Beijing 100084, PR China.

E-mail ludiannan@mail.tsinghua.edu.cn

ABSTRACT

Laccase is a kind of copper-containing oxidase found in plants and fungi. It is unique capability in catalysing oxidation of phenols using molecular oxygen attracted immense efforts in exploring its applications in textile, papermaking, wastewater treatment and so on. However the poor activity resulted from the lagged electron transfer among the cooper ions in the active site of laccase hinders the above mentioned applications of laccase.

Here we present a method using electric filed to enhance the intramolecular electron transfer of laccase. In the present study, laccase from *Trametes versicolor* (TvL) was adsorbed on a mesoporous carbon nitride (CN) matrix consisting pyridine and benzene rings interconnected by nitrogen atoms. The immobilized laccase was coated onto Au electrode using PVA as agglutinant, giving rise to a novel CN-Lac/PVA/Au electrode. It is shown that CN-Lac/PVA/Au was highly active for O_2 reduction, requiring an over-potential as

low as 0.013 V to generate an current density of 103.1 μA/cm^2. The structural characterization indicates the protuberances at the CN internal surface acted as molecular wires to connect the active site of laccase and Au electrode.

To test of workability of the CN immobilized laccase in the electric field in the absence of oxygen, oxidization of 2,2'-azinobis-(3-ethylbenzothiazoline-6-sulfonate) (ABTS) was performed at pH 5.5 and 25°C under anaerobic condition. While laccase catalysed oxidization of ABTS scarcely proceeded (<0.2% within 1800 s) in the absence of electric field, the application of electric field enabled a continuous oxidation at a conversion rate of 32 μmol/min. Reproducible ABTS oxidation yield as well as cyclic voltammetry were obtained in 5 cycles. The continuous oxidation by laccase in the absence of oxygen, we believe, is attributed to the timely electron transfer by electricity to the active site of laccase through the nanowires inside the CN matrix and holds great promise for unprecedented applications.

465071 Rare Residues Mutation Approach for Changing Specificity of Antibody

Abstract:

Glioma is a type of tumor formed in brain, and mutations at a functional arginine residue of isocitrate dehydrogenase 1/2 (IDH1/2) in glioma have been reported. In the case of IDH1, mutations of the arginine 132 residue (R132) of IDH1 in glioma have shown good prognosis of patients [1]. Therefore, the R132-mutated IDH1s can be a promising prognostic marker in glioma. Antibodies specific for the R132-mutated IDH1 are powerful tools for glioma diagnosis, however developing antibodies to all of the IDH1 mutants is challenging. Recently, multi-specific anti-mutated IDH1 monoclonal antibody, termed MsMab-1, has been generated [2]. This antibody showed specific binding to four R132-mutated IDH1 (R132H, R132S, R132G and R132L), enabling detection of multiple IDH1 mutations by a single antibody. In this study, we aim to develop an antibody that can detect all of the R132-mutated IDH1s by improving multi-specificity of MsMab-1 by directed evolution.

We first constructed a single chain fragment of variable region (scFv) format of MsMab-1 antibody that allows us to perform high throughput functional analysis of MsMab-1 mutants. We transformed *E. coli* with the vectors encoding scFv of MsMab-1. The specificity of MsMab-1 scFv against various R132-mutated IDH1 was measured by sandwich ELISA. As a results, MsMab-1 scFv showed specificity against four R132-mutated IDH1, indicating that MsMab-1 scFv retained the multi-specificity of its parental antibody, MsMab-1 IgG.

Next, we attempted to improve multi-septicity of MsMab-1 scFv by a site-specific mutagenesis. Because MsMab-1 antibody has an unique molecular recognition mechanism, we focused on unusual amino acid residues of MsMab-1 compared to other antibodies. We prepared all mutants at these amino acid residues, respectively. High throughput functional analysis revealed that Cys 47 residue in VH was critical for multi-specificity of MsMab-1. Unexpectedly, a C47N mutant showed improved affinity and multi-specificity. These results suggest that rare residue in variable domains influences specificity of antibody.

Together, we succeeded in preparing MsMab-1 scFv and found that rare residues influenced the specificity of MsMab-1. We are currently working on further improving multi-specificity by introducing mutations at multiple rare residues.

[1] H.Yan *et al.*, *New England Journal of Medicine*, **360**, 763-73, 2009.

[2] Y. Kato, *Brain Tumor Pathology,* **32**, 3-11, 2015.

465316 Antisense RNA Exclusively Regulate Key Genes in Anaerobic Fungi

Abstract:

Anaerobic fungi resident in the digestive tracts of large mammalian herbivores degrade crude plant material through the expression and regulation of powerful cellulolytic enzymes. These enzymes are strongly expressed on demand with specific activities tuned to the architecture of the available substrates. Thus, elucidating the specific regulatory patterns of these genes may reveal more efficient strategies of converting complex plant materials to the simple sugars needed for microbial biofuel production. Here, we use strand-specific RNASeq and genomics to examine the transcriptional regulatory patterns of 3 non-model species of anaerobic fungi: *Anaeromyces robustus, Neocallimastix californiae,* and *Piromyces finnis*. In this work, we sequenced the genomes of these organisms and collected their transcriptomes when grown on a collection of monosaccharides, disaccharides, oligosaccharides, and complex plant material. Mapping these transcripts to the genome revealed that non-coding antisense RNA formed up to 30% of the transcriptome. These antisense RNA target genes involved in a number of cellular processes, including lignocellulose hydrolysis. Antisense RNA arose early in gut fungal genomes with transcripts targeting genes involved in RNA processing and protein expression sharing a common ancestor across all three fungal genera. Similarly, antisense RNA is critical to the regulation of genes involved in protein expression, and DNA replication and repair as revealed by their functional enrichment in fungal genomes and convergent evolution. More importantly, antisense regulation is not redundant â€“ genes targeted by antisense RNA differentially expressed under various substrates are typically not transcriptionally regulated. That is, seemingly constitutive genes were, in fact, regulated exclusively via antisense RNA. Collectively, this work demonstrates the central nature of antisense RNA regulation in anaerobic fungi and highlights the importance of antisense RNA in understanding gene regulation.

465391 Antibiotic and Physical Treatments Shift Multispecies Staphylococcal Biofilm Growth Behaviors and Structures

Abstract:

We report that the growth behavior and structure of multispecies staphylococcal biofilms was sensitive to antibiotic and physical treatment conditions. *Staphylococcus aureus* and *Staphylococcus epidermidis* commonly adhere to medical device surfaces (i.e. orthopedic implants) and subsequently form bacterial biofilms that can contribute to infection and virulence. However, little is known about the growth behavior and structure of these strains in multispecies communities. Here we establish that *S. aureus* has a higher prevalence than *S. epidermidis* in unstressed growth conditions. We report the structural changes of single and multispecies staphylococcal biofilms during 18-hours of surface colonization and biofilm formation. Finally, we assess the variations in multispecies staphylococcal biofilm growth behavior while undergoing antibiotic (vancomycin) and physical (pH, temperature) treatments. Understanding variations in multispecies staphylococcal behavior in different growth environments is important for guiding treatments that disrupt multispecies biofilms at infection sites.

465653 Nemaflex: A Microfluidic Tool for Measuring Muscle Strength in *C. Elegans* Across Lifespan

Abstract:

Maintenance of physical fitness is essential for an individual's health and well-being. In fact, a decrease in physical fitness indicates an increased risk of hospitalization and a higher mortality risk, especially in elderly populations. While there are many aspects of physical fitness, a decline in muscle strength correlates with poor physical performance. Loss of muscle strength alone is a prognostic indicator for a variety of disorders including dynapenia, cancer, and neuromuscular diseases. Although it is straightforward to record muscle strength in people, it is not feasible to conduct whole life studies looking at the impact of genetics on muscle health. Promisingly, prospective life-long studies can be accomplished in the roundworm *C. elegans*. The nematode body wall muscles have similarities to the human muscle and strikingly also deteriorate with age much like in humans which makes *C. elegans* a premier genetic model for muscle strength investigations.

Building on recent advances in microfluidic tools for mechanobiology, in this study we report a soft deformable micropillar based device, NemaFlex, for quantifying muscle strength in *C. elegans*. Animals crawl through the pillars pushing them, allowing extraction of local forces from pillar displacements. It is expected that the forces fluctuate depending on animal's behavior, velocity, body shape and position with respect to pillar, making quantitation of animal strength thus far elusive. Driven by the need to anchor NemaFlex for high throughput muscle strength assays, we develop a robust experimental protocol and analysis workflow for quantifying maximum strength in *C. elegans* and its genetic mutants. We also configure NemaFlex for recording strength virtually from 'womb to tomb' providing insights into how muscle strength changes during development and declines with age.

We find that forced contractions, such as induced by an acetylcholine agonist, show the same maximum strength as the untreated animals suggesting NemaFlex quantitates true strength. We tested mutants with neuronal defects (*unc-17*) and impaired sarcomeres (*unc-52* and *unc-112*) and observe changes that verify the neuromuscular origin of strength. We also address how animal size affects muscle strength, allowing accurate comparisons among worms at different developmental stages or among mutants with size variations. We profile strength across the lifespan for individual worms, and our results show that strength increases 7-fold from the young adult followed by a sharp late-age decline—providing the first direct evidence of muscle strength loss due to aging, similar to dynapenia in humans. In summary, NemaFlex is a powerful tool to conduct prospective life-long investigations of muscle strength in *C. elegans*.

465667 Lipidomic and Transcriptomic Biomarkers for Diagnosis of Nonalcoholic Fatty Liver Diseases

Abstract:

In humans, accumulation of fat in the liver and the associated inflammation, over many years, can results in what is referred to as nonalcoholic fatty liver diseases (NAFLD). NAFLD consists of three stages, namely, steatosis, (nonalcoholic) steatohepatitis (or NASH) and cirrhosis. Steatosis is characterized by accumulation of fat and some inflammation and can be reversed (to normality) by controlling diet and lifestyle changes. NASH is characterized by substantial inflammation and may or may not be reversible in which case it progresses to cirrhosis or liver cancer. Since progression of NAFLD may be reversible during the steatosis and NASH stages, it is important to be able to accurately diagnose the stage of NAFLD. With that objective, in this work, we have developed a two-stage linear discriminant analysis (LDA) based approach to identify lipidomic and transcriptomic biomarkers to identify (classify) among people at distinct stages of NAFLD and normal people (Gorden et al., 2015).

Using lipidomic and transcriptomic data from about 90 human liver-tissue samples, we developed a LDA-based classifier. We found that about 75-85 variables (genes or lipids) are needed to accurately classify among all the four classes/conditions (normal, steatosis, NASH and cirrhosis) whereas about 20 genes or lipids were enough to classify between steatosis and NASH samples. These findings are in contrast to our initial expectation to be able to find a small number of biomarkers, for example, 5 or less genes or lipids, for accurate classification. Two likely reasons for the need of a larger number of biomarkers for complete diagnosis are: (1) relatively much larger variability in the measurements in tissue-level samples as compared to that in cell culture, and (2) the complex and systems-level nature of the disease. Towards the latter possibility, we carried out KEGG pathway and GO-term enrichment of the genes differentially regulated among the four classes or between steatosis and NASH. We found that pathways such as focal adhesion and ECM-receptor interaction were enriched in NASH vs. steatosis comparison suggesting increased inflammation during the progression from steatosis to NASH. Fibrosis pathway and collagen-related genes were

enriched in the overall comparison, indicating their role during the late stage NASH and cirrhosis. Using PLINK, we also analyzed single nucleotide polymorphism (SNP) data from the samples. The well-known SNP marker in PNPLA3, rs738409, showed a strong statistical significance with a p-value of 1.7E-6 between normal and cirrhosis samples, but not between NASH and steatosis samples. We are continuing further integrated analysis of the gene-expression and SNP data. Overall, we conclude that progression of NAFLD involves many genes, lipids and pathways and only a systems-based classification and analysis approach will lead to accurate/reproducible diagnosis and mechanistic understanding of NAFLD.

Acknowledgements: This study were primarily supported by the National Institutes of Health (NIH) glue grant GM U54069338 to the LIPID MAPS consortium. We would also like to acknowledge the National Science Foundation (NSF) collaborative grant STC-0939370, and NIH R01 grants HL087375-02, HL106579 and HL108735.

References:

Gorden, D. L., D. S. Myers, P. T. Ivanova, E. Fahy, M. R. Maurya, S. Gupta, J. Min, N. J. Spann, J. G. McDonald, S. L. Kelly, J. Duan, M. C. Sullards, T. J. Leiker, R. M. Barkley, O. Quehenberger, A. M. Armando, S. B. Milne, T. P. Mathews, M. D. Armstrong, C. Li, W. V. Melvin, R. H. Clements, M. K. Washington, A. M. Mendonsa, J. L. Witztum, Z. Guan, C. K. Glass, R. C. Murphy, E. A. Dennis, A. H. Merrill, Jr., D. W. Russell, S. Subramaniam and H. A. Brown, â€œBiomarkers of NAFLD Progression: A Lipidomics Approach to an Epidemicâ€, *Journal of Lipid Research*, **56**(3), 722-736, 2015.

Corresponding author: Shankar Subramaniam, Department of Bioengineering, University of California San Diego, 9500 Gilman Drive, La Jolla, CA, 92093-0412, Phone: (858) 822-0986, E-mail: shankar@ucsd.edu.

Combinatorial Engineering *Yarrowia Lipolytica* As a Platform for Production of Long Chain ω - Hydroxy Fatty Acids from Renewable Feedstock

AIChE®
2016 Annual Meeting, San Francisco

Ali Abghari and Shulin Chen

Bioprocessing and Bioproduct Engineering Laboratory, Department of Biological Systems Engineering, Washington State University

ali.abghari@wsu.edu

1. INTRODUCTION

Bio-based production of long chain monomers can steer chemical industry from utilizing non-renewable hydrocarbon feedstock to renewable carbohydrates and by-products.

Long chain omega-hydroxy (ω-hydroxy) fatty acids (LCHFA) have many commercial uses as building blocks for production of:

- poly(ω-hydroxy fatty acids) and green synthetic fibers of excellent quality
- cosmetic intermediates, anticancer medicals, polyketides antibiotics
- macrocyclic lactones
- lubricants and adhesives

in chemical, food, cosmetic and bioplastic industries.

Y. lipolytica is an industrially important microorganism. This safe organism has strong capability for accumulation and degradation of lipid and lipid precursors.

It can utilize a wide variety of carbon sources including sugars from pretreated cellulosic biomass and glycerol from biodiesel industry (Abghari and Chen 2014).

It is also versatile for P450-catalyzed biotransformation of hydrophobic compounds (Mauersberger, 2013).
This platform has been studied before for biodegradation of oil to long chain oxygenated fatty acid monomers (Waché et al., 2006; Gatter et al., 2014).

2. CURRENT TECHNOLOGIES AND LIMITATIONS

LCHFA can be chemically synthesized:

- reductive cleavage of olefinic fatty acids
- reduction of dicarboxylic acid

Problems:
harsh reaction conditions, use of hazardous reagents, less selective, various byproducts formation, costly purification, environmentally costly, dependency on crude/plant oils and higher green-house gas (GHG) emissions thereof

Limited commercial availability and high price have prevented LCHFA widespread growth/usage in many applications.

A bio-based process for the direct one-step biosynthesis of LCHFA from renewable unrelated carbon sources will be safer, milder, more selective, and relatively cost effective.

The current bio-process faces:

- Use of pathogenic yeast microorganism (e.g., *C. tropicalis*) (Lu et al., 2010)
- Dependency on non-renewable feedstock (e.g., crude/plant oil and low solubility of hydrophobic feedstock

3. BIOSYNTHESIS OF LCHFA FROM RENEWABLE FEEDSTOCK

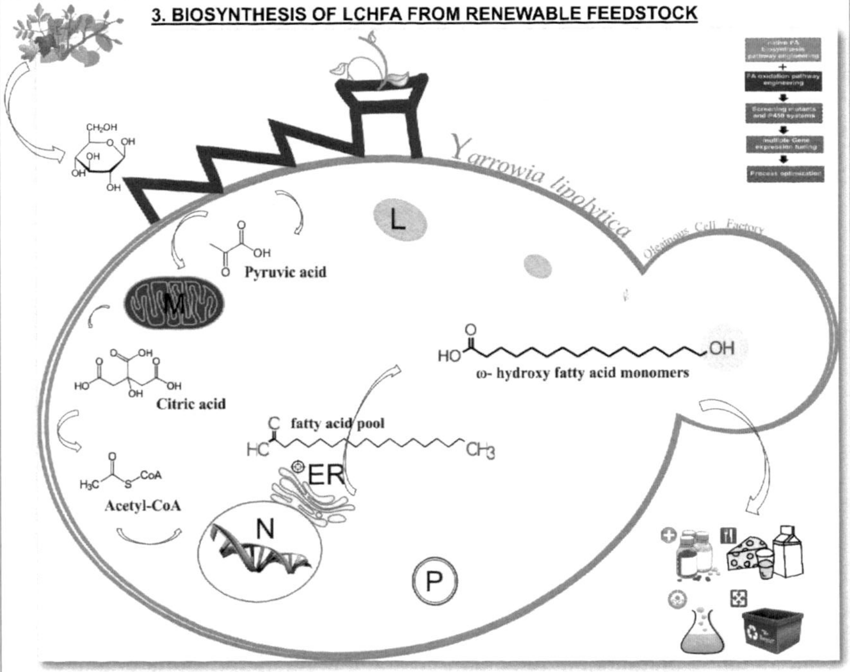

4. RESEARCH STRATEGIES

➢Enzymes catalyzing the bioconversion of fatty acids to LCHFA are cytochrome P450 monooxygenases, the hydroxylation activity of yeast CYP52 toward C_{12}–C_{18}fatty acids, is higher than the bacterial ones

➢Engineering FA biosynthesis pathway to create pool of long chain fatty acid (e.g., FA-C16) as precursor. This pool can be created by elimination of non-essential FA utilization pathways (storage lipid), disruption of FA activation and degradation pathways

➢Formulation of culture media for induction of FA biosynthesis pathway

➢Engineering ω -oxidation pathway as well as disruption of β-oxidation pathways to selectively hydroxylate endogenously synthesized FAs and to prevent further degradation of precursors and products, respectively

➢Screening of potential native P450 monoxygenases for selected *YIALK* genes in combination with native P450 reductase as well as potential mutant strains

5. RESUTLS

GC-MS analysis of the products

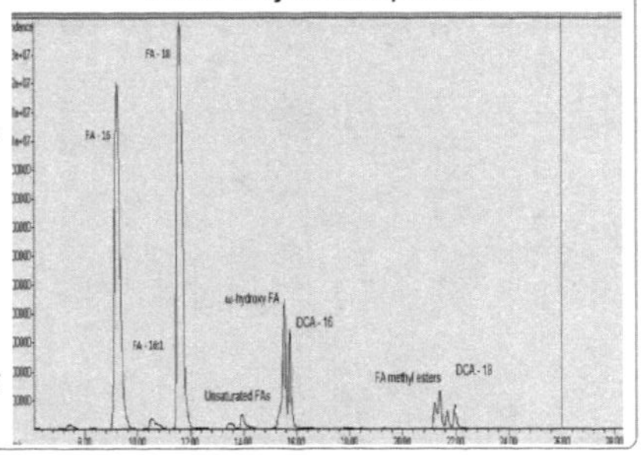

6. FINDINGS AND CONCLUSIONS

The presentation employs a simple ONE STEP fermentation using renewable feedstock (sustainable raw materials) and/or industrial waste products (e.g., glycerol) procedure rather than multiple chemical and/or biocatalytic processes for production of the LCHFA monomer

This unique biosynthetic pathway performs engineered *Y. lipolytica* as whole-cell biocatalyst for *de novo* production of LCHFA-16

Combinatorial engineering FA metabolism pathway and FA oxidation pathway resulted in production of the monomers at minimum titer of 220 mg/L under nitrogen limiting conditions of cultivation

6. FUTURE WORKS

Despite some improvements, production of our target ω-hydroxyl fatty acid needs more improvement in space time yields for commercial production

Enhancement of yield and productivity by following strategies is recommended:

- Protein engineering of selected enzyme and co factor engineering for rational increase of NADPH pool
- Identification of all native contributory genes to ω-oxidation pathway for further genetic manipulation and selectivity

Last, direct use of non-detoxified lignocellulosic hydrolysates and crude glycerol will be beneficial

7. SELECTED REFERENCES

1. Abghari, Ali, and Shulin Chen. "Yarrowia lipolytica as an oleaginous cell factory platform for production of fatty acid-based biofuel and bioproducts."*Frontiers in Energy Research* 2 (2014): 21.
2. Gatter, Michael, et al. "A newly identified fatty alcohol oxidase gene is mainly responsible for the oxidation of long-chain ω-hydroxy fatty acids in Yarrowia lipolytica." *FEMS yeast research* 14.6 (2014): 858-872.
3. Lu, Wenhua, et al. "Biosynthesis of monomers for plastics from renewable oils." *Journal of the American Chemical Society* 132.43 (2010): 15451-15455.
4. Mauersberger, Stephan. "Cytochromes P450 of the Alkane-Utilising Yeast Yarrowia lipolytica." *Yarrowia lipolytica*. Springer Berlin Heidelberg, 2013. 227-262.
5. Waché, Yves. "Production of dicarboxylic acids and flagrances by Yarrowia lipolytica." *Yarrowia lipolytica*. Springer Berlin Heidelberg, 2013. 151-170.

8. ACKNOWLEDGMENTS

❖We acknowledge Financial support from:

Graduate and Professional Student Association Travel Award

WASHINGTON STATE UNIVERSITY

❖Scientific support from:

Dr. Scott E Baker (EMSL) and **Dr. Catherine Madzak (INRA)**

❖ Some mutant strains from

Dr. Michael Gatter (TUD) and

❖ Technical support for Chromatography Analysis from:

Mr. Jonathan Lomber (WSU) and **Mr. James G Wallis (WSU)**

WASHINGTON STATE UNIVERSITY

466386 Controlling Glycolytic Flux Regulation for Production of *N*-Acetylglucosamine in *Saccharomyces Cerevisiae*

Abstract:

We previously demonstrated that the production of N-acetylglucosamine (GlcNAc) in *Saccharomyces cerevisiae* was enabled by overexpressing an allosteric regulation-free *GFA1* mutant and the haloacid dehalogenase-like phosphatase YqaB. Wefurther improved GlcNAc production with controlling glycolytic flux. Rructose 2,6-bisphosphate (F26BP) was known as a modulator for the activity of phosphofructokinase 1 (PFK-1), the rate limiting enzyme of glycolysis. Disruption of PFK-2 was attempted to abolish synthesis of F26BP and to increase GlcNAc production by reducing glycolytic flux, but no significant change of GlcNAc with glucose as a sole carbon source. However, when galactose was used as a sole carbon source, GlcNAc production was significantly increased in the strain without PFK-2, suggesting significant reduction of glycolytic flux is needed for improving GlcNAc production. Controlling glycolytic flux of *S. cerevisiae* can be applied to improve production of carbohydrate derivatives.

466680 Silkworm Expression System Suitable for High-Throughput Screening of Antibody Fragment

Abstract:

Phage display method is the technique to select the antibodies with affinity for a target molecule such as membrane protein from the library of the phages displaying a large variety of antibody fragments on the surfaces. However, to identify the antibody with high binding affinity and structural stability, many antibody fragments as possible as we can select from the library should be prepared as recombinant protein to analyze the affinity and stability without phage, and the rapid protein preparation method for achieving the analysis should be prepared. Here, we focused on the protein production by means of silkworm. Because this protein production system can produce relatively large amount of proteins in small scale with high cellular density, it may prepare a variety of antibody fragments at a time for correct analysis. We develop process for preparing a variety of antibody fragments to obtain the antibody with affinity and stability.

We attempted to produce single chain fragments of variable region (scFv), where heavy chain of variable region (VH) and light chain of variable region (VL) are linked via a peptide linker. First, we inserted the scFv gene from the antibody with affinity for CD3 on lymphocyte cell and with affinity for epidermal growth factor receptor (EGFR) on cancer cells into baculovirus vectors and we transduced silkworm pupae with the vectors. The pupas were smashed and then ultracentrifuged. From the supernatant after ultracentrifugation, the scFv were purified by means of immobilized metal affinity chromatography and size exclusion chromatography. Finally, we measured the binding affinity of scFvs for each target molecule by means of flow cytometry and surface plasmon resonance.

In this study, 3 and 8 kinds of scFvs were prepared from anti-CD3 and anti-EGFR antibody, respectively. The scFvs were purified from the supernatant after ultracentrifugation, although the expressed amounts were varied. We confirmed that 8 out of 11 scFvs bound to each target molecule. From these results, we concluded that scFvs can be

prepared from silkworms.

465975 Conformational Change of 2-(2 Hydroxyphenyl)Benzenesulfinate Desulfinase during Catalysis

Abstract:

Increasingly stringent regulation of sulfur oxide emissions and environmental stewardship necessitates more effective fossil fuel desulfurization technologies. While traditional catalytic means of desulfurization effectively remove simple sulfur compounds, more complex thiophenic molecules remain intact; these thiophenic molecules now account for the majority of sulfur emissions from liquid transportation fuels. Biodesulfurization via enzyme catalysis has the potential for highly specific, rapid thiophenic desulfurization occurring at ambient temperature and pressure. The 4-step catabolic pathway converts dibenzothiophene (DBT), a common crude oil contaminant, into the sulfur-free molecule 2-hydroxybiphenyl (2-HBP) without the disruption of carbon-carbon bonds. 2-hydroxybiphenyl desulfinase (DszB), the rate-limiting enzyme in this biocatalytic process, is capable of selectively cleaving carbon-sulfur bonds. Accordingly, fundamental understanding of the molecular mechanisms of DszB must be developed. Based on crystallographic evidence, we hypothesize that DszB undergoes an active site conformational change associated with the catalytic mechanism. Moreover, we anticipate this conformational change is responsible, in part, for enhancing product inhibition. Rhodococcus erythropolis IGTS8 DszB was recombinantly produced and purified via Escherichia coli BW25113 to test these hypotheses. Activity and the resulting conformational change of DszB in the presence of 2-HBP were tested. The activity of recombinant DszB appears comparable to the natively expressed enzyme and is inhibited via competitive binding of the product, 2-HBP. Using circular dichroism, we observed conformational changes in DszB upon introduction of product, 2-HBP.

467014 Characterization of an Evolved Carotenoids Hyper-Producer of *Saccharomyces Cerevisiae* through Bioreactor Parameter Optimization and Raman Spectroscopy

Abstract:

We have previously developed an evolutionary engineering approach for enhancing heterologous carotenoids production in an engineered *Saccharomyces cerevisiae* strain, and isolated several carotenoids hyper-producers from the evolved populations. In this work, we optimized β-carotene production of the parental and one of the evolved carotenoids hyper-producers (SM14) in bench-top bioreactors. We found that maintaining a low pH and increasing the carbon-to-nitrogen ratio (C:N) from 8.8 to 50 in standard YNB medium resulted in a higher β-carotene production level at 25.52 ± 2.15 mg β-carotene g^{-1} [dry cell weight] in the carotenoids hyper-producer. We hypothesized that the increased carotenoids production in the evolved hyper-producer and the increase in beta-carotene production at a high C:N ratio are due to increased cellular lipid contents. In this study, we demonstrate that Raman spectroscopy is capable of monitoring β-carotene production in these bioreactor cultures. Raman spectroscopy is a powerful tool because it is adaptable to large-scale fermentations and can return results in near real-time. Raman spectroscopy was successfully used to return the relative lipid compositions and protein content of the parental and SM14 strains at two different C:N ratios in the bioreactor. The data showed a higher total lipid content in SM14 compared with the parental strain and that an increased C:N ratio resulted in significant increase in total fatty acid content of both strains, strongly supporting our hypothesis. The data suggests a positive correlation between the yield of β-carotene per biomass with total lipid content of the cell, suggesting one mechanism for the hyper-production of β-carotene in SM14 to be related with increased total lipid content.

467383 Crystallisation of Intact Monoclonal Antibody anticd20

Abstract:

Silica based nanotemplates with tuneable mesoporosity and surface chemistry, have been synthesised to facilitate protein crystallisation.[1] These nucleants act as a substrate for nucleation for macromolecular monoclonal antibodies (mAbs), which have greater complexity and segmental flexibility than many proteins. The class of anti cd20 mAbs is one of the most important therapeutics used in the treatment of lymphoproliferative disorders in the last 30 years.[2] Rituximab is a well known and widely used anti cd20. It has been crystallised previously both intact and as mAb fragments. Unfortunately, regardless of how similar anti cd20 mAbs may be, not all crystallise under the same solutions conditions.

Porous nanotemplates, tuned to the appropriate size, have been used to crystallise mAb anti cd20 by creating a spherulite seed stock. Templates have shown to significantly increase the number of crystals produced compared to crystallisation void of the heterogeneous nucleants. Stock solutions of these seeds were used for crystallisation and showed significant improvement in crystal quality. Moreover, seeding in different crystallisation conditions lead to various crystals with different morphologies.

References:

1. U.V. Shah, D.R. Williams, and J.Y.Y. Heng, â€œSelective Crystallization of Proteins Using Engineered Nanonucleantsâ€,Cryst. Growth Des.,(2012), 12(3), 1362â€"1369.
2. Lim, S.H., et al., *Anti-CD20 monoclonal antibodies: historical and future perspectives.* haematologica, 2010. **95**(1): p. 135-143.

467388 Site-Specific DNA Methylation Using Engineered dCas9-Methyltransferases

Abstract:

Heritable changes in gene expressions via epigenetic modifications are paramount to the normal functioning of the host and its phenotype. In particular, cytosine methylation has been shown to regulate gene expression in eukaryotes to facilitate vital cellular processes such as cellular differentiation and embryonic development. High levels of methylation at promoter regions are often associated with transcriptional silencing. One can thus envision that a bio-tool for promoter- specific methylation may aid in gene repression of overexpressed oncogenes or disease-linked genes during malignancy, such as leukemia, atherosclerosis, and Alzheimerâ€™s disease The typical strategy for targeting DNA methylation involves end-to-end fusion of a DNA binding domain (DBD), such as zinc fingers and TALE, to a cytosine DNA methyltransferases (MTase), typically bacterial enzymes or human DNMT3a. The two main drawbacks of this approach are (1) methylation occurs at non-target sites since the MTase domain remains active to methylate these sites when the DBD is not interacting with DNA and (2) new sequence recognition requires DBD redesign. Our solution to the first problem is to use the target site for methylation as a template for assembling the active form of an otherwise ineffective split methyltransferase (sMTase). This strategy has been shown to result in very specific targeting in bacteria (~80% methylation at the target site and â‰¤0.8% methylation at other CpG sites) using zinc fingers as the DBD. The second problem can be addressed by the CRISPR-Cas9 genome engineering tool. A key feature of nuclease- null Cas9 (dCas9) as a DBD is its ability to bind specific sites defined by the single-stranded guide RNA (sgRNA) sequence. A carefully selected sgRNA allows for highly specific targeting without the requirement for DBD redesign. An ideal dCas9-sMTase would have many exemplary features: the precision of true single-site targeting of methylation, ease of targeting via simple design of sgRNA, and multiplexed targeting of several sites via multiple sgRNAs. The later advantage addresses the concern that single site methylation might not be sufficient to have a desired phenotypic effect. To this end, we designed fusion proteins of dCas9 and a sMTase. The dCas9-sMTase provides high precision and control over CpG methylation in bacteria and human cells.

467232 Batch Production of Coenzyme Q By Metabolic Engineered *Escherichia coli* strains

Abstract:

Coenzyme Q (CoQ) plays an important role as an electron transporter in the respiratory chain. It is formed from a benzoquinone ring and an isoprenoid chain of a specific length depending on the organism. CoQ10 has been used in the treatment of different diseases including Parkinsons, Alzheimer and cardiovascular diseases. In addition, it is used as a dietary supplement and in cosmetic applications due to its important antioxidant property. *Escherichia coli* produces CoQ8 naturally but it is able to produce CoQ10 when an heterologous decaprenyl synthase is expressed. *E. coli* is easy to culture and relatively easy to modify genetically which makes it suitable for the development of an industrial-scale process. In a previous work, we constructed strains unable to produce demethylmenaquinone (DMK) and menaquinone (MK), compounds that compete for both chorismate, precursor of the benzoquinone ring, and the isoprenoid chain. In addition, mutant strains unable to produce enterobactin, high affinity siderophore, synthesized from chorismate, were also constructed. These strains where designed as platforms for the generation of novel CoQ-producing strains. In the present work, the Batch production of CoQ was evaluated in these strains using different carbon sources, such as glucose, fructose, succinate and glycerol, in a Lab-Scale Bioreactor.

467498 Engineering Red Blood Cell-Based Biosensors for Physiological Monitoring

Abstract:

Cell-based therapies have a wide range of applications ranging from cancer immunotherapy to regenerative medicine. A promising emerging frontier of this field is the development of engineered red blood cells (eRBCs) for therapeutic and diagnostic applications. RBCs have exceptionally long circulation times (around 120 days â€" far longer than synthetic vehicles), lack DNA (and thus are safe), and can be loaded with drugs, proteins, or other cargo. Technologies that enable one to engineer RBCs to act as biosensors, performing specific functions in vivo, could serve unmet diagnostic and therapeutic needs. In particular, new technologies are required for non-invasive, routine monitoring for pathogen exposure (e.g., in the context of first responders) and for other â€œactionableâ€ analytes (e.g., markers of inflammation post-surgery).

In this project, we are developing eRBC biosensors that generate a fluorescent output upon detection of the analyte of interest. Ultimately, eRBC biosensor fluorescent output may be monitored non-invasively using established technologies for fluorescent imaging of the retina. Using this simple imaging technology, a patient could perform regular self-analysis and enable real time, high frequency monitoring outside clinical settings, none of which is possible with existing technologies requiring specialized equipment, trained personnel, and/or sample collection. Thus, such biosensors that enable the detection of actionable analytes could benefit exposed personnel by accelerating the initiation of treatment (perhaps before obvious symptoms present) and reduction of further exposure risks when possible.

As a first step towards the goal of building eRBC biosensors, we designed and evaluated a novel biosensor strategy that is suitable for achieving biosensing in eRBCs, which lack DNA and thus require a readout other than gene expression. Towards this end, we engineered a novel cell-surface receptor protein in which ligand binding induces receptor dimerization, which then facilitates reconstitution of an intracellular split fluorescent protein. Importantly, our strategy involves modification of RBC-resident proteins, since retention of membrane proteins during RBC maturation is a tightly regulated and an incompletely understood process. We comparatively evaluated a range of biosensor architectures that implement the proposed mechanism, enabling us to identify biosensor designs and design features that successfully conferred significant ligand-induced generation of fluorescent output. We also evaluated and implemented strategies for improving biosensor performance, including minimization of background fluorescence and enhancing fold-induction upon exposure to ligand. This crucial proof-of-principle demonstration establishes a foundation for developing eRBC biosensors that could ultimately address an unmet need for non-invasive monitoring of physiological signals for a range of diagnostic applications.

467417 Production of Ethanol and Butanol from CO_2 and H_2 By Clostridium Aceticum

Abstract:

Clostridium aceticum is a homoacetogenic bacteria which converts CO_2 and H_2 to acetate without losing any carbon, thus resulting in a higher yield and titer than fermentation from heteroacetogenic bacteria such as *Clostridium ljungdahlii* and *Clostridium carboxidivorans*.
In this study, *Clostridium aceticum* DSM 1496 (ATCC 35044) was used as a host to produce ethanol and butanol by overexpressing aldehyde ferredoxin oxidoreductase (AOR) from *Clostridium carboxidivorans*, and aldehyde/alcohol dehydrogenase 2 (adhE2), thiolase (thl), butanol dehydrogenase (bdh) and BCS cluster (includingcrotonase (crt), putative butyryl-CoA dehydrogenase (BCD), electron-transfer flavoprotein (etf), and 3-hydroxybutyryl-CoA dehydrogenase (hbd) genes) from*Clostridium acetobutylicum* ATCC 824. Ethanol titer of 8.13 g/L was obtained in *C. aceticum* overexpressing adhE2, and 4.86 g/L in *C. aceticum* overexpressing AOR. Butanol production was low in *C. aceticum* overexpressing adhE2, thl and BCS, but increased significantly in *C. aceticum* overexpressing bdh, thl and BCS. These results suggest that *C. aceticum* is a promising heterologous host for ethanol and butanol production from CO_2.

467516 New Approaches in Engineering Somatic Embryogenesis in Loblolly Pine Suspension Cultures

Abstract:

Many industries, including agriculture and healthcare, require efficient methods for replication of plants with optimal traits. The loblolly pine (*Pinus taeda*) is a valuable crop in the timber industry, occupying 30 million acres of U.S. land, and breeding efforts aim to produce a crop with ideal phenotypic traits, including superior growth and wood quality. One method to large-scale clonal crop propagation is somatic embryogenesis (SE), the process through which aggregated plant cells undergo differentiation *in vitro,* resulting in a germination-competent embryo. We seek engineering strategies in the earliest stage of growth to improve yield and more efficiently use resources. There are three stages of growth and development that lead to the production of embryos: 1) aggregated cells that form embryonic suspensor masses (ESMs) are grown and scaled up in maintenance cultures; 2) ESMs are plated on solid media to initiate SE; and 3) embryos are separated from the ESM and germinated to generate plants. However, the process is not fundamentally understood, leading to large, unpredictable variability in embryo yield, a number only determined 8-12 weeks downstream of SE initiation. Here, we present on a variety of process manipulations (both biochemical and biophysical) we have evaluated and their effects on SE process outcomes and performance. We have demonstrated that induction of stress-related pathways through exogenous addition of plant stress hormones prior to moving a culture into development improves the rate of SE, and have used this knowledge to explore a means for more efficient embryo production. We have separated cellular communities at different stages of development within maintenance cultures and determined how the different populations progress to form embryos. We will also comment on recent advances in culture protein modification without genetic transformation, by inhibiting or supplementing endogenous arabinogalactan proteins (AGPs), which we have found to significantly influence SE in loblolly pine. We will present the influence of these treatments on embryo yield as well as the molecular response of the cultures, including culture stress (determined by phenolic content), growth (using the Coulter Counter), as well as total and specific protein biomarkers. These are the first studies to consider process engineering as a simple and cost effective means to improve the overall feasibility of SE on both an academic and industrial scale.

467530 On the Improvement of Immobilized Metal Affinity Chromatography (IMAC): Preparation, Characterization and Validation of a Superporous Agarose Monolith

Abstract:

On the improvement of Immobilized Metal Affinity Chromatography (IMAC): preparation, characterization and validation of a superporous agarose monolith.

E.M. Martín del Valle*, M. Elviro, , M.A. Galán

*Corresponding author: E. Martín del Valle. Department of Chemical Engineering, University of Salamanca P/Los Caidos, 37008 Salamanca, Spain.

Mail: emvalle@usal.es Phone: +34 923 294479

Purification of biomolecules is a key step in the biotechnology industry as it often represents the major manufacturing costs. Affinity chromatography is a high performance purification step with high selective separation as it uses a reversible binding between biomolecules wanted to purify and a specific ligand joined to a chromatographic support.

Immobilized metal ion affinity chromatography (IMAC) is a type of affinity chromatography based in the affinity of some side chain of amino acids for metal ions bound to a metal-chelating agent immobilized on a chromatographic matrix. Since the interaction between the immobilized metal ions and the side chain of amino acids has a reversible character, it can be utilized for adsorption and then be recovered using nondenaturing conditions.

In the last years, new kinds of chromatographic supports have emerged to improve the chromatographic properties and industrial scale-up of the known matrix, as higher mass transfer, and specially a lower pressure drop.

Monolithic superporous agarose beds were prepared by an emulsion method prior to use. The agarose matrix is formed by a homogeneous distribution of macropores. This bed is first used as support in catalase and BSA adsorption and purification by IMAC chromatography using Cu^{2+} as metallic ion. The monolithic matrix was used in order to achieve high superficial area and low pressure drop.

Furthermore, catalase and BSA adsorption equilibrium and elution were studied. The obtained results were compared in order to determine the influence of the adsorbed protein in adsorption and elution processes. The matrix was reused fourteen times and it can be concluded there is not any lose of yield, and any regeneration of Cu^{2+} was required.

In summary, a monolithic superporose agarose have been developed and used in catalase and BSA purification by IMAC. This support had high superficial area available for protein adsorption, it had improved the amount of protein adsorbed, it had decreased the needed time for adsorption and elution and it had been reused fourteen times without any losing of yield neither Cu^{2+} regeneration. Furthermore, influence of the kind of target compound was evaluated in order to understand better the chromatographic process inside the monolith, necessary to scale up.

467648 Transient Production of a Recombinant Anthrax Receptor Fusion Protein in *Nicotiana Benthamiana* Plant Cell Suspension Culture

Abstract:

Faster production of experimental protein therapeutics or vaccines at a large-scale is needed to respond to emerging infectious disease outbreaks or bioterrorist threats. We are investigating transient protein expression in plant cell suspension culture as a new platform that could be used to rapidly manufacture large quantities of novel protein therapeutics. Plant cells are being explored as an alternative to mammalian expression systems because they can also produce complex proteins, but enhance product safety and efficacy in some cases. Our system uses genetically engineered *Agrobacterium tumefaciens* to mediate transient protein expression in *Nicotiana benthamiana* plant cell suspension culture. A key advantage of our system is that these *Agrobacterium* constructs could be produced in a few weeks while stable transgenic plant or animal cell lines take months to develop and screen. By co-culturing recombinant *Agrobacterium* and *N. benthamiana* in suspension, we have produced an anthrax toxin receptor Fc fusion protein at levels up to 0.7 mg per kg of plant cell fresh weight after 9 days of co-culture. However, reduced intracellular protein levels and evidence of plant cell lysis have also been observed after co-culture with *Agrobacterium.* To increase recombinant protein expression levels in this system, we are testing various strategies to reduce effects on plant cell health associated with *Agrobacterium* infection.

468100 Biochemical Characterization of the *N*-Acetylglutamate Kinase (ArgB) from *Lactobacillus Plantarum*

Abstract:

Arginine is one of the twenty standard amino acids and considered a conditionally essential amino acid. Most mammals can produce arginine *in vivo*, but it is generally not sufficient to meet nutritional needs. Arginine is important because of the beneficial health effects it offers to mammals, specifically in the immunological, neurological, and cardiovascular systems. In addition, arginine plays an important role in embryogenesis, implantation of blastocysts, placental cell growth, and nutrient transfer to the fetus. Arginine has also been used to help rescue compromised pregnancies.

Probiotics, specifically *Lactobacillus*, offer the opportunity to deliver arginine-producing bacteria to mammalian gastrointestinal tracts. Arginine biosynthesis in prokaryotes is encoded by the *arg* operon and involves eight enzymatic steps starting at glutamate. The second step is catalyzed by a *N*-acetylglutamtate kinase, encoded by the *argB* gene, that phosphorylates an acetylated glutamate molecule. Based on literature, arginine exhibits feedback inhibition toward the ArgB enzyme, which is one of the key methods in how the cell regulates the production of arginine in many organisms. In order to increase the production of arginine, this feedback inhibition must be fully understood and characterized.

The *argB* gene was isolated from *Lactobacillus plantarum*, cloned into the pET32a(+) plasmid with a C-terminal His_6 tag, and transformed into *E. coli* BL21(DE3) for expression. The resultant recombinant proteins were recovered from cell lysates by affinity chromatography using a Ni-NTA column and subjected to full biochemical characterization. In future work, results will be used to relieve arginine feedback inhibition of ArgB and enable the construction of arginine-producing *L. plantarum*.

467932 Bactericidal Surface Chemistry That Enhances Implant Biointegration

Abstract:

Implants for years have been a vital tool in the medical field, but these implants often integrate poorly with the recipient tissue which can eventually lead to the implant failure. Bacterial infection is another major concern associated with the implantation procedures. Implant failure requires the removal surgery which is costly and invasive. In order to address this important issue, we propose a multifunctional surface chemistry that enables the implant surface to integrate with the surrounding tissue and at the same time suppresses bacterial growth. The main feature of this surface chemistry is the presence of nanopatterns of cell adhesive ligands (RGD peptide) on top of the bactericidal coating. The silane chemistry was used to coat the glass substrate with the quaternary amine-based antibacterial molecules. Nanopatterns were created by the adsorption of gold nanoparticles, onto which cell adhesive RGD peptides were immobilized. The surface chemistry was characterized by X-ray photoelectron spectroscopy (XPS) and field-emission scanning electron microscopy (FE-SEM). The antibacterial property of the surface was tested by culturing E.coli with the surface for 4 hours and measuring its proliferation. Human mesenchymal stem cells (MSCs) were cultured on the surfaces, and the differentiation into osteoblasts was observed over two weeks. The differentiation of the cells were characterized using alkaline phosphatase assay (ALP), qRT-PCR, and fluorescence microscopy. These surfaces have the potential to advance the use of medical implants by improving implant integration and reducing the risk of bacterial infection associated with the implantation procedures.

467963 Assessment of Bioseparation Technology Options for Bio-Based Chemicals Generated from Microbial Cultures

Abstract:

Assessment of bioseparation technology options for bio-based chemicals generated from microbial cultures

Kirti M. Yenkie, WenZhao Wu, Christos T. Maravelias*

Department of Chemical and Biological Engineering,

University of Wisconsin-Madison.

Abstract Background The alarming climatic changes like global warming and constantly declining fossil reserves have established the need for alternative sources which are environmentally benign and sustainable. Production of chemicals from microbial sources is one such alternative that has potential to generate carbon neutral cycle (Fresewinkel et al., 2014). However, along with satisfaction of these crucial requirements their production must also be commercially viable. The separation of bio-based chemicals contained in process streams from microbial cultivations prove to be most challenging and can contribute to more than 70% of the total production costs. Thus, the economic feasibility is dependent upon the efficient synthesis of separation systems. Separation synthesis requires evaluation of several alternative technologies performing similar tasks, since their suitability is affected by variations in parameters like separation unit efficiency, input stream characteristics and desired product specifications. A sensitivity study involving potential variations in these parameters can provide a general guideline for technology selection while synthesizing separation systems. Previous work in this area has focused on the performance of individual technologies and has provided some general guidelines regarding their applicability to microalgae harvesting (Molina Grima et al., 2003) and biofuels (Brennan and Owende, 2010). However, a simultaneous and detailed quantitative analysis for synthesis and evaluation of bioseparation systems is not available. Methodology We develop a bioseparation superstructure framework involving four stages: (i) cell treatment, (ii) product phase isolation, (iii) concentration and purification and (iv) refinement. Each stage comprises of technology alternatives for tasks like biomass harvesting, cell disruption and product recovery based on their separation principles. Then, we generate reduced superstructures for specific product categories based on their physicochemical properties (solubility, density and volatility), physical state (solid or liquid), localization (extracellular or intracellular) and intended use (commodity or specialty). We generate a base case by assuming nominal values for parameters like plant capacity, product titer, desired purity, technology parameters, cost of raw materials, chemicals and separating agents from relevant literature (Choi and Lee, 1999; Mooibroek et al., 2007) and process simulation packages (SuperPro Designer) for economic assessment. We have a combinatorial optimization problem with an objective to minimize the overall bioseparation costs. The solution provides the cost distribution in each bioseparation stage and identifies the technologies selected. However, technologies performing intended tasks and their suitability for a specific product can vary depending on stream characteristics (biomass titer, product concentration), technology performance indices (efficiencies) or matching parameters (like mass separating agents contributing to raw material costs or energy separating agents contributing towards utility requirements) and final specifications (product purity and recovery). Hence, we perform an optimization based sensitivity analysis by varying these critical parameters and study the effects on technology selection as well as process economics. Results

To illustrate, we present some results for the product category of intracellular(IN)-insoluble (NSL)-heavy(HV)-solid(SLD)-commodity(CMD) chemical. The stage-wise cost contribution analysis predicts (Fig 1A) that stage I-cell treatment is the key cost driver, followed by stage II-product phase isolation and stage IV-refinement while stage III-concentration and purification is absent for the base case. We select the task of biomass harvesting in stage I, which can be performed by sedimentation, centrifugation and filtration, for sensitivity analysis. The results for variation in performance indices of centrifugation (efficiency) and filtration (retention factor) on total separation cost ($/kg product) can be seen in Figure 1B. The contour lines are horizontal in the region where filtration is selected and vertical where centrifuge is selected. Threshold values when there is a shift in technology selection is denoted by the white lines. We observe that total separation cost can vary from 3.12 to 8.44 $/kg. Thus, maximum improvement that can be achieved by selecting a suitable biomass harvesting operation is ~32% when compared to the base case (shown by the black point in Figure 1B).

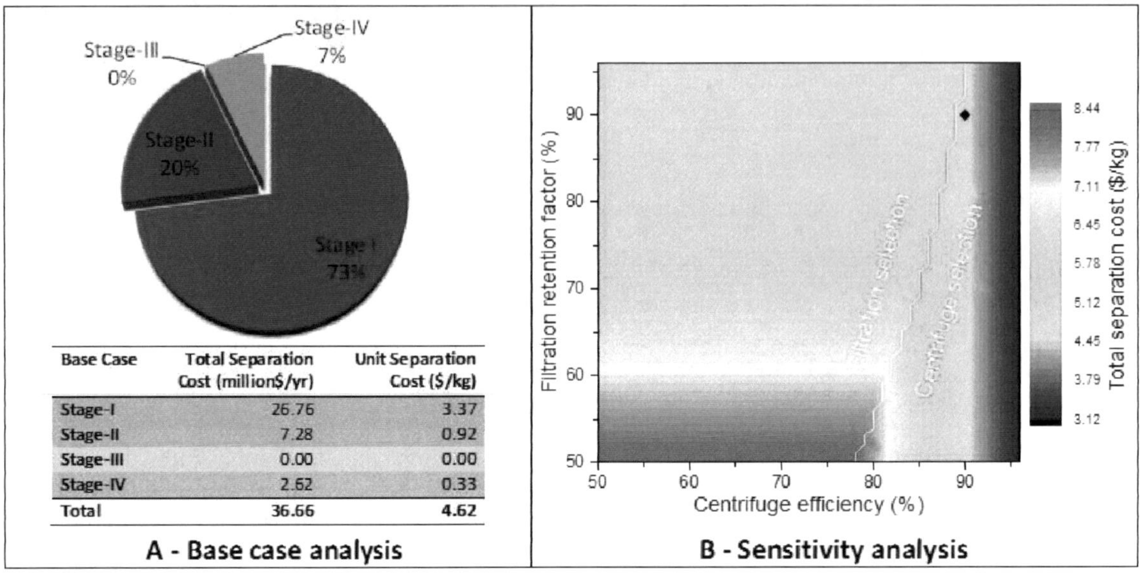

Figure 1 Results for the product category of intracellular (IN)-insoluble (NSL)-heavy (HV)-solid (SLD)-commodity (CMD) chemical. (A) Stage-wise cost contribution analysis for the base case. (B) Sensitivity analysis for biomass harvesting task in stage I by varying performance index of centrifuge (efficiency) and filtration (retention factor). Conclusions We studied separation systems for the recovery of chemicals produced via a range of bio-conversions. We believe that such an analysis will prove valuable in (1) selecting the chemicals that can be produced economically using bioconversions, and (2) designing separation processes for bio-based chemicals. The threshold values determined by our analysis will help in deciding which technology shall be more promising for a particular task when the information about relevant parameters are known apriori. Sometimes, a particular technology selected in a previous stage also influences the technology selection in the following stages. Thus, we can make an informed decision regarding which technologies should be placed in conjunction. Most importantly, the proposed framework help us identify some promising research directions in the area of separations, a topic that has received limited attention despite its high impact on the economics of biomass-to-fuels/chemicals strategies. References

Brennan, L., Owende, P., 2010. Biofuels from microalgae—A review of technologies for production, processing, and extractions of biofuels and co-products. Renew. Sustain. Energy Rev. 14, 557–577. doi:10.1016/j.rser.2009.10.009

Choi, J., Lee, S.Y., 1999. Factors affecting the economics of polyhydroxyalkanoate production by bacterial fermentation. Appl. Microbiol. Biotechnol. 51, 13–21. doi:10.1007/s002530051357

Fresewinkel, M., Rosello, R., Wilhelm, C., Kruse, O., Hankamer, B., Posten, C., 2014. Integration in microalgal bioprocess development: Design of efficient, sustainable, and economic processes. Eng. Life Sci. 14, 560–573. doi:10.1002/elsc.201300153

Molina Grima, E., Belarbi, E.-H., Acién Fernández, F.G., Robles Medina, A., Chisti, Y., 2003. Recovery of microalgal biomass and metabolites: process options and economics. Biotechnol. Adv. 20, 491–515. doi:10.1016/S0734-9750(02)00050-2

Mooibroek, H., Oosterhuis, N., Giuseppin, M., Toonen, M., Franssen, H., Scott, E., Sanders, J., Steinbüchel, A., 2007. Assessment of technological options and economical feasibility for cyanophycin biopolymer and high-value amino acid production. Appl. Microbiol. Biotechnol. 77, 257–267. doi:10.1007/s00253-007-1178-3

467994 Characterization of Formaldehyde-Inducible Promoter with Massively Parallel Partitioning and Sequencing

Abstract:

Transcription factor-promoter pairs can be used to control expression of a desired gene based on a chemical input. Characterization of sequence-function relationships of DNA-protein interactions in transcription factors is key for understanding inducible regulator-promoter pairs. Nucleotide resolution quantitative analysis of these interactions enables the design of tunable synthetic promoters with varied and desirable properties.

Applications where engineered pathways produce toxic intermediates (e.g. formaldehyde) require dynamic internal pathway balancing to avoid growth and production inhibition. The formaldehyde-inducible promoter (Pfrm) transcribing the *frmRAB* operon is repressed in the presence of FrmR, the product of the first gene in the operon. FrmA and FrmB encode the glutathione-dependent formaldehyde dehydrogenase and S-formylglutathione hydrolase, respectively, which together oxidize formaldehyde to carbon dioxide. While the general locus of the Pfrm-FrmR interaction has been identified, the consensus sequence has not been previously elucidated.

Here, we characterize the *in vivo* properties of the Pfrm/FrmR including the Hill cooperativity coefficient, sensitivity, and dynamic range for the wild type and *Î"frmR* strain against varying formaldehyde concentrations using a fluorescent reporter plasmid and flow cytometry analysis. The constructed reporter system can induce more than 30-fold change in expression rate and is sensitive to formaldehyde concentrations ranging from 1 to 1000 ÂµM, where transcriptional noise limits the lower end and toxicity the upper.

To elucidate the sequence-strength and sequence-repression relationships of the promoter regulator pair, we screened and partitioned a library of mutated promoters using FACS and analyzed the resulting activity-based binned populations with high-throughput sequencing. We quantitatively describe the effect of mutations at single nucleotide resolution on promoter strength and efficacy of the repressor. This process enables a generalizable method for construction of custom tunable inducible promoter systems for refactoring pathways for precise metabolic flux control.

This work was supported by the US DOE ARPA-E agency through contract no. DE-AR0000432 and the NIH through the NRSA F32GM109617.

468636 Effect of Sucrose Concentration on Levan Molecular Weight from Different Bacteria

Abstract:

EFFECT OF SUCROSE CONCENTRATION ON LEVAN MOLECULAR WEIGHT FROM DIFFERENT BACTERIA

Alvaro G. Garcinuño, Antonio Tabernero, Jose M. Sanchez-Álvarez , Miguel A. Galan, Eva M. Martin Del Valle

Department of Chemical Engineering. University of Salamanca.

Levan is biopolymer formed by furanoses units (fructose) that shows a great potential for different applications such as industry, cosmetics, medicine or nanotechnology.

Levan has been traditionally obtained from *Zymomonas mobilis*, a gram negative bacterium, by culturing with high sucrose concentration. An enzyme located in its cell wall (Levansucrse) catalyzes the sucrose breakdown and fructose obtained is polymerized in order to build levan polymer. Levan has been proposed as a good polymer for nanoparticle formation. Particularly, due to its ability for nanoparticle formation by self-assembly; because of its hydrophobic and hydrophilic character.

One crucial factor for nanoparticle formation is related with polymer molecular weight. Moreover, it has been reported that molecular weight is also important for its antitumor activity [1]. For that reason, the main objective of this work is controlling levan molecular weight by modifying the culture medium.

Specifically, the effect of sucrose concentration on levan synthesis (and its molecular weight) from two no-studied bacteria (*Bacillus atrophaeus* and *Acinetobacter nectaris*, gram positive and gram negative respectively) is studied. It is found that *A. nectaris* growth and levan production are affected by sucrose concentration; wheras *B. atrophaeus* growth and levan production are not affected by modifying that parameter.

On the other hand, levan molecular weight from *A. nectaris* is considerably modified by sucrose concentration in culture medium (from 2000 kDa to 8000 kDa depending on sucrose amount), whereas levan from *B. atrophaeus* has a constant value around 50 kDa.

Studies with glucose (a by-product from sucrose hydrolysis) were carried out in order to study why that difference happens. It was found that glucose inhibits levansucrase in Gram negative bacteria (*A. nectaris*) and has no effect on levansucrase from Gram positive bacteria (*B. atrophaeus*). That inhibition could explain why polymer from Gram negative is longer than the obtained from Gram positive. Bacterial peptidoglycan wall acts as a barrier for starting new polymer chain, so that; polymerization has to continue with the chain that is being synthetized. These results allow controlling levan molecular weight by modifying sugar amount in culture medium.

References:

[1] Calazans G, Lima R, de França FP, Lopes CE. (2000). Molecular weight and antitumour activity of *Zymomonas mobilis* levans.

468375 A Computational Fluid Dynamics Model to Predict Turbulent Flow Damage in Medical Devices

Abstract:

Turbulence is a random, highly three dimensional, chaotic, irregular and multiscale flow condition that results in strong vorticity, high rate of mixing, with pressure and velocity variations occurring over a wide range of time and length scales. [1] Rapidly moving impeller blades of heart devices cause highly disturbed turbulent flow in/near these devices. It is commonly accepted that turbulence contributes to RBC damage. One of the most common and widely studied effects is hemolysis. While an increase in hemolysis is observed when cells are exposed to turbulent stresses, [2] the structure of turbulence in proximity to the blood cells and the fundamental mechanism by which cells are injured remain unclear. [2-4] Therefore, understanding and predicting the effect of turbulent stresses on erythrocytes is a major concern when designing prosthetic heart devices. [2-4]. In this work, a computational technique (using the CFD software FLUENT) that employs Reynolds-Averaged Navier-Stokes models of turbulence (k-Îµ and k-Ï‰ SST models) to simulate the classic Couette flow, the capillary flow and the jet flow hemolysis experiments was used. The simulations predicted the rate of dissipation of turbulent kinetic energy and the size of corresponding turbulent eddy structures. Idealized structures were assumed to be spherical and to have sizes that correspond to the KLS of the flow at different locations of the devices. The study investigated the relation between the surface area of the KLS structures and predictions of cell damage following a procedure detailed elsewhere [5]. Extensive measures like total eddy areas for specific KLS were used to provide a prediction of hemolysis. Depend on the results of three systems, regression analysis was performed and a new model was proposed to predict hemolysis in turbulent flow that takes into account the complexity of turbulence by giving varying weight to eddies of different sizes. Results showed that dissipative eddies comparable or smaller in size to the RBCs are related to hemolysis, and that hemolysis corresponds to the surface area of eddies that are associated with KLS smaller than about 10 µm [5]. This is roughly the length scale of an RBC. The model was applied to results from three systems. Reasonable agreement has been obtained for flow fields and exposure times of these three distinctly different experiments. Predictions are also to be compared to the

experimental findings on release of a centrifugal blood pump that has been identified as a case study through the critical path initiative of the U.S. Food and Drug Administration (FDA) [6].

References

[1] Pope, S. B., 2000, Turbulent flows, New York, USA.

[2] Kameneva, M. V., Burgreen, G. W., Kono, K., Repko, B., Antaki, J. F., and Umezu, M., 2004, "Effects of turbulent stresses upon mechanical hemolysis: experimental and computational analysis," ASAIO J, 50(5), pp. 418-423.

[3] Aziz, A., Werner, B. C., Epting, K. L., Agosti, C. D., and Curtis, W. R., 2007, "The cumulative and sublethal effetcs of turbulence on erythrocytes in a stirred-tank model," Ann. Biomed. Eng., 35(12), pp. 2108-2120.

[4] Bludszuweit, C., 1995, "Three-dimensional numerical prediction of stress loading of blood particles in a centrifugal pump," Artif. Organs, 19(7), pp. 590-596.

[5] Ozturk, M., O'Rear, E. A., and Papavassiliou, D. V., 2015, "Hemolysis related to turbulent eddy size distributions using comparisons of experiments to computations," Artif. Organs, 39(12), pp. E227-E239.

[6] US Food and Drug Administration, 2013, "Computational fluid dynamics: an FDA Critical Path Initiative project," https://fdacfd.nci.nih.gov.

468579 Motility Is Necessary to Enable Efficient Intratumoral Dispersion of *Salmonella*

Abstract:

Small molecule cancer therapies are limited by diffusion. This prevents the therapies from reaching the entire tumor in high concentrations. Moreover, these diffusional limitations are one reason for drug resistant cancer relapse.

Facultative anaerobic bacteria, like *Salmonella,* on the other hand, are not restricted by diffusion since the bacteria can swim and colonize tumors 10,000 fold more than any other organ. But, the bacteria predominantly accumulate in quiescent and necrotic but not actively dividing tissue. This prevents *Salmonella* from delivering cancer therapies to a tumor uniformly. *We hypothesize that highly motile Salmonella can colonize tumors uniformly, migrate far away from tumor vasculature and colonize actively dividing tissue.*

To test this hypothesis, we transformed an arabinose inducible flhDC gene circuit into non-pathogenic *Salmonella*. We overexpressed flhDC to investigate the effect on aqueous motility. Mice were used to determine how highly motile *Salmonella* dispersed in *in vivo* tumors.

Overexpression of flhDC enabled a significant increase in aqueous motility. Upon arabinose induction, the *Salmonella* demonstrated a 50% overall increase in aqueous motility. Highly motile *Salmonella* populations had fewer non-motile bacterium when compared to control ($P<.01$). While 60% of control bacteria exhibited motilities between 0-15 micrometers/second, only 20% of flhDC induced *Salmonella* exhibited the same range of motility. Moreover, 85% of flhDC induced bacteria exhibited aqueous motilities between 16-30 microns/second compared to only 20% in the control.

Salmonella with overexpressed flhDC were also smaller than a control strain of *Salmonella* ($P<.01$). While the control

Salmonella was approximately 10 μm in length, flhDC overexpressing *Salmonella* were 4 μm in length. This finding was consistent with other results reporting that aqueous motility is inversely proportional to bacterial size. This finding further confirms that flhDC overexpression strongly influences bacterial lifestyle and motility.

The same flhDC overexpressing *Salmonella* was administered to mice along with a less motile control. Histological staining detected extravasated, highly motile *Salmonella* within tumors. This finding demonstrated that highly motile *Salmonella* could circumvent dispersion limitations and be an effective cancer drug delivery vehicle.

Engineering *Salmonella* to disperse uniformly after extravasation will improve drug distribution in a tumor. By developing a bacterial drug delivery vehicle that disperses, cancer therapies can be administered to the entire tumor. This can potentially limit drug resistance and relapse of the disease.

468921 Effect of Subtle Differences in Amino Acid Sequences on the Aggregation of Insulin Fragments

Abstract:

Insulin aggregates at low pH and high temperature. The aggregates have a signature intermolecular beta sheet structure which is a characteristic of amyloid structures. Insulin is a highly conserved hormone exhibiting minor differences in the amino acid sequence from species to species. For example, human and bovine insulin differences are limited to four amino acids. In a previous study we found that human and bovine insulin exhibit very different aggregation kinetics. Both lag times and rate constants were different. Moreover, dynamic light scattering and FTIR experiments showed that both insulins seem to follow different aggregation pathways as shown by different populations of aggregate sizes and different secondary structure change. Still, mature fibrils were practically identical. To advance our knowledge about this we decided to study fragments of these two insulins that contain the sequence where the differences exist. The fragments were synthesized in solid phase using Fmoc chemistry. Liquid/gas interfaces were provided by stirring whereas solid/liquid interfaces were provided by adding liposomes of different chemistries. Fibrillation was studied as a function of pH, temperature and the presence of air/liquid and liquid/solid interfaces. An experimental design was made to determine the main factors that contribute to fibrillation. A Thioflavin T fluorescence assay was performed to monitor the time evolution of the aggregation process. FT-IR (Fourier Transform Infrared Spectroscopy) was used to monitor the changes in secondary structure and the presence of fibrils was further confirmed by transmission electron microscopy. The exploratory experiments showed differences in the kinetics of both fragments with the bovine one aggregating faster; the same trend is observed with the entire hormone. The best conditions were pH 2.5, a temperature of 60°C under stirring conditions. These conditions were used to further explore the differences between the aggregation phenomena of both fragments by increasing the sampling time and by including DLS measurements. Different mutations in the non-conserved amino acids were explored to evaluate the aggregation mechanism.

468943 Selection of Methylxanthine-Responsive Aptamers

Abstract:

RNA has been known to possess the abilities of catalysis and molecular recognition. In fact, the theophylline (1,3-methylxanthine) aptamer has found great utility in synthetic biology, with the generation of many theophylline-dependent riboswitches. The theophylline aptamer is selective against caffeine and other related methylxanthines, but a single nucleotide mutation can change the selectivity, resulting in an aptamer that recognizes 3-methylxanthine. However, there are currently no riboswitches or aptamers for any other methylxanthine. The objective of this study was to develop new RNA aptamers to recognize theobromine (3,7-dimethylxanthine) and paraxanthine (1,7-dimethylxanthine) through iterative rounds of *in vitro* selection. We first generated a pool of RNA molecules containing 40 random nucleotides affixed to the stem II loop of the hammerhead ribozyme. The initial C-G base pair on the stem II was left in place in order to retain ribozyme activity. After an initial negative selection step to remove ribozymes that self-cleaved in the absence of ligands, the remaining uncleaved molecules were incubated in a positive selection reaction with ligand molecules (i.e., theobromine or paraxanthine). The 5â€™ cleavage products were purified and used as a template for subsequent rounds. After 8-10 rounds of allosteric selection, we found aptamers specific for theobromine and paraxanthine. These aptamers are expected to provide new tools to control gene expression and select for cells capable of producing high-value methylxanthines.

469183 Using Raman Spectroscopy in Cell Culture Process Development

Abstract:

In recent years, the United States Food and Drug Administration (FDA) and the European Medicines Agency (EMA) have stressed the need of better process understanding and control in the area of pharmaceutical development, manufacturing and quality assurance, summarized in the Process Analytical Technology (PAT) initiative. Lately, Raman spectroscopy has emerged as a promising online monitoring and control tool in bioprocesses and, particularly, in cell cultures for different process variables. The advantages of this technology are the invasive and nondestructive manner, the simultaneously detection of multiple process variables as well as the online operation mode and real-time information release.

Although Raman spectroscopy is used very often as a monitoring tool at production stage, its use in process development has been very rare. The major challenges are large variabilities in process conditions, media products or cell lines, which might lead to rather small data sets based on similar process conditions, needed for an appropriate model calibration. Since the benefits of Raman spectroscopy can be also useful in the development stage, this work presents several ideas, experimental results and recommendations to overcome those challenges in process development.

Different fed-batch and perfusion cultures, varying in cell lines, products and platform media, were monitored online. Besides classical parameters such as glucose, lactate, titer and viable cell density, also parameters such as amino acids and quality attributes, for example aggregates and fragments, showed promising prediction results. In addition, diverse spiking strategies were performed to artificially increase the calibration range as well as the quality of the prediction models. Based on D-optimal designs, spiking was performed for the variables glucose, lactate, ammonium and viable cells in different background media. The different spiking strategies included synthetic cell-free solutions of the components of interest in media, semi-synthetic samples containing cells and different components using a special miniaturized spiking equipment as well as spiking directly into the bioreactor.

A central role of this work was in the evaluation of advanced modeling techniques to derive optimal results from the obtained spectral data. Those techniques included an in-depth analysis of the pretreatment procedure, wavelength selection and outlier elimination with regards to robust calibration and prediction of the components of interest. In particular, the application of a genetic algorithm for selecting specific wavelengths to build a generic model showed promising results. Such generic model can be applied independent of the cell line, process platform and medium and, hence, is very attractive for the utilization at the development stage.

469051 Production of n-Butanol and Ethanol from Biodiesel-Glycerol By Engineered *Clostridium Formicoaceticum*

Abstract:

With concerns about the high oil price and greenhouse gas emissions, biofuels have gained large attention from industry and government. Nowadays, more and more research has focused on the development of sustainable bioprocesses for bio-butanol production. n-Butanol, a four-carbon primary alcohol ($C_4H_{10}O$), is a significant industrial chemical and has a great potential to be used as a superior biofuel. Especially, compared with ethanol, n-butanol can be used as an ideal substitute of gasoline because of its low water solubility, high energy density, and low vapor pressure. Unfortunately, bio-butanol production using the conventional acetone-butanol-ethanol (ABE) fermentation process is not economically feasible because of the co-production of other metabolites, including CO_2, acetone, acetate, and butyrate, resulting in a low butanol yield. In this study, we focuses on a homoacetogen, *Clostridium formicoaceticum*, which naturally produces acetic acid as the sole fermentation product from a carbon source, such as fructose, lactate and glycerol, with a high product yield of >90% (w/w) without losing any carbon to CO_2. Based on the analysis of recently sequenced whole genome, *C. formicoaceticum* has all the genes needed for glycerol utilization and Wood-Ljungdahl pathways, suggesting that *C. formicoaceticum* can utilize glycerol and CO_2 as carbon sources to produce acetate. In this study, we introduced *thl*, *hbd*, *crt*, *bcd-etfB-etfA*, and *adhE2* genes from *Clostridium acetobutylicum* into *C. formicoaceticum* and investigated ethanol and n-butanol production from crude glycerol present in biodiesel wastewater by the engineered mutant. The result showed that *C. formicoaceticum* is a promising heterologous host for biofuel production from cheap and abundant industrial byproducts and its use can further reduce CO_2 emission.

469483 Classification and Quantitative Analysis of Normal and Apoptotic Cells from Fluorescence Microscopy Images

Abstract:

Fluorescence microscopy is a well-developed tool to study in vitro behaviour of cells. However, microscopy experiments can generate a great amount of images of cells with varying image qualities (Waters 2009). The manual quantification and analysis of these data is time consuming. Hence, accurate and automatic analysis of cells images such as Chinese Hamster Ovary (CHO) cells can be very useful.

Mammalian cells are prone to apoptosis (programmed cell death), which is a key metabolic event that can restrict the growth of cells and decreases the productivity in a bioreactor (Rulter, Spearman & Braasch 2014). The accurate detection of apoptotic cells will help identifying the critical factors that trigger apoptosis. This knowledge may be used for delaying apoptosis and potentially increase the productivity (Taatjes, Sobel & Budd 2008).

Morphological changes in cells are highly indicative of the occurrence of apoptosis (Henry, Hollville & Martin 2013). For example, shrinkage and blebbing of the cytoplasmic membrane are found to be significant characteristics of apoptotic cells, which may cause cells to lose normal, smooth and circular shapes. Blebbing during apoptosis is generally associated to swell of the cell membrane into spherical bubbles. Hence, microscopic observation of morphological changes can be used to discern normal from apoptotic cells. However, cells may exhibit highly variable values of these morphological measures due to the dynamic nature of apoptosis.

This work presents a new image processing and quantitative analysis method that can automatically differentiate apoptotic from normal cells, while maintaining the computational time at a reasonable level. The proposed method involves three consecutive steps: (i) a coarse segmentation that can be used to identify the number of cells in a given image of cells; (ii) a fine segmentation step to detect the boundaries of cells and to identify particular morphological features related to these boundaries; and (iii) a support vector machine (SVM) based classification model that uses the morphological features identified in the fine segmentation step (step ii) to distinguish apoptotic cells from normal cells.

References

Waters, JC 2009, 'Accuracy and precision in quantitative fluorescence microscopy', *The Journal of Cell Biology*, vol 185, no. 7, pp. 1135-1148.

Rulter, M, Spearman, M & Braasch, K 2014, 'Monitoring cell growth, viability and apoptosis', in R Portner (ed.), *Animal cell biotechnology, method and protocols*, Springer, Hamburg, Germany.

Taatjes, DJ, Sobel, BE & Budd, RC 2008, 'Morphological and cytochemical determination of cell death by apoptosis', *Histochemistry and Cell Biology*, vol 129, no. 1, pp. 33-43.

Henry, CM, Hollville, E & Martin, SJ 2013, 'Measuring apoptosis by microscopy and flow cytometry', *Methods*, vol 61, no. 2, pp. 90-97.

469543 Identifying NAD-Dependent Methanol Dehydrogenases for Synthetic Methylotrophy

Abstract:

Methylotrophy describes the ability of microorganisms to utilize reduced one carbon compounds, such as methane and methanol, as growth and energy sources. The recent discovery of large natural gas reserves has prompted considerable interest in utilizing these compounds as substrates or co-substrates with sugars in industrial fermentation of fuels and chemicals, as higher biomass and product yields are expected from these more reduced substrates. Native obligate methylotrophs are not suitable as industrial microorganisms since they produce few metabolites, lack genetic engineering tools, and have methanol dehydrogenase (MDH) enzymes that cannot conserve NADH for metabolite production. Therefore, the development of synthetic methylotrophy is of considerable interest. Herein, we discuss the selection of suitable MDH enzymes for engineering nonnative methanol metabolism in *Escherichia coli*, a well-developed host for industrial fermentation.

Methanol oxidoreductase enzymes are classified by the cofactor associated with methanol oxidation. Generally, methylotrophic yeasts utilize methanol oxidases, Gram-negative methylotrophic bacteria utilize pyrroloquinoline quinone-dependent MDHs and Gram-positive methylotrophic bacteria utilize NAD-dependent MDHs. NAD-dependent MDH enzymes offer the best selection towards synthetic methylotrophy since methanol oxidation supplies reducing equivalents in the form of NAD(P)H to drive downstream metabolite-production pathways. Of these NAD-dependent MDH enzymes, those from thermophilic *Bacillus methanolicus* and *B. stearothermophilus* have been well characterized *in vitro*. Of more importance, however, is heterologous *in vivo* activity of these enzymes in *E. coli*, which we demonstrate. We further discuss how to overcome the challenges involved with these MDH enzymes, including unfavorable thermodynamics of methanol oxidation and broad substrate selectivity. We report a high-throughput screening technique that we developed to identify improved MDH enzymes through protein engineering. This technique takes advantage of the native *E. coli* formaldehyde-responsive transcription factor/promoter system (*frmR-Pfrm*), derived from the formaldehyde detoxification *frmRAB* operon. We constructed a product-responsive reporter strain to isolate improved MDH variants via fluorescence-activated cell sorting (FACS). This reporter strain enables *in vivo* quantification of MDH activity via fluorescence detection upon methanol addition. We demonstrate that this technique successfully isolates MDH variants based on *in vivo* methanol oxidation activity.

SUPPORTED by the US DOE ARPA-E agency through contract no. DE-AR0000432.

469499 Fluxomic Topology of Cyanobacteria: Rigidity and Plasticity

Abstract:

Synechocystis sp. strain PCC 6803 is a model cyanobacterium that has been widely used as photo-biorefinery chassis. Like fast-growing *E.coli*, this species contains a complete TCA cycle that is predicted to be highly active for glucose fermentation in absence of photosynthesis. Via ^{13}C-metabolic flux analysis and isotope dilution mass spectrometry for metabolite pool measurements, we observe that *Synechocystis* 6803 has a minimal flux through the TCA cycle (relative flux < 5% of glucose uptake), while its pentose phosphate pathway operates in a cyclic mode for complete glucose oxidation under dark condition. We further demonstrate that cyanobacteria can adapt its metabolism towards different light and carbon sources via flexible fluxes through sugar phosphate pathways and ATP charges, while its fluxes and metabolite pool sizes in the TCA cycle are maintained at very low levels. For example, its acetyl-CoA pools remain 5~10 folds lower than *E.coli*under phototrophic or heterotrophic conditions, minimizing cell growth and acetate overflow capability. Flux analysis also reveals a 6.9 mmol/g/h of non-growth-associated ATP maintenance requirements for *Synechocystis* 6803 and the essentiality of oxidative phosphorylation for ATP generations under light insufficient conditions. The results explain why most cyanobacteria are unable to grow anaerobically with glucose under dark or low light conditions. In summary, enzyme machineries and thermodynamic favorability constrain cyanobacterial fluxome for optimal biosynthesis with organic carbon sources. Lessons from this study also include *Synechocystis* 6803 as a good chassis for the production of chemicals derived from its sugar phosphate pathways rather than its TCA cycle.

469593 Development and Characterization of Collagen Substrates to Study Antimicrobial Peptide Tethering Using QCM-D

Abstract:

The success of chronic wound scaffolds is severely threatened by infection, but current methods to manage infection are toxic, impede tissue regeneration, and may promote antimicrobial resistance. Antimicrobial peptides (AMPs) are promising as alternative chronic wound therapeutics because of their function in the innate immunity of several species, including broad-spectrum antimicrobial activity and immunomodulatory functions, and their unique mechanisms leading to a low likelihood of promoting antimicrobial resistance. Unfortunately, clinical application of AMPs is hindered by their fast degradation and toxicity at high concentrations *in vivo*. Tethering, or binding AMPs to surfaces has been shown to increase stability and reduce toxicity. Previously, we studied AMP tethering onto inorganic and metallic surfaces using quartz-crystal microbalance with dissipation (QCM-D), a real-time method for characterizing mass and film rigidity changes on a surface, and coupled it with antimicrobial assays. However, QCM-D methods for studying AMP tethering onto relevant biopolymer surfaces have not been developed. Our overall goal is to develop a QCM-D platform to study AMP tethering onto collagen, a prevalent biopolymer in wound scaffolds, and assess AMP bioactivities. For this, we chose human-derived AMP LL37 because it has well-characterized antimicrobial and wound healing activities, and modified it with a collagen binding domain (CBD-LL37) for binding specificity. Here, we characterized adsorbed collagen films using QCM-D in terms of thickness, concentration, substrate, morphology and crosslinking, studied real-time CBD-LL37 binding to these films, and developed *ex situ* antimicrobial assays to study bound CBD-LL37 bioactivity. These results demonstrate QCM-D versatility in studying AMP-biopolymer tethering for development of antimicrobial scaffolding materials.

469571 Graphene-Based Microfluidics for Serial Microcrystallography

Abstract:

The idea of coupling microfluidic technology with advanced protein crystallography techniques for *in situ* analysis is an area where significant advances can be made. Microfluidic platforms have the benefit of not only enabling experiments at small volumes, but also create an environment free of inertial or convective effects while providing exquisite control over local conditions and gradients. A significant problem of microfluidic-based crystallography is the background noise caused by device materials, which may obscure the signal from weakly diffracting micro-crystals. Graphene is a single-atom-thick material that has a negligible scattering effect on X-ray and should be the best option available to solve this problem. Also its remarkable mechanical strength and impermeability to gas make it suitable to be integrated into microfluidic devices.

We have successfully validated the feasibility of graphene as a component in an ultra-thin microfluidic chip, demonstrating a significant reduction in background noise and a subsequent enhancement in the quality of the observed diffraction signal. Furthermore, graphene acts as a diffusion barrier within our microfluidic devices, creating a sample environment that is stable over the course of weeks. We are currently utilizing these platforms for the structure determination of caspase7, a member of the cysteine-aspartate family of proteases that are implicated in cellular apoptosis, and halorhodopsin, a light-sensitive membrane protein commonly utilized in optogenetics. Our goals include the collection of radiation-damage free structural data and the time-resolved structural analysis of key reaction intermediates for these targets.

469751 Chitin-Binding Proteins Induce Biomacromolecular Complexation: Inspirations from Insect Cuticle for Hierarchical Self-Assembly

Abstract:

Insect cuticle, the primary component of the exoskeleton, is one of the most widespread materials in nature. It is a valuable model for the development of biomimetic materials because its moduli can vary by six orders of magnitude or more. The exceptional properties of insect cuticle are hypothesized to arise from a combination of covalent and non-covalent interactions among proteins, catechols, chitosan and chitin nanofibers, including protein-catechol crosslinking, generation of catechol-derived microparticles and protein-metal ion interactions. Here we examine the interactions between novel structural proteins that we have identified in the beetle *Tribolium castaneum* and their interactions with chitin, chitosan and metal ions. The goal of this work extends from understanding the *in vivo* roles of specific interactions to development of new biomaterial design motifs.[1]

Two abundant cuticle proteins in the elytra of *T. castaneum* that we named CPR27 and CP30 were the foci of this study. CPR27 has a conserved sequence of amino acids first hypothesized by Rebers and Riddiford to bind chitin. In this work, we establish direct evidence of this protein-chitin binding by using an active microrheology technique, coupled with fluorescence and bright-field microscopy. While our results from active microrheology showed that addition of CPR27 to aqueous chitosan solutions caused a 2-fold decrease in viscosity, simultaneous visualization of the solution microstructure shows the formation of micron-sized particles. Together these results indicated that CPR27 complexed with chitosan as hypothesized to form micron-scale structures. Furthermore, by using fluorescently-labeled chitosan, our simultaneous fluorescence images confirm the presence of clusters of chitosan, suggesting that the protein CPR27 serves as a nucleation agent to induce the complexation of the polyelectrolyte. However, varying the concentration of the protein over several orders of magnitude does not alter the viscosity drop, suggesting that the RR interactions are the first step in inducing complexation of chitosan. In contrast, CP30, which does not contain the chitin-binding sequence, displayed no evidence of complexation. The role of quinone-crosslinking of both proteins with the catechol N-Î²-alanyldopamine (similar to that observed in mussel adhesion) was also examined. Microparticle formation was observed in solutions containing protein, catechol and the oxidative enzyme laccase. Understanding the interactions involving these cuticle proteins suggest new motifs that could be used in the design of new composite materials. The rational design of recombinant proteins following these principles, with specific covalent and non-covalent interactions with polysaccharides or ions, inspired by insect cuticle, may lead to biomaterials with enhanced mechanical properties.

REFERENCE

Â [1] Lomakin, J.; Huber, P.A.; Eichler, C.; Arakane, Y.; Kramer, K.J.; Beeman, R.W.; Kanost, M.R.; Gehrke, S.H. *Biomacromolecules* (2011) 12, 321.

469605 Antibacterial Activity of Catechins Against *Aggregatibacter Actinomycetemcomitans*

Abstract:

Antibacterial Activity of Catechins against *Aggregatibacter actinomycetemcomitans*

Virulence factors are molecules produced by pathogenic bacteria that allow them to establish a niche within the host. These molecules include adhesions, enzymes, and toxins, among many others. Our lab is focused on inhibiting the action of toxins for the development of antivirulence strategies to treat bacterial infections. The gram-negative bacterium *Aggregatibacter actinomycetemcomitans* (*A.a.*) produces numerous virulence factors, including a leukotoxin (LtxA), which selectively kills host immune cells. It has been reported that catechins, polyphenols extracted from green tea leaves, have antibacterial properties against *A.a.* However, the mechanism by which catechins interact with both the bacteria and mammalian cells has not yet been uncovered. In this project, six common catechins were used: (-) catechin (C), epicatechin (EC), epigallocatechin (EGC), gallocatechin gallate (GCg), epigallocatechin gallate (EGCg) and epicatechin gallate (ECg) to correlate inhibition of LtxA activity with the catechins¡¯ ability to alter the host membrane properties. In addition, the effect of the six catechins on bacterial membrane properties was examined. Our results indicate that the galloylated catechins significantly alter membrane fluidity in both model membranes and target cell (THP-1) membranes, while non-galloyated catechins have little effect on membrane fluidity. These changes appear to be caused by the catechins alone, as there are no changes in the cell membrane cholesterol composition after incubation with the catechins. Two of these galloyated catechins, EGCg and GCg, inhibit the ability of LtxA to kill target cells, while the non-galloylated catechins do not, suggesting that the ability of the catechins to alter target cell membrane fluidity plays a role in their inhibition of toxin activity. Our current work is focused on better understanding this relationship as well as investigating the effects of the catechins on bacterial cell membrane properties to better define their antibacterial properties.

469625 Transcriptome-Level Signatures in Gene Expression and Gene Expression Variability during Bacterial Adaptive Evolution

Abstract:

Antimicrobial resistance is a worldwide healthcare issue of increasing concern, as pathogens have evolved immunity to all currently available treatments. By decoding the intrinsic mechanisms that bacteria use to gain tolerance, it may be possible to avert the continued emergence of new resistances. Landmark bacterial evolution studies have eloquently uncovered trends in genetic diversity during adaptation; incremental genomic changes undoubtedly promote heritable resistance, and studies that focus on the genome elucidate *what* specific mutations enable resistance to a stress. Contrastingly, less is known about *how* resistance is achieved â€" namely, the regulatory responses promoting heterogeneity and underlying stepwise increases in resistance. Emerging evidence suggests that cellular phenotype has a strong non-genetic basis, influenced by shifts in gene expression, which provide immense flexibility when cells are exposed to a stressful environment. Here, we probe the regulation of adaptive resistance by investigating transcriptome profiles of *Escherichia coli* upon adaptation to the disparate toxins ampicillin, tetracycline, and n-butanol. We analyze transcriptome data for both traditional differential gene expression as well as more unconventional analysis focusing on gene expression variability. Our results highlight the occurrence of extensive gene expression heterogeneity across adapted populations, including shifts in both gene expression level and gene expression variability, then filters that information to arrive at a small set of genes with conserved signatures across diverse adapted bacterial populations. To validate the biological relevance of the observed changes, we synthetically perturb gene expression using CRISPR-dCas9. Manipulation of select genes in combination with antibiotic treatment promotes greater heterogeneity in MIC and an increased degree of antibiotic tolerance, with no observed changes in mutation rate, which suggests that these genes may provide a useful set of targets for novel therapeutics that hinder adaptation mechanisms. Overall, this work provides evidence for a complex non-genetic response, encompassing shifts in gene expression and gene expression variability, which contributes to adaptive resistance.

469919 Modeling the Aggregation and Oxidation of Therapeutic Proteins

Abstract:

Therapeutic proteins are the most rapidly growing class of pharmaceuticals for use in diverse clinical settings including cancer, chronic inflammatory diseases and infectious diseases. However, these proteins tend to degrade due to several mechanisms such as aggregation and oxidation during production, formulation or storage. This degradation can reduce their biological activity or cause immunological response in the patients. Here we discuss in silico molecular modeling tools to assess the risk of degradation from aggregation and oxidation, and to identify the regions of the protein responsible for them. Aggregation is assessed using Spatial-Aggregation-Propensity (SAP) model which identifies the hydrophobic patches on the protein that are prone to aggregation. Oxidation of methionine residues is assessed using molecular simulation models based on solvent accessible area and 2-shell water coordination number. We apply these models on several model proteins and therapeutic candidates and demonstrate that they are highly accurate based on comparison with experimental data. These models aid in assessing the risk of protein degradation early on in the drug development cycle and enable further protein engineering to improve stability.

470040 Conformational Landscape of Mutations in the N-Terminus of the a$\beta_{1\text{-}42}$ Monomer

Abstract:

The only naturally occurring variant demonstrated to protect against developing Alzheimer's disease is a hydrophilic mutation located in the N-terminus of amyloid-β (A2T). Current hypotheses regarding A2T's protective effect are: (1) reduced processing from Aβ's parent protein, the amyloid precursor protein (APP), and (2) differential biophysical properties compared with the wild type Aβ that affect its reactivity and clearance. Specifically, it has been shown this N-terminal modification alters conformational landscape, aggregation kinetics, aggregate morphology, and long term potentiation of $A\beta_{1\text{-}42}$. Conversely, a hydrophobic mutation (A2V) in the same position has been shown to give different results in the conformational landscape and aggregation kinetics of $A\beta_{1\text{-}42}$. Here, we perform a systematic evaluation of mutations with varying degrees of hydrophobicity and hydrophilicity to interrogate the N-terminus' effect on the conformational landscape of the $A\beta_{1\text{-}42}$ monomer, using explicit solvent atomistic replica-exchange molecular dynamics simulations. To analyze these variants we examined contact (2D heat) maps of their conformational landscape to estimate the thermodynamic stability of the various conformations adopted by the variants. These results further implicate the importance of the N-terminus in the biophysical properties of $A\beta_{1\text{-}42}$.

470077 High Throughput Cell Screening Via Luminescent Nanoparticles in a Microfluidic Device

Abstract:

With the understanding that the molecular backbone of genetic diseases and cancers is the result of lower than expected success rates, the need to screen and tailor treatments to each patient is becoming more vital. Droplet microfluidic devices have garnered significant interest in recent years due to their ability to encapsulate single cells in aqueous droplets for high throughput screening. Recent innovations with droplet microfluidics have included the addition of downstream trapping arrays to collect the aqueous droplets and perform dynamic analysis on intracellular enzyme activity and the cellular response to therapeutics. While these trapping arrays can assess a heterogeneous population of cells, they are currently limited in their ability to test multiple inputs (e.g., different doses or combinations of chemotherapeutics) in a single device. Therefore, it is important to develop this droplet â€˜barcodingâ€™ screening technique that allows for the simultaneous measurement of multiple inputs across a heterogeneous population of cells to perform high throughput, personalized diagnostics. In this work, upconversion luminescence nanostructures are incorporated into a droplet microfluidic trapping array to barcode individual droplets as a means of performing real-time, dynamic analysis of the cellular response in single cells to different treatment protocols.

Specifically, this work focuses on the viability of cells in the presence of the upconversion nanoparticles, $NaYF_4$ core-shell particles. The rare earth emitter ions (Er^{3+} or Tm^{3+}) are located in the core while the Yb^{3+} sensitizer is doped in the shell layer. The core particles are synthesized using a hydrothermal growth process with growth control between 40-500 nm. Afterwards, a shell layer is deposited using a sol-gel method for shell thickness up to 30 nm. The particles were shown to have an upconversion luminescence emission in the blue, green, and red spectral region upon excitation with 980 nm light allowing for the barcoding of three different treatment protocols. The ratio of intensity of the different emission peaks were engineered through the dopant type, concentrations, and location within the core-shell structure. As a proof-of-concept, combinations of one, two, and three different types of nanoparticles were encapsulated in aqueous droplets in the array to demonstrate the ability to selective tag individual droplets. In order to determine the applicability of the nanoparticles as a means for droplet barcoding, viability studies were performed using different sized nanoparticles across a range of concentrations in several model cancer cells lines including HeLa, OPM2, THP-1, and HCT116 cells. This work provides the technological first step for a novel, antibody-free method to barcoding droplets to increase the throughput and utility of microfluidic droplet trapping arrays as a technique for personalized medicine.

469656 Engineering Short-Chain Fatty Acid Production in Probiotic Bacteria

Abstract:

Short-chain fatty acids (SCFAs), especially butyric acid, have many roles in the human gut, affecting immunomodulation, cell differentiation and apoptosis. In addition, butyrate is the preferred carbon source for colon cells. Butyrate is normally produced in the colon by butyrogenic anaerobic bacteria that respond to dietary fiber or other prebiotics. However, diets rich in low-fiber, highly processed foods can result in low gut butyrate levels, which can lead to inflammation and other bowel diseases. In these disease states, populations of butyrogenic bacteria are also decreased.

Using a CRISPR-Cas based genome engineering method, we have engineered a model probiotic organism, *E. coli* Nissle 1917, to produce butyrate in both aerobic and anaerobic culture conditions. Further genetic optimization has been completed to increase anaerobic butyrate production, mimicking conditions in the gut. Additionally, to facilitate a high-throughput approach to butyrate production, we have developed a biosensor that responds to intracellular butyrate concentrations with a fluorescent reporter.

Implications of probiotic butyrate delivery on cellular health and microbial community makeup in higher order model systems, such as nematodes, mice, and intestinal stem cells will also be discussed.

470277 A Comprehensive Mutagenesis Study to Improve Yeast Cytosine Deaminase for Enzyme/Prodrug Therapy

Abstract:

Yeast cytosine deaminase (yCD) is an enzyme commonly used in gene directed enzyme/prodrug therapy (GDEPT) for cancer. GDEPT is a two-step process whereby the gene encoding a prodrug-converting enzyme is selectively delivered to tumor cells, followed by systemic administration of a nontoxic prodrug. Consequently, only cells that express the prodrug-converting enzyme (i.e. tumor cells) will be capable of converting the prodrug to its toxic form and induce cell death. One challenge of GDEPT is the inefficient transduction of tumor cells to carry the gene encoding the prodrug-converting enzyme, which limits the efficacy of this therapy. This work aims to address this issue by improving yCD activity against its prodrug, 5-fluorocytosine (5-FC), thus increasing cytotoxin production. While several studies have examined the effects of mutations at specific residues in yCD, a yCD variant with significantly improved activity against 5-FC has yet to be discovered. We performed a comprehensive mutagenesis study on yCD to improve its activity against 5-FC. Using inverse PCR, we created a comprehensive site-saturation mutagenesis library comprised of approximately 3000 yCD mutants. These mutants were subjected to selection for retained catalytic activity and subsequently screened for increased toxicity of 5-FC to *E. coli*. Mutants (and combinations thereof) that conferred increased 5-FC toxicity in our assays were further characterized to determine the mechanism for the increased sensitivity.

470377 *In Vitro* Characterization of a Novel MMP-2 Activatable Peptide for Photoacoustic Lifetime Imaging of Tumor Tissue

Abstract:

Many molecular imaging agents suffer from uptake and retention in healthy clearance organs. Due to constant signal emission by the contrast agent, this accumulation results in high off-target signal in these organs which significantly inhibits image specificity and sensitivity and hinders imaging of tumors near these organs. Additionally, many of the contrast agents used are handicapped by either poor resolution due to tissue-related scattering (fluorescence imaging), or require expensive equipment and radioactive agents (PET/SPECT). To address these issues, we have created an activatable peptide probe that only generates signal in the presence of the cancer biomarker matrix metalloproteinase 2 (MMP-2). This probe utilizes the novel photoacoustic lifetime imaging (PALI) modality, which decreases background signal, increases imaging depth penetration, lowers costs, and avoids ionizing radiation associated with current molecular imaging methods.

There are two key components to this probe: a dual-state imaging contrast agent, and a biocompatible, activatable motif. The contrast agent for the PALI imaging modality is methylene blue, which was shown to exhibit quenching of the PA signal lifetime upon dimerization but a long lifetime as a monomer. This enables differentiation between the inactive and MMP-2-activated states. The activatable motif used is a peptide design – adapted from the literature (Levi, *et al. J Am Chem Soc.*2010) – which will induce dimerization of MB to quench the monomer signal until cleavage of the peptide by MMP-2. The peptide contains poly(glutamic acid), an MMP-2-cleavable peptide, and poly(arginine) segments with methylene blue at each terminus. The ionic segments align in a 'zipper', which dimerizes the attached methylene blues. The zipper and associated methylene blues dissociate upon cleavage of the MMP-2 recognition peptide.

The initial peptide design, Ac-Keeee[Ahx]PLGLAGrrrrrK (E4R5), displayed a $k_{cat}/K_m = 1.2 \times 10^5\ M^{-1}s^{-1}$, comparable to previously published values for similar constructs. Flash photolysis and PALI demonstrate dimerization and effective quenching of the lifetime signal. A systematic series of additional peptides of varying designs of zipper and methylene blue linker were synthesized and tested for their dimerization efficacy and activation rates. Within the E4R5 peptide, the addition of glycine (GE4R5G) or 6-aminohexanoic acid (XE4R5X) between zipper and methylene blue decreased the dimerization efficacy (ratio of dimer/monomer absorptivities, $\varepsilon_{613\ nm}/\varepsilon_{668\ nm}$) from 2.57 ± 0.03 to 2.40 ± 0.01 and 2.09 ± 0.01, respectively. Shorter and longer zipper sequences (E2R3, E3R4, E4R4, E4R5 and E5R6) were shown to dimerize the attached methylene blue at equivalent efficacies. Initial activation rates of the various zipper constructs at 10 µM peptide, 0.2 ng/µL MMP-2, and 37 °C decrease with increasing zipper length from 269 ± 19 nM/min for E2R3 to 207 ± 12 nM/min for E5R6, demonstrating a potential increase in the strength of the interaction between the two ionic peptides and their increasing length. Notably, the E4R4 and E4R5 peptides have indistinguishable rates of 221 ± 12 nM/min and 226 ± 8 nM/min, respectively, potentially demonstrating the initial design of N poly(arginine) to N+1 poly(glutamic acid) does not affect peptide activity compared to equivalent amino acid numbers. The extended linker peptides (GE4R5G and XE4R5X) also showed decreased rates of 178 ± 13 nM/min and 173 ± 20 nM/min, compared to E4R5 at 226 ± 8 nM/min. Ongoing studies, including additional peptide design comparisons and evaluation of PALI signal before and after activation will be discussed. Additional results to be shared include evaluating the *in vitro* cellular activation and internalization of select peptides, as well as their *in vivo* activation via murine tumor models.

470382 Engineering E. coli for Methanol Metabolism

Abstract:

Due to the volatility of sugar prices, growing concerns in the "food vs fuel" debate, and the recent relatively low cost of natural gas, methanol has emerged as a promising feedstock for bio-based manufacturing of fuels and chemicals. Organisms that natively efficiently metabolize methanol are not generally amenable to metabolic engineering, owing to poorly developed genetic tools, and an incomplete understanding of their physiology, biochemistry and regulation. By contrast, *E. coli*, the workhorse of metabolic engineering, does not natively consume methanol. Here we present progress in engineering *E. coli* to grow and produce a variety of target molecules toward from methanol. Using isotopically labeled methanol, we demonstrate the ability of an engineered strain to assimilate methanol into central carbon metabolism. We further demonstrate the improvement of the assimilation rate through rational strain engineering based on sensitivity analysis of an experimentally parameterized kinetic model. Finally, we report on the effects of methanol supplementation on growth rate, yield, and product titer. The results lay the foundation for the development of an *E. coli* strain that can be "dropped in" to existing bioprocesses to enable the utilization of the alternative substrate.

470118 Liquid-Phase Synthesis of Low-Cost, Low-Polydispersity, Periodically-Sequenced Polypeptides

Abstract:

One of the main limitations to modern peptide synthesis is the relatively high cost of the products relative to the reagents. Liquid-phase synthesis offers the possibility of forming low-cost polypeptides, but the condensation reaction results in high polydispersities when forming high molecular weight polypeptides. To overcome this, we engineered and synthesized amino acid dimers that are both amphipathic and water-soluble. When polymerized, these dimers form repeat sequences in polypeptides with alternating hydrophilic/hydrophobic side groups, and are thus designed to develop a periodicity conducive to the formation of ß-sheets. We then decrease polydispersity of the growing polypeptide chains by modifying the kinetics of growth through transport-limited chain elongation. Our experiments show that in the absence of a surfactant interface, standard bulk-phase condensation polymerization occurs. In contrast, in the presence of surfactant, the amphipathic character of the polypeptides results in the transport-limited elongation as the polypeptide partitions into micellar assemblies. This partitioning, taking place during chain growth, serves to narrow the polydispersity of our periodically-sequenced polypeptide. We quantify polypeptide size and assembly using multi-angle light scattering and mass spectrometry and define evolving secondary structure using circular dichroism. Our results show that peptides grown in the presence of surfactant micelles show significantly enhanced self-assembly and narrowed polydispersity indices. We conclude that the transport-limited chain elongation polymerization in a surfactant-laden solution can be used in the manufacture of low-cost, self-assembling polypeptides. Finally, purified polypeptides were compressed on a Langmuir-Blodgett trough and imaged using Brewster angle microscopy, yielding surface pressure data characterizing the ß-sheets on an air water interface.

470588 Rapid in Vitro Screening of Proteins with Non-Natural Amino Acids

Abstract:

The incorporation of non-natural amino acids (NNAA) can introduce novel functional groups into proteins with site-specific precision and control, with significant applications in basic science and protein engineering. However, NNAA can decrease protein expression and functional activity depending on the location within the protein â€“ a process that is not yet completely understood and difficult to predict. Therefore, practical applications often necessitate a time-consuming optimization of NNAA location by individual expressions, purification and evaluations for many variant proteins. Such optimization could be accelerated by a high-throughput approach. Here, we introduce a highly customizable in vitro system using PCR products as expression templates for rapidly expressing and screening NNAA-containing proteins without cloning or purification procedures. We utilize this technology to simultaneously screen 24 different T4-Lysozyme mutants with different NNAA incorporation sites in a matter of few days, and further explore the effects of NNAA location on expression and activity.

470646 Role of Dimension and Spatial Arrangement on the Activity of Coupled Biocatalytic Reactions on Scaffolds

Abstract:

Despite broad interest in engineering enzyme cascades on surfaces (*i.e.,* for multi-step biocatalysis, enzyme-mediated electrocatalysis, biosensing, and synthetic biology), there is a fundamental gap in understanding about how the local density and spatial arrangement of enzymes impacts overall activity. In this work, the dependence of the overall activity of a cascade reaction on geometric arrangement and density of enzymes immobilized on a two-dimensional scaffold was elucidated using *de novo* kinetic Monte Carlo simulations. Simulations were specifically used to track the molecular trajectories of the reaction species and to investigate the turnover frequency of individual enzymes on the surface under diffusion-limited and reaction-limited conditions for random, linear striped, and hexagonal arrangements of the enzymes. Interestingly, the simulation results showed that, under diffusion-limited conditions, arrangements with the shortest distance between sequential enzymes were favored at high enzyme surface densities. Conversely, under reaction-limited conditions, the spatial arrangement of enzymes did not impact the overall rate. These results, which have widespread practical implications, suggest that random immobilization of the enzymes yields similar rates to enzyme arrangements that may be achieved using sophisticated approaches to patterning, thereby simplifying the engineering of artificial biocatalytic cascades.

470663 Solution Phase Characterization of [Fe-Fe] Hydrogenase O_2 Sensitivity

Abstract:

[FeFe] hydrogenases can provide large specific activities for sustainable, biological production of H_2 but are highly O_2 sensitive. We describe a new approach to provide quantitative assessment of O_2 sensitivity by using an assay employing ferredoxin $NADP^+$ reductase (FNR) to transfer electrons from NADPH to hydrogenase *via* ferredoxins (Fds). Hydrogenase inactivation is measured during H_2 production in an O_2-containing environment. An alternative assay uses dithionite (DTH) to provide reduced Fd. This second assay measures the remaining hydrogenase activity in periodic samples taken from the NADPH-driven assay. The second assay validates the more convenient NADPH-driven assay which also better mimics physiological conditions. During development of the NADPH-driven assay and while characterizing the *Clostridium pasteurianum* (Cp) [FeFe] hydrogenase, CpI, we detected significant rates of direct electron loss from reduced Fds to O_2. However, when decay in H_2 production rate is only assessed after O_2 addition, both assays indicate first order hydrogenase inactivation with rate constants insensitive to initial hydrogenase concentration. A protein fusion between Cp ferredoxin (CpFd) and CpI mediated by a 15 amino acid linker provides a higher specific activity than CpI alone (suggesting more effective electron delivery) and also was more O_2 tolerant. A longer linker, however, provided no activity or O_2 tolerance benefit. We suggest that this precise, solution phase assay for [FeFe] hydrogenase O_2 sensitivity and the insights we provide constitute an important advance toward the discovery of the O_2 tolerant [FeFe] hydrogenases required for photosynthetic, biological H_2 production.

470683 Nicking Mutagenesis: A Plasmid-Based Single-Pot Saturation Mutagenesis Method

Abstract:

The coupling of synthetic biology to protein mutational studies (i.e. directed evolution or deep mutational scanning) has greatly expanded the utility of these methods[1,2]. A core component of these experiments is the generation of a mutational library. Error prone PCR and cassette mutagenesis are the most common strategies but suffer from limitations in scalability, deficient codon sampling, and imprecise control over the number of mutations introduced. Based on these limitations, we developed Nicking Mutagenesis, a robust and accessible method for the construction of high quality, user-defined mutational libraries. Nicking Mutagenesis is a single day, single pot method using routinely prepped plasmid dsDNA as an input substrate. In this talk I will describe the overall method and demonstrate the efficacy on multiple systems. I will present data for saturation mutagenesis at multiple positions and single-site saturation mutagenesis libraries containing all possible single codon substitutions for full-length genes. Efficacy of the method will be presented by assessing library coverage using deep sequencing, revealingÂ 100% coverage of all possible single non-synonymous mutations (2840 total) and 100% of all possible programmed codon mutations (8946 total), with over 60% of the library containing exactly one mutation. Resources for portability of the method, including plasmid sharing and protocol capture will also be discussed.

1. Currin, A., Swainston, N., Day, P. J. & Kell, D. B. Synthetic biology for the directed evolution of protein biocatalysts: navigating sequence space intelligently. *Chem. Soc. Rev.* **44,** 1172â€"1239 (2015).

2. Fowler, D. M. & Fields, S. Deep mutational scanning: a new style of protein science. *Nat. Methods* **11,** 801â€"807 (2014).

470793 Inhibitors for Allergic Reactions to Drugs and Other Small Molecules

Abstract:

Penicillin is one of the most influential compounds in the history of human healthcare and it saves countless lives a year, but this drug is not always a suitable choice. For penicillin and countless other drugs that are heavily relied upon in healthcare, allergic responses are a constant concern that restricts their usage. In particular, type I hypersensitivity reactions, also known as Immunoglobulin E (IgE) mediated hypersensitivities, are the most problematic due to their ability to cause life-threatening anaphylaxis reactions. During these reactions, allergen specific IgE molecules crosslink when exposed to a carrier protein that has been conjugated with several drug molecules, triggering mast cell degranulation and therefore allergy symptoms. Our laboratory has developed targeted inhibitor molecules that specifically bind allergen specific IgEs monovalently, and thereby prevent any crosslinking from drug haptenized proteins, consequently preventing type I hypersensitivity reactions. We engineered a versatile molecular design that can potentially be adaptable in the synthesis of inhibitors for any drug or small molecule allergens. We used the design to synthesize inhibitors for penicillin G and dansyl chloride. Here, we have demonstrate that these molecules are specific only to their target IgE, form long lasting interactions, and prevent cellular degranulation from haptenized proteins both *in vitro* and *in vivo*. These inhibitors will potentially prevent life-threatening allergic reactions to drugs, and improve patient outcome.

470862 Specific siRNA-Target Interactions Influence siRNA Activity

Abstract:

Short-interfering RNAs (siRNAs) direct targeted post-transcriptional gene knockdown through the native eukaryotic RNAi pathway. siRNAs are widely used in research settings and several siRNA-based therapies are progressing through clinical trials. Still, siRNA design is not straightforward; this is due to an incomplete understanding of the RNAi mechanism and how chemically synthesized siRNAs interact with the various proteins of the RNAi pathway. Here we investigated how the 5' terminal nucleotide, siRNA hybridization stability, and siRNA-mRNA interactions impact siRNA strand loading and activity in cultured HeLa cells. We found that low hybridization stability at the 5' terminus of the intended guide strand (Nucleotides: 1-2) correlates with greater siRNA loading. Post-loading, low stability siRNA-mRNA interactions within the seed region of the guide strand (Nucleotides: 3-4) and high stability interactions near the 3' terminus (Nucleotides: 17-18) correlate with greater siRNA activity. Further, our results indicate that 5' terminal nucleotide sequence impacts the activity of an siRNA post-loading. In this presentation, we will discuss these results and how these parameters influence the design of highly active siRNAs.

471203 Nanofiltration for Mitigation of Upstream Viral Contamination Risks from Cell Culture Media and Feed

Abstract:

Adventitious viral contamination of production cells of commercial drugs can cause severe economic consequences due to losses of drug product batches and market share, costs associated with decontamination, and fines for regulatory non-compliance. A number of measures have been implemented in manufacturing processes to mitigate the risk of viral contamination of biopharmaceutics, which include testing of production cell lines, raw materials, process intermediates and final products, and inclusion of downstream process steps capable of reducing, removing, and inactivating viruses. Nonetheless, virus entry into the process through the cell culture medium preparation remains a possibility. Technology for upstream processing such as high temperature short time (HTST) processing for prevention of viral infection in cell culture is practiced by a small number of biotech companies. Retentive viral filtration (nanofiltration) of cell culture media and feeds is another key viral contamination risk mitigation technology, which has been implemented in GMP Manufacturing by a number of biopharma companies. In this study, small-scale nanofilters from three nanofilter vendors were evaluated for BI production and feed media, and the results with four CHO cell lines show no significant negative impact on cell culture performance. The nanofilter from one of the vendors demonstrated promising potential for implementation at 12K manufacturing scale and is being evaluated at pilot scale.

471031 *Bioactive BDNF Expression Using Brevibacillus Choshinensis*

Abstract:

Neurotrophins are essential growth factors in the growth, differentiation and survival of neuronal cells, playing a wide variety of roles in the central and peripheral nervous systems. Recombinant neurotrophins are used for research and therapeutic purposes, however, the cost of neurotrophins, in particular brain-derived neurotrophic factor (BDNF), can be limiting to both researchers and clinicians. The cost of BDNF is related to difficulties in the production of protein with the correct cystine-knot configuration. Here we investigated the recombinant expression of BDNF in the *Brevibacillus choshinensis,* exploring the effects of nutrients, temperature, and culture time on expression of bioactive BDNF. Of the variables examined, we found that lower temperature (25Â°C vs. 30Â°C) had the greatest impact on the expression bioactive BDNF. This proof-of-concept study demonstrates that we can produce ~200 Î¼g/L of bioactive BDNF. Further, BDNF was purified in a single step. This presentation will discuss the use of *B. choshinensis* to produce BDNF as well as methods to further optimize expression.

471038 Finding Diamonds in the Rough, Genome Mining of Fungi for Human Therapeutics

Abstract:

Natural products (NPs) are major sources for drug discovery. Due to increasing drug resistance to existing scaffolds and dwindling pipeline of new drug leads, our needs to uncover novel NPs and generate chemical diversity are becoming ever more important. Traditional NP discovery is achieved through bioactivity-guided isolation of NPs produced by microbes. With the exploding amount of genome information available for microorganisms, the direct mining of new biosynthetic gene clusters (BGCs) that can lead to new chemical structures and biological activities is a promising approach. In my thesis project, I proposed a highly efficient strategy, target-guided mining (TGM), to bridge this gap between activity-guided NP isolation and genome-guided NP discovery. I am aiming to combine my knowledge in chemistry and biology to develop new strategies to discover promising drug molecules that may benefit human health. There are three aims to be pursued: 1. Develop bioinformatics methods and TGM to identify BCGs that may produce NPs which target enzymes that are relevant in human diseases; 2. Produce the NPs encoded in the silent BCGs using synthetic biology strategies and validate the proposed enzyme targets; 3. Optimize the activities of the compounds through synthetic and biosynthetic modifications.

471210 Using Population Balance Models to Learn about the Evolution of Tumor Populations

Abstract:

Tumors can grow via cell division, can shrink via the attack of chemotherapy or the immune system (lately enabled by a new class of checkpoint inhibitor antibodies) and can spawn metastases via the breakoff and implantation of cells from existing tumors. Different tumors inside an individual patient each have elements of statistical variation in how they undergo these processes due to the occurrence of random mutations. However, these processes are still rate processes. We consider an ensemble of tumors from one or many patients and, in a type of model not previously used in this field, propose a population balance model approach to study the time evolution of the populations of tumors of all sizes in this ensemble. The goal of such a model is to obtain the probability of finding tumors within a specified size range at any given time. Traditional mathematical models produce a growth trajectory for an individual tumor based on either the fit of its early growth to an arbitrary mathematical form or by the use of average growth rates. This is inadequate because tumors that are the most important to a patientâ€™s prognosis are the largest, not the average size tumors. The proposed model combines the processes described above in a way that produces a distribution of tumors sizes rather than a single trajectory and whose results relate to the probability of finding tumors of each size as a function of time. These processes mathematically combine to produce an advection-diffusion equation in tumor-size space, which makes intriguing patient-relevant predictions. For example, with somewhat size-dependent growth and/or death parameters garnered from experiment, the model predicts that a patient whose tumor cell-division and cell-death rates are balanced can survive with a stable tumor load for a long time until a tumor emerges that grows rapidly and can quickly become fatal. It also illustrates how a patient can relapse after surgery even if all tumors large enough that their growth rates exceed their cell-death rates are surgically removed. These patient-relevant predictions arise very naturally from the model and seem to describe cohorts of hard-to-explain patients.

We test our model on existing hepatocellular carcinoma data and make predictions as to the evolution of this disease with and without therapy. We present our new data from zebrafish that are genetically engineered to be stripe-less and nearly clear, and which have been inoculated with a green fluorescent protein-labeled malignant melanoma cell line. We follow the evolution of both the primary tumor and the formation and growth of its metastases and compare these data to our model. Our long-term goal is to try to explain confounding patient cohorts, potentially propose new treatment strategies and, eventually, to use patient-specific information to guide treatment and to potentially predict likely time-to-recurrence.

471078 Discovery and Characterization of Novel Plant Natural Products through Biosynthetic Clusters

Abstract:

Plant natural products (PNPs) have been a rich source for discovery of pharmaceutical and antimicrobial compounds. While the onset of next-generation sequencing has dramatically increased the number of sequenced genomes, providing an invaluable resource for pathway discovery of natural products, the number of new natural product-derived or -inspired drugs per year has remained approximately constant over the past 34 years (Newman et al). In order to help close the gap between availability of genomic information and discovery of novel PNPs, herein we have developed a high-throughput platform for discovery of PNPs by exploiting the genomic clustering of biosynthetic genes. Though biosynthetic clustering is a common phenomenon in bacteria, until recently it has been largely overlooked in plants; however, with growing interest for new ways to discover novel PNPs, more and more biosynthetic clusters have been shown in plants (Nützmann et al). Through genomic mining and transcriptomic co-expression analysis, our platform assembles and filters potential clusters to a small list of top candidates. These candidates are then combinatorially tested for new PNPs in a plant transient expression system using untargeted metabolomics. Our platform combines rapid computational and experimental processes for the assembly, filtering, and characterization of potential biosynthetic clusters to the end of PNP discovery.

Newman, D.J.; Cragg, G.M. Natural products as sources of new drugs from 1981 to 2014. *J. Nat. Prod.* **79**, 629-661 (2016).

Nützmann, H.W.; Osbourn, A. Gene clustering in plant specialized metabolism. *Curr. Opin. Biotechnol.* **26**, 91-99 (2014).

471183 Accurate Structure Prediction of CDR H3 Loops Enabled By a Structure-Based C-Terminal Constraint within Rosetta

Abstract:

The adaptive immune system in vertebrates is capable of raising antibodies against a countless number of antigens. More recently, however, engineered antibodies have been used as therapeutic molecules and biosensors. The source of these antibodies varies across specific applications. In order to optimize specific modes of interactions, rational engineering techniques must be developed. Rational engineering of antibodies requires accurate structural models, but crystallization is not always practical or even possible. Additionally, expressing a large library of mutants in order to assess the energetic implications of specific mutations is time consuming, resource intensive and, in some cases, technically challenging. Computational methods, namely antibody homology modeling, are poised to enable the realization of rational design.

RosettaAntibody's approach to modeling is to break the structure into eight distinct structural components: the heavy- and light-chain frameworks; CDR loops L1–3; and CDR loops H1–3. Because the non-H3 CDR loops adopt canonical conformations, accurate backbone conformations for them can usually be found in known structures. RosettaAntibody exploits this by selecting templates from curated structural databases by BLAST bit-score for CDRs L1–3, H1 and H2 and the framework regions. Each structural component is defined such that they have overlapping residues that can then be superposed to create a grafted model. An initial VH–VL orientation is also selected from databases, and the grafted heavy and light chains are each superposed to the corresponding chain in the orientation template. After this, the CDR H3 loop is modeled de novo while sampling the VH–VL orientation.

We have presented the performance of RosettaAntibody in Antibody Modeling Assessment II (AMA II). With few exceptions, RosettaAntibody selects templates for the framework regions and the non-H3 CDR loops with sub-angstrom RMSDs from the native structure. The most difficult aspect of antibody homology remains accurately predicting the VH–VL orientation and the CDR H3 conformation.

A large majority of CDR H3 loops have a C-terminal kink, and in AMA II we found that producing low-RMSD models required filtering out non-kinked H3 conformations. However, the scores of the kinked structures was higher than some of the extended structures that Rosetta produced. In response to these findings, we developed new geometric parameters that describe the kink.

Because the CDR H3 loop lies at the interface between the heavy and light chains, incorrect VH–VL orientations can frustrate identifying correct CDR H3 conformations. In the time that has elapsed since AMA II was conducted, progress has been made in predicting VH–VL orientation from sequence by training a random forest model on a set of "fingerprint" residues at the VH–VL interface using ABangle's six degree-of-freedom description of orientation. Similarly, effort has been made to develop a CDR H3-specific loop modeling routine, but successful predictions require extremely accurate atomic coordinates for the rest of the FV, which may make these tools better-suited for refining crystal structures with poor electron density around the CDR H3 loop than for homology modeling.

De novo loop modeling has endured as a challenging problem in part because of the large number of degrees of freedom that need to be sampled, as well as the challenges associated with accurately ranking different structures that may appear to be very similar when using a coarse-grained measurement such as RMSD. Additionally, side-chain interactions may play key roles in stabilizing observed loop conformations, potentially complicating low-resolution searches. Complicating the task even further is the most common source of the reference coordinates: crystal structures. Crystals are extremely crowded environments in which each protein molecule is surrounded by several by others; this may or may not influence the observed conformation within the asymmetric unit. Without the existence of a crystal structure of the same protein in more than one distinct crystal form, it cannot be determined if these "crystal contacts" perturb the conformation of any region of the protein.

Similarly, another complication of loop modeling is the search for a single set of coordinates. Proteins in physiological conditions are not completely rigid, and estimating the conformational entropy of a loop requires supplying a model to describe the describe the modes of flexibility accessible to the loop. Nevertheless, the possible existence of multiple degenerate-energy conformations cannot be dismissed.

In this study, we use the parameters defined in our previous work on CDR H3 structures to constrain the kink during the course of a simulation. To limit the uncertainty in the crystallographic coordinates, we have constructed a set of extremely high-resolution H3 loops. Given the high degree of confidence in the atomic coordinates, computed RMSD values are also better-defined. The constraint is tested by predicting H3 conformations on the crystal framework structure across the set of benchmark structures. Finally, to test the utility of the constraint, we also assess the ability to dock an antibody with a modeled H3 loop and CDR H3 modeling on a homology modeled framework.

The method is integrated into the computational protocols implemented within Rosetta for predicting the three-dimensional structure of an antibody from sequence and subsequently docking the antibody to its cognate antigen. The approach employed is robust; antibody modeling leverages the existence of canonical loop conformations to graft large segments from experimentally-determined antibody structures as well as energetic calculations to (1) minimize canonical loops, (2) docking methodology to refine the relative orientation of the VH and VL domains, and (3) de novo loop structure prediction to model the elusive CDR H3 loop. Similarly, to alleviate model uncertainty, antibody–antigen docking resamples CDR loop conformations (either through minimization or explicit refinement) in the binding context and can use a set of input models to represent an ensemble of conformations for the antibody, the antigen or both.

Our methods are freely available to academic users and can be run fully-automated on a user's computer or via the ROSIE web server, or semi-automated on a user's computer.

471345 Site-Specific Immobilization of Antibody on Gold Nanoparticle Surface for One-Step Diagnosis of Prostate Specific Antigen with Dynamic Light Scattering

Abstract:

Small dimensions of gold nanoparticles (AuNPs) necessitate the oriented immobilization of antibodies to conserve antibodiesâ€™ antigen binding activity for proper function. In this study, we used four different methods to immobilize antibodies on AuNP surface; i) UV-NBS, ii) EDC-NHS, iii) disulfide reduction for thiol conjugation chemistry, and iv) physical adsorption; and compared the efficiency of these methods in terms of antigen binding activity, detection sensitivity and selectivity. Two different clones of anti-PSA monoclonal antibodies, B731M and B728M, specific to different epitopes on Prostate Specific Antigen (PSA) were chosen to functionalize on 5 nm AuNPs. Once the mixture of two different bivalently active antibodies on AuNPs recognize PSA, they form large clusters depending on the concentration of PSA in a sample; size of those clusters can be easily measured by dynamic light scattering (DLS) allowing selective detection of the antigen. Our results demonstrate that the limit of detection (LOD) for PSA utilizing antibody functionalized AuNPs via the UV-NBS method was significantly lower than the other immobilization methods, with the LOD of 0.1 nM PSA. The UV-NBS method yielded AuNP surfaces with significantly enhanced antigen recognition capabilities and improved antigen detection sensitivity with high level of selectivity when compared to other commonly used AuNP functionalization methods including: EDC-NHS, disulfide reduction for thiol conjugation chemistry, and physical adsorption. Consequently, this study demonstrates a one-step PSA detection method that yields highly sensitive, selective and rapid results.

471674 Expression and Isolation of a Precious Metal Binding Peptide Useful for Peptide-Directed Nanoparticle Synthesis

Abstract:

The need for low-cost nanoparticle (NP) materials is of great importance within the context of new technologies for catalysis. Processes used today to produce peptide-directed NP morphologies are not scalable in an economical way primarily due to the challenges imposed by the use of a biological molecule. The purpose of this work is to address this challenge thorough the development of a cost effective means of peptide manufacturing.

The poster describes the cloning, fermentation, and isolation of a 45 amino acid long metal binding peptide, accompanied by an assessment of its ability to direct NP synthesis and morphology. A synthetic gene encoding a fusion protein of the metal binding peptide with Green Fluorescent Protein was constructed by overlap extension PCR, introduced into an arabinose inducible vector and transformed into *Escherichia coli*. High cell density fed batch culture was used to prepare extracts containing the fusion protein, with varying degrees of purity correlated to NP morphology. Finally, NP synthesis was characterized by electron microscopy.

471690 Scaleable Assembly and Characterization of Fungal Natural Product Gene Clusters

Abstract:

Fungi have been an important source of medically relevant natural products (NPs) since the inception of biochemical engineering in the 1940s. Recent advances in DNA sequencing have revealed that the biosynthetic potential of fungal genomes is much deeper than early investigators realized. However, due to difficulties of culturing and genetically engineering many fungi, and the fact that many NP biosynthetic gene clusters (BGCs) are not expressed under laboratory conditions, progress to experimentally characterize the products of these gene clusters has been slow.

Here we describe the realization of a pipeline for the high-throughput characterization of fungal NP biosynthetic gene clusters. Our pipeline encompasses:

- bioinformatic tools for identification and prioritization of BGCs, and tools for sequence editing (e.g. intron removal) for improved expression in S. cerevisiae
- an assembly platform that combines commercial gene synthesis and yeast homologous recombination for the cost-effective construction of large (> 20 kb) DNAs
- high-throughput untargeted LC-MS based metabolomics screening of recombinant yeast cultures for the rapid detection and prioritization of potential novel compounds

With this pipeline, we successfully detected potentially novel NPs in four out of six previously unstudied fungal BGCs, and have structurally characterized a subset of the BGC-associated compounds. Our pipeline shows how high-throughput synthetic biology tools can facilitate the rapid discovery and biosynthetic production of complex chemical scaffolds of potential pharmaceutical relevance.

471600 Spore Surface Display of a Laccase to Enhance Its Robustness

Abstract:

Cost-effective deconstruction of lignocellulose is a major challenge to produce biofuel and other valuable fermentation products from renewable biomass. Large amounts of enzymes are needed due to the recalcitrant structure of lignocellulose, and the enzymatic degradation is often carried out under nonideal conditions owing to the pretreatment of the biomass. Inexpensive production of robust cellulolytic enzymes is critical to develop economically feasible biomass degradation processes. Extremophiles, growing under harsh conditions, provide a good resource and have been extensively explored to find robust enzymes. In this work, a thermostable laccase from thermophilic bacterium *Geobacillus thermoglucosidasius* was characterized and functionally displayed on the spore surface of *Bacillus subtilis*. It belongs to a newly discovered family of laccases, DUF152 family, and exhibits excellent activity and thermostability. The half-life of the purified laccase is about 3 hours at 80 °C. It was displayed on the spore surface to further enhance its robustness and reduce the production cost, because spores are able to survive various harsh environments and sporulation is well-established protocol and no protein purification is needed. A few spore surface proteins were explored as the anchor and different linkers were tested. The displayed laccase was shown to have a half-life at 80 °C about 1.5 times of the purified protein and improved solvent resistance. Co-expression of the protein together with the anchor-laccase fusion protein is critical to maximize the activity of dimers displayed on the spore surface. Overall, this work showed that spore surface display is a cost-effective protein expression and display platform that can enhance enzyme stability and complex protein such as multimers can be also functionally displayed on the spore surface.

471731 Sequence Specific Antisense Inhibitors of Non-Traditional Antibiotic Pathways for Eliminating Multidrug-Resistant Bacteria

Abstract:

Multidrug-resistant (MDR) infections are a pressing concern to global health which is made worse by the lack of new antibiotics being introduced. The overuse and misuse of current antibiotics has created pan-drug resistant bacteria to which there are no longer efficacious antibiotics. These current antibiotics are typically limited to three main pathways in bacteria: DNA replication and cell growth, protein biosynthesis, and cell wall biosynthesis. Developing antisense therapeutics, which enable sequence specific targeting of genes, allows us to take advantage of the many non-traditional antibiotic pathways to both uncover novel antibiotic targets and elucidate potential combination therapies which capitalize on the unexplored bacterial pathways. Using *Escherichia coli* as a model organism we created peptide nucleic acid based antisense molecules which target essential genes in non-traditional antibiotic pathways including metabolism, cell signaling, stress response, and gene regulation via antisense inhibition. Although the antisense molecules were designed against essential genes in *E. coli*, these molecules demonstrate therapeutic potential against multiple pathogenic bacteria including in clinical isolates of *E. coli*, *Klebsiella pneumoniae*, and *Salmonella typhimurium*, thus highlighting the potential to create broad-spectrum yet gene specific antibiotics. The clinical isolates used in this study are highly resistant to most classes of antibiotics, yet when the antisense molecules are used in combination with traditional antibiotics we observe a strong synergistic effect that significantly inhibits cell growth greater than either mono-therapy. We are able to re-sensitize the clinical isolates to Clinical Laboratory Standards Institute (CLSI) levels of antibiotics using the antisense inhibitors of non-traditional antibiotic pathways. Our findings highlight the potential utility of both targeting novel targets in non-traditional antibiotic pathways as well as the promise of combination therapies taking advantage of the novel targets in bacteria.

471797 Hybrid Promoter Engineering for the Development of Novel Promoters Inducible By Lignin Derived Phenolics

Abstract:

Lignin valorization is viewed as a key for the development of a cost effective lignocellulosic biorefinery, and synthetic biology tools would play an important role in the construction of an efficient chassis towards this goal. In this work, we have employed hybrid promoter engineering approach for the construction of higher strength phenolics inducible promoters. The hybrid promoters were constructed by replacing the spacer region of an endogenous promoter, PemrAB that was naturally inducible by phenolics. The strength of the promoters was interrogated by using *mCherry* as the reporter, and the engineered promoters were found to be stronger than the native promoter. In the presence of vanillin, the strains RIF02, RIF03 and RIF04, with the engineered promoters P_{vtac}, P_{vtrc}, and P_{vtic} had their fluorescence increase by 4.6, 3.0, and 1.5 fold respectively in comparison to RIF01 that had the native promoter PemrAB. The promoters were also observed to be highly specific towards vanillin and further promoter engineering efforts can increase the specificity and promoter strength beyond the current levels. Flow cytometry was employed to identify heterogeneity if present in the cell population. We observed the emergence of a sub-population constituting the healthy and dividing cells especially in the cultures that were induced with 5mM vanillin. The size of this sub-population remained about the same regardless of the promoter employed and it lay between 24% and 27% of the total cell population. The analysis also uncovered that this smaller sub population of cells were primarily responsible for the increased expression that was observed with the engineered promoters. The study also reveals that to achieve higher density of the heterologous proteins, efforts must be laid not only in increasing the expression level of genes but also in maintaining a healthy cell population in the presence of phenolics.

471924 Integrated Electrochemical-Biological Systems for the Production of Fuels and Chemicals from CO_2

Abstract:

One of the challenging processes in producing fuels or chemicals from CO_2 has been efficiently and selectively producing C_4-C_6 and higher multifunctional organic compounds from CO_2. In efforts to overcome these challenges a system that combines the faster kinetics of inorganic CO_2 reduction catalysts with highly selective microbial metabolism is being designed and investigated. It is known that electrochemical synthesis of small molecules which requires only 2-electron reduction steps such as H_2 or formate is able to convert electrical energy into chemical energy with high efficiency. On the other hand, it is also known that certain microbial organisms have the capacity to synthesize higher multifunctional organic compounds from simple C_1 or H_2 precursors with high selectivity but generally have low efficiency. This collaborative research is a unique study of the synergy of novel hybrid microbial-electrochemical platforms for the production of sustainable fuels from CO_2. The initial stage of this research involves the development of an experimental platform which allows for continuous and sustainable electrochemical CO_2 reduction on metal catalysts as well as continuous growth of microbial culture from select products of CO_2 reduction. Methods for the analysis of the products from the electrochemical CO_2 reduction as well the microbial culture has been developed from existing methods which utilize Gas chromatography, H^1NMR, and HPLC. This research involves the design and investigation of three main systems for the integration of electrochemical CO_2 reduction systems and cultures of microorganisms. The first design is focused on coupling these systems via the transfer of the gaseous products CO and H_2 along with unreacted CO_2 to the microbial culture. This was done by using established methods for electrochemical CO_2 reduction and continuous microbial culture growth. The second system design is a further integrated system where the microbial culture and the electrochemical cell is separated via an anion membrane. This is focused on the production of fuels by microorganisms which can metabolize CO_2 reduction products such as formate, while reducing the amount of contamination of the electrode from the microbial culture. The third system to be investigated is a fully integrated system where the electrochemical cell with specific metal catalysts are directly in the microbial culture. This investigation brings together electrochemical engineering and microbial science to utilize the efficiency of electrochemical CO_2 reduction to 2-electron products and the selectivity of specific microorganisms to metabolize these 2-electron products to produce fuels.

471939 Loop-Mediated Isothermal Amplification (LAMP): An Insight into Reaction Mechanism and Application in Point-of-Care Diagnostics

Abstract:

Analysis of nucleic acids is an essential part of both qualitative and quantitative diagnostics. Although polymerase chain reaction (PCR) based diagnostic techniques are well validated, the associated turnaround time, complex instrumentation and thermal cycling of reagents makes it suboptimal for point-of-care applications. Loop-mediated isothermal amplification (LAMP) is an alternate nucleic acid amplification method offering simple and cost-effective diagnosis of infectious diseases by eliminating thermal cycling and tedious sample preparation steps. It uses auto-cycling strand displacement synthesis with the aid of 6 specific primers. However, the intermediate reaction products are intricate stem loop DNA constructs (DNA hairpins), which may or may not contain the target sequence. This often makes LAMP a semi-quantitative analysis method suitable for yes/no type detection.

Although LAMP is increasingly used for diagnostic testing, current softwares for LAMP are limited to primer design, and unlike PCR, there is no known mechanistic description of the complete LAMP reaction. Here we present a kinetic model that explores all reaction intermediates and their evolution in respective reaction pathways. The kinetic parameters are estimated from literature and statistical mechanics calculation. The agent based model predicts around 15 side products along with the double loop DNA structure responsible for the autocatalytic amplification reaction. Some of these side products without the target sequence can lead to misrepresentation of the true amplification signal rendering LAMP as a semi-quantitative technique. The fundamental insights gained from our model are then used to determine optimal reaction parameters and efficient primer design. We leverage our new understandings to demonstrate multiplexed assay detection via the quencing of unincorporated amplification of signal reporters (QUASR) technique, enabling us to detect bright, unambiguous fluorescence signals with a simple smartphone based fluorimeter for point of care applications.

Sandia National Laboratories is a multi-program laboratory managed and operated by Sandia Corporation, a wholly owned subsidiary of Lockheed Martin Corporation, for the U.S. Department of Energyâ€™s National Nuclear Security Administration under contract DE-AC04-94AL85000.

471950 Understanding the Metabolism of *Scheffersomyces Stipitis* during the Transition from Aerobic Growth to Fermentation Condition

Abstract:

Xylose fermentation is essential in converting lignocellulosic material to biofuels in order to make an economically feasible process. *Scheffersomyces stipitis* is a native strain with the highest known capability of converting xylose to ethanol by fermentation. Since *S. stipitis* is a respiratory yeast strain, the oxygen availability is a critical parameter to consider for ethanol optimization due to the redox balance. Many other studies showed the effects of oxygenation conditions on ethanol yield and productivity; however, these studies have not look at the metabolism shifts occurred during the transition stage, where the oxygen condition changed from aerobic to oxygen-limited conditions. In order to systematically understand the metabolic states of *S. stipitis* during this transition period, we conducted experiments under controlled chemostat to achieve aerobic growth with no product and by-products, then it was switched to fermentation conditions by switching the oxygen condition from aerobic to oxygen-limitation. During this transition, we took samples frequently to capture the rapid changes of genome expression and metabolites. To doing this, we manipulated carbon-to-oxygen ratio (C/O2) to reach desired metabolic states of aerobic and oxygen-limited conditions. Different ranges of C/O2 ratio define different phenotypes of S. stipitis to produce different product and by-products. During the transitions stage, the experiments showed that the specific rates of substrate consumption, products and by-products profiles were changed within 1.5 hours to switch their metabolic states. This study allows us to understand the mechanisms of shifting metabolism of S. stipitis.

472044 Production and Purification of Oxidation-Resistant Alpha-1 Antitrypsin By Transient Expression in *N. Benthamiana*

Abstract:

Alpha-1 antitrypsin (AAT) is an acute phase protease inhibitor. Currently, the only available treatment for patients with COPD caused by AAT deficiency is replacement therapy, requiring weekly infusions of human plasma-derived AAT. While AAT has shown potential when inhaled to treat both emphysema and cystic fibrosis, the oxidation susceptibility of methionine residues at positions 351 and 358 in the reactive site loop have led to difficulties in maintaining AAT activity and stability when aerosolized. Using a novel cucumber mosaic virus-based inducible transient expression system, we have produced a biobetter plant recombinant AAT (prAAT) in *Nicotiana benthamiana*. This variant replaces the oxidation-susceptible methionine residue at position 358 with a valine residue to increase resistance to oxidation.

N. benthamiana plants were vacuum infiltrated as detached leaves with two strains of *Agrobacterium tumefaciens*: one carrying the biobetter AAT gene and one carrying a gene for the tomato bushy stunt virus p19 viral RNA silencing suppressor driven by a constitutive cauliflower mosaic virus 35S promoter. Leaves were incubated in humidity chambers at 20 °C in the dark for 6 days before flash freezing in liquid nitrogen, grinding, and protein extraction in a 20 mM Tris, 150 mM NaCl, and 0.01% (v/v) Tween-80 AAT stability buffer. Purification was achieved using Alpha-1 Antitrypsin Select Affinity Chromatography (GE Healthcare) with a 58 ± 8 % yield of active AAT. Activity of purified AAT was confirmed via permanent binding as seen in an enzyme-linked immunosorbent assay, a residual elastase activity inhibitory assay, and a band shift in a Western blot after incubation with excess porcine pancreatic elastase, however, multiple prAAT bands were also observed in all purified extracts. These bands may be due to glycosylation variants or degradation.

Purified prAAT was tested against an analytical standard (Calbiochem) and Prolastin-C (Griffols), a common

therapeutic formulation of human AAT, to determine its activity and oxidation resistance. For the purified prAAT, 41 ± 11% of the total AAT was active (able to inhibit porcine pancreatic elastase as determined by a residual elastase activity inhibitory assay) while 72 ± 6% of Prolastin-C AAT was active. Upon exposure to oxidation conditions with 48.9 mM H_2O_2 for 60 minutes at room temperature, prAAT was found to retain 100 ± 20% of its original anti-elastase activity, while Prolastin-C retained 14 ± 5% anti-elastase activity and analytical standard AAT retained 12 ± 8% anti-elastase activity. Future work will focus on increasing the percentage of total prAAT that is active during recovery and purification.

471733 Production Improvement of Butyrylcholinesterase (BChE) in Rice Cell Lines

Abstract:

Toxic nerve agents have been used as chemical weapons since they can disrupt the mechanism by which nerves transfer signals in the nervous system. Toxic nerve agents act by inhibiting the enzyme acetylcholinesterase which is responsible for the breakdown of acetylcholine (ACh) in the synapse. ACh gives the signal for muscles to contract, preventing them from relaxing so the intoxicated person will lose control of body function which can lead to death.

Some antidotes are being produced to combat poisoning by toxic nerve agents. One of them is the human blood protein, butyrylcholinesterase (BChE), which acts as a bioscavenger for organophosphorous nerve agents. BChE belongs to the family of serine hydrolases and can form a tetramer from four identical subunits (each 85 kDa) interacting with each other via a four-helix bundle at the C-termini. Although the tetramer has longer circulatory half-life, each monomer is fully active independent of its oligomerization state. Recombinant BChE has been produced in several expression systems such as mammalian cells (Lockridge et al.1997; Parikh et al. 2011; Xue et al. 2011; Ilyushin et al. 2013), silkworm (Wei et al. 2000), milk of transgenic goats (Huang et al. 2007), insect cells (Brazzolotto et al. 2012), tobacco leaves (Schneider et al. 2014; Geyer et al. 2010) and most recently in rice cell culture (Corbin et al. 2016). However, the expression of active BChE is not easily obtained.

Our approach is the production of recombinant BChE in rice cell culture using the rice alpha amylase 3D (RAmy3D) expression system. Plant cell culture is being used as a platform for the production of biopharmaceuticals because of its safety and cost-effectiveness in the manufacturing and purification of recombinant proteins. BChE has been produced successfully in rice cell culture using the rice RAmy3D (Corbin et al. 2016). The RAmy3D system allows the production of the recombinant protein under the control of an inducible promoter which is activated by sugar starvation and a signal peptide targets the protein for secretion. However, accumulation of the BChE protein into the culture medium has been challenging, perhaps because of an inefficient excretion of the recombinant BChE presumably due to the large size of the protein or perhaps due to BChE degradation by proteases.

In this study, 8 stable transgenic rice cell lines were analyzed for the expression of BChE. The quantity of recombinant BChE produced was measured at the 1, 3, 6, 9, 12 and 15 days post-induction (dpi) in both the culture media (secreted) and within the plant cell aggregate (homogenization). The time course analysis revealed that at 6 dpi the major amount of the protein was found in the callus while for the secreted protein the highest level was obtained between 6 to 9 dpi depending on the cell line.

In order to enhance the accumulation of the secreted protein in the media culture some biopolymers were added. Gelatin, polyethyleneglycol (PEG) and polyvinylpyrrolidone (PVP) were added to the culture medium at two different concentrations 10 g/L and 15 g/L, respectively. The mechanism of action of the biopolymers as stabilizers is not clear yet however a hypothesis is that they are target of proteases rather that the recombinant protein and it helps to block the degradation of the protein of interest (Kwon et al. 2002). Our results revealed that the best results were achieved by the 16-10 cell line and the results discussed below are regarding this line. The control used was the cell line induced with the media without sugar and 0 g/L biopolymer. The levels of protein at 6 dpi in the media reached up to 35 µg BChE/ g fresh weight (FW) calli (32 fold compared to control) when 10g/L gelatin was added. At the concentration of 15 g/L gelatin results were similar to 10 g/L gelatin at 6 dpi the levels detected reached 32 µg BChE/ g fresh weight (FW) calli (28 fold compared to control).

Interestingly with the addition of 10 g/L PEG, 9 dpi was the best day of production of BChE in the medium reaching levels up to 45 µg BChE/ g FW calli (21 fold compared to control). In the case of the addition of 15 g/L PEG, it produced 17 Âµg BChE/ g FW calli at 9 dpi (8 fold compared to control).

On the other hand, PVP did not help to increase the level of BChE found in the culture medium. In fact the levels were similar or lower than the control.

In the calli homogenization it was observed that the addition of either 10 g/L or 15 g/L gelatin or PEG didnâ€™t increases the levels of BChE detected, suggesting that the effect of the biopolymer is just in the excreted protein . All results mentioned were statistically significant ($p<0.05$).

In summary, these data suggest that the best biopolymers used were 10 g/L gelatin and 10 g/L PEG. These results were confirmed by Western blot analysis under denaturing conditions and the monomer of 85kDa was observed in the media culture starting at 6 dpi. Protease activity analysis is being done to establish if the improvement of the BChE accumulation in the media culture is because of the protease enzymes present in the media. Moreover, glycan analysis will identify the glycosylation pattern in either secreted or cell associated BChE.

471738 Immobilized BDNF Mimetic Peptides to Combat Secondary Injury Following TBI

Abstract:

Traumatic Brain Injury (TBI) is a debilitating condition for which there currently exists no fully restorative treatment. A primary, mechanical trauma results in instantaneous cell damage and death. Little can be done to address this damage once it has occurred, but a secondary, inflammatory injury cascade persists for weeks and months following the initial insult. Elevated concentrations of free radicals, glutamate induced excitotoxicity, and other comorbidities greatly exacerbate the initial damage and effects of injury. Despite the hostile, complex nature of the secondary injury environment, the timing offers a potential window for treatment. Promising research efforts to protect healthy tissue following TBI have included the delivery of neurotrophic factors, such as brain derived neurotrophic factor (BDNF). BDNF can provide neuroprotection and improve neuron survival following TBI, but it is rapidly cleared in vivo. Recently, short peptides that mimic the activity of native BDNF have been identified. Our laboratory has routinely created functional biomaterials by covalently grafting small molecules and peptides to collagen. We propose to immobilize these BDNF fragment peptides (IKRG) to a polymer matrix for sustained presentation of neurotropic cues at the site of TBI to combat the effects of secondary injury. Herein we present our initial efforts using collagen as a carrier molecule. BDNF fragment peptides were covalently grafted to lysine residues on the collagen backbone via 1-ethyl-3-(dimethylaminopropyl) carbodiimide (EDC). A mixture of IKRG peptide, EDC, and collagen is reacted overnight at 4Â°C, dialyzed to remove un-grafted peptide, lyophilized, and reconstituted at 3mg/mL in 0.02N acetic acid. Peptide grafted collagen was used to treat cultures of cerebellar granule neurons (CGNs) following simulated secondary injury in vitro. CGNs were isolated from day 6 rat pups and plated onto poly-L-lysine coated tissue culture well plates at 2,000,000 cells/mL. Cultures were treated with Cytosine Î²-D-arabinofuranoside (Ara-C) after 48 hours to limit the proliferation of astrocytes. At DIV 9, secondary injury was modeled by treating cultures with 300μM or 500μM glutamate for one hour after which glutamate containing media was removed. Fresh media containing recovery treatments was added for a period of 24 hours. Immediately following recovery treatments, cellular metabolic activity was evaluated with a MTT assay. Recovery treatments consisted of different ratios of IKRG grafted collagen and native collagen at a net concentration of 0.33mg/mL in culture media. These were compared to untreated and uninjured controls. Following injury with glutamate, CGN cultures demonstrated ~60% reduction in metabolic activity (Untreated 300μM: 45.0 ±13.4; 500μM 33.5 ±11.9). After treatment with native collagen, metabolic activity further decreases (Native Collagen 300μM: 32.9 ±6.2; 500μM 22.2 ±5.5). Treatment with IKRG grafted collagen resulted in a nearly two-fold increase in metabolic activity as compared to native collagen (IKRG-Collagen 300μM: 54.6 ±19.5; 500μM 44.3 ±10.4), but represented only a slight increase as compared to the untreated condition. This suggests that collagen may be having unintended, negative consequences on neuronal viability that is masking the full potential of these IKRG peptides. To realize the full potential of immobilized, BDNF mimetic peptides we are investigating PEG based polymers as a bio-inert alternative for a carrier molecule. Our hope is that immobilization to an alternative polymer will allow us to deliver IKRG peptides without the confounding effects of collagen.

472161 Efficacy of Antibacterial Agents on *Staphylococcus Aureus* biofilms Based on the Hydrodynamic Conditions and Nutrient ConcentrationsÂ of the Growth Environment

Abstract:

Biofilms arise when microorganisms attach to a surface of their host and form a bio-layer. Growth of these biofilms and their ability to resist antimicrobial agents are the main cause of numerous tenacious and chronic bacterial infections. *Staphylococcus aureus* is one of the most common causes of bacterial infections and has the ability develop biofilms easily. Bacteria within biofilms tend to be hundreds of folds more resistant to antibiotics than planktonic bacteria. Biofilm development can be affected by several factors such as nutrient concentrations, hydrodynamics conditions (fluid shear stress and shear rate), types of bacteria, and cell-cell communication (quorum sensing). This project aims to determine the isolated and combined effects of nutrient concentration and shear rate on the application of antibacterial drugs. We hypothesized that the combined effect of nutrient concentration and shear rate applied to the bacterial biofilm would have statistically significant influence on the ability of antimicrobial drugs to degrade bacterial biofilms. To test this hypothesis, biofilms were grown under different nutrient concentrations (2X, 1X, 0.5X, 0.1X) of TSB and fluid shear rates (ranging from 10 s^{-1} to and 300 s^{-1}) conditions in the presence or absence of known antibacterial agents. Biofilm cultures of *S. aureus* were grown under continuous flow of TSB nutrient to determine the impact of hydrodynamic conditions on the growth of bacteria. The cultures were grown at 37°C and exposed to various fluid flow rates under the control of a BioFlux 1000z flow system (Fluxion Biosciences Inc., South San Francisco, CA). The acquired images from the experiment were then analyzed to determine the area coverage. From this study we can conclude that the application of antibiotics is greatly affected by the shear force where the biofilm was developed under. The study suggests that the efficacy of antibiotics is greatly affected by the shear forces present within the growth environment and, to a lesser extent, by nutrient strength. These results may be used to predict the efficiency of antimicrobial dosage regimens against pathogens prone to biofilm formation under hydrodynamic

milieus.

472214 Applying Standardized Three-Dimensional Mechanoculture to Multi-Well Plates

Abstract:

Introduction: Although multi-well plates mark the pinnacle standard for in vitro cell culture studies, they are essentially limited in emulating 3D and dynamic extracellular milieu in the body that are critical in restoring cellular functions. Various scaffolds and bioreactor designs have been introduced to reduce this gap, and their biological significance has been evidently demonstrated. Yet, these platforms have been sparingly accepted in biological experiments and pharmaceutical screenings partly because the benefits of using multi-well plates (e.g. reproducible, quantitative and observable experiments with affordable cost) outweigh the advantages of 3D bioreactor cultures. Here, we introduce an integrative approach to realize a standardized 3D mechanoculture in a multi-well plate using a hydrogel scaffold and a magnetically responsive floating disc.

Materials and Methods: 3D polyacrylamide hydrogel scaffolds with controlled microporous structure (D=250-300μm) and defined thickness (H=1mm) were prepared following listed method [1]. A magnetically responsive floating disc was fabricated by attaching a medical grade stainless steel ball (D=4.76mm) in a cylindrical (D=12mm, H=3mm) PDMS slap containing air sac (V=0.1mL). An array of neodymium magnets (D=H=9.5mm) were placed on an acrylic stage. A pneumatic piston was used to adjust the distance between the floating disc and magnet-acrylic plate. As the pump pushed the magnet-acrylic plate up, the discs move to the bottom of the well in response to the magnetic field, and generate shear and mechanical stimulus. An acrylic O-ring (D=15.4mm) was placed on the bottom of some wells to prevent the floating disc from striking the bottom. The frequency of pneumatic actuation was controlled by Arduino Uno (Adafruit). Human Bone Marrow Stromal Cells (hBMSCs) and human prostate cancer cells (PC-3) were used for proof of concept studies. The confluent areas in each well were progressively analyzed using the open source program ImageJ.

Results and Discussion: hBMSCs response to three distinct culture conditions were analyzed: disc only, disc with O-Ring, and control. Optical clarity of the disc allowed us to image the cells without the risk of airborne contaminants from open-lead operation. BMSC in the disc and O-Ring well have much more densely aligned cellular morphology in the direction of the applied shear force compared to that of the control well. Physical interaction between the disc and the well plate may have caused some cellular damage. A Similar phenomenon was observed for PC-3 cells, thus conclusive that scaffolds further increased the complexity of in-vitro culture systems by offering 3D matrices. Current efforts are in directing stimuli to the 3D scaffolds.

Conclusion: We demonstrated a simple and versatile approach to apply 3D and mechanically active culture in a multi-well plate. Our modular approach allows simulating specific structure and geometry of extracellular matrices as well as shear and mechanical stimuli of a diverse range of tissues/organs. We envision our multi-well plate based 3D bioreactor design will facilitate practical implementation of 3D mechanoculture to broader research communities.

472312 Maximizing P-Glycoprotein Expression in the Presence of Certain Therapeutic Compounds

Abstract:

Alzheimer's disease (AD), the most common form of neurodegenerative disorder, affects over 5 million people and is the 6th leading cause of death in the United States. AD is characterized by an accumulation of amyloid-β protein (Aβ) in the brain. Currently under study as a therapeutic approach is the transport of excess Aβ out of the brain through the single endothelial cell thickness of the blood-brain barrier. P-glycoprotein (P-gp), an ATP binding cassette transporter located on the apical side of the blood-brain barrier, has been shown to transport the oligomeric aggregates of Aβ. While monomeric Aβ is inert, oligomeric Aβ exhibits neurotoxicity and leads to the formation of Aβ fibrils that deposit as amyloid plaques, which are characteristic of AD brain. However, formation of oligomeric Aβ may be important to its clearance from the brain.

Previously, a correlation was demonstrated between low levels of P-gp expression at the blood-brain barrier and high levels of Aβ aggregate deposits around the cerebrovascular. Also, it has been recently discovered that the presence of certain forms of Aβ in the cells of the blood-brain barrier can reduce P-gp levels through the ubiquitin-proteasome degradation pathway. These findings identify P-gp as a novel therapeutic target for AD. This study seeks to identify therapeutic compounds that will increase the expression levels of P-gp at the blood-brain barrier in order to mitigate the accumulation of Aβ by allowing for increased clearance of Aβ from the brain.

Madin-Darby Canine Kidney (MDCK) epithelial cells continuously express the multidrug resistance 1 (MDR1) gene responsible for the presence of P-gp. These cells are often used in the study of P-gp due to the availability of a MDCK-MDR1 cell line, a MDCK cell line transfected to express additional P-gp via the MDR1 gene, that can be used for comparison when up-regulating P-gp expression in standard MDCK cells. Cells were treated with three prospective therapeutic compounds: verapamil, caffeine, and rifampicin. Verapamil is a known P-gp substrate recently suggested to up-regulate P-gp expression. Rifampicin, an antibiotic used to treat tuberculosis and leprosy, and caffeine have been shown in epidemiological studies to reduce dementia symptoms. Following treatment, alteration of P-gp expression was determined using confocal microscopy to visualize cell membrane location of P-gp and SDS-PAGE with Western blotting to quantify P-gp expression. Results highlight the prospective therapeutic potential of these compounds.

472367 Waste Water Treatment Using Field Deployable Macroalgal Tissue Culture

Abstract:

Macroalgae can remove nitrates, phosphates and heavy metals from contaminated waters as well as break down recalcitrant compounds, e.g. trinitrotoluene (TNT), while synthesizing halogenated compounds which have antibiotic, antifungal and antitumoral properties. The macroalgae also produce high levels of compounds relevant to biofuels, such as terpenes and cyclic sesquiterpenes. The metabolic pathways that allow the synthesis and breakdown of cyclic compounds may also enable the capability to remediate phenolics or other wastes from agricultural and petrochemical industries. Hence, initially, we aim to produce preliminary data to develop a field-deployable macroalgal tissue culture system for the bioremediation of surface waters in a "waste to value" process that will simultaneously produce bioactive compounds or biofuels. Further, we propose a field-deployable, "biogenic packed bed," high cell-culture density system and determine nutrient, metal and phenolic removal capacities. Therefore, this study has investigated a high density cultivation system and quantified the

ability to remove nutrient loads from agricultural wastewater. Additionally, because the biomass itself can also be used as a fertilizer or animal feed, the objective of the study is directly related to the water-energy- food nexus.

Keywords: Photobioreactor, Macroalgae, Waste water treatment, Biofuel, Terpenes, Bioactive compounds, Bioremediation

472465 Memory of DNA Methylation in Live Cells Exposed to Environmental Chemicals

Abstract:

Exposure to heavy metals, such as lead (Pb), at an early development stage is known to be affililiated with adverse health outcomes at a late life stage. In spite of decades of efforts, the seach of molecular markers that characterize early-stage heavy metal exposure does not have any successful hits. Comparative transcriptomics study suggests that Pb exposure can trasiently alters gene expression pattern but the observed changes do not persist. Furthermore, gene sets that contribute to disease development at a late-life stage have little overlap with those identified at an early stage right after heavy metal exposure. In spite of accumulating evidence suggesting the existence of long-term cellular memory arising from previous Pb exposure, the exact molecular traits of "memory" remains unknown. Accumulating literature and our own study suggest that epigenetic changes of specific gene sets can be the major composing unit of long term memory formed due to Pb exposure. The goal of this study is thus to identify specific epigenetic traits that are induced upon Pb exposure and subsequently persist through cell cycles and divisions. We examined human cell lines exposed to Pb at different dose and duration using ChIP-qPCR assay, bisulfite sequencing along with the live-cell imaging tools developed in our laboratory. We identified several epigenetic changes that arise from Pb exposure and characterized their short- and long-term stability using a coarse-grained signaling model framework. The identified epigenetic traits can be used as a potential biomarker for characterizing individual health risk arising from previous Pb exposure.

472247 Characterization of Conjugated Polyelectrolytes As Molecular Sensors for the Detection of Amyloids in Neurodegenerative Diseases

Abstract:

Misfolding and aggregation of proteins is a central pathogenic event in neurodegenerative disorders such as Alzheimer's and Parkinson's diseases. Currently, there lacks a reliable sensor platform for detecting the range of protein aggregates involved in disease etiology, particularly the pre-fibrillar aggregate conformations that are the most neurotoxic due to a lack of molecular probes that could selectively and differentially target different protein aggregate conformations. We have shown that a class of synthetic polyelectrolytes, oligo *p*-phenylene ethylenes (OPEs), show promise as molecular sensors that can selectively detect fibrillar conformations of two model proteins, lysozyme and insulin. Compared to existing probes that are of limited clinical use, e.g., Thioflavin-T based probes, OPEs offer many distinct advantages, including versatile and highly tailorable structural and chemical properties, and most notably the multiple modes by which OPEs respond to interactions with ligands. In this study, we test the sensing capability of two leading OPE compounds against aggregates of a series of disease-relevant proteins, including the amyloid-beta peptide and tau protein implicated in Alzheimer's disease, islet amyloid polypeptide in diabetes, and the Huntingtin protein in Huntington's disease. Changes in the photophysical properties of the OPEs including absorbance, fluorescence excitation and emission were characterized in the presence and absence of amyloid protein aggregates. Our results indicate that the OPE sensors exhibit non-protein specific, large fluorescence enhancements in the presence of protein aggregates. Preliminary data of OPE staining of brain tissue sections of Alzheimer's mouse models indicate that the OPEs has the potential to specifically stain amyloid deposits in the brain. This project contributes towards our goal of developing a novel class of sensors for the early detection and tracking of neurodegenerative disorders such as Alzheimer's disease, which will not only help us understand the diseases better, but will also lead to early diagnosis and therapies to prevent and treat these devastating diseases.

472580 The Use of Microbiota to Degrade Hydrocarbons

Abstract:

Contamination of soil and water in Mexico, due to oil spills, leaks clandestine milkings, by poor plumbing or accidents, is a problem with serious effects, since these impacted areas are no longer available for agricultural activities and become potential sources of groundwater contamination. Some fractions from crude oil have toxic and carcinogenic effects in animals and humans, so by removing them we can transform them into harmless compounds this is an important goal in the decontamination process. In the process of degradation, have found through the capacity of a microbiome, that were extracted from an agricultural soil site in southeastern Mexico contaminated with petroleum hydrocarbons, through bioremediation study. Water samples contaminated with commercial diesel and light oil with 30°API were prepared. Samples were inoculated with growing microbiome, analyzing the concentration of hydrocarbons by gas chromatography (GC-FID), UV-vis and IR-ATR, microbial growth by plate count spectroscopy. degradation efficiencies 95% for water samples with oil and 98 contaminated% for water samples were obtained contaminated with diesel.

472601 Improving Dynamic Regulation of Biodiesel Production in *E. coli*

Abstract:

Dynamic regulation is frequently employed in nature to optimally allocate cellular resources for efficient substrate utilization. Dynamic regulation is utilized as an advanced metabolic engineering strategy to improve production of a bioproduct by employing metabolite responsive feedback of gene expression. The dynamic nature of these feedback control schemes require a sensor, which detects metabolite concentration in real-time, and an actuator, which regulates gene expression based on the sensor measurements. Bacterial transcription factors have been used for this purpose, as they are sensor-regulators by design. This approach has been implemented for the production of fatty acid ethyl ester (FAEE) biodiesel in *E. coli* with great success. Based on a tractable non-steady model of this pathway, we propose to improve the dynamically regulated FAEE production by: 1) altering the ethanol-producing enzymes for more balanced pathway flux, and improving the dynamics of the regulator; 2) increasing the expression of the regulator protein; and 3) engineering a reversible ethanol pathway to enable reutilization of excess ethanol. These changes result in better carbon distribution and bring the pathway into better balance, allowing the dynamic nature of dynamic control schemes to be utilized. The strategies employed here are likely to be generally applicable to metabolic pathways rather than merely specific to this one.

472904 *in Vitro* DNA Assembly for Marker-Less Gene Replacement in *Bacillus Subtilis*

Abstract:

Bacillus subtilis, the best-characterized Gram-positive bacterium, is the model organism for studying Gram-positive pathogens such as *Bacillus anthracis* and biotechnologically important species such as *Bacillus licheniformis* and *B. subtilis* for the production of proteins and food additives. It is also a model organism for studying cell biology including sporulation, bacterial chromosome replication, cell development and differentiation. Along with the development of functional genomics, *B. subtilis* is now one of the most advanced model organisms for systems biology study and host for synthetic modules. Despite of its importance and wide use as research or industrial host, genetic engineering of *B. subtilis* is not very efficient. Though a genetic toolkit is available for modifying its genome, almost all of them rely on *E. coli-B. subtilis* shuttle vectors and genetic modification takes much longer time compared with *E. coli*. In addition, a resistance marker is often created for either gene deletion or insertion. The limitation of available resistance markers and long turnover time significantly impede the research related to *B. subtilis*. In this work, we developed an *E. coli* independent method for marker-less gene deletion or insertion. Four pieces of DNA for deletion or five pieces for insertion, including the homologous fragments and a resistance marker, are assembled using Gibson assembly and transform *B. subtilis*. Double crossover homologous recombination followed by linearizing the chromosome at a designed I-sceI site will force the removal of the resistance marker. Compared with previous method, our design does not require cloning in *E. coli* and this greatly shortens the overall process. In addition, the efficiency of resistance marker removal is 100% and all colonies obtained have the desired design, which is much better than a 50% from a previously reported method, and no further experiment is needed for confirmation. We have demonstrated this method by deleting the *amyE* gene or integrating a *gfp* expression operon into the *amyE* site. It takes about one day to get the deletion or insertion and another 2-3 days to remove the resistance marker, which is comparable to genome modification in *E. coli*. The development of the *E. coli* independent marker-less gene replacement method will greatly facilitate the genome editing in *B. subtilis* and expedite relevant research including cell biology study, strain development, and system/synthetic study.

473657 Comparing the Predictive Accuracy of Translation Rate Models Across 16,685 mRNA Sequences

Abstract:

Precise control over protein expression is required when engineering biological systems such as genetic circuits or metabolic pathways for chemical production. Modifying the ribosomal binding sites (RBS) of the encoding mRNA is one of the most commonly employed and effective strategies used for tuning protein expression levels. Early on, biological engineers achieved this by using previously characterized RBS parts or designing *ad hoc* RBS sequences, both of which often resulted in unpredictable protein expression in various genetic contexts. To address this challenge, a number of models of translation initiation have recently been published including the RBS Calculator (versions 1.0, 1.1, 2.0), UTR Designer, RBS Designer, and EMOPEC. These models attempt to map the complex relationship between sequence and expression, but do so with varying biophysical descriptions of the interactions between the 30S ribosome and the mRNA. These differences reveal the ongoing debate over what the definitive determinants are that control translation initiation. Here, we present a systematic model comparison on a diverse 16,685 mRNA sequence-expression database, and the result of that analysis, a definitive list of determinants that regulate translation initiation. We used a combination of sequence structure categorization, model error analysis, and RNA folding dynamics simulations to identify that the rate and efficiency of mRNA folding into low free energy structures can modulate the rate of translation initiation by over 1,000-fold â€“ a phenomena not properly accounted for in any of the existing models. With these new insights, we look to develop a more accurate biophysical model of translation initiation to enable more rapid and efficient design of genetic systems.

474084 Transdermal Delivery of Traditional Chinese Medicine Via a Microfabricated Zeolite Microneedle

Abstract:

Traditional Chinese medicines (TCM) have more than 3000 years of history and include many bioactive phytochemicals derived from plants. Decoction and oral administration remain common extraction methods for TCM with inconvenience and low efficacy. The possibility of reformulating TCM for transdermal delivery is attractive and microneedle could provide an avenue for effective administration of TCM. This work explores the delivery of ferulic acid (FA) derived from TCM, *Danggui* (*Angelica sinensi*) by microneedle. Ferulic acid is shown to be an antioxidant with pro-apoptotic effects on cancer cell. The microneedle was made of pure silica zeolites that have good hardness and strength, and can create channels through skin with minimal peripheral damages. The delivery of ferulic acid was monitored in an *in vitro* experiment using pig skin as model skin surface to measure the efficiency of the microneedle transdermal delivery system.

473982 Enhancer Control of Transcriptional Bursts

Abstract:

Transcription is episodic, consisting of a series of discontinuous bursts. Using live imaging methods and quantitative analysis, we examine transcriptional bursting in living *Drosophila* embryos. Different developmental enhancers positioned downstream of synthetic reporter genes produce transcriptional bursts with similar amplitudes and duration, but generate very different bursting frequencies, with strong enhancers producing more bursts than weak enhancers. Insertion of an insulator reduces the number of bursts and the corresponding level of gene expression, suggesting that enhancer regulation of bursting frequency is a key parameter of gene control in development. We also show that linked reporter genes exhibit coordinated bursting profiles when regulated by a shared enhancer, challenging conventional models of enhancer-promoter looping.

450409 Modulation Effect of Acidulated Human Serum Albumin on Cu^{2+}-Mediated Amyloid Î² -Protein Aggregation and Cytotoxicity

Abstract:

Alzheimer's disease (AD), a prevalent neurodegenerative disease, is characterized by dysfunction of memory. The main histofpathological hallmark of AD is senile plaques formed by amyloid β-proteins (Aβ) in the hippocampus and cortex. In addition, elevated levels of metal ions in plaques, mainly Cu^{2+} and Zn^{2+}, indicate that they play a crucial role in Aβ aggregation in vivo. It has been proven that metal ions not only induce the generation of off-pathway Aβ aggregates but also lead to the formation of reactive oxygen species (ROS), which are harmful for nerve cells. Cerebral acidosis is another common complication of AD. Under mildly acidic conditions, Cu^{2+}-Aβ species have a higher tendency to generate neurotoxic aggregates. In this work, we fabricated acidulated human serum albumin (A-HSA) to mitigate Cu^{2+}-mediated $A\beta_{42}$ aggregation and cytotoxicity at pH 6.6. Extensive experiments showed that A-HSA could alter the pathway of Cu^{2+}-mediated $A\beta_{42}$ aggregation and protect SH-SY5Y cells from cytotoxicity and oxidative damage induced by Cu^{2+}-$A\beta_{42}$ species. Noticeably, stopped-flow fluorescence analysis revealed that A-HSA changed the Cu^{2+}-$A\beta_{42}$ coordination mode from complex I to complex II on the millisecond timescale, which avoided the formation of aggregation-prone Cu^{2+}-$A\beta_{42}$ aggregates. Based on the findings, we proposed a mechanism of the modulation effect of A-HSA on inhibiting Cu^{2+}-mediated $A\beta_{42}$ aggregation and cytotoxicity under the mildly acidic condition.

458860 Designed Ankyrin Repeat Proteins in Chimeric Antigen Receptor Engineered T Cells

Abstract:

Designed Ankyrin Repeat Proteins in Chimeric Antigen Receptor Engineered T Cells

Elizabeth Siegler[1], Yu Jeong Kim[2], Pin Wang[3]

[1]Department of Biomedical Engineering, University of Southern California, Los Angeles, CA 90089

[2]Department of Pharmacology and Pharmaceutical Sciences, University of Southern California, Los Angeles, CA 90089

[3]Mork Family Department of Chemical Engineering and Materials Science, University of Southern California, Los Angeles, CA 90089

Cancer immunotherapy has the potential to achieve long term remissions by harnessing the patient's own immune system to attack cancer cells. Chimeric antigen receptors (CARs) are engineered proteins that offer a method of reprogramming autologous immune cells to recognize and target cancer cells. CAR-engineered T cells are not MHC restricted and can be engineered to recognize specific tumor-associated surface antigens (TAA). Typically, this recognition is due to a single chain variable fragment (scFv) which is derived from the desired antibody and fused to intracellular signaling domains. Antigen binding to the scFv triggers a signaling cascade that activates the CAR-T cell against the TAA-expressing cancer cell.

CAR-T therapy has been largely successful in hematological cancers, but has faced many setbacks in treating solid tumors. Often this is because CAR-T cells become exhausted and lose their cytotoxic capabilities after long exposures to TAA. Many factors are thought to contribute to CAR-T exhaustion, but one important factor may be due to crosslinking of the scFvs on the CAR-T cell surface. scFvs tend to aggregate, which can stimulate the CAR-T cell even in absence of the appropriate antigen. Prolonged stimulation leads to eventual exhaustion and compromised function, including diminished cytokine release and decreased cytotoxicity.

Antibody mimetic proteins also are capable of binding to specific antigens. One such protein family is designed ankyrin repeat proteins (DARPins), which are based off naturally occurring ankyrin proteins. In general, these proteins are smaller than scFvs and consist of two or three repeat units of 33 amino acids each. Constructs with specific binding affinities, sizes, or other desired qualities can be generated from DARPin libraries. DARPins are more stable and less prone to aggregation compared to scFvs and are being explored as an alternative to scFv use in CARs. In particular, we are comparing two Her2-binding DARPin-CARs to a traditional anti-Her2 scFv CAR.

We have selected two DARPins, G3 and 929, to compare to the scFv 4D5. G3 is the smallest of the three Her2 targeting proteins, which may make it more easily expressed during lentiviral transduction. G3 also has the higest binding affinity (k_d = 90 pm), which may make the CAR more sensitive to Her2. We are also testing the DARPin 929, which binds to a different Her2 domain than either 4D5 or G3. Most anti-Her2 scFvs, including trastuzumab-derived 4D5, bind to Her2 domain IV, which is partially obstructed when Her2 dimerizes. 929 binds to domains I-III, which may be more easily accessible than domain IV.

To date, our group has performed side-by-side *in vitro* testing of G3, 929, and 4D5 CAR-T cells. Each of these antibody binding domains were cloned into 3rd generation CD28-41BB-CD3ζ CARs. CAR surface expression, interferon-γ release, and cytotoxicity are comparable between the three CARs. However, T cell exhaustion is not readily observable *in vitro.* Our next steps include *in vivo* testing, in which we predict the DARPin CARs will be superior to the scFv CAR due to decreased aggregation and exhaustion.

KEYWORDS: cancer therapy, cancer immunotherapy, chimeric antigen receptors (CARs), designed ankyrin repeat proteins (DARPins)

References

1. Chang, Z.; Silver, P.; Chen, Y., Identification and selective expansion of functionally superior T cells expressing chimeric antigen receptors. J Transl Med. 2015, 13 (161), 1-16.
2. Chmielewski, M.; Hombach, A.; Abken, H., Antigen-specific T-cell activation independently of the MHC: chimeric antigen receptor-redirected T cells. Front Immunol. 2013, 4 (371), 1-5.
3. Münch, R.; Mühlebach, M.; Schaser, T.; Kneissl, S.; Jost, C.; Plückthun, A.; Cichutek, K.; Buchholz, C., DARPins: An efficient targeting domain for lentiviral vectors. Mol Ther 2011, 19, 686-693.
4. Riddell, S.; Srivastava, S., Engineering CAR-T cells: design concepts. Trends Immunol 2015, 36 (8), 494-502.

459651 Enhancing *in Vitro* Macrophage Drug Delivery Efficiency Via Co-Localization of Cells and Drug-Loaded Microcarriers in 3D Ultrasound Standing Wave Field

Abstract:

Introduction: Since macrophages are intrinsically involved in inflammatory process which is a key driver for onset and/or progression of various diseases such as cancer, tuberculosis, and HIV infection, strategies aimed at targeting the macrophages have gained increasing attention in the field of drug/gene delivery over the past decade. However, macrophages have long been reputed to be difficult targets because of limited efficiency of molecular transfer and rapid degradation of internalized chemicals and/or biologicals caused by the innate endolysosomal system that have seriously hampered the macrophage-related applications in clinic. Although some success has been achieved through the use of synthetic drug vehicles and/or utilization of physical approaches (e.g., electroporation and sonoporation), several drawbacks, such as insufficient transfection rates, serious cell damage, and/or lack of scalable capacity still remain obstacles for their practical use. To circumvent these issues, a synthetic molecular transfer system involving use of drug-loaded poly(lactic-co-glycolic acid) (PLGA) microspheres and ultrasound standing wave field (USWF) was explored in this study.

Materials and Methods: The calcein-AM-loaded PLGA microparticles (CAPMs) were fabricated by single oil-in-water emulsification in association with a solvent evaporation approach. The calcein-AM with expression of green fluorescence (GF) was served as the model drug. To investigate the efficacy of cellular uptake of CAPMs under USWF exposure, canis macrophages (DH82 cells) were exposed to 1-MHz USWF with output intensity of 0.5 W/cm^2 for 0, 1, 3, 5, 10, and 15 min, respectively, following wash twice with PBS and incubation at 37°C for an additional 24 h. The drug delivery efficiency which was represented by the intensity of green fluorescence expressed (i.e., relative fluorescence units; RFUs) was detected using a fluorescent microscope and quantitatively analyzed through flow cytometry.

Results and Discussion: Based on the phase-contrast microscopic and SEM detections, the produced CAPMs remained intact spheroids without deformation after the fabrication process including centrifugation and vacuum-aid filtration. The mean size of the CAPMs was 2.68 ± 0.07 µm in which > 90% of the particles were in the range of 1 – 10 µm based on the Stable Light Scattering analysis. The mean surface charge of the CAPMs was -91.8 ± 2.82 mV, and the encapsulation efficiency of the calcein-AM was 76.8 ± 3.2%. Based on the delayed and prolonged GF expression in the CAPMs-treated DH82 macrophages, our data showed that the PLGA microspheres were able to protect the encapsulated calcein-AM molecules from enzymatic digestion in the phagocytic endolysosomal system and thus the effect of GF expression can be extended. This is particularly important for macrophage drug delivery because mostly the exogenous molecules are often quickly degraded by the phagocytic endolysosoms. Through the fluorescent microscopic and flow cytometric analyses, our results showed that both DH82 macrophages and CAPMs can be quickly brought to acoustic pressure nodes within 20 sec under USWF exposure, and were consequently aggregated throughout the time course. In this study, the efficacy of cellular uptake of CAPMs enhanced with increased USWF exposure time that a three-fold augmentation ($P < 0.05$) was obtained after 15 min of USWF exposure. We further demonstrated that the enhanced CAPM delivery efficiency was mainly contributed by the co-localization of cells and CAPMs due to USWF exposure, rather than from sonoporation..

Conclusions: In summary, the developed USWF-mediated microparticulate delivery approach provides a feasible means for macrophage drug delivery.

460794 Chemotaxis-Based Mesenchymal Stem Cell Migration in 2D Microfluidic Maze

Abstract:

The migration of cells is an important event that involves in embryonic morphogenesis, wound repair, and cancer invasion. Most migration studies focus on cancer and immune cell, whereas for tissue manufacturing stem-cells are more relevant. Yet, their migratory behavior is studied a lot less in comparison. Part of the reason for this is simply because stem cells are slower, and thus single cell studies require lengthy experiments with live imaging and incubation on the microscope (challenging). Another reason is that tissue engineering studies are frequently done by seeding cells on 3D scaffolds, in which real-time non-invasive observation is not possible. As a result, most of the analysis is done by performing bulk differentiation and proliferation assays on crushed samples (i.e., a single time point at the end of the experiment. Therefore here we utilize the most advanced aspects of cell cancer/immune migration studies and apply them specifically to mesenchymal stem cells (MSCs) - the most prominent cell type used tissue engineering. The few MSC migration studies we have found, so far have not considered heterogeneous-yet-well-defined chemoattractant gradient profile (the drivers of migration that the cells usually encounter *in vivo*). Hereby we endeavor to explore the long-term migration of MSCs in 2D microfluidic maze platforms which represent the complexity of the tissue geometry *in vivo* with long paths, short paths, and dead ends in which a multitude of cell directional decisions are possible through different chemical gradient scenarios. The migration of MSCs is chemotactically induced by diffusion-based concentration gradients of platelet-derived growth factor-BB (PDGF-BB) and fetal bovine serum (FBS). The gradients are maintained stable for more than 48 hours by means of slow diffusion of the chemoattractants. The whole cell migration process in the microfluidic device is monitored and imaged in real time using an inverted microscope (IX83, Olympus) installed with a custom culturing chamber. The stability of the chemical gradient is tracked by fluorescent 20 kDa Dextran-FITC while the initial concentration gradient of PDGF-BB and FBS is computationally predicted by COMSOL modeling basing on their molecular weights. Migratory cell decisions are quantified in terms of percentage of cells migrating across long path, or short path, or being trapped in dead ends. Different maze dimensions as well as various chemotactic profiles affecting the migration of cells are also investigated. For instance channel size must be large enough for the cell to be able to pass through, while small enough for it to be able to grab onto the walls; while at the same time, the chemoattractant gradient steepness is also affected by the maze geometry. The ultimate end-product of our work is a microfluidics platform custom-tailored to migration studies relevant to tissue engineering-like systems.

460934 Multi-Layer Agent-Based Modeling of Vascularized Bone Regeneration within a Porous Biomaterial Scaffold with Growth Factor Gradient

Abstract:

The success of engineered bone tissue regeneration requires the synergistic combination of biomaterial scaffolds, growth factors, and osteogenic cells. Biomaterial scaffolds serve as a temporary matrix for cell proliferation and extracellular matrix deposition, as well as a template for vascularization. The formation of a sufficient and proper vascular network is critical for the survival of the regenerated tissues. Exogenous growth factors released from the scaffold, including osteogenic growth factors (e,g, bone morphogenetic protein 2, BMP2) and angiogenic growth factors (e.g. vascular endothelial growth factor, VEGF), are often required to promote the osteoblastic differentiation of stem cells and angiogenesis, respectively. The interactions between blood vessel formation, growth factor release, bone cell growth, and scaffold structure properties are often difficult to investigate systematically with experimentation alone. Computational models provide powerful tools to simulate the process of bone tissue regeneration within biomaterial scaffolds, and capable to test hypothesis and strategies to enhance bone tissue regeneration in an organized manner. Due to the complexity and heterogeneity of the whole process of bone tissue regeneration, conventional empirical and mechanistic models are sometimes inadequate to describe the underlying biological process. Agent-based modeling (ABM) is well suited for modeling multi-cellular biomedical systems where different type of cells, molecules, and organelles can be represented as autonomous, interacting software agents while allowing an easy integration with continuous models for more realistic environment description.

We have previously developed an ABM to investigate angiogenesis (Mehdizadeh et al., 2013) and osteogenesis (Bayrak et al., 2014) in a porous, degradable biomaterial scaffold. In addition, a biomaterial system that creates gradients of growth factors within the porous scaffold was developed to accelerate scaffold vascularization (Akar et al.,

2015). The current work extends and modifies the ABM to simulate the interaction between stem cells and growth factors. Both endogenous and exogenous growth factor release are considered to investigate the effects on bone tissue regeneration. The multi-layer ABM is utilized to investigate optimum growth factor gradient properties and scaffold geometrical structure to form clinical size vascularized bone tissue.

Reference:

Bayrak, E. S., Mehdizadeh, H., Akar, B., Somo, S. I., Brey, E. M. & Cinar, A. Agent-Based Modeling Of Osteogenic Differentiation Of Mesenchymal Stem Cells In Porous Biomaterials. Engineering In Medicine And Biology Society (EMBC), 2014 36th Annual International Conference Of The IEEE, 2014. IEEE, 2924-2927.

Mehdizadeh, H., Somo, S., Bayrak, E. S., Brey, E. M. & Cinar, A. 2013. Three-Dimensional Modeling of Angiogenesis In Porous Biomaterial Scaffolds. *Biomaterials,* 34, 2875-2887.

Akar, B., Jiang, B., Somo, S.I., Appel, A.A., Larson, J.C., Tichauer, K.M. and Brey, E.M., 2015. Biomaterials with persistent growth factor gradients in vivo accelerate vascularized tissue formation. *Biomaterials*, *72*, pp.61-73.

460849 Computational Modeling of Therapeutic Protein Distribution in the Rabbit Eye

Abstract:

Sight-threatening diseases such as age-related macular degeneration and diabetic retinopathy are currently treated by intravitreal injection of anti-VEGF antibodies drugs. To ensure optimum therapeutic benefits, drug concentration should be within or above certain concentration in target tissues. For this purpose, a computational 3D model based on the anatomically accurate rabbit eye dimensions was developed to simulate spatio-temporal evolution of protein distribution in ocular tissues after intravitreal injection. The model considers the convective-diffusive transport processes along with drug elimination in choroidal sink and episcleral veins. It also includes flow of aqueous humor generated in ciliary body within ocular tissues. Previous studies indicated the high dependence of drug distribution by diffusivity coefficients in vitreous. So in order to obtain accurate picture of drug clearance in different ocular tissues, it is necessary to have the accurate drug transport properties. To do this, apparent diffusion coefficient of fluorescein-conjugated protein in fresh rabbit Retina and choroid-RPE was determined in a custom-built diffusion apparatus. The proteins used in this study---- Avastin, Lucentis, Eylea, lysosoms and BSA--- are of different molecular weight and some of them are commonly used drugs for the treatment of ocular diseases. Fluorescein probes concentration was determined by fluorometry. Diffusion coefficient and mobility of proteins in fresh Rabbit vitreous was estimated using fluorescence recovery after photo-bleaching (FRAP), a microscopy based method. Model validated with experimental data for suprachoridal injection of fluorescein labeled Avastin. Comparison of the result for diffusion coefficient through RPE-choroid, retina and vitreous between different proteins showed that it decreased with increasing size as expected. The model simulation indicates such difference has a significant effect on the clearance pattern inside the eye.

461631 Effect of Fatty Acid Compounds in Persister Formation of *Escherichia coli*

Abstract:

Persisters are a dormant subpopulation of bacterial community that exhibits significant tolerance to antibiotics without genetic modification, because slow cell division and protein synthesis in the dormant state results in insensitivity to antibiotics targeting these steps. Once the level of antibiotics drops, persister cells revive by becoming active cells. Thus, persister cell formation of bacteria is problematic in controlling infectious diseases. Little is known about the mechanisms of persister formation and revival, and thereby there is few effective means to eradicate persister cells. In this study, we investigate different types of fatty acid compounds to see how they affect the survival of *Escherichia coli* under antibiotic treatment. Short chain fatty acids including butyric acid, propionic acids, and valeric acids are known for antibacterial activity by damaging cell membranes which is a different mechanism from common antibiotics. We observed that high concentration (20 mM) of fatty acids arrests cell growth and decrease the level of pH in the *E. coli* culture, while low concentration (<5 mM) of fatty acids does not affect cell growth but has a less effect in pH decrease. When the *E. coli* culture was treated with ampicillin (100 μg/mL) along with short chain fatty acids both high and low concentrations, persister cell formation was increased significantly similar to the cell survival level of un-treated *E. coli*. We are working on figuring out how cells increase their survival under the presence of both antibiotic and fatty acid antibacterial. We will further assess effect of fatty acids on pathogenic bacteria such as enterohemorrhagic *E. coli* and *Pseudomonas aeruginosa*. This study opens the development of novel strategies for combating pathogenic bacteria and controlling infectious diseases.

462538 Inkjet Printed Emulsions and Microparticles for Single-Cell Analysis

Abstract:

The field of droplet or emulsion microfluidics has seen rapid expansion in recent years, enabling novel techniques to address challenging problems in biology such as single-cell analysis. Droplet microfluidcs typically relies upon microfluidic channels with flow-focusing or T-junctions to break up an aqueous stream into a stream of monodisperse droplets surrounded by an immiscible oil phase, with each droplet (typically ranging from picoliters to a few nanoliters in volume) serving as a discrete reaction chamber. While numerous variations upon droplet generation devices have been described, almost all rely upon fluid mechanical effects to acheive periodic breakup of the aqueous stream into uniform droplets. This can be quite effective but is ultimately dependent upon precise generation of highly stable flows. In practice this can be difficult to acheive reliably, and almost invariably results in some stabilization time for each experiment, and furthermore presents challenges with coating devices for compatibility with immiscible liquid phases.

We describe an elegant alternative in which electrical signals rather than fluid mechanics are used to acheive extremely regular, periodic generation of droplets. This is acheived using a commercial, off-the-shelf piezoelectric inkjet dispenser suspended above a stirred bath of surfactant-laden oil. Stable and precise droplet generation, at frequencies up to 2 kHz, and with sizes somewhat tunable in the 30-50 micron range, is acheived at the flip of a switch in the "drop-on-demand" mode. Larger droplets (80-100 microns) are produced at 10-20 kHz using the jet breakup mode by applying a constant head pressure to the fluid. Since the aqueous phase (in the inkjet dispenser) and the oil phase are not directly in contact, the two phases can contain incompatible components such as reactive monomers. We demonstrate the use of this system to produce standard water-in-oil emulsions with a variety of hydrocarbon oil and fluorinated oil phases, as well as core-shell microcapsules, and photopolymerized hydrogel microbeads. We demonstrate applications to genetic analysis of single microbial cells. We will discuss practical considerations of the inkjetting setup, and we will discuss unique advantages as well as disadvantages to the inkjetting approach.

461811 Measuring Adhesion Between Uropathogenic *E. coli* and Bladder-Epithelial Cells

Abstract:

Bacterial adhesion to host cells is often a first step in the infection process. For example, uropathogenic *Escherichia coli*, the major causative agent of urinary tract infection, bind to host bladder-epithelial cells and initiate cell invasion. This triggers a subsequent pathogenic cascade characterized by recurrent infection. There is currently growing interest in developing new antimicrobials that, instead of targeting bacterial survival and placing high selective pressure for drug-resistant mutations, target mechanisms promoting infection such as binding to host cells. This new therapeutic strategy requires a detailed understanding of bacterial adhesion. To address this issue, we used *in vitro* methods to examine the factors that contribute to the adhesion of *E. coli* to the bladder epithelium, including both traditional tissue-culture based assays and novel mechanical measurements. Specifically, we used an adapted live-cell rheometer to make direct physical measurements of bacterial adhesion to a monolayer of bladder epithelial cells. This method allowed us to quantify the extent to which extracellular bacterial components affect adhesion. Our results indicate that an extracellular amyloid fiber commonly produced by uropathogenic stains of *E. coli*, termed curli, increase adhesion to bladder epithelial cells and may therefore play a role in the infection process. In the future, we plan to use these techniques to assess the effectiveness of various small molecules in blocking bacterial binding to host cells.

463028 Perfluoro-Treated Bioresorbable Polylactic Acid for Coronary Artery Stents

Abstract:

Perfluoro-treated Bioresorbable Polylactic Acid For Coronary Artery Stents

The permanent presence of rigid metallic coronary stents in the vessel wall can cause stent failure over long times. This failure may be in part related to hyperplasia stimulated by the stent and mechanical inhibition of physiologic vasomotion around it. Bioresorbable stents, in which their composition allows gradual breakdown, are an emerging and novel treatment for improving stent long-term outcomes. Poly(lactic acid) (PLA) has been the predominant polymer used for making bioresorbable stents. This polymer is fully resorbable and it degrades through a bulk erosion process through hydrolysis of ester bonds, with resulting lactic acid and oligomers being metabolized into CO_2 and water.

Although bioresorbable vascular scaffolds have various advantages over the long term, they are associated with higher thrombosis than permanent metallic stents. To address this issue, we modify the surface of PLA with a perfluoro compound facilitated by surface activation using RF plasma. Fluoropolymers have been extensively used in blood contacting materials, such as blood vessel replacements due to their lower thrombogenicity. In addition, our previous studies with fluoropolymers have shown low platelet reactivity.

Here, we will discuss the synthesis and blood compatibility of our perfluoro-treated PLA substrates. The compositions of plasma-treated surfaces determined by electron spectroscopy for chemical analysis (ESCA) will be studied. Also, contact angle measurements, cell cytotoxicity and the degradation profile of our treated polymers will be presented. Finally, relevant blood compatibility parameters, including plasma protein adsorption, platelet adhesion and morphology, will be evaluated. A hypothesis will be presented where tight binding of adsorbed albumin by fluoropolymers accounts for its success in blood-contacting applications.

462614 Integrative Meta-Modeling Ranks RTK Signaling and Identifies Connection Between Nuclear Translocation and Extracellular Ligand Concentrations

Abstract:

Integrative meta-modeling ranks signaling strength across eight RTKs and finds that RTK signaling is dependent on the extracellular ligand concentration

Introduction:

Current therapeutics seeking tyrosine kinase receptor (RTK) control are ineffective and met with drug resistance [1]; however, controlling RTK signal transduction would allow treatment of many pathologies, including cancers [2] and vascular diseases [3]. This challenge underlies a continuing need to increase our understanding of RTK signaling to cell response. Indeed, while signal transduction pathways have been established for many membrane receptors [4]–[7], these pathways are often established for individual RTKs, accounting for mechanisms specific to a particular receptor. As such, insights gained for a specific RTK have never been compared to other RTKs. In order to achieve signal transduction control for any RTK, a systematic analysis identifying how cell physiology generally mediates RTK signaling is needed. Moreover, this systematic analysis delineating how cell physiology mediates signal transduction, in general, would serve as a "signaling template" of signaling fundamentals that could be tuned to account for receptor-specific spatiotemporal dynamics [9-10]. Such an analysis would also improve and enhance knowledge of signal mapping and facilitate research on receptor-based signaling control, both of which are critical in ultimately being able to treat pathological conditions. Here, we engineer such a signaling template and used it to meta-analyze signaling of eight receptor tyrosine kinases: EGFR, FGFR1, IGFR1, PDGFRα, PDGFRβ, VEGFR1, VEGFR2, and Tie2 (Table 1). This computational meta-analysis delineates the complex endocytosis mechanisms underlying intracellular RTK signaling in general, by examining how cell (compartment volume, trafficking kinetics and pH) and ligand-receptor physiology (ligand/receptor concentration and interaction kinetics) determine signaling.

Materials and Methods: We analyze RTK signaling by computationally modeling ligand-receptor interactions and subsequent receptor internalization and trafficking using MATLAB Simbiology. While undergoing trafficking, compartment pH dynamically regulates ligand-receptor interaction kinetics. Using this understanding, we quantitated receptor phosphorylation, a post-translational modification, associated with each endocytic compartment, to weigh the signaling contribution from each intracellular compartment. We take a data-driven approach towards model development, mining ligand-receptor kinetics via surface plasmon resonance, receptor concentrations via qFlow (quantitative) cytometry and western blot, ligand concentrations via ELISA, and compartment volumes and trafficking kinetics via microscopy. We also account for pH-dependent ligand-receptor interaction kinetics across the compartments. With this computational model, we rank order signaling across RTKs and examine how RTK parameters (Table 1) direct the receptor signaling strength. We also conduct a correlation analysis, assuming a lognormal fit, between the physiological RTK parameter values given by each RTK (Table 1) to membrane-based and nuclear-based RTK signaling in Origin. Here we focus on membrane- and nuclear-based RTK signaling, as these highlight the initial (membrane) and final (nucleus) compartments that receptors are trafficked throughout.

Results and Discussion: *The extent of absolute signaling is dependent on the RTK.* With this computational signaling template, we quantify the integrated response, the total receptor phosphorylation over time, at each compartment. We find that receptor signaling primarily occurs within endocytic vesicles, comprising > 43% of total receptor signaling within the cell for these eight RTKs. Conversely, we found that membrane signaling is relatively low, providing < 1% of total receptor signaling within the cell across these eight RTKs, indicating that nearly all receptor phosphorylation throughout the cell occurs intracellularly. While these eight RTKs follow the trend of receptor signaling primarily occurring within endocytic vesicles with low membrane-based receptor signaling, we found that absolute receptor signaling is highly variable across the eight RTKs. Indeed, among the eight RTKs, PDGFRβ has the largest absolute membrane signaling at a level $3.1 \cdot 10^3$-fold greater than FGFR1, which has the lowest absolute membrane signaling (Table 1). By analyzing the three RTK-specific parameters - receptor concentration [R], ligand concentration [L], and ligand-receptor dissociation constant [Kd] (Note [Kd] = k_{off}/k_{on}) - we analyze the complex concentration, [R][L]/[Kd], across several RTKs. Indeed, FGFR1 has the lowest complex concentration, while PDGFRβ has the highest—due to its very high [R]. Overall, our meta-modeling ranks RTK signaling strength: PDGFRβ > IGFR1 > EGFR > PDGFRα > VEGFR1 > VEGFR2 > Tie2 > FGFR1.

Table 1. Experimentally obtained parameters (receptor concentration, interaction kinetics, and extracellular ligand concentration) were pulled from literature for the eight meta-analyzed RTKs.

	Ang2-Tie2	EGF-EGFR	FGF2-FGFR1	IGF1-IGFR1	PDGFAA-PDGFRβ	PDGFBB-PDGFRα	VEGFA-\
Receptors/cell	$1.8 \cdot 10^3$	$5.0 \cdot 10^4$	$2.8 \cdot 10^4$	$2.5 \cdot 10^4$	$5.3 \cdot 10^4$	$5.1 \cdot 10^3$	2.7·
k_{on} ($M^{-1}s^{-1}$)	$6.0 \cdot 10^3$	$3.0 \cdot 10^7$	$9.6 \cdot 10^4$	$2.7 \cdot 10^5$	$8.8 \cdot 10^3$	$7.8 \cdot 10^6$	3.0·
k_{off} (s^{-1})	$6.1 \cdot 10^{-4}$	$3.8 \cdot 10^{-3}$	$5.9 \cdot 10^{-3}$	$1.2 \cdot 10^{-3}$	$1.5 \cdot 10^{-4}$	$7.6 \cdot 10^{-3}$	1.0·
Ligand in Serum (pg/mL)	1865	917	2.2	$1.65 \cdot 10^5$	1769	8506	16
Membrane-receptor integrated response (p-Receptor·time)	$8.9 \cdot 10^2$	$7.7 \cdot 10^3$	$1.5 \cdot 10^1$	$1.1 \cdot 10^4$	$4.7 \cdot 10^4$	$2.3 \cdot 10^3$	1.6·
Nuclear-receptor integrated response (p-Receptor·time)	$8.0 \cdot 10^3$	$3.6 \cdot 10^5$	$2.6 \cdot 10^2$	$4.2 \cdot 10^5$	$7.5 \cdot 10^5$	$8.1 \cdot 10^4$	5.8·

Complex concentration determines the extent of nuclear translocation. Our meta-analysis also reveals that the extent of nuclear translocation varies significantly across the eight RTKs (Table 1). This is evidenced by nuclear signaling, which we find ranges between 3.3% - 27%, given by FGFR1 and EGFR, respectively, of the total receptor signaling within the cell. By analyzing the three RTK-specific parameters, we predict that, like membrane-based RTK signaling, nuclear-based RTK signaling is determined by the complex concentration. Indeed, EGFR has the highest complex concentration, while FGFR1 has the lowest, among the eight RTKs. Therefore, the extent of nuclear-based RTK signaling is also dependent on the RTK and can be predicted by the complex concentration.

The signaling given by a RTK can be tuned with the extracellular ligand concentration. Our results show that absolute receptor signaling is variable among these eight RTKs. To understand which individual RTK parameter best regulates signaling within a RTK, we alter three RTK parameters and observe membrane (Fig. 1A) and nuclear (Fig. 1B) signaling. We find that membrane RTK signaling is well regulated by [Kd] and [L] but not by [R]; increasing the ligand-receptor on-rate (k_{on}) or [L] three orders of magnitude above the physiological concentrations (Table 1) exponentially increases membrane signaling, while increasing the ligand-receptor off-rate (k_{off}) abrogates membrane RTK signaling (Fig. 1A). Conversely, nuclear RTK signaling is regulated by [L] and [R], but not [Kd]; increasing [R] eight orders of magnitude above physiological concentrations (Table 1) increases nuclear signaling from 3% to 22% of the total cell signaling – increasing [L] the same amount increases nuclear signaling to 24% of the total cell signaling (Fig. 1B). Like signaling within a single RTK, we find that [L] strongly predicts nuclear signaling across the eight RTKs (Fig. 1C), whereas [R] (Fig. 1D) and [Kd] (Fig. 1E) are weak predictors. Indeed, we find that increasing [L] one order of magnitude increases nuclear RTK signaling 3.2-fold (Fig. 1C). We also find that [R] and [Kd] have low weight, while [L] has high weight, in mediating membrane-based RTK signaling. Overall, these results suggest that [L] is the only RTK parameter capable of regulating both membrane- and nuclear-based RTK signaling, in addition to [L] being the RTK parameter that best mediates RTK signaling at these two compartments.

Conclusions: We introduce a new integrative, data-driven, computational approach that provides a "signaling template" for understanding RTK receptor signaling. This is an important step, because our template can be applied to develop new therapeutic approaches targeting specific RTK signaling, optimizing treatment for many pathologies, including cancers [9] and vascular diseases [10]. We also arrive at important findings with this computational approach: We also arrive at important findings with this computational approach: receptor signaling can best be regulated by controlling the extracellular ligand concentration, whereas altering the membrane receptor concentrations will have negligible effect on receptor signaling. This finding has significant implications for drug delivery, suggesting that therapies inhibiting membrane receptor signaling alone will be ineffective, and should, instead, target intracellular receptors or extracellular ligands.

References: [1] Corcoran C. *Methods Mol Biol* (2015) 1233:169. [2] Niepel M. *Sci Signal* (2013) 294:ra84. [3] Murphy E. *Circ Res* (2011) 109:687. [4] Maruyama IN. *Cells* (2014) 3:304. [5] Seshacharyulu P. *Expert Opin Ther Targets* (2012) 16:15. [6] Kabbani N. *Proteomics* (2008) 8:4146. [7] Arish M. *Biochimie* (2015) 113:111. [8] Mukherjee S. *Circ Res* (2006) 98:743. [9] Pálfy M. *Trends Cell Biol* (2015) 22:447.

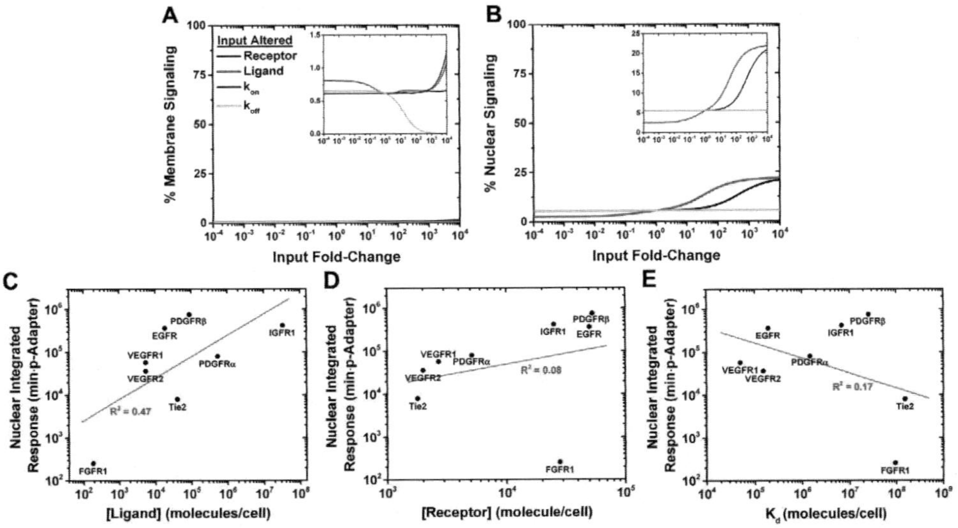

Figure 1. The effect of the individual RTK parameters on (A) membrane-based and (B) nuclear-based RTK signaling, using Ang2-Tie2 as the representative RTK. A correlation analysis was performed across the three individual RTK parameters: (C) the receptor level [R], (D) the ligand-receptor dissociation constant [Kd], and (E) the extracellular ligand concentration [L]. The correlation analysis used the physiological values for each RTK parameter given each RTK type (Table 1).

463206 Pharmacometabonomics Approach for Early Prediction of Neuropathy

Abstract:

Vincristine (VCR) is a core chemotherapeutic drug administered to pediatric Acute Lymphoblastic Leukemia (ALL) patients during induction phase. In a subgroup of population, it leads to Vincristine Induced Peripheral Neuropathy (VIPN), which is the dose-limiting toxicity. For few patients, VIPN is severe with long-term effects. Even though VCR has been used in chemotherapy for more than 50 years now, predictors and mechanism of VIPN induction are unclear. Historically, an empirical cap of 2 mg has been kept for dosage. Due to this, a population not susceptible to VIPN receives sub-therapeutic treatment, while another population may experience severe neuropathy. It is of interest to find predictors/biomarkers which can discriminate populations based on VIPN, before the treatment starts.

Few trends in VIPN incidence have been observed. Caucasians experience VIPN more than African-Americans. CYP3A5 high expressers experience lesser VIPN, though, no clear relation between VIPN and CYP3A4/5 expression has been established. Recently, VIPN was found to be associated with a SNP in the promoter region of CEP72 gene, which encodes a centrosomal protein, involved in microtubule formation. We use pharmacometabonomics approach to find metabolites as biomarkers which can predict VIPN. In a retrospective pilot study done on 12 ALL patients, metabolites were found to be differentially expressed in high and low VIPN patients. We plan to conduct a prospective study of 77 pediatric ALL patients administered with VCR as the primary chemotherapeutic drug for treatment. Fasting serum samples are being collected at 3 times points: before induction, one week after induction and at the end of a month. An attempt will be made to find metabolites differentially expressed in pre-dose samples of low and high VIPN patients. Presently, retrospective non-fasting samples of pediatric ALL patients are used for a preliminary study. These samples were collected at three time points: day 8, day 29, +6 months. Feature selection machine learning techniques have been used to identify biomarkers in a high-dimensional data set of metabolite profiles of patients. Samples at time point +6 months are used for identification of biomarkers, primarily.

We have future plans of exploring what relevant pathways the identified metabolites lie in. We further plan to bring mechanistic modeling into the picture to provide biological insight to the problem. For this purpose, we plan to study VCR metabolism, with an assumption that liver cells have a goal while metabolizing VCR.

463433 Conjugation in *Escherichia coli* biofilms on Poly(dimethylsiloxane) Surfaces with Microtopographic Patterns

Abstract:

Bacterial biofilms are less susceptible to antimicrobials and play an important role in the spread of antibiotic resistance due to close cell-to-cell contact. As an important surface character, topography has been shown to affect bacterial-surface interactions and biofilm formation. To understand if surface topography can also affect horizontal gene transfer in biofilms, we prepared poly(dimenthylsiloxane) (PDMS) surfaces with 20 μm by 20 μm, 50 μm by 50 μm, and 100 μm by 100 μm protruding square-shaped patterns (all features are 10 μm tall). Biofilm formation and associated conjugation on these surfaces were studied using fluorescently labeled donor and recipient strains of ***Escherichia coli***. The results demonstrated that surface topography with an inter-pattern distance larger than 10 μm can promote biofilm formation with the majority of cells attached on the side of protruding features. This location was also found to be a "hot spot" for biofilm-associated conjugation due to the increase in cell density. Furthermore, a motility (***motB***) mutant of ***E. coli*** exhibited defects in biofilm formation and conjugation, which was recovered by complementing the ***motB*** gene on a plasmid. These results suggest that bacterial motility is important for biofilm-associated conjugation, and optimizing surface topography is a potential strategy for the design of antifouling materials.

463578 Plasma Immunoglobulins Drive Corona-Induced Negative Adhesion of Drug Carriers in Human Blood Flow

Abstract:

Plasma immunoglobulins drive corona-induced negative adhesion of drug carriers in human blood flow

Daniel J. Sobczynski and Omolola-Eniola-Adefeso The plasma protein corona has been identified to have deleterious effects on nanoparticle targeting efficiency to reactive substrates as well as human blood flow *in vitro*. Although protein size has been implicated in corona-induced negative targeting effects, the role of specific proteins in this process remains largely unknown. This work explores the role of immunoglobulin (Ig) proteins on prescribing the impact of the protein corona on inducing negative adhesion of PLGA drug carriers in blood flow via a parallel plate flow chamber assay. It is observed that binding of PLGA particles is restored upon depletion of immunoglobulins, in particular, IgG and IgA, however this is shown to depend on the human donor. Interestingly, no dependence of corona-derived effects on plasma donor Ig concentration is observed. Overall, this study reveals that the impact of the protein corona on PLGA drug carrier targeting efficiency is dominated by high abundant immunoglobulin proteins in blood plasma.

464434 Cells Sense and Respond to the Surface Tension of Biomaterials

Abstract:

Cells physically interrogate their extracellular environment to inform decision making related to cell proliferation, differentiation, migration and other critical processes. In addition to its biochemical properties, the physical properties of the extracellular matrix, including its stiffness, are key inputs into signaling networks. Recent experiments have illustrated that solid surface tension can have a dominant role in the mechanical behavior of soft materials. Here we test whether cells are able to sense material surface tension and whether surface tension is a dominant physical cue that regulates cell response to soft biomaterials. We find that cells interacting with soft materials with appreciable surface tension primarily sense and respond to surface tension and not the bulk moduli of the materials. Our results are consistent with theory that predicts that solid surface tension can dominate over elasticity at cellular length scales. Even on very soft substrates (E=100 Pa), if the surface tension is appreciable, cells assemble robust adhesion structures, spread over large areas, and migrate efficiently. Remarkably, we find that cells can even spread and migrate on liquid biomaterials that have high surface tension. On these materials, cells migrate through a highly collective, swarming-type motility that is significantly faster than migration on solid biomaterials. Together, our results indicate that material surface tension is an important criteria for the design of biomaterials for cell growth and morphogenesis in tissue engineering.

464874 Fabrication of Biodegradable Spheroidal Drug Carriers with Modified Emulsion Solvent Evaporation Technique

Abstract:

Fabrication of biodegradable spheroidal drug carriers with modified emulsion solvent evaporation technique

Hanieh Safari and Omolola Eniola-Adefeso

Rod-shaped particles have recently gained interest for drug delivery applications evidenced by their reduced immune clearance and improved adhesion efficiency in blood flow relative to spheres. Oil-in-water emulsion solvent evaporation has recently been applied to fabricate poly(lactic-co-glycolic) acid (PLGA) micro-sized spheroids. This method benefits from its simple set up, scalability and large production rate and adaptability to different biodegradable polymers compared to other techniques. However, this method has been limited to rods with the major axis size of greater than 10 μm and minor axis size of greater than 2.0 μm that risks blood vessel occlusion. In this study, we developed a two-step fabrication system enabling stretching of smaller sized droplets by separating the droplet formation and stretching steps. In addition, careful tuning of the oil phase solvent and surface-active molecule is critical to spheroid yield. With our two-step method, we were able to scale down the size of the rods to major axis of 3.0 μm and minor axis of 500 nm, which is equivalent to the volume of 1.0 μm spheres. The rods fabricated with this method have the capability to be used both *in vitro* and *in vivo* to study the effect of shape on the efficiency of biodegradable drug carriers.

464978 Differentiation of Neural Stem Cells into Neurons By Visible Light Induced Electrical Stimulation

Abstract:

Neural tissue engineering aims to regenerate irreversibly damaged neural systems (either peripheral or central nervous systems) by differentiation of neural stem cells into neurons on three dimensional scaffolds by providing various chemical and physical cues. One physical cue that has received much attention is electrical stimulation by an external electric potential on conducting substrates such as graphene. One potentially useful method of achieving electrical stimulation of neural stem cells is by free electron generation on the surface of titanium oxide (TiO2) nanoparticles by light irradiation. However, the band gap of TiO_2 nanoparticles is in the UV range which can be harmful to cells and has many limitations in clinical applications. The goal of this research is to develop a novel method to achieve visible light-triggered differentiation of human neural stem cells (hNSCs) into neurons. We have discovered that TiO_2 grown on polydopamine (PDA) coated substrates exhibits an absorption peak in the visible light range. The surface chemistry was confirmed by various surface characterization methods including XPS, SEM and FT-IR. Using this substrate, we investigated the effect of YIGSR peptide on hNSC adhesion, surface free electron generation on TiO2 by visible light irradiation, and preferential differentiation of hNSCs into neurons. The results from this research open up a new possibility for implantable neural tissue engineering scaffolds that can be activated by light irradiation.

465752 Cytometric Measurement of Caspase-3 Activation Using Forster Resonance Energy Transfer (FRET) Between Peptide-Linked Bioprobes

Abstract:

Background

Fluorescence lifetime measurements with flow cytometry provide high throughput and quantitative information about molecules within cells based on fluorescence intensity decay times. This time-resolved cytometry method is not commercially available but can be introduced onto any cytometry system, as our laboratory has done. Fluorescence lifetime measurements are helpful for a range of applications including the measurement of Forster Resonance Energy Transfer (FRET). In this contribution we screen large cell populations with our cytometry instrument in order to identify the induction of apoptosis (programmed cell death) by quantifying the activity of caspase-3 with a FRET assay.

Methods

A FRET based bioprobe is engineered with green fluorescent protein as well as a caspase-3 recognition sequence. Our modified GFP, which is the FRET donor is connected to an exogenous fluorophore (based on spectral properties: AlexaFluor® Dye, Thermo Fisher Scientific Inc.), which acts as the acceptor in the FRET pair. A Caspase-3 recognition peptide linker site connects the FRET pair such that during the induction of apoptosis, caspase-3 is present and available for cleavage of the peptide. Our measurements include excitation and emission spectra before and after caspase-3 activation with a fluorimeter. Additionally we detect fluorescence lifetime changes using our flow cytometer before and after caspase-3 activation.

Results

Our analysis of the emission spectra confirm the presence and loss of FRET prior and post caspase-3 treatment respectively. We calculated the emission ratios representing maximum of dye emission to maximum of modified GFP emission. Furthermore the observed change in fluorescence lifetimes measured before and after caspase-3 activation help us discriminate the loss of FRET accurately.

Conclusions

Fluorescence lifetime measurements on a flow cytometer is a powerful method to provide a quantifiable way to study apoptosis in mammalian cells grown in a monolayer culture. This assay permits us to determine quantities of caspase-3 through the measurement of FRET as well as loss of FRET.

466901 E-Selectin-Mediated Rolling and Firm Adhesion of Pancreatic Cancer Cells in Shear Flow

Abstract:

E-Selectin-mediated Rolling and Firm Adhesion of Pancreatic Cancer Cells in Shear Flow

Daniel J Shea[1], Yi Wai Li[1] and Konstantinos Konstantopoulos[1]

[1]Department of Chemical and Biomolecular Engineering, Johns Hopkins University

Introduction: Metastasis is a highly-regulated, multistep process, in which tumor cells migrate from the primary tumor and enter the circulatory system where they interact extensively with host cells before extravasating and colonizing a distal organ. Extravasation of a tumor cell from a blood vessel first requires the rolling and arrest of the cell to the vessel wall through the formation of specific bonds [1,2]. Our lab recently identified the sialofucosylated surface protein Podocalyxin (PODXL) to be overexpressed in metastatic pancreatic cancer cells and to bind to E/L-selectin [3]. PODXL-E-selectin mediated rolling velocities were found to depend on single bond kinetics and could be predicted by the single receptor-ligand *off*-rate and the number of bonds [4]. We looked to build on this and develop a quantitative understanding of how selectin-ligand tethering facilitates firm adhesion at elevated levels of shear stress. We also determined the rate-limiting parameters in this cascade of events, such as the lengths of coated regions with selectins and hyaluronic acid, necessary to support downstream firm adhesion. The quantitative analysis of the selectin-ligand binding kinetics of known pancreatic cancer surface proteins in the presence of shear flow will enable us to further our understanding of the metastatic process.

Materials and Methods: Using standard microfabrication principles we patterned a glass slide with E-selectin and hyaluronic acid (HA) patches with lengths varying from 10-160μm in geometrically defined regions with a gap distance between the two regions of between 30-120μm (**Fig. 1A**). Using a microfluidic chamber, PODXL expressing metastatic pancreatic cancer cells (Pa03c) were then perfused over the patches and all cells were tracked using a modified code package produced by Eric M. Furst (University of Delaware). Using a custom Matlab program, we identified both the rolling and firm adhesion regions and separate these regions into ten binned sections of increasing patch length. The program then analyzes each cell track with respect to these regions and indicates the position and velocity of the cell, while also determining if the cell binds in either the rolling or firm adhesion regions.

Results and Discussion: Experiments were performed with Pa03c pancreatic cancer cells using devices coated with E-selectin (rolling) and HA (firm adhesion). Cells were tracked as they pass over the rolling, gap and firm adhesion regions, respectively (**Fig. 1A**). Cells interacting with E-selectin showed a clear rolling behavior. The extent of interacting cells decreased with shear stress. However, it increased with longer E-selectin patch lengths due to increased ligand-selectin contact times (**Fig. 1B**). We determined that cells bind infrequently to hylaronic acid-coated regions, however the binding frequency increases when a cell first rolls on E-selectin (**Fig. 1C**). Interestingly, for rolling cells, firm adhesion is most frequent when cells roll on shorter E-selectin patch lengths (<40μm), likely because these cells are bound immediately before entering the gap thus having a slower velocity. Together these data indicate that rolling on E-selectin can facilitate firm adhesion with HA and can do so even at small E-selectin patch lengths.

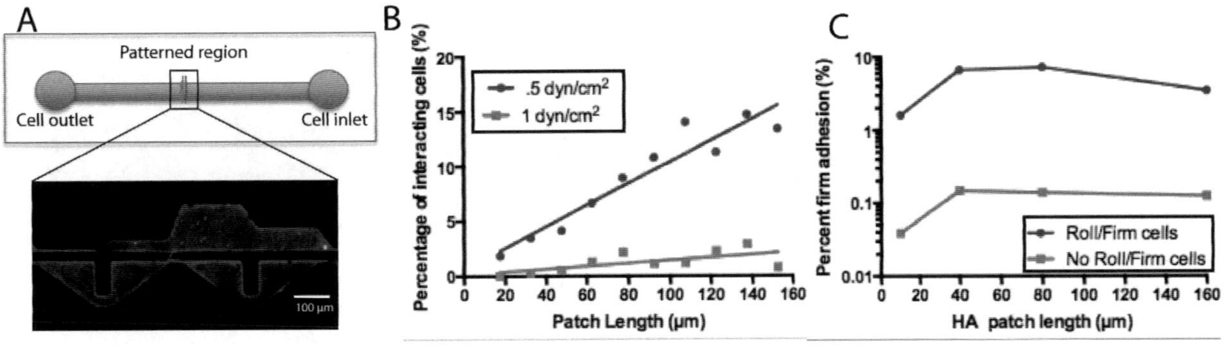

Figure 1. (**A**) Device schematic with rolling region (red) and firm adhesion region (green). (**B**) Frequency of rolling events at increasing patch lengths. (**C**) Frequency of firm adhesion events for cells that roll (blue) or that do not roll (red) before firmly adhering. 30μm gap.

Conclusions: This work impacts a broad range of areas including cell-biophysics, engineering and cancer research and is critically important as it advances the knowledge of receptor-ligand processes as they pertain to cancer metastasis. Knowledge of the properties of receptor-ligand-mediated tumor cell adhesion, in physiologically relevant environments is critical to the design parameters necessary to engineer sensors, to target biological entities based on recognition, and to design molecules to interrupt adverse or pathological adhesion events.

References: (1) Cheung, L. S. L. *et al., Phys Biol*, 2011, *8*(1), 015013. (2) Stroka, K. M., and Konstantopoulos, K. *Am J Physiol Cell Physiol*, 2014, *306*(2), 98-109. (3) Dallas, M. R. *et al., Am J Physiol Cell Physiol*, 2012, 303, C616–624. (4) Shea, D. J. *et al., Oncotarget*, 2015, *6*(28), 24842.

468076 NANOG Rejuvenates the Impaired Extracellular Matrix Expression in Senescent Cells and Thus Restores the Decreased Mechanical Properties of Engineered Tissues

Abstract:

Extracellular matrix molecules (ECM) such as elastin and collagen are critical in regulating the arterial mechanical properties. Elastin, imparts the elastic properties to the vascular wall whereas collagen regulates the strength of tissues in general and the arterial wall in particular. Notably, aging is accompanied by significant reduction in the elastic properties of the vascular wall mostly due to loss of elastin fibers. Similarly, although crosslinking of old collagen increases, new collagen production decreases with aging. As a result, aging is characterized by stiffer vessels, which are prone to aneurysms. Hence, there is a need to develop strategies to restore the impaired deposition of ECM. Here we report that ectopic expression of the pluripotent factor, NANOG, in aged myogenic progenitors could restore the impaired ECM production and increase the mechanical properties of aged engineered tissues.

To study the effect of NANOG on aged cells, we applied two widely established models of aging i.e. replicative senescence of human Mesenchymal Stem Cells (MSC), and myofibroblasts derived from Hutchinsonâ€™s Guilford Progeria Syndrome (HGPS) patients. Human Mesenchymal Stem Cells (MSC) were induced to senescence by serially passaging. To express NANOG, cells were transduced with a tetracycline-inducible virus, which allowed expression of NANOG only in the presence of Doxycycline (Dox). Upon senescence, cells were treated with DOX (LPN) and the production of ECM was evaluated and compared to Early passage (EP) and senescent cells (Late Passage or LP). Similar experiments were also performed with HGPS derived cells.

Our results show that senescence reduced the production of both Collagen 3 and Elastin. In agreement, tissues engineered using senescent cells showed inferior mechanical properties. Interestingly, NANOG restored the capacity of senescent cells to express ECM molecules and increased the mechanical properties of engineered tissues. Next we performed RNA-seq to identify the mechanism mediating the effects of NANOG on ECM production. We found that NANOG altered the expression of genes that are part of the TGFbeta pathway, which is known to be necessary in ECM production. In agreement, we discovered that NANOG restored the activity of the TGFbeta pathway that was impaired in senescent cells, whereas inhibition of the TGFbeta pathway with either shRNA (SMAD2, SMAD3, SMAD4) or chemical inhibitor (SB431542) abolished the effects of NANOG. Conversely, overexpression of SMAD2 and SMAD3 increased ECM production.

In conclusion, our study provides a novel strategy to restore the impaired production of ECM, which may have broad applications in tissue regeneration and anti-aging treatments.

468059 Extending the Dynamic Range of Nanosensors

Abstract:

Recent years have shown the incredible versatility and utility of polymeric optode-based nanosensors. Sensors have been developed for a wide variety of ionic analytes (e.g. sodium, potassium, pH) as well as nonionic analytics (e.g. glucose) and have been applied for challenging research fields such as detection of sodium sparks in live cardiomyocytes. This field of research has yielded new knowledge on cellular metabolism and function. Additionally, these sensors can be applied in vivo, through implantation in or under the skin in animals; showing potential as a research tool for continuous monitoring of analyte concentrations without the need for direct samples to be taken from the animal.

However, one of the key limitations of these sensors is the dynamic range is often limited relative to the desired range for physiological application. There are two key causes for this limitation: the optode mechanism and the available selection of chromoionophores. Relative to a similar technology, ion-selective electrodes, the pH based mechanism inherently limits dynamic range based on the pKa of the chromoionophore. In this work we control the dynamic range of these nanosensors through altering the formulation of the optode components. This results in nanosensors with tunable and extended dynamic ranges â€“ a necessary step in measurement cases where concentrations change by more than one order of magnitude.

468389 Targeted Alpha-Particle Therapy of Triple Negative Breast Cancer Using 'sticky' Liposomes

Abstract:

Statement of Purpose: The goal of this project is to design a new class of intravenously-administered lipid-nanoparticles (sticky liposomes) loaded with DNA damaging agents (the alpha-particle generator Actinium-225, Ac-225) to enable selective targeting of Triple Negative Breast Cancer (TNBC) cells expressing low or too low antigen densities. Alpha-particle emitters typically cause DNA double-strand breaks and, in combination with receptor-mediated targeting, could result in significant radiobiological effects. The key to this unique treatment is to use drug-carrying liposomes whose surface phase-separates to form 'sticky patches' when in proximity to tumor cells.

To better understand the material properties and mechanism of action of these nanoparticles, the kinetics of receptor binding, controlled drug release as well as transport properties of our â€˜stickyâ€™ liposomes will be assessed using monolayer cell culture as well as 3-D multicellular tumor spheroid models. By optimizing our liposome properties in these 3-D spheroid systems, which mimic small avascular tumors and micrometastases, we increase the likeliness of successful in vivo efficacy.

Methods: HER2-targeting sticky liposomes were loaded with Ac-225. The sensitivity to alpha particle emitters of various TNBC cell lines (some bearing the BRCA-1 mutation) was evaluated in vitro. Dose response studies to sticky liposomes loaded with Ac-225 were conducted on TNBC cells expressing variably low levels of HER2 receptor copies and were compared to HER2-overexpressing breast cancer cells.

Passive diffusion of fluorescently labeled phase-separating lipid nanoparticles into 3-D tumor spheroids was analyzed over time. Content release of these drug carriers into tumor spheroids was also monitored via fluorescence, using calcein as a therapeutic analogue.

Results: Liposome loading efficiency with Ac-225 was greater than 80% with stable retention in the presence of cells for over 24 hours. BRCA-1 mutated cell lines showed greater sensitivity to alpha particle irradiation. Initial experiments indicate the promising killing efficacy of HER2-targeting sticky liposomes loaded with Ac-225 towards TNBC cells when compared to radiolabeled antibodies or uniformly functionalized liposomes.

Conclusions: Initial results show promise in the capability of these Ac-225 loaded sticky liposomes to selectively target and kill cancer cells that are currently not targetable by reported functionalized nanoparticles.

468084 *motB* Is Involved in the Response of *Escherichia coli* to Material Stiffness during Biofilm Formation

Abstract:

Bacterial biofilm formation is a leading cause of chronic infections, which lead to nearly 100,000 deaths and 28-45 billion dollars of cost each year in the U.S. alone. The chemical and physical properties of substratum surfaces play important roles in biofilm formation. We recently reported that the stiffness of poly(dimethylsiloxane) (PDMS) affects bacterial adhesion and the physiology of attached cells. In this study, the automated contour-based tracking package for *in vitro* environment (ACT*IV*E) developed in the Henderson Lab was validated for tracking bacterial cells. By comparing the motility of attached *Escherichia coli* cells on PDMS surfaces with different Youngâ€™s moduli (0.1 MPa for soft and 2.6 MPa for stiff) using ACT*IV*E, the attached cells were found to be more mobile on stiff surfaces during the early stage of biofilm formation than those on soft surfaces. To further investigate if motility is important to bacterial response to material stiffness, *Escherichia coli* RP437 and its isogenic mutants of motility (*motB*), flagella (*fliC*) and type I fimbriae (*fimA*) were compared for adhesion on soft and stiff PDMS. The *motB* mutant exhibited defects in response to PDMS stiffness (based on cell counting and tracking with ACT*IV*E), which was recovered by complementing the *motB* gene on a plasmid. In comparison, mutation of *fliC* and *fimA* did not show significant effect. Collectively, the findings from this study suggest that *Escherichia coli* cells can actively respond to material stiffness during biofilm formaiton, and *motB* is involved in such response.

468746 Computational Modeling of Tuberculosis Granuloma Activation

Abstract:

Tuberculosis (TB) is one of the most common infectious diseases and deadliest diseases worldwide. It is estimated that one-third of the world's population is infected with TB, and 1.5 million TB-related deaths were reported in 2014. TB is spread by aerosol droplets containing Mycobacterium tuberculosis (Mtb). The Mtb bacteria enter through the respiratory system and are attacked by the immune system in the lungs, primarily by aveolar macrophages. The bacteria are clustered and contained by the macrophages into cellular aggregates called granulomas. These granulomas can hold the bacteria dormant for long periods of time, even decades, in a condition called latent TB. However, the bacteria persist and can be activated when the granulomas are compromised by other immune response events in a host, such as cancer, HIV, or aging. The activation and subsequent spread of bacteria leads to active TB disease. It is difficult to study the activation process in humans because those with latent TB are asymptomatic and are often undiagnosed. Current animal models all have limitations. Computational and mathematical models can be useful tools for inexpensively conducting short- and long-term in silico experiments with multiple, interacting factors and can aid in generating and testing hypotheses. Several previous computational and mathematical models have been developed to describe the infection or granuloma formation stages of TB. No computational approach has been proposed considering the dynamics of matrix metalloproteinase 1 (MMP-1) regulation and its impact on TB activation. MMP-1 dysregulation has been recently implicated in TB activation through experimental studies, but the mechanism is not well understood. Animal and human studies currently cannot probe the dynamics of activation, so a computational approach is proposed to fill this gap. The overall objective of the study is to predict TB cavity formation (a hallmark of activation) in response to the dynamics of MMP-1 dysregulation. We will report on the status of development of mathematical and computational tools to test the hypothesis that the dynamics of MMP-1 regulation play a key role in the transition from latent TB to active TB.

468678 A Study of the Effects of Heat Therapy through Thermomechanical Coupling on Action Potential Properties and Spiral Wave Dynamics in a Cardiac Electromechanical Model

Abstract:

The heartbeat is triggered by electrical impulses that travel along a special pathway through the cardiac muscle and initiate its contraction. Abnormal electrical activity, such as spiral waves, in cardiac tissue have been largely linked to the onset of cardiac arrhythmias, such as Ventricular Fibrillation (VF) [1], which lead to cardiac arrest and results in over 300 000 deaths annually in the United States. In its turn, contraction of the heart causes cardiac tissue deformation which feeds back on the wave propagation and affects electrophysiological properties via mechanoelectrical feedback (MEF) [2]. Electrical-excitation and contraction of the heart can be linked by an electromechanical model of cardiac tissue. The heat therapy has been used in medicine. When a hyperelastic material is subjected to heat, it deforms. The hyperelastic material [3] model has been used to describe the mechanical behavior of passive myocardium. The therapeutic effects of heat include decreasing joint stiffness; increasing the extensibility of collagen tissues, etc.

In this work, we introduce a thermo-electromechanical model for cardiac tissue, which is provided by thermomechanical coupling. Mathematically, this mechanism of thermomechanical coupling is based on the theory of thermoelasticity [4]. The electrical activity is represented by the Luo-Rudy [5] model, and the mechanical properties are described using the Mooney-Rivlin material response [3]. The active tension that couples the electrophysiological model with the cardiac mechanics model is generated using the Niederer-Hunter-Smith [6] model which is the most advanced model, including all features contained in other models. We apply the model to study how thermomechanical coupling affect the action potential shape, restitution properties, and dynamics of spiral waves. We show that deformation caused by heat transfer through the thermomechanical coupling affect these properties via MEF.

To the authors' best knowledge, this is the first work that studies the effects of coupling the heat with the soft tissue mechanics of the heart on the properties and dynamics of action potential. This will serve to demonstrate that heat therapy can be pursued in the design of therapeutic strategies and algorithms for implantable devices capable of preventing or treating life-threatening cardiac arrhythmias, VF for example, where the electrical therapy is currently the most widely used treatment.

[1] M. E. Zevitz, Ventricular fibrillation [online]. eMedicineÂ http://www.emedicine.com/med/topic2363.htm (2004).

[2] M. J. Lab, Mechanoelectric feedback (transduction) in heart: concepts and implications, Cardiovasc.Res, vol.32, p. 3-14, 1996.

[3] A. Hazim, Y. Belhamadia, and S. Dubljevic, Control of cardiac alternans in an electromechanical model of cardiac tissue, Comput in Biol and Medicine, vol. 63, p. 108-117, 2015.

[4] R. B. Hetnarski, â€Žand M. R. Eslami, Thermal Stresses: Advanced Theory and Applications. Springer 2009.

[5] C. Luo, and Y. Rudy, A model of the ventricular cardiac action potential. Depolarization, repolarization, and their interaction, Circ Res, vol. 68, p. 1501-1526, 1991.

[6] S. A. Niederer, P. J. Hunter, and N. P. Smith, A quantitative analysis of cardiac myocyte relaxation: a simulation study, J Biophysics, vol. 90, p. 1697-722, 2006.

468816 Nanoparticle-Loaded Hydrogel Patches on Cells for Drug Delivery

Abstract:

Cell surface patches provides a potential approach to deliver therapeutics to targeted tissues. The goal of this work is to form a drug-loaded patch on the surface of cells without altering the capacity for cell-substrate interactions on non-coated regions. Our approach is to synthesize polymer patches on cell surface via PEG-diacrylate (PEGDA) photopolymerization and photomask-mediated photolithography. These PEGDA patches are formed using an eosin/tertiary amine initiation system with 530 nm light. The fundamental limits of feature size is probed through experimental analysis of pattern geometry, system geometry, prepolymer composition, and prepolymer viscosity. The patterning was initially studied on a glass microarray where the reaction environment is carefully controlled, and <10 micrometer features are easily attained. PEGDA polymer formation showed that high intensity, short time illumination had the best chance to accurately generate the desired PEGDA pattern. Additionally, this reaction/diffusion system was modeled using coupled differential equations in Matlab. Trends in the experimental system are directly mapped to the experimental system, and the model is used to guide the experimental approach for patterning on cells, where high intensity light is not feasible for cell function.

To demonstrate cell surface labeling, A549 cell surfaces were loaded with eosin (initiator) through sulfo-NHS coupling or through antibody interactions. Sub-10 micrometer nanoparticle-loaded patches are successfully deposited on cells, and factors promoting cell viability and function are discussed. The release of doxorubicin from nanoparticles incorporated into these coatings is quantified through doxorubicin fluorescence. The release profile is not influenced by the coating thickness, and the kinetics of release in PBS are consistent with that of the free nanoparticles in solution.

468984 Engineered Photocrosslinkable Protein Coatings Improve Osseointegration of Orthopaedic Implants

Abstract:

As the U.S. population continues to age, the use of orthopedic implants is projected to grow rapidly in the coming decades. With failure rates of orthopedic implants around 10%, combined with the millions of new implants a year, implant failures will lead to an ever-increasing burden on patients and the healthcare system. Developing new technologies to improve the long-term osseointegration of orthopaedic implants to prevent premature failure is paramount in addressing the growing cost of implant failures. We present an elastin-like engineered protein (ELP) material that incorporates a bioactive RGD sequence to promote osteoblast and mesenchymal stem cell (MSC) adhesion to facilitate rapid bone growth and implant integration. The recombinant ELP polymer is covalently modified with a photocosslinkable moiety to enable facile spin-coating or dip-coating of any implant geometry prior to covalent crosslinking to form a stable coating on the implant surface. ELP coatings are shown to remain intact after implants have undergone simulated dental and orthopaedic insertion and removal. Using the MG63 osteoblast-like cell line, ELP coated implants showed 80% adhesion of cells after 24 hours compared to 38% for an uncoated Ti6Al4V sample. Additionally, MG63s showed a significant increase in bone mineral deposition at all time points up to two weeks, demonstrating the osseointegrative capabilities of the coating. Human mesenchymal stem cells (hMSCs) similarly showed an increase in bone mineral deposition on ELP implant surfaces compared to uncoated controls. The ELP coating also accelerated hMSC differentiation into osteoblast-like lineages, as evident by a spike in alkaline phosphotase activity at 7 days, compared to 14 days for the control. Finally, our coatings improve functional recovery of dental implants, as demonstrated by a significant increase in removal torque of ELP coated implants after one week. We have shown an increase in key *in vitro* markers that point to early osseointegration of ELP coated implants, and verified that through functional recovery tests. By engineering a bioactive peptide sequence into a durable protein coating for orthopaedic implants, we have been able to create a more favorable microenvironment for tissue regeneration and implant recovery.

469256 Quantification of Alkaline Phosphatase Activity in Single Algal Cells Using a Microfluidic Device

Abstract:

Harmful algal blooms (HABs) are the result of the overabundance of naturally occurring algae due, in part, to an excess of nutrients such as nitrogen and phosphorus. While measuring nitrogen and phosphorus content in the water column can provide information about their respective availability, this analytical metric does not account for how much of each nutrient is fixed by the algae. Conversely, measuring alkaline phosphatase (AP) activity provides a more direct metric for how much phosphorus is being fixed by algae in the bloom. AP is expressed on the surface of algae and is responsible for converting the dissolved organic phosphorus to inorganic phosphorus required by algae. One limitation to using AP as a metric for dissolved phosphorus in a bloom is that it requires several washing and centrifugation steps confining the assay to a lab and resulting in a 1-2 week lag time between sample acquisition and analysis. Microfluidics devices however can be used to drastically decrease the time required to run such biochemical assays while using a fraction of the reagents coupled with their ability to serve a portable readouts of cellular behavior. The goal of this study was to develop a microfluidic device capable of single cell capture and isolation followed by directly measuring AP activity in the trapped algal cells. The device consisted of an array of 15 Î¼m semi-circular cups capable of trapping individual *Chlamydomonas reinhardtii* cells followed by detection of AP activity using a commercially available fluorescent biomolecule with fluorescent microscopy. The device eliminated the need for the washing and centrifugation steps as the introduction and removal of all reagents was achieved by changing the inlets to the device. To induce AP activity, *C. reinhardtii* cells were starved in phosphorus free 1X TA buffer followed by resuspension in phosphorus-rich 1X TAP prior to initiation of the AP assay. A range of increasing phosphorus concentrations were tested which correlated to increasing levels of AP activity, all performed in a fraction of the time in the device. In order to demonstrate the utility of this device, similar experiments were performed examining AP activity in the small centric diatom *Thalassiosira*, with similar results of increasing AP activity correlating to increasing dissolved organic phosphorus levels. Thus, the microfluidic device allows for facile single cell analysis and

the ability to incorporate all the steps in the biochemical assay in a single device. This technology exhibits the potential for portability needed to measure AP activity in HAB samples collected during cruises without the need for long-term storage and analysis in a lab.

469568 Substrate Physical Cues Regulated Fibroblasts Sensing Carbon Nanotubes

Abstract:

While the rapidly evolving nanotechnology has shown promise in electronics, energy, healthcare and many other fields, there is an increasing concern about the adverse health consequences of engineered nanomaterials. It has been shown that inhaled carbon nanotubes (CNTs) rapidly enter the lung interstitium to stimulate collagen production and induce progressive interstitial lung fibrosis in weeks in mice. The lung interstitium displays interrelated framework of nanoscale fibrous proteins. The lung has stiffness ranging from £ 5 kPa for native lung tissue to 25 – 100 kPa for fibrotic tissues. The majority of current *in vitro* models lack characteristics of *in vivo* microenvironment, leading to the cell behavior deviated from *in vivo* phenotypes and responsiveness. We hypothesize that substrate nanotopography and stiffness cues can affect toxic response of cells to nanomaterials. We thus engineered polydimethylsiloxane (PDMS) nanotopographies of different shapes and dimensions and polyacrylamide (PAAm) gels of various stiffnesses, and investigated how the physical cues influenced fibrogenic response, including cell spreading, proliferation, collagen I production and reactive oxygen species generation of normal human lung fibroblasts to multi-walled carbon nanotubes (MWCNTs). The results show that the substrate physical cues can regulate cell spreading and intracellular tension, and consequently affect toxic response of the fibroblasts to the CNTs. When the cells are less spread, the cellular sensitivity to the CNTs is enhanced. This study highlights the important role of cell-substrate interactions in fibrogenic response of the fibroblasts to nanomaterials, and contributes to development of physiologically relevant *in vitro* models for nanotoxicology study.

469590 Using Green ’ s Functions to Model Electrokinetic Flow in Tumor Cells for Tumor Treating Fields (TTF) Therapy

Abstract:

Tumor treating fields (TTF) therapy has recently emerged as a viable treatment for Glioblastoma Multiforme (GBM). This therapy uses frequency specific alternating current (AC) applied electrical fields to interfere with cell division by disrupting mitosis through prevention of tubulin polymerization. This treatment, however, remains unpredictable and a comprehensive mechanistic understanding of its effects on cells has yet to be achieved. The mathematical model presented here studies electroosmotic flows inside a single tumor cell during different cell cycle phases. Idealized descriptions of the cell geometries are employed with the electrokinetic hydrodynamics (EKHD) approach to obtain electroosmotic velocity profiles within the tumor cell under the influence of direct current (DC, limiting case) and AC electrical fields. The governing equations were solved both analytically, using Greenâ€™s Function approaches, as well as numerically. Various physiochemical and geometrical parameters were identified and varied to study their effects on the flow profiles, and it was found that the ratio between the electrical and viscous contributions to the flow had the largest impact on the profiles.

469675 Adhesion-Based Tumor Cells Capture Using Nanotopography

Abstract:

Circulating tumor cells (CTCs) shed from primary tumors, transport through the blood stream to distant sites, and cause 90% of cancer deaths. Although different techniques have been developed to isolate CTCs for cancer detection, diagnosis and treatment, the heterogeneity of expression of the target antigen and the significant size variance in CTCs limit clinical applications of antibody- and size-based isolation techniques. Cell adhesion using nanotopography has been suggested as a promising approach to isolate CTCs independent of surface marker expression or size of CTCs. However, the influence of nanotopography configuration (geometry and dimensions) on CTC capture efficiency has not been fully investigated. This study examined capture performance of several cancer cell lines of different types, surface marker expression and metastatic status on nanotopographies of various geometries and dimensions. The cancer cells exhibited differential capture performance with an efficiency up to 52% on the nanotopographies without antibody conjugation. Compared with flat surfaces and isotropic, discrete nanopillars, anisotropic nanogratings favored cancer cell adhesion, thus improving the capture efficiency. The cancer cells were less sensitive to the height range studied. By optimizing the nanotopography configuration and performing appropriate surface modification, the CTC capture efficiency was further improved. This study is expected to shed light on optimization of nanotopography to improve adhesion-based CTC capture technique.

469962 Polyacetal As a Novel Drug Carrier for Targeted Cancer Therapies

Abstract:

There has been an increasing focus in modern cancer research on the development of materials that can efficiently deliver therapeutics to a pre-identified solid tumor target. Due to their unique properties, stimuli responsive polymers (SRPs) have been of particular interest. One such SRP is a polyacetal-based copolymer (PAC). PAC shows both remarkable temperature response that is linearly dependent on composition and pH dependent degradation properties at physiological pH ranges; making it one of the most desirable materials for targeted cancer therapy. Here, we discuss the fundamental origins of the lower critical solution temperature (LCST) behavior, exhibited by this polymer. Our results indicate that the observed LCST scales linearly with the number of carbon and oxygen atoms in the polymer repeat units, allowing for precise control over the LCST. Furthermore, due to its unique pH dependent breakdown mechanism, the molecular purity of the encapsulated drug is retained post degradation. Therefore, we design PAC to include cancer therapeutics in its polymer-backbone, utilizing strategies to modify step-growth polymerization to obtain, for the first time, temperature-responsive main-chain drug conjugates. The temperature response in these main-chain drug conjugates allow for effective delivery of therapeutics to the tumor site, followed by acid-hydrolysis in acidic local tumor environments, to release pristine drug-molecules directly at the tumor site. These results strongly suggest that PAC possess both active targeting and degradation mechanisms that will greatly enhance the design of drug delivery vehicles.

470066 Photoreversible Gel Patterning with Site-Specifically-Modified Proteins to Direct Cell Fate in 4D

Abstract:

As microenvironmental reconfigurations govern many important biological processes, tunable *in vitro* culture platforms that recapitulate such dynamic phenomena would be invaluable for fundamental studies in stem cell biology, as well as in the eventual engineering of functional human tissue[1]. Preliminary efforts have exploited photochemical reactions to tether peptides spatially within hydrogels[2–4]. While such approaches have proven successful in directing 3D cell physiology, the realized biological control has been confined to relatively simple cellular functions (*e.g.*, adhesion, proliferation). To govern more complex decisions of fate, a system that enables dynamic presentation of *full-length proteins* would be of great interest[5]. However, proteins are commonly recognized to be as delicate as they are powerful; careful consideration must be given to the immobilization chemistry and precise site of protein tethering to ensure sustained stability and activity. Here we present a robust synthetic strategy enabling the *user-defined immobilization and subsequent release of proteins in a site-specific manner within a 3D cell culture platform*, thereby preserving protein function to modulate intricate cellular behavior including stem cell differentiation, protein secretion, and cell-cell interactions.

1. DeForest, C. A. & Anseth, K. S. Advances in Bioactive Hydrogels to Probe and Direct Cell Fate. *Annu. Rev. Chem. Biomol. Eng.* **3,** 421–444 (2012).

2. DeForest, C. A., Polizzotti, B. D. & Anseth, K. S. Sequential click reactions for synthesizing and patterning three-dimensional cell microenvironments. *Nat. Mater.* **8,** 659–664 (2009).

3. DeForest, C. A. & Anseth, K. S. Cytocompatible click-based hydrogels with dynamically tunable properties through orthogonal photoconjugation and photocleavage reactions. *Nat. Chem.* **3,** 925–931 (2011).

4. DeForest, C. A. & Anseth, K. S. Photoreversible Patterning of Biomolecules within Click-Based Hydrogels. *Angew. Chemie Int. Ed.* **51,** 1816–1819 (2011).

5. DeForest, C. A. & Tirrell, D. A. A photoreversible protein-patterning approach for guiding stem cell fate in three-dimensional gels. *Nat. Mater.* **14,** 523–531 (2015).

470453 Regulation of Insulin Secretion in Pancreatic β-Cells with an Optogenetic Approach

Abstract:

Regulation of blood glucose is dependent on proper secretion of insulin by pancreatic β-cells. Cyclic AMP (cAMP) is a key activator of the release of hormone through the increase in Ca^{2+} influx via voltage-dependent channels. Optogenetics has been widely implemented to control biological processes with light. Here, we hypothesized that a photoactivatable adenylyl cyclase (PAC) can be employed to modulate intracellular cAMP in β-cells. To test this hypothesis, the PAC gene was delivered to cultured mouse insulinoma β-cells, which were subsequently illuminated with light and the levels of cAMP and secreted insulin were assessed under various conditions. The PAC gene was delivered with an adenoviral vector and its efficient expression were confirmed by western blot analysis and fluorescence microscopy. Various multiplicities of infection were screened to achieve high transduction efficiency without compromising cell viability. The level of cAMP peaked within 10 minutes of light stimulation and returned to baseline in less than 20 minutes after terminating illumination. A concomitant increase was observed in secreted insulin. The results demonstrate the potential of the optogenetic system described here for the precise and efficient control of insulin secretion by pancreatic β-cells and may facilitate the development of technologies for the treatment of metabolic diseases such as diabetes.

470740 Cytoplasmic Stiffness in Migrating Cells at the Interface of a Chemical/Mechanical Gradient

Abstract:

In order to migrate, cells must exhibit a certain degree of anisotropy resulting in a well-defined leading and trailing edge. Cell polarization and the redistribution of cytoskeletal components are key steps in cell migration. Therefore, we seek to investigate how substrate rigidity and chemical composition could affect cellular cytoplasmic stiffness. Our goal is to correlate anisotropy in cytoplasmic stiffness to migratory speed and directions. Such studies will provide fundamental insights into wound healing, cancer cell metastasis and in developmental biology. We have investigated the changes in cytoplasmic stiffness of fibroblasts on chemical-only, mechanical-only and dual chemical-mechanical gradient substrates developed previously in our group[1]. These studies allow us to make correlations between substrate rigidity, protein concentrations, cytoplasmic elasticity, and migration.

Polyacrylamide hydrogels exhibiting chemical, mechanical and opposing dual mechanical/chemical gradients were designed according to previously established methods[1]. Cytoplasmic stiffness measurements were conducted on BALB/c 3T3 fibroblasts using a Veeco Bioscope II AFM equipped with a heated stage. The AFM was configured with a Nikon TE-2000U optical microscope allowing for precise positioning of the AFM cantilever over the cellular region of interest, as well as conducting fluorescence microscopy of cytoskeletal components. Force distance curves were collected from different regions such as nuclear, peri-nuclear and lamellipodia, leading and trailing edges of the cell. Correlations to cell cytoplasmic stiffness and the organization of the actin cytoskeleton will be made by imaging fibroblasts labeled with CellLight® Actin-GFP (Life Technologies) through confocal microscopy immediately preceding AFM measurements.

For our preliminary studies, AFM measurements have been taken on fibroblasts seeded on collagen-coated glass coverslips or polyacrylamide (PAAM) hydrogels. The Young's modulus (YM) of the PAAM gels was either 45 or 120kPa. The elastic modulus was obtained from fitting force- distance curves with the

Sneddon's modified Hertz model. Cells were indented up to 1µm and the resulting force distance curves were fit for indentations up to 10% of the cell's thickness to avoid substrate effects. Force distance curves obtained from single cells were collected within a minute's timespan to minimize interactions between the cell and AFM cantilever. We have observed YM ranging from 1-20kPa with as much as a 12-fold increase in cytoplasmic stiffness between the perinuclear and peripheral regions of fibroblasts seeded on rigid glass coverslips. When fibroblasts were cultured either on the 45 or 120kPa PAAM gels, similar trends were observed. On the 45kPa gels the decrease in YM between the stiff periphery and soft perinuclear regions was approximately 2-fold. However, on the 120kPa gels, the corresponding decrease was approximately 6-fold. These trends demonstrate that the substrate rigidity modulates the cytoplasmic stiffness of cells. More importantly, substrate rigidity appears to play an important role in the anisotropy of cellular cytoplasmic stiffness.

We will correlate anisotropies in cytoplasmic stiffness to cytoskeletal and focal adhesion protein expression. The use of controlled gradient substrates enables us to separate multiple influences to cellular locomotion. These studies will allow us to gain a greater understanding of the complex interplay between the cytoskeleton, substrate properties and migratory processes.

References: 1.) Jain G, ACS Biomater. Sci. Eng., 2016; 8:621-631. 2.) Solon J, Levental I, Biophys. J., 2007;93:4453-61.

471390 Free Surface Electrospinning of Microemulsions Containing Vitamin E

Abstract:

Free surface electrospinning (FSE) of microemulsions is a novel method proposed to increase solubility of poorly soluble active pharmaceutical ingredients (API) and to eliminate batch powder processes that challenge the pharmaceutical industry today. An oil-in-water (o/w) microemulsion containing vitamin E, El-35 surfactant (castor oil), ethyl butyrate, water and ethanol was investigated. Turbidity of the microemulsion was measured at varied stir time and sonication time to determine its stability. The five-component microemulsion stabilized at 1 hour stir time and 45 seconds sonication time. Polyethylene glycol (PEG) and polyvinylpyrrolidone (PVP) were used as *excipient.* A ratio of 9 wt. % PEG and 6 wt. % PVP relative to the aqueous phase was found to be the effective ratio to produce high quality desirable mats. Scanning electron microscopy (SEM) confirmed uniformity and amorphicity of the fibers. The role of EL-35 surfactant and the release characteristics of nanofibrous mats were also examined using high performance liquid chromatography (HPLC). Dissolution of fiber mats containing vitamin E in acetate buffer solution reached 100% within 10 minutes indicating a high release rate when EL-35 served as the surfactant. Moreover, the mats showed higher release characteristics compared to casted film emulsions. Productivity of FSE was determined to be 0.26 mg/min per centimeter of electrode length, significantly higher than conventional needle based electrospinning.

470676 Investigating Macrophage Plasticity and Migration in a 3D Wound Healing Model

Abstract:

Cellular interactions with neighboring cells and their microenvironments are critical to wound healing. Specifically, in a 3D environment matrix metalloproteinases (MMPs) and tissue inhibitor of metalloproteinases (TIMPs) enable cellular migration and remodeling of the extracellular matrix (ECM), while cytokine secretions modulate phenotypic functions. Despite the critical roles played by macrophages and fibroblasts during wound healing, changes in macrophage activation, protein expression and the corresponding changes in fibroblast function are not fully understood. We have designed 3D and 2D macrophage-fibroblast co-culture systems that mimic different phases of the wound healing process. The ratios of macrophages to fibroblasts were varied to simulate early and late stages of wound healing. We investigate the effects of matrix elasticity, EC composition and cellular interactions in 3D and 2D system on macrophage plasticity, migratory behavior, cell-cell adhesion molecules, and the secretion of MMPs and TIMPs.

Collagen gels exhibiting Youngâ€™s modulus values of 6 or 36kPa were assembled to mimic early and late-stage wound environments. Co-cultures of RAW 264.7 macrophages and BALB/c 3T3 fibroblasts were cultured within (3D) or on the surface (2D) of these gels. Time-lapse images obtained on a confocal microscope were used to track cells in 3D. Immunofluorescence stainings were conducted to detect cell-cell adhesion molecules and ELISA assays were conducted to measure the concentration of pro- and anti-inflammatory cytokines, MMPs and TIMPs.

Macrophages co-cultured with fibroblasts showed a strong tendency to converge on and adhere to fibroblasts in both 2D and 3D systems. In 2D cultures, cells migrated independently or in aggregates of cells. Migratory speeds increased 28.5% for individual macrophages and 101.2% for cell aggregates at the 1:1 ratio when compared to the macrophage monocultures on 6kPa gels. This trend held for cells on 36kPa gels, however migratory speeds were reduced.

Surprisingly, macrophages in 3D monocultures often migrated at speeds up to 30Î¼m/h, similar to their 2D counterparts. We are currently analyzing migration in 3D co-cultures as well as investigating the signaling molecules responsible for attractions between fibroblasts and macrophages. As we have observed strong adhesion between macrophages and fibroblasts, we have examined the expression of intercellular adhesion molecule 1 (ICAM-1), a mediator of inflammatory cell recruitment. Fibroblasts expressed ICAM-1 across all conditions. In macrophage monocultures on 6kPa gels, almost all macrophages expressed ICAM-1 at the cell periphery and points of contact between neighboring cells. Macrophages cultured on the stiffer 36kPa gels showed decreased expression compared to those on the 6kPa gels. The expression of ICAM-1 by macrophages was significantly reduced when fibroblasts were added at a 1:1 ratio. These trends suggest that the presence of fibroblasts may be inhibiting the inflammatory profile of macrophages.Â Interestingly, decreases in ICAM-1 expression in macrophages could be correlated to a reduction in the concentration of tumor necrosis factor alpha (TNF-a). The reduction of this inflammatory cytokine between mono- and co-cultures of macrophages with fibroblasts (1:1 ratio) was approximately 50%.

We have developed a platform to study the complex interplay between two cell types that play critical roles in wound healing. Significant changes were observed between 2D and 3D migration of co-cultures of these cell types. The co-cultures can be expanded to include different cells types and tissue environments enabling the investigations into cell-cell interaction in 3D migration.

471513 Production of Self-Assembled Aggregates of Human Mesenchymal Stem Cell in the Wave bioreactor^{TM} for Cell Therapy

Abstract:

Human mesenchymal stem cells (hMSCs) are primary candidates in cell therapy and regenerative medicine, and have been tested in clinical trials for a wide range of diseases. Recent studies showed that hMSC have natural ability to self-assemble into three-dimensional (3D) aggregates that enhance their therapeutic functions with augmented multi-lineage differentiation potential, migration ability, secretions of anti-inflammatory and angiogenic factors, and resistance to ischemic conditions post-transplantation. Presently, various laboratory methods have been developed for hMSC aggregates production, including manual hanging drops, centrifugation with microfabricated surface, low attachment surface treatment, thermal lifting, and microfluidic technologies. However, these method have limited scalability and/or poor control in aggregate size, and can not meet the required production in clinical trials. This study is to produce size-controlled aggregates of hMSCs in a scalable bioreactor for cell therapy.

The WAVE bioreactor^{TM} provides a repeated back and forth gentle wave motion that accelerates cell-cell collisions and boosts hMSC aggregation at low shear stress. We first investigated the influence of rocking angle (3Ëš, 6Ëš, and 9Ëš) and rocking speed (10, 15, 20 rpm) on aggregation kinetics and size distribution. hMSC aggregates were formed in an ultra-low attachment 6-well plate on a WAVE bioreactorTM base at a certain rocking condition for 18 h to 72 h. Aggregate size was converted by the projected area of aggregates based on the light microscope images with an assumption that the aggregates are spherical in shape. Since the fluidic shear stress increases the cell-cell collision but decreases the binding kinetics upon cell-cell contract, the shear stress significantly impacts the aggregation process and aggregate size distribution. Computational fluid dynamics analysis using COMSOL Multiphysics indicates that the maximum shear stress was proportional to the rocking angle and speed under a two-variable linear regression, and the rocking angle had a more significant effect than the rocking speed. Aggregate size distribution was also found to be

significantly dependent on the cell-cell collision frequency as alteration in cell seeding density and incubation time influence aggregate size distribution. Therapeutic functional tests supported that hMSCs derived from engineered aggregates in WAVE bioreactor^{TM} have significantly higher expression of stem cell genes and secretion of anti-inflammatory factors compared with those from monolayer culture.

471638 Development and Physicochemical Characterization of Acetalated Dextran Aerosol Particle Systems Designed for Deep Lung Delivery

Abstract:

Pulmonary aerosols have attracted increasing attention for the treatment of lung diseases, as they are capable of delivering therapeutics directly and efficiently to the lung. As the result, pulmonary delivery systems exhibit advantages including increased local drug concentration, reduced side effects, rapid onset of action due to the enormous surface area and plentiful capillary vessels in the lung, and avoidance of the first-pass metabolism of the liver. Polymers with biocompatibility and biodegradability are used as excipients to impart sustained release to aerosol formulations, in which acetalated dextran (Ac-Dex) exhibits promising potential in various applications.

Acetalated dextran is an acid-sensitive, biodegradable, and biocompatible polymer prepared in a one-step reaction by reversibly modifying dextran with acetal groups. This modification reverses the solubility properties of dextran from hydrophilic to hydrophobic, making it possible to form polymeric particles using standard emulsion or nanoprecipitation techniques. In contrast to other commonly used polymers such as poly (lactic-co-glycolic acid) (PLGA), polylactic acid (PLA), and poly-Îµ-caprolactone (PCL), Ac-Dex exhibits attractive properties suitable for the controlled release of the payloads. By controlling the reaction time during the formation of Ac-Dex, the ratio of cyclic acetal groups with a slower degradation rate to acyclic acetal groups with a faster degradation rate can be adjusted. As a result, the degradation rate of resulting Ac-Dex can be tuned from hours to months to suit various applications. Moreover, the acid-sensitivity of Ac-Dex enable it to degrade faster in lower pH environments such as lysosomes in macrophage or tumor cells, resulting in faster release of drug in such environments, while controlling the release before reaching them. Furthermore, Ac-Dex degrades into neutral by-products, which avoids undesirable changes of the micro-environmental pH in the body. At last, Ac-Dex shows promising potential of targeted delivery, as the dextran chain can be further modified with a variety of functional moieties, which enhances the efficacy of therapeutic delivery to the targeted site.

There is no known comprehensive study on using Ac-Dex as an excipient for dry powder aerosol particulate formulations. In this study, we aimed to develop and characterize pulmonary delivery systems based on spray-dried Ac-Dex particles with capabilities of: (1) delivering particles to the deep lung, (2) targeting particles to a desired location, and (3) releasing therapeutics from particles in controlled rate. Two types of Ac-Dex with rapid (5min) or slow (3h) degradation rate were synthesized. Nanocomposite microparticle (nCmP) and microparticle (MP) system were successfully formulated using both kinds of Ac-Dex as excipients and curcumin (CUR) as model drug. MP were approximately 1.5 Î¼m with wrinkled surfaces, while nCmP also exhibited wrinkled surfaces and disassociated into 200 nm nanoparticles on being dissolved in water. 3h Ac-Dex particles exhibited slower CUR release rates than 5min Ac-Dex particles. Release of CUR from nCmP systems was finished within 24 hours, while the release from the MP system can last for several days. All nCmP and MP system exhibited desirable aerodynamic diameters, which are suitable for deep lung delivery. Overall, the designed Ac-Dex aerosol particle systems have the potential of targeted and effective delivery of therapeutics in to the deep lung.

471714 Incorporating Time Constraints in Model Based Control of siRNA Delivery

Abstract:

Incorporating Time Constraints in Model Based Control of siRNA Delivery

Small-interfering RNA (siRNA) is a therapeutic RNA interference (RNAi) agent with strong potential for the treatment of a wide range of diseases that are associated with undesirable gene expression [1]. Safe and effective targeted delivery of siRNA therapeutics to cells in the human body remains a main barrier to realize its clinical potential. Several modelling approaches have been used in the field of gene therapy; however, a comprehensive approach that integrates all the key pharmacological issues into a framework that is applicable for *in vivo* conditions is still lacking. This contribution presents a model-based control study of non-viral siRNA delivery that describes the dynamics of the delivery process and takes into account the main pharmacological issues such as efficacy and toxicity, as well as the influence of cell division on intracellular transport.

In this work, pharmacokinetic (PK) and pharmacodynamic (PD) modelling approach was developed based upon the experimental analysis of siRNA delivery that had been measured in the intracellular compartments, which is reported in the literature [2]. The PK model was used to describe intracellular transport processes responsible for siRNA delivery during *in vitro* cell transfection, and modified to incorporate the infusion rate of siRNA therapeutics for *in vivo* conditions. The PD model was constructed to describe the correlation between siRNA concentration and its pharmacological effects. The developed modeling and control framework provides integration of the pharmacokinetic and pharmacodynamic models, which is then used for computation of optimal siRNA delivery profile. The optimization formulation is set up to control the siRNA delivery system in the presence of different practical constraints while simultaneously considering the key pharmacological issues such as cytotoxicity. The impact of cell division was evaluated by incorporating time constraints into the optimal control framework in order to achieve the desired therapeutic effect before the cell division takes place. The siRNA delivery optimal control problem was formulated and solved while considering different cell doubling times to account for uncertainty in the pace of cell cycling. The obtained results suggest the optimal dosage injection rate and optimal cellular exposure of siRNA in order to exert maximum therapeutic effect during the therapy while simultaneously preserving cell viability. The solution of the constrained optimization problem indicates how these profiles are affected by the trade-offs between efficacy and toxicity.

[1] WANG, J., LU, Z., WIENTJES, M. G. & AU, J. L. S. 2010. Delivery of siRNA Therapeutics: Barriers and Carriers. AAPS Journal, 12, 492-503.

[2] ZHOU, C. G., ZHANG, Y., YU, B., PHELPS, M. A., LEE, L. J. & LEE, R. J. 2013. Comparative cellular pharmacokinetics and pharmacodynamics of siRNA delivery by SPANosomes and by cationic liposomes. Nanomedicine-Nanotechnology Biology and Medicine, 9, 504-513.

471814 Convenient Generation and Detection of Oxygen Gradients to Investigate Hypoxia-Induced Effects on Cancer Cells

Abstract:

Recent studies have shown that hypoxia is a critical factor in the induction of drug resistance and angiogenesis in cancer cells. However, most current studies are performed either at the atmospheric levels (21% O_2) or utilize O_2 controlled incubators to maintain a defined O_2 stress, conditions that do not adequately reflect the microenvironment of solid tumor tissues. Recently, microfluidic approaches have also been proposed to effectively generate O_2 gradients but involve complex gaseous connections or indirect oxygen detection through secondary byproducts. Here, we demonstrate a novel approach for passive generation of linear oxygen gradients within microfluidic channels with simultaneous detection capability of real-time O_2 concentrations. This platform consists of just a single source of O_2 enriched and an O_2 depleted media to establish long lasting linear gradients ideal for investigating the hypoxic effects on cancer cells. The approach is based on the split and recombine strategy commonly utilized to create passive chemical gradients. The diffusion of dissolved oxygen through PDMS is prevented by creating a thin (5-10 μm) glass coating on the top three sides of the microchannels while excluding the bottom side to allow real-time detection through integrated Platinum (II) octaethylporphyrin ketone (PtOEPK) sensors. PtOEPK is dissolved in polystyrene-toluene (7% w/w) solution at 1 mg/ml concentration and spin-coated on a glass slide. The PDMS device is subsequently bonded to the oxygen-sensing layer after applying the three-sided sol-gel coating. Compared to the uncoated channels, the coated device exhibits superior dynamic oxygen sensing ability, with significantly larger changes in the detection intensity. Calibration studies reveal that the glass coated device follows a linear Stern-Volmer relationship, with a superior $I_0:I_{100}$ ratio as compared to bare PDMS. Fluorescent imaging confirmed successful generation of an oxygen gradient at flow rates of 10 $nLmin^{-1}$ and 0.1 $\mu Lmin^{-1}$consistent with the COMSOL simulation results. This is the first study where gaseous concentration gradients have been established through passive diffusion

without requiring active gaseous connections or cytotoxic chemicals. This unique approach offers versatile applications such as the creation of parallel oxygen and chemical gradients to investigate effects of hypoxia induced drug resistance on cancer cells.

Keywords: Hypoxia, oxygen sensing, microfluidics

471851 Development of a Virus-like Particle for Targeted Delivery of Therapeutic Biomolecules

Abstract:

The treatment of diseases through systemic delivery of therapeutics is often limited by harmful off-target effects. Despite advances in medicine, many diseases â€“ including cancer, Parkinsonâ€™s disease, and Tay-Sachs disease â€“ are still in need of a novel treatment with reduced side effects.

To address this problem, we are developing a nanoparticle-based delivery system to load a therapeutic cargo and deliver it specifically to the intended site-of-action. This system will increase the therapeutic index while reducing side-effects caused by the typical systemic delivery. In addition, using a nanoparticle to deliver the therapeutic allows us to use cargo, such as nucleic acids and proteins, which would not ordinarily reach the intended tissue.

Here we used advances in cell-free protein synthesis and synthetic biology to engineer a Virus-Like Particle (VLP) - a hollow, non-infectious protein-based nanoparticle - derived from Hepatitis B to efficiently load nucleic acids and proteins and specifically deliver the cargo to the targeted cells. Cell-free protein synthesis is a powerful platform that harnesses the capabilities of natural biological systems without using living cells. This allows us to efficiently express properly folded viral proteins and incorporate non-natural amino acids. These non-natural amino acids provide for easy conjugation points using the Copper(I)-Catalyzed Azide-Alkyne Cycloaddition reaction and allows us to functionalize the surface of the VLP with targeting ligands. The ligands used are antibody fragments or aptamers with high affinity towards receptors overexpressed on the targeted cells, allowing the VLP to enter the cells via receptor-mediated endocytosis. Furthermore, the VLP has been engineered such that it will disassemble once inside the cytosol, allowing the cargo to be delivered. These advantages, combined with extensive engineering of the viral protein itself to allow concurrent cargo loading and capsid assembly, will allow us to develop an ideal delivery vehicle.

471855 Engineering of Virus-like Particles for Targeted Delivery of Cancer Drugs

Abstract:

Prostate cancer is the second leading cause of cancer death in American men, and chemotherapy has been limited because of harsh side effects caused by the systemic delivery of cancer drugs including docetaxel. Virus-like particles (VLPs) designed for targeted drug delivery to prostate cancer cells can help increase the docetaxel efficacy and reduce side effects. VLPs are hollow and non-infectious nanoparticles derived from the viral capsid, and they have been engineered to be used as targeted drug delivery vehicles by loading therapeutic cargo inside the capsid. The Swartz lab has engineered Hepatitis B core VLPs to be stable during manufacture and storage, but to lose stability under cytoplasmic conditions. This allows the drug cargo to be released inside the cells. Targeting molecules, such as antibody fragments that bind to the cell surface markers of prostate cancer, can be used to direct the VLPs to prostate cancer cells to reduce the undesired side effects. The Swartz lab has developed a process to functionalize the VLP surface with these ligands using non-natural amino acid incorporation and cell-free protein synthesis. This work focuses on recent efforts for the effective loading and retention of docetaxel inside the VLP. For effective loading, we introduced several different C-termini that extend into the VLP interior, on which different types of charged and hydrophobic cancer drugs will adsorb by electrostatic and hydrophobic interactions. We further engineered the VLPs to constrict the pores for the better retention of the cargo during the circulation. The final delivery vehicle functionalized with the targeting molecules will provide improved efficacy and reduced side effects compared to the systemic administration of docetaxel.

471943 Biodegradable Proteolipobeads for Dynamic Ligand Display in Cell Microenvironments

Abstract:

Biomaterials and biomedical research are often focused on the re-creation of complexities found in cells and cellular microenvironments. We hypothesize that biomembrane-microsphere assemblies are a viable biomimetic means to present ligands/bound factors to stem cells and mimic cellular communication in the stem cell niche. Supported lipid bilayers (SLBs) are ideally suited for the study of biomembraneâ˘'biomembrane interactions and for the biomimicry of cell-to-cell communication, allowing for surface ligand displays that contain laterally mobile elements. However, the SLB paradigm does not include three-dimensionality and biocompatibility. As a way to bypass these limitations, we have developed a biodegradable form of microsphere SLBs, also known as proteolipobeads (PLBs), using PLGA microspheres. Microspheres were synthesized using solvent evaporation and size selected with fluorescence activated cell sorting (FACS). Biomembranes were covalently tethered upon fusion to microsphere supports via short-chain PEG spacers connecting membrane-integrated Î±-helical peptides and the microsphere surface, affecting membrane diffusivity and mobility as indicated by confocal FRAP analysis. Membrane heterogeneities, which are attributed to PLGA hydrophobicity and rough surface topography, are curtailed by the addition of PEG tethers. This method allows for the presentation of tethered, laterally mobile biomembranes in three dimensions with functionally embedded attachment peptides for mobile ligand displays.

472363 Aggregate-Derived Human Mesenchymal Stem Cells Improve Cell Survival and Neurological Function Recovery in Rats Mcao Stroke Model

Abstract:

Introduction

In the US, stroke is the primary cause of severe disability and the third leading cause of death. To date, the only FDA-approved drug for stroke is tissue plasminogen activator (tPA), a thrombolytic agent that has limited benefits to patients. Originally isolated from bone marrow as the progenitor cells responsible for the repair and regeneration of mesenchymal tissues, human mesenchymal stem cells (hMSCs) have emerged as an important cell source in stroke treatment due to their trophic effects that promote endogenous tissue repair by forming a molecular milieu. However, hMSCs have low cell survival and secretory function post-transplantation, which significantly reduced their therapeutic potency. Cellular damages induced by the microenvironment at the stroke lesion site, such as ischemia, high level of reactive oxygen species (ROS) and inflammation, challenge the survival and function of transplanted hMSCs. While studies have demonstrated that in vitro hMSCs preconditioning, such as hypoxia and aggregation culture, could significantly improve hMSCs survival, migration, and secretory function in vitro, it remains to be determined whether the in vitro beneficial effect could be reflected after transplantation. The objective of this study is to investigate the effects of hMSCs aggregation preconditioning on cell survival and injury recovery in a rat MCAO stroke model. We hypothesize that PI3K/Akt anti-apoptotic pathway activated by aggregation preconditioning play a significant role in enhancing hMSCs survival post-transplantation, and further enhance the trophic effects and neurological function recovery.

Materials and Methods

hMSC aggregates were formed by seeding cells into ultra-low adherent surface plates. After dissociation, aggregate-derived hMSCs were labeled with Bangs iron particles and transplanted to MCAO stroke SD rats by I.V. injection. MRI scanning and behavioral tests were performed for each group.

Survival of aggregate-derived hMSCs was also assessed in vitro under the treatment of ischemia and hydrogen peroxide. To investigate the role of PI3K/Akt anti-apoptosis pathway during aggregation, several PI3K/Akt modulators were used to treat aggregate-derived hMSCs and 2D culture cell. Genes involved in PI3K/Akt pathway were also tested by RT-PCR.

Results

MRI image of the stroke rats with transplantation of aggregate-derived hMSC revealed low cell clearance after one week compared to control group. Aggregate-derived hMSCs lasted longer at lesion site after transplantation. Lesion recovery and neurological function were improved. Histological images showed cell presence and lesion sites after 4 weeks. In vitro ischemic or exogenous ROS challenge of aggregate-derived hMSCs showed higher survival rate compared to the cells from 2D culture. The enhanced survival of aggregate-derived hMSCs was correlated with increased expression of several anti-apoptosis genes involved in PI3K/Akt anti-apoptotic pathway, including Akt, PDK-1, Bcl-2, ERK1/2. PI3K inhibitor could reduce the survival of aggregate-derived hMSCs, which confirmed the significant role of PI3K/Akt signaling pathway in the enhancement of cell survival of aggregate-derived hMSC.

Conclusions

The results of this study demonstrate that aggregation preconditioning of hMSCs enhanced cell survival and functional recovery post-transplantation in MCAO stroke rats. PI3K/Akt pathway is activated in hMSCs aggregates and plays an important role to enhance anti-apoptosis and survival under ischemic and high level ROS conditions. Thus, 3D aggregation culture of hMSCs could be a non-genetic preconditioning method to potentiate hMSC properties for stroke cell therapy.

472493 Leukocyte Purification and 3D Migration from a Drop of Whole Blood

Abstract:

The directional migration of leukocytes in response to soluble chemokine gradients is the basis of multiple physiological and pathophysiological processes including wound repair, immune response, and several other homeostatic phenomena. Current *in vitro* approaches developed to characterize the behavior of primary leukocyte cells rely on fractionation procedures to obtain white blood cells from whole blood (e.g. centrifugation, chemical lysis, or fluorescence assisted cell sorting). While these techniques efficiently deplete unwanted cell populations, they are prone to activating leukocytes and altering their native phenotypes. Furthermore, due to the low abundance of leukocytes in whole blood, these procedures can be time consuming, laborious, and may not be suitable for small sample volumes. Recently, microfluidic approaches have also been utilized to achieve highly precise fractionation of whole blood. Using these devices, researchers have been able to analyze the migratory and chemotactic behavior of leukocytes along two dimensional surfaces. To adequately mimic the three-dimensional in vivo environment, we have developed an integrated platform that combines whole blood sorting and hemodynamic focusing to enhance leukocyte-endothelial cell interactions for subsequent chemokine stimulated adhesion and transendothelial migration (TEM). Once through the endothelium, the leukocytes can migrate towards a chemokine gradient that is generated through a biomimetic three-dimensional extracellular matrix. The unique design of this device preferentially limits migration to single cells allowing high resolution and real-time observations of native cell behaviors. We have characterized the separation technique using finite element simulations, and verified the efficiency using fluorescent microparticles with radii similar to red blood cells and leukocytes. Furthermore, we have recently demonstrated the efficiency of our single cell migration device (S.A. Roberts, Anal. Chem. 2016). While we utilize this platform primarily to study leukocyte TEM behavior, the integration of cell isolation with 3D migration makes the design highly versatile making it suitable for many applications, including liquid biopsies and pharmaceutical screenings.

472722 High-Throughput Combinatorial Drug Screening of *Caenorhabditis Elegans* Using Droplet-on-Demand Technology

Abstract:

The nematode *Caernorhabditis elegans* (*C. elegans*) is an excellent model organism for research on neurodegenerative disorders. The ability to assay worm behavior in response to multiple cues is important for drug screening to assess potential synergistic or interferential effects. However combinatorial screening is difficult because of the complexity of sample preparation and costs of the drugs. Droplet microfluidic technology can address the cost by reducing volume consumption by several orders of magnitude. However, handling *C. elegans* in droplets remains challenging due to the animalsâ€™ ability to locomotion and interaction with structures on-chip.

Here we present a platform to enable high-throughput combinatorial screening of drugs for *C. elegans*. Animals are encapsulated in individual droplets and ordered in an array for imaging. The chambers are designed to have a trapping capacity over a wide range of volume, enabling sequential loading of droplets containing worms and stimuli, and to release the droplets on demand. We optimized structured wall design to obtain 100% trapping efficiency. Chemical reagents are mixed on chip to create combinatorial stimuli in an automated way via the actuation of valves and then delivered to the worms. Upon merging of droplets, worm responses are recorded, then the array is flushed, freeing space for the next batch. Videos are subsequently analyzed, screening for specific behaviors. To demonstrate screening capability, we studied worm responses to an anesthetics (tetramisole). Thrashing frequency as a function of drug concentration was analyzed over time to quantify paralysis, using a custom MATLAB program. We demonstrate strong increase of paralysis efficiency with increased drug concentration.

This platform provides a robust method to perform high-throughput combinatorial screening of chemicals in *C. elegans*, using droplets on demand to create mixtures while minimizing reagent consumption. We envision this platform to be determinant in discovering drugs or deciphering genetic basis of neurodegenerative disorders.

472732 Microfluidic Generation of 3D Micro-Hydrogel Constructs for Cell Encapsulation and Cell-Cell Interaction Studies

Abstract:

Stem cells hold immense potential in tissue engineering applications due to their ability to differentiate into multiple cell types based on their cellular niche. In order for them to successfully differentiate and proliferate ex-vivo, it is imperative to design a micro-environment which is able to reasonably mimic the mechanical and biochemical native cell environment. We conduct in vitro studies to better understand the effect of different hydrogel matrices and intercellular interactions on the differentiation and proliferation of Human Bone Marrow Mesenchymal Cells (HBMSCs). We use droplet microfluidics to encapsulate HBMSCs using photo-curable gelatin methacrylate (GelMA) and alginate hydrogels to study the effect of the encapsulating microhydrogel matrix on the proliferation and differentiation of HBMSCs. Further, we describe a strategy to co-culture HBMSCs with Endothelial Cells (ECs), in 3D compartmentalized hydrogel micro-particles in order to study the paracrine signaling pathways that exist between these cells and their effect on the differentiation potential of the HBMSCs. This system offers confinement of different cell types, isolation of paracrine from juxtacrine signaling, high surface area for transport, biocompatibility, control over cell ratios and good experimental statistics. Initial results show this platform is suitable for the long term culture of these cell types and detection of interactions between them.

472709 Effect of Spinning Time and Speed on Separation of Bacteria from Blood for Microfluidic Identification of Bacteria

Abstract:

Bacterial infections in the blood stream (sepsis) can cause severe illness and death if untreated. Even more dangerous is sepsis caused by antibiotic resistant bacteria, which in some cases has a 50% mortality rate. Current clinical methods to diagnose the microorganism and its antimicrobial resistance profile involve culturing the blood, which adds many hours of a time-consuming step in the diagnosis procedure. Considering the fact that the situation of the patient will worsen as the time passes, it is crucial to develop fast diagnosis techniques. The first step in rapid diagnosis using a microfluidic system is to separate the bacteria from the blood. Spinning the blood sample in a hollow disk will produce a centrifugal force acting on all the particles present, including blood cells and bacteria. The difference between the size and density of the bacteria and the blood cells will cause these components to move with different sedimentation velocities. If spinning speed and time are optimized, cells will sediment more quickly than the bacteria, the majority of which will remain in the plasma. This will lead to rapid and practical separation of the bacteria from blood cells. We have conducted experiments using human blood spiked with *E. coli* spinning in a hollow disk. By varying the spinning time and speed, we have been able to optimize the separation of the bacteria from blood and recover about 50 percent of the bacteria spiked in blood within minutes, which is sufficient for identification. This is the first and perhaps most important step in rapidly identifying the antibiotic resistance of organisms in sepsis. The separated bacteria are injected into a microfluidic system employing genomic identification processes.

472804 Detection of Epithelial-to-Mesenchymal Transition in Breast Cancer Cells Using Endoplasmic Reticulum Stress

Abstract:

Epithelial-to-mesenchymal transition (EMT) is a complex remodeling of cell signaling pathways that activates metastatic abilities in cancer cells. Detection of EMT is essential for evaluation and management of cancer progression in patients. EMT is commonly detected using cell morphology changes and EMT markers expression, however, real-time detection of EMT in live cells is challenging. Endoplasmic reticulum (ER) stress is due to accumulation of unfolded proteins in the ER lumen, and has been related to EMT in cancer. Here, we describe the previously unknown role of ER stress in EMT in breast cancer cells, and evaluate the use of ER stress as an indicator of EMT. We have observed that induction of EMT in mammary cell lines using TGF-Î^{2}1 is accompanied by a transient increase of ER stress, as measured by immunostaining of GRP78 marker. Surprisingly, specific inhibition of ER stress using 4-PBA during EMT induction does not affect early EMT marker ZEB1, however, it completely blocks the induction of late EMT markers E-cadherin and Vimentin. Considering the essential role of ER stress in EMT in mammary cells, we hypothesized that measuring the transient ER stress increase after TGF-Î^{2}1 treatment would a good indicator of EMT. Thioflavin T (ThT) is an inexpensive fluorescent compound capable of quantitatively detecting ER stress in live cells. We observed that ThT can be used conveniently to detect the transient increase of ER stress during EMT in live breast cancer cells at the single cell level. We expect that this new inexpensive and easy-to-use EMT detection method will help the investigation of EMT in breast cancer, in laboratory and clinical settings.

472978 Impact of Membrane Cholesterol on Macrophage Biomechanics

Abstract:

Monocytes/macrophages play a critical role in the pathogenesis of atherosclerosis, and cholesterol has been shown to greatly influence the development and progression of the disease, Although cholesterol is known to affect the various stages of atherosclerotic process from monocyte recruitment to atherosclerotic foci to foam cell formation, the mechanisms are not well understood. Plasma membrane contain majority of the total cellular content, and is the entry point of dietary cholesterol that infiltrate the macrophages. We hypothesized that cellular cholesterol affects macrophage recruitment and transmigration by altering the biophysical properties. To test this hypothesis, we investigated the effect of cholesterol depletion and enrichment on membrane stiffness, lipid raft and receptor clustering, cytoskeletal deformability, and its consequences on monocyte recruitment, transmigration and chemotaxis. We observed that a decrease in cholesterol levels resulted in an increase in membrane fluidity and uniformity in lipid raft distribution but a decrease in overall cellular deformability as measured by atomic force microscopy, and increase in actin polymerization and cell spreading. The changes in the distribution of adhesion receptors CD44 followed the lipid raft distribution. As a consequence, we observed that cholesterol-depleted cells rolled slowly and more uniformly on E-selectin coated surfaces under shear flow. We also observed that a decrease in cholesterol levels resulted in a decrease in the transmigration of monocytes to chemotactic gradient triggered by MCP-1, due to a redistribution of the cognate receptor, CCR2. These effects were reversed when the cells were enriched with cholesterol. Together, our data suggests that the biophysical regulation of monocyte/macrophage behavior by membrane cholesterol levels may influence cellular interactions that lead to atherosclerosis.

452849 On the Small-Angle Scattering Signature of Monoclonal Antibodies

Abstract:

Among biotherapeutics, the high efficacy and specificity of monoclonal antibodies (mAbs) make them key in controlling different type of illnesses such as cancer, allergies, and autoimmune diseases. However, small changes in the protein sequence and formulating conditions may lead to physical instabilities such as protein phase separation and aggregation. These stability concerns constitute some of the main causes of impurities during biotherapeutic manufacturing, as well as may bring immunogenicity issues in patients. Understanding how different factors such as solution conditions, sequence identity, and protein concentration affect the behavior and stability of mAbs is critical for the rational design of new therapeutics. In that regard, small-angle scattering (SAS) experiments have been proven to be powerful methods to evaluate the stability of proteins over a wide range of length-scales (1–100nm) and formulation conditions. However, quantifying the behavior of proteins via SAS generally requires reliable theoretical models in order to relate scattering profiles at different length-scales to changes in protein structure, protein concentration, and solution conditions. In the case of mAbs, the use of such models results challenging due to their inherent anisotropy, as well as their high interdomain flexibility. In this work, a minimal, flexible molecular model was developed to evaluate the behavior of mAbs in solution and their corresponding SAS profiles, which coarse-grains antibodies based on the large domains (e.g., Fc and Fabs). Comparison against detailed models (e.g., fine-grain and atomistic representations) at dilute conditions showed that this minimal model effectively captures the main SAS signatures of mAbs. Unlike those detailed models, the simplicity of the working representation permits to evaluate the effect of high-concentration non-idealities such as molecular crowding on the behavior of mAbs. The results indicate there are significant structural changes as a consequence of high protein concentrations (>200mg/ml), where mAbs transition into closed, globular-like conformations. Finally, analysis of the effect of interdomain interactions provides a phenomenological phase boundary for different states that cause mAb instabilities, which include formation of micelles, gelation, and liquid-liquid phase separation.

455721 *Agrobacterium Tumefaciens* Production to Enable the Large-Scale Transient Expression of Recombinant Proteins in Plants

Abstract:

Production of recombinant proteins through *in planta* transient expression offers an alternative to conventional microbial and mammalian cell culture systems. This platform is particularly appealing because of its rapid and relatively low-cost implementation and its ease of scale-up. *Agrobacterium*-mediated transient expression in plants has been shown to be a viable method of producing functional human therapeutic proteins, vaccines, and commercial enzymes. Large-scale transient expression of recombinant proteins in plants is a developing area, and studies are underway to optimize the stages of the process. One area that has not been examined is the fermentation of *Agrobacterium* for the process at large-scale. We investigated the effects of growth conditions including temperature, pH, and media on *Agrobacterium* growth kinetics and gene transfer capability, with the goal of identifying optimal conditions for growing *Agrobacterium* at large scale. Growth temperature was found to affect bacterial growth rate but not gene transfer capability, and a growth temperature of 28°C was selected as optimal. Growth in Lysogeny Broth (LB) and Yeast Extract Peptone (YEP) media was examined and subsequent transient gene expression was measured. A defined media was developed and optimized for growing *Agrobacterium* and growth and gene transfer capability with this media was found to be comparable to LB and YEP media. Growth of *Agrobacterium* strain C58C1 pTFS40 in LB, YEP, and defined media resulted in maximum specific growth rates of 0.36 ± 0.01, 0.37 ± 0.03, and 0.33 ± 0.01 h^{-1} and maximum biomass concentrations of 1.9, 3.6, and 3.9 grams dry cell weight per liter after 12, 16, and 20 hours, respectively. It was demonstrated that direct infiltration with *Agrobacterium* in diluted growth media was an effective method of inducing transient expression. Batch fermentation of *Agrobacterium* was scaled up to benchtop (5 L) scale with the three types of media. Finally, production was successfully scaled up to a 100 L working volume bioreactor.

456207 High-Yield Productino of Ginsenosides 20(S)-Rg3, Rk1, and Rg5 from Ginseng Extract By Hydrothermal Reaction

Abstract:

Ginseng has been widely used as an herbal medicine because it has many bioactivities such as immune system modulation, antistress activity, and antihyperglycemic activity. Moreover, its anticancer activity has been found recently. These bioactivities are generally attributed to ginsenosides which are main ingredients of ginseng. In unprocessed ginseng, intact ginsenosides such as Rg_1, Re, Rf, Rb_1, Rc, Rb_2, and Rd are major compounds. The intact ginsenosides can be converted to transformed ginsenosides through deglycosylation and/or dehydration. Because of fewer sugar moieties, the transformed ginsenosides are more bioavailable than intact ginsenosides. Furthermore, it was revealed that the transformed ginsenosides have special bioactivities which intact ginsenosides do not have. Therefore, there have been many researches to increase the yield of the transformed ginsenosides, with three processes commonly utilized. The most typical process is steaming, which is used to make red ginseng and sun ginseng. Other processes are operated under application of acidic catalysts or microorganisms. However, these conventional methods are frequently sluggish and complex. The steaming process takes over 3 h, and biotransformation process takes longer as it includes incubation. Shorter reaction times can be achieved for the acid catalyst process, but the acid leads to corrosion of the reactor. Besides, additional step that makes the process complex is needed to separate the acid. In this study, to overcome the disadvantages of these conventional processes, hydrothermal process was introduced. The hydrothermal process consisted of two steps. First step was to extract ginseng, and second step was to carry out hydrothermal reaction of the ginseng extract. Hydrothermal reactions are either homogeneous or heterogeneous chemical reactions performed in the presence of a solvent above room temperature and atmospheric pressure in a closed system. When the solvent of a hydrothermal reaction is water, as the temperature increases to the critical temperature, the ion product of water increases, whereas the dielectric constant decreases. According to the definition of ion product, the increased ion product means higher concentrations of H^+ and OH^-; hence, water with a higher ion product can be used as a good medium for acid- or base-catalyzed reactions. Moreover, with decreasing dielectric constant, water can dissolve organic materials like other organic solvents. Therefore, same effects can be achieved by hydrothermal reaction in water without using organic solvents and catalysts. Because of these properties, biomass and biowaste which traditionally needed acid catalysts or organic solvents have been hydrothermally converted to more valuable chemicals through hydrolysis and/or dehydration. As the transformation of intact ginsenosides to more valuable ginsenosides occurs through similar reaction mechanisms, we introduced the hydrothermal process to obtain transformed ginsenosides. The purpose of this study was to increase the yields of the transformed ginsenosides, such as 20(S)-Rg_3, Rk_1, and Rg_5 and find optimized operating conditions. For comparisons, red ginseng and sun ginseng were prepared by conventional steaming. The maximum yields for the hydrothermal reaction were 2.5â^'25 times higher than those for steaming. This was because that the hydrothermal reaction was a homogeneous reaction while the steaming was a heterogeneous reaction. As reactants co-exist in the same phase for a homogeneous reaction, the chemical reaction, heat transfer, and mass transfer occur more efficiently than those in a heterogeneous reaction. Additionally, the antioxidant activities were increased 1.6â^'4 fold for the hydrothermal reaction. Because the increase in the antioxidant activities was not as large as that in the yields, the yields of the three transformed ginsenosides may not be the principal factor influencing the antioxidant activities. Considering yields and antioxidant activities together, the optimized operating conditions for the hydrothermal process were temperatures over 140 °C for tens of minutes.

The Effect of Labisia Pumila Plant Extract on Biological Changes and Morphological Modifications of Various Human Skin Cells

Rozaliana Abdul Karim, UTM, Johor, Malaysia
Harisun Yaakob, UTM, Johor, Malaysia*
Mohamad Roji Sarmidi, UTM, Johor, Malaysia
Mukrish Hanafi, UTM, Johor, Malaysia
Noor Azwani Zainol, UTM, Johor, Malaysia
***Siti Nor Azlina Bt Abd Rashid**, UTM, Johor, Malaysia*

Abstract

Labisia pumila (LP) or also known as Kacip Fatimah, is a type of traditional herb that has been used from generation to generation to cure various illnesses and currently was reported having a good antiaging effect to the skin. However, an extensive research of this plant efficacy has not yet been studied into details especially in terms of biochemical and morphological changes of human skin cells. This study aimed to investigate the anti-photoaging efficacy ofLPextract on UVB-irradiated human skin cells and to observe the biochemical and morphological changes of the cells. In this study, human skin fibroblast cells (HSF1184) and melanoma cells (B16F1) were exposed to UVB irradiation and cultured in serum free media with LP extract. After UVB exposure, the changes on biochemical were investigated via cell viability assay. The cell viability result showed that LP extract at concentration below than 1 x10^{-5} mg/mL does not giving any toxicity effect to the HSF118 cells but a bit higher concentration of LP extract at 1 x 10^{-4} mg/mL for B16F1 cells showed LP extract caused significantly decreased of apoptosis activity compared to the control. Finally, themorphological changes of the cells were evaluated by microscopic observation using SEM (Scanning Electron Microscopy) and TEM (Transmission Electron Microscopy). The microscopic observation through SEM indicated that, at the higher concentration of LP extract at 1 mg/mL for HSF1184 and B16F1 cells, all cells were shrinking and died; whereas through TEM this concentration had caused the cells mitochondria swollen and contained disorganized of crystae which indicated the cells were programmed to death. Thus, this study revealed that both biochemical and morphological changes of the UVB-irradiated human skin cells after the treatment can be used to confirm the anti-photoaging effect of *L. pumila*.

Keyword: *Labisia pumila*, morphological, biochemical, anti-aging, SEM, TEM

Introduction

A huge number of Malaysian medicinal plants have long been used as traditional remedies from one generation to another up until today. The information of the plant materials that have been used as medicines since ancient times can be found in history books, old literature, pharmacopoeias, archaeological finds (Jantan, 2004) and other research publication. This fact is supported by the holy Quran and Bible, saying about 20 and 125 plants have been used as medicines to treat various diseases (Musselman, 1999). Out of more than 20,000 species of vascular plants in Malaysia, about 10%, or approximately 2,000 species, have been documented to have medicinal qualities (Herbal Medicine

Research Centre, 2000). Thus, these have made most Malaysian nowadays seeking an option of consuming natural herbal remedies rather than synthetic medicines.

Labisia pumila (*L. pumila*) or also known as Kacip Fatimah is one of the most popular herbal plants among Malay community which contain many bioactive compounds that are beneficial to human vitality. Recent study has shown that this plant contained several compounds include flavonoids, phenolic acids, benzoic acids and also cinnamic acids (Chua *et al*., 2011). Flavonoids and phenolic acids have been proven containing antioxidant effect, which is very essential to promote healthy skin and provide skin protection from UV harm. As known, UV rays can enhance reactive oxygen species (ROS) generation in skin cells, which indirectly initiate the intracellular and extracellular oxidative stress; thus, they cause wrinkles and atypical pigmentation formation to the skin. These photo-induced aging skin symptoms can be prevented by practicing effective approach such as the use of antioxidants (Masaki, 2010). The same study by Chua *et al*. (2011) also reported that the presence of antioxidant activity in aqueous *L. pumila* extracts can help the skin against reactive oxygen species (ROS) by scavenging the free radical activity.

Oxidative stress that is initiated by ROS generation is an essential factor in modulating skin alterations, especially those caused by UV exposure and aging. The aging effect of excessive sun exposure on skin has been shown in the earliest recognition on the necks of sailors who suffer Cutis rhomboidalis nuchae, which includes the changes of textural and pigmentary of the skin when compared with the necks of nuns. Other consequence of chronic sun exposure is Favre-Racouchot disease or also known as modular elastosis, a condition where cysts and comedons form on the exposed parts of the skin (Giacomoni and Rein, 2004). Preliminary experiments with some antioxidants such as co-enzyme Q10, vitamin A and E, ferulic acid and silymarin which act as anti-aging compound, have been revealed able to repair long-term damage skin from environmental influences and also promote skin self-repair, which have been applied in topical skin care formulation (Pinnell, 2003). Thus, treatment using some antioxidants should be effective to enhance resistance to oxidative stress and improve skin aging (Masaki, 2010)

Recent study has demonstrated that all varieties of *L. pumila* plant extracts contain bioactive compounds that possess antioxidant activity (Karimi *et al*., 2011). Besides, phytomolecules such as gallic acid contained in *L. pumila* extracts has been reported able to scavenge free radicals from skin cells, including a sun protector factor (SPF) of 15 or higher, and resulting in giving protection to the skin from wrinkles and leading to a healthy younger skin complexion (Mukherjee *et al*., 2011). Thus, this study was conducted to investigate the efficacy in term of biochemical and morphological changes when the *L. pumila* var. *alata* (LPva) plant extract, which acts as anti-photoaging agent, is applied on the UVB-irradiated Human Skin Fibroblast (HSF1184) and melanoma (B16F1) cells.

Materials and Methods

Cell Culture Preparation

Human skin fibroblast (HSF1184) cells were cultured in Modified Essential Medium (MEM) while melanoma cells (B16F1) were grown in Dulbecco Modified Essential Medium (DMEM) with 10% of fetal bovine serum (FBS) and 1% of Pennicillin-Streptomycin antibiotics. The cells were then incubated in 37°C containing 5% of CO_2 incubator. For experiments, fibroblasts cells from passages 5 to 8 were used while melanoma cells from any passage can be used. The cells took about 96 hours to reach confluency (Freshney, 2005).

Biochemical Analysis

UVB Irradiation and Treatment

The fibroblasts cells were seeded at 2×10^5 cells/well density in 96-well plate. The cells were then cultured in MEM up to 70-80% confluence and starved for 24 hours in serum free MEM and washed with phosphate buffer saline (PBS). UVB irradiation was performed at a wavelength of 312 nm and 80 mJ/cm^2 as measured with SX-312 research radiometer. UVB-irradiated cells were then cultured in serum free MEM with or without LPva extract or ascorbic acid for 24 and 48 hours. B16F1 cells were grown at a density of 2×10^4 cells/well in 96-well plate and cultured in DMEM for 24 hours. The medium was then removed and substituted with PBS. UVB irradiation exposure was performed to the cells at 312 nm and 80 mJ/cm^2. After that, the cells were cultured in serum free DMEM with or without the addition of LPva extract or ascorbic acid for 24 and 48 hours.

Cytotoxicity Test via MTT Assay

The effect of LPva extract on cell viability recovery was examined using MTT (3-(4,5-dimethylthiazol-2-yl)-2,5-diphenyltetrazolium bromide) assay. A serial dilution of LPva extract in media without serum was performed in order to determine the effect of LPva extract concentration towards the cells cytotoxicity. After the cells were treated with LPva extract and incubated for 24 and 48 hours in CO_2 incubator, 10 µL of MTT (0.33 g/L) solution (5 mg/mL dissolved in PBS) was added into each well and incubated again for 4 hours at 37°C. After that, both media and the MTT solution were removed from all wells, and 200 µL of DMSO were added into all wells in order to dissolve the formazan product. The absorbance measurement was taken at 570 nm wavelength using a microplate reader (Choi *et al.*, 2010).

Morphological Analysis

The morphologicals of the cells before and after being treated with LPva extract were fixed and observed using Scanning Electron Microscopy (SEM) and Transmission Electron Microscopy (TEM) at an initial magnification of 100×. The micrographs were then analyzed. The structures, numbers and sizes of intramembrane particles of the cells were next evaluated on micrographs taken at higher magnification (×1,000).

Results

Cytotoxicity Studies

From the MTT assay results as shown in Figure 1 and 2, the optimum concentrations of LPva extract for HSF1184 cells treatment were 5.00E-05 mg/mL and also 1.00E-05 mg/mL but for B16F1 cells the optimum concentrations of LPva extract were 5.00E-04 mg/mL and 1.00E-04 mg/mL. However, the first two concentrations which are 1.00E+00 mg/mL and also 5.00E-01 mg/mL were found to be toxic to both of the cells.

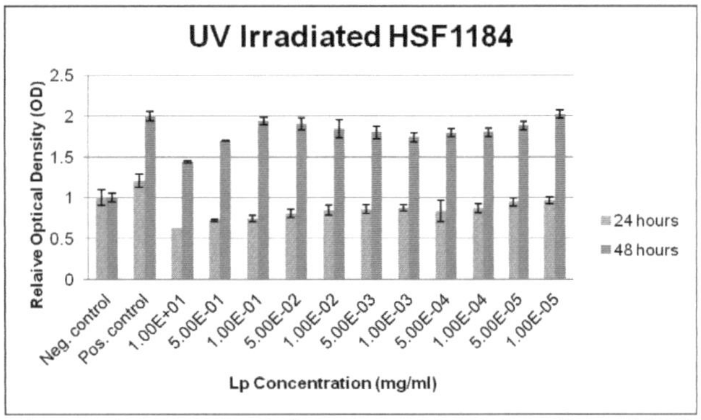

Figure 1: Cytotoxicity test of UV-irradiated HSF1184 cells.

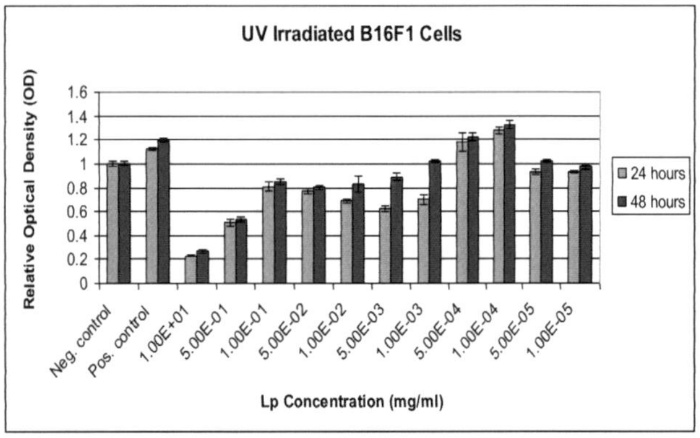

Figure 2: Cytotoxicity test of UV-irradiated B16F1 cells.

Cells Observations using Scanning Electron Microscope (SEM)

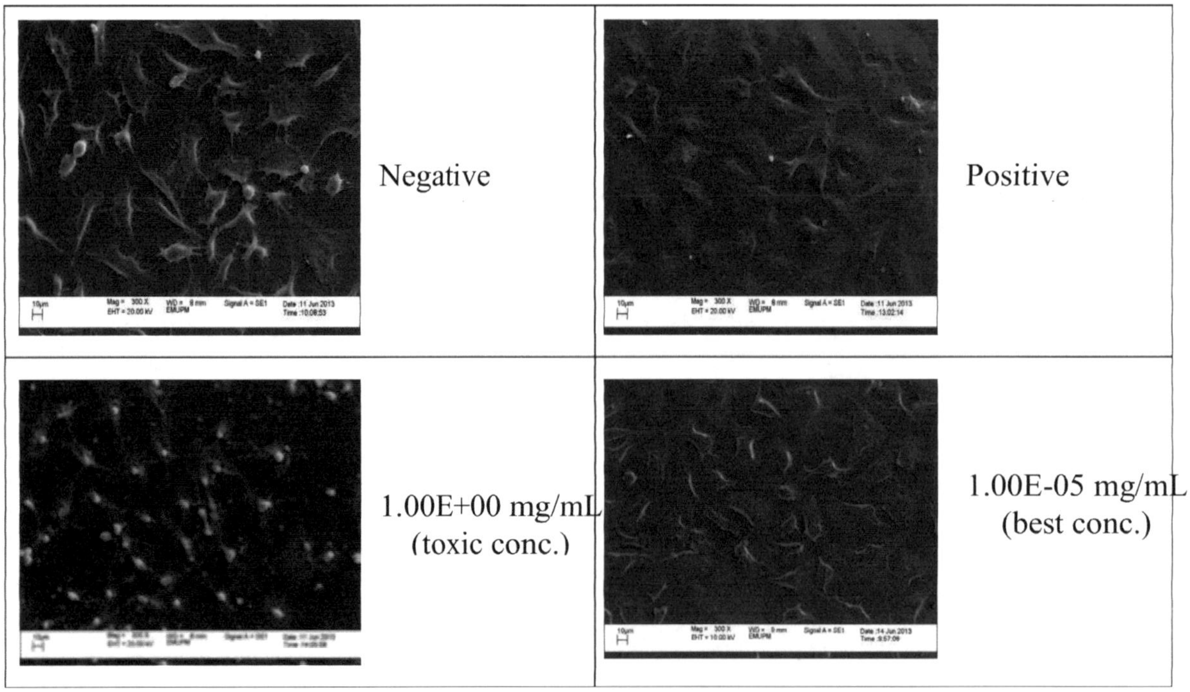

Figure 3: Cells Observation Using SEM (UVB-irradiated HSF1184 cells).

From the observation of UVB-irradiated HSF1184 cells after being treated with LPva extract and also incubated for 48 hours as shown in Figure 3, a lot of cells shrinked and a few cells with large nucleus invaginations for the negative control. For the positive control, these cells were seen elongates back to normal after being incubated. The morphological changes in this positive control were seen almost identical with the cells treated with both of the best LPva concentrations. However, for the 1.00E+00 mg/mL toxic concentrations, all of the cells were seen died after being incubated for 24 and 48 hours.

Cells Observations using Transmission Electron Microscope (TEM)

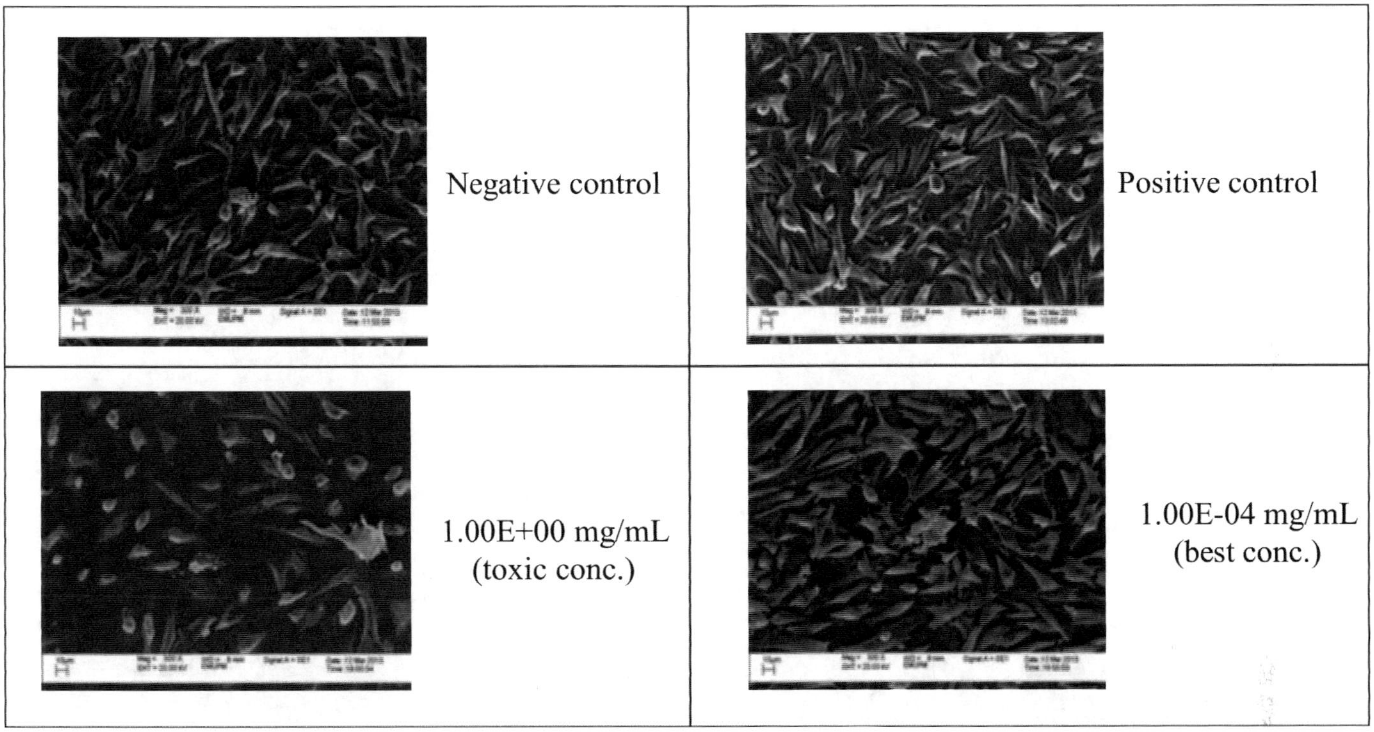

Figure 4: Cells Observation Using SEM (UVB-irradiated B16F1 cells)

The micrographs of UVB-irradiated B16F1 cells after being treated with LPva extract and also incubated for 48 hours as shown in Figure 4, some of the cells were observed died and the cells proliferation has been decreased and thus has reducing the accumulation of the adherent cells. Most of the cells survived were slightly shrinking when incubated for 48 hours.

Cells Observations using Transmission Electron Microscope (TEM)

The Transmission Electron Microscopy (TEM) micrographs analysis for HSF1184 cells as can be seen in Figure 5 and 6 supported the micrographs analysis viewed by using SEM. The toxic concentrations showed that the cells were died while the cells were managed to live healthily in the best concentrations. This evidence has been supported by the micrographs of mitochondria for each of the samples. The mitochondria for the toxic concentrations were swollen and have disorganized of cristae which indicates the cells were programmed to death. While the normal conditions of mitochondria for the best concentration of Lp extract showed that the cells were alive. However, for the B16F1 cells

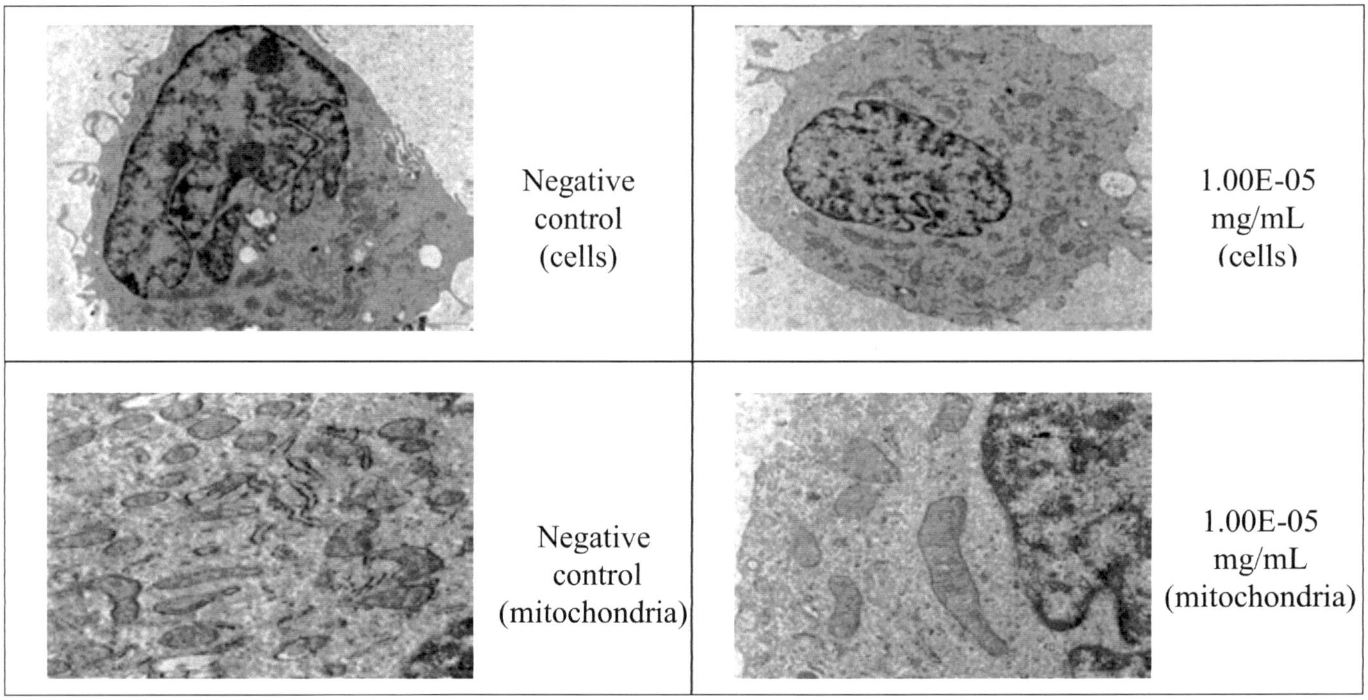

Figure 5: Cells and Mitochondria Observation Using TEM (UVB-irradiated HSF1184).

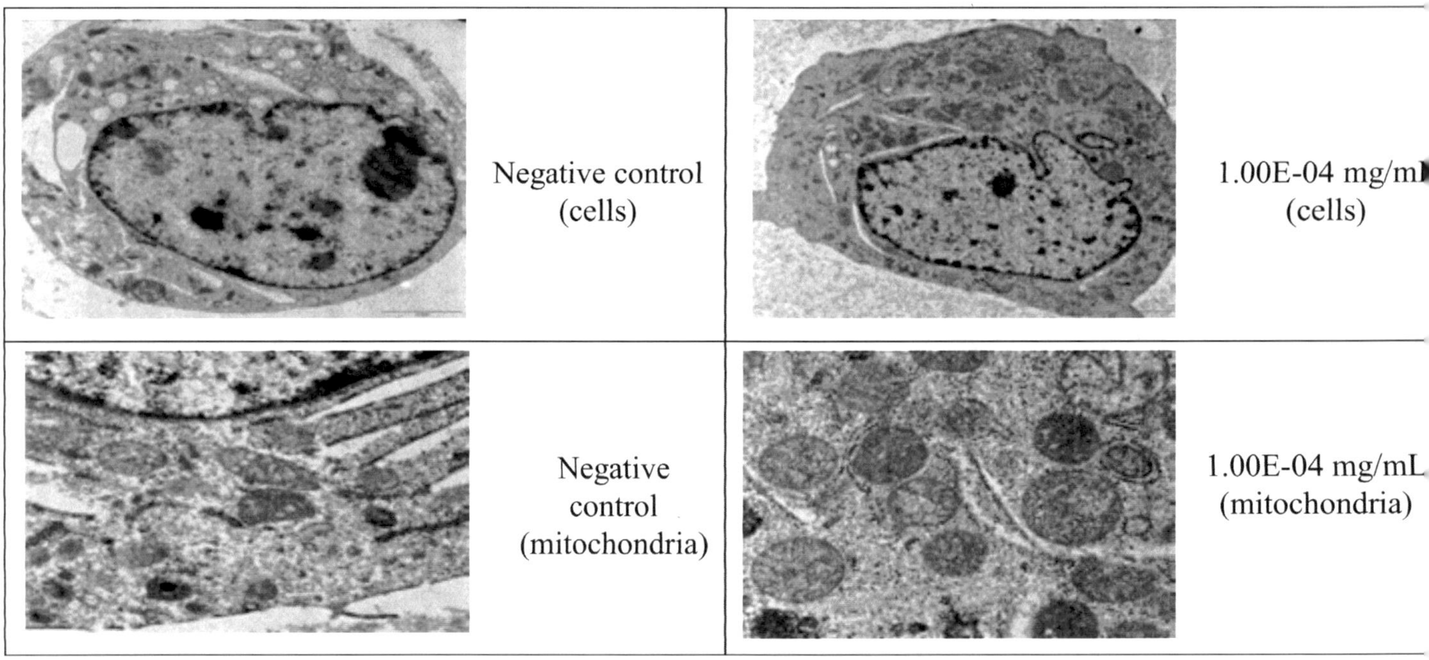

Figure 6: Cells and Mitochondria Observation Using TEM (UVB-irradiated B16F1).

micrographs analysis showed that all of the cells structures for the control, the toxic and best concentrations were almost identical. This may be due to the cells characteristics which are already abnormal, cancerous cells compared to the normal cells.

Discussions

Recent studies have showed that *L. pumila* extract was able to stimulate cells regeneration after being exposed to UVB irradiation as well as increasing the collagen synthesis in the UVB-irradiated cells. This is due to the presence of chemical compounds called antioxidant that can be able to fight against free radicals which can cause various human ailments (Galati and Brien, 2004). The antioxidant activity presence will scavenge the various free radicals and thus will help to restore the skin firmness and elasticity. Besides, *L. pumila* also contains flavonoids, phenolic acids and also resorcinols that have antioxidant and also anti-cancer activities (Tapiero *et. al.*, 2002) which can slower down the aging process. Previous studies also have shown that *L. pumila* var. *alata* contains some phytochemicals such as flavonoids, ascorbic acid, beta-carotene, anthocyanin and also phenolic content which can help to restore the decrement of collagen synthesis of the UV-irradiated skin cells to normal after being treated with the Lp extract (Norhaiza *et. al,* 2009).

The cytotoxicity test was used to examine the effect of LPva extract on cell viability recovery via the MTT assay. The cytotoxicity study was conducted using series of concentrations of LPva extract in order to identify which concentration are toxic and best for the cells to grow. According to the MTT assay data, the concentrations of 1.00E+00 mg/mL and also 5.00E-01 mg/mL have given the highest toxicity level on all of the three cells. However, the percentage of the cell viability has increased when the concentrations of the LPva extract has been decreased from 1.00E-01 mg/mL to 1.00E-05 mg/mL. Based on IC_{50}, the concentrations range were safe to be applied to the cells as they did not bring any harm towards all of the cells. However, the optimum concentrations need to be identified in order to proceed for the Electron Microscopy viewing. Thus, the best concentration for HSF1184 cells were 5.00E-05 mg/mL and also 1.00E-05 mg/mL while for the B16F1 cells, the best concentration were 5.00E-04 mg/mL and 1.00E-04 mg/mL.

The graphical data clearly proved that the both cells showed growth stimulatory effect, and cell regeneration process occurred when being treated with the best concentration of LPva extract, but suppressed to cell death when treated with the toxic concentration. These results also showed that the LPva extract had a strong antioxidant activity comparable with the ascorbic acid when applied to the UVB-exposed to the cells. Besides, the LPva extract was also observed able to provide significant protection to the cells from the damage caused by the UV radiation.

The SEM micrographs analysis showed the different morphologies exhibited by the cells as a result of different treatments applied on the UVB-irradiated cells. The micrographs showed that there were a lot of shrunk cells, which had large nucleus invaginations, but only a few flattened cells were observed in the negative control. This morphology was exhibited by the cells maybe due to the stress caused by the UVB radiation and the scarcity of the nutrient supply in order for the cells to be back to normal condition again, since media without serum was used in this experiment. Meanwhile, for the positive control, the amount of cells that contained flattened cells were observed to be more than in the negative control, and the cells showed indistinct cell borders which indicated normal morphology of the cells. At the toxic concentration of LPva, showed that all the cells underwent apoptosis process. The dead cells were observed shrinking and blebbing from the cell surface with the nucleus pinched off of the cells. However, at the best concentration of LPva extract for the UVB-irradiated cells, showed that all the cells managed to survive and the cells morphology was as identical as shown in the positive control. This evidence supports that the best concentration of LPva extract can restore the cells' integrity and resilience back to normal as well as reducing the likely of the cells to undergo apoptosis process.

Excessive exposure to UV radiation has long been thought overwhelming and reducing antioxidant supply, thereby leading to a state of oxidative stress (Fuchs, 1998). This will eventually result in skin morphological changes due to the connective tissue damage which will then lead to wrinkles, laxity and leathery appearance. According to Varani *et al.* (2004), fibroblasts in photoaged skin are elongated and collapsed. However, as found in this study, after the cells were treated with LPva extract, the cells were observed restored back to their normal spindle-shaped

The TEM micrographs analysis showed at Figure 5 and 6 supported the micrographs analysis that had been viewed by using SEM. The cells managed to live healthily in the best concentration. This evidence was supported by the mitochondria micrographs of the cells for each of the samples. The mitochondria for the toxic concentration were observed inflammed and underwent membrane lysis, leading them to burst completely with the cells, indicating the cells died severely. Meanwhile, the mitochondria established regular morphological structures in the positive control, and the best concentration of LPva extract showed that the cells survived. The mitochondria of the negative control, on the other hand, were found swollen and contained disorganized of cristae which indicated that the cells were under stress condition after being exposed to the UVB radiation and were programmed to cell death.

The dense nucleus was shown by the negative control and also the best concentration of LPva applied to the cells. However, the dense nucleus of the cells with folded nuclear membrane was expressed by the cells in negative control due to the stress caused by the UVB exposure. However, the nucleus of the cells in the toxic concentration of LPva already burst together with the cells. The slightly folded nucleus of the best concentration of the LPva extract showed that the cells were restored to normal condition and were under a recovery process

From the overall observations on the cells treated with LPva extract, it was confirmed that the plant extract has the anti-photoaging effect on UVB-irradiated HSF1184 and B16F1 cells. The LPva extract has also been observed and proven able to increase the growth stimulation, cell regeneration and cell photoprotective effects, as well as enhancing collagen and also elastin expressions. These properties of LPva extract are a hallmark of which will lead the LPva to be a novel anti-photoaging cosmetic ingredient, and it can be the 'new antiaging material' in the near future.

Acknowledgements

We would like to thank you the Institute of Bioproduct Development (IBD) and Faculty of Chemical and Energy Engineering, Universiti Teknologi Malaysia for providing us cell culture facilities and helping us dealing with the cell culture works.

References

1. Choi, H., Kim, D., Kim, J. W., Ngadiran, S., Sarmidi, M. R., & Park, C. S. (2010). Labisia pumila extract protects skin cells from photoaging caused by UVB irradiation. Journal of Bioscience and Bioengineering, 109: pp. 291–296.
2. Chua, L. S, Abdul Latiff, N., Lee, S. Y., Lee, C. T., Sarmidi, M. R., Abdul Aziz, R. (2011). Flavonoids and phenolic acids from Labisia pumila (Kacip Fatimah). Food Chemistry 127:pp. 1186–1192.
3. Freshney, R. I. (2005). Freshney's culture of animal cells: A manual of basic technique (5th Edition). Wiley Publication.
4. Fuchs J. (1998). Potentials and limitations of the natural antioxidants RRR-alpha tocopherol, L-ascorbic acid and beta-carotene in cutaneous photoprotection. Free Radic. Biol. Med. 25: pp. 848-873.
5. Galati, G. and Brien, P. J. (2004). Potential toxicity of flavonoids and other dietary phenolics: Significance for their chemopreventive and anti-cancer properties. Free Radic. Biol. Med. 2004, 37: pp. 287-303
6. Giacomoni, P. U., Rein, G. (2004). A Mechanistic Model for the Aging of Human Skin. Micron 35: pp 179–184.
7. Herbal Medicine Research Center IfMR. (2000). Compendium of Medicinal Plants Used in Malaysia. Kuala Lumpur: Aktif Bestari Sdn. Bhd.
8. Jantan, Ibrahim. (2004). Medicinal Plant Research in Malaysia: Scientific Interests and Advances. Jurnal Sains Kesihatan Malaysia 2(2):27-46.
9. Karimi, E., Jaafar, H. Z. E., Ahmad, S. (2011). Phenolics and flavonoids profiling and antioxidant activity of three varieties of Malaysian indigenous medicinal herb Labisia pumila Benth. Journal of Medicinal Plants Research Vol. 5(7), pp. 1200-1206.
10. Masaki, H. (2010). Role of antioxidants in the skin: Anti-aging effects. Journal of Dermatological Science 58:85-90.
11. Mukherjee, P. K., Maity, N., Nema, N. K., Sarkar, B. K. (2011). Bioactive Compounds from Natural Resources against Skin Aging. Journal of Phytomedicine 19:64-73.
12. Musselman, L. J. (1999). Holy Pharmacy. Modern Medical Uses of some Plants of the Quran and the Bible: Its Relation to Biodiversity.
13. Norhaiza, M., Maziah, M. and Hakiman, M. (2009). Antioxidative properties of leaf extracts of a popular Malaysian herb, Labisia pumila. Journal of Medicinal Plant Research, 3:217–223
14. Pinnell, S. R. (2003). Cutaneous photodamage, oxidative stress, and topical antioxidant protection. J. Am. Acad. Dermatol., 48:1-19.
15. Tapiero, H., Tew, K. D., Nguyen, B. G., Mathé, G. (2002). Polyphenols: Do they play a role in the prevention of human pathologies? Biomed. Pharmacother. 56:200-207.
16. Varani, J., Schuger, L., Dame, M. K., Leonard, C., Fligiel, S. E. G., Kang, S. (2004). Reduced fibroblast interaction with intact collagen as a mechanism for depressed collagen synthesis in photodamaged skin. J. Invest. Dermatol. 122:1471-1479

Improvement of the Olive Oil Quality with Olive Leaf Extract through Response Surface Technology

Selin Şahin, Ezgi Sayım and Mehmet Bilgin*

Istanbul University, Engineering Faculty, Department of Chemical Engineering, 34320 Avcilar, Istanbul, Turkey

*Corresponding Author E-mail: selins@istanbul.edu.tr
Phone: +90-212-4737070 (17679), Fax: +90-212-4737180

Introduction

Olive oil was enriched by adding high added value olive leaf extract. After extracting the dried and ground olive leaves with the assistance of homogenizer, the dried extract was partially dissolved into the oil to increase the shell-life of the oil. A face central composite design (FCCD) through response surface methodology (RSM) was used to investigate the effects of enrichment conditions (extract content, time and mixing speed) on the responses, total phenolic content (TPC) and oleuropein concentration of the enriched olive oil. Furthermore, antioxidant activity of the oil was determined by 2,2′-azino-bis-(3-ethylbenzothiazoline-6-sulfonic acid) diammonium salt (ABTS) method. Additionally, oxidative stability of the enriched oil was assessed by the Rancimat method. Total carotenoid content (TCC), peroxide value (PV), tocopherol and chlorophyl were also measured, respectively. Finally, pure and enriched oil were compared according to those quality parameters.

Results and Discussions

In this study, quick, simple and cost-effective method, homogenizer assisted extraction (HAE) method was utilized in order to obtain olive leaf extract. Optimization studies of enrichment conditions for olive oil were evaluated according to the maximum TPC and oleuropein concentration of the oil, since oleuropein is the major compound contributing to the total phenols in the leaf (Figure1).

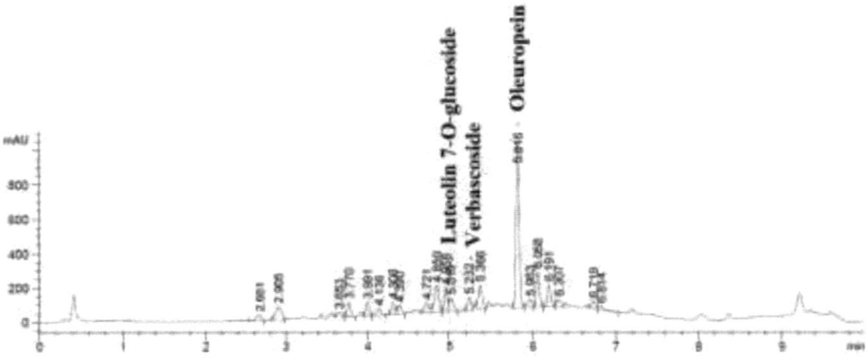

Figure 1. An HPLC chromatogram of the olive leaf extract obtained by HAE at the optimum conditions

A face central composite design (FCCD) was performed with three variables to explore the effect of variables on the response (Table 1).

Table 1. Values of the independent variables and their coded forms with their symbols employed in RSM for optimization of enrichment

Independent Variables	Units	Symbol of the variables	Coded Levels		
			-1	0	1
Mixing speed	rpm	X_1	4000	7000	10000
Time	sec	X_2	30	60	90
Solid mass	ppm	X_3	1000	1250	1500

20 experimental runs were chosen randomly by Design Expert 9.0.6 as seen in Table 2.

Table 2. FCCD of the independent variables (X_1, X_2, X_3) and experimental results for the TPC and oleuropein*

Run number	Independent variables			Responses	
	X_1 (rpm)	X_2 (sec)	X_3 (ppm)	TPC (ppm)	Oleuropein (ppm)
1	10000	30	1500	249±5.19	26.36±1.19
2	4000	30	1500	258±4.29	26.08±1.30
3	7000	30	1250	160±3.09	23.54±0.19
4	4000	30	1000	158±1.25	20.62±1.01
5	7000	60	1500	156±2.01	18.00±1.33
6	7000	60	1000	137±1.01	24.25±2.19
7	10000	90	1500	196±4.44	13.26±1.00
8	7000	90	1250	176±2.98	11.07±0.92
9	7000	60	1250	173±5.00	14.17±1.10
10	10000	30	1000	221±5.02	25.37±1.15
11	4000	60	1250	158±2.21	12.00±0.19
12	7000	60	1250	150±1.14	14.00±0.99
13	10000	60	1250	177±2.91	17.33±1.15
14	7000	60	1250	143±3.33	13.50±0.97
15	7000	60	1250	133±4.01	13.90±0.66
16	4000	90	1500	184±2.81	19.02±0.21
17	10000	90	1000	157±3.03	17.33±1.01
18	7000	60	1250	113±0.98	14.00±1.30
19	4000	90	1000	139±1.15	15.96±0.99
20	7000	60	1250	128±2.17	13.94±0.25

*Data are expressed as the mean (n=9) ± S.D.

Optimization of enrichment conditions for olive oil

The models derived for TPC and oleuropein were found significant ($p<0.05$) to display the relationship between the response and independent variables according to the ANOVA results of Design Expert 9.0.6. The ANOVA result also showed that the experimental data had correlation coefficients (R^2) of 0.8193 and 0.8909 with the calculated models, accounting for the 81.93 and 89.09 % of the results, respectively. The quadratic models for TPC (Y_1) and oleuropein (Y_2) were derived as given in the equations below:

$$Y_1 = 140.98 + 10.30X_1 - 19.40X_2 + 23.10X_3 - 3X_1X_2 - 9.75X_1X_3 - 5.50X_2X_3 + 25.05X_1^2 + 25.55X_2^2 + 4.05X_3^2 \quad (1)$$

$$Y_2 = 14.61 + 0.60X_1 - 4.53X_2 - 0.081X_3 - 1.18X_1X_2 - 1.45X_1X_3 - 0.93X_2X_3 - 0.98X_1^2 + 1.66X_2^2 + 5.48X_3^2 \quad (2)$$

In order to explore the potential relationships between the variables, three dimensional (3D) response surface plots (Figures 2 to 4) are constructed according to the quadratic Eqs.(2) and (3).

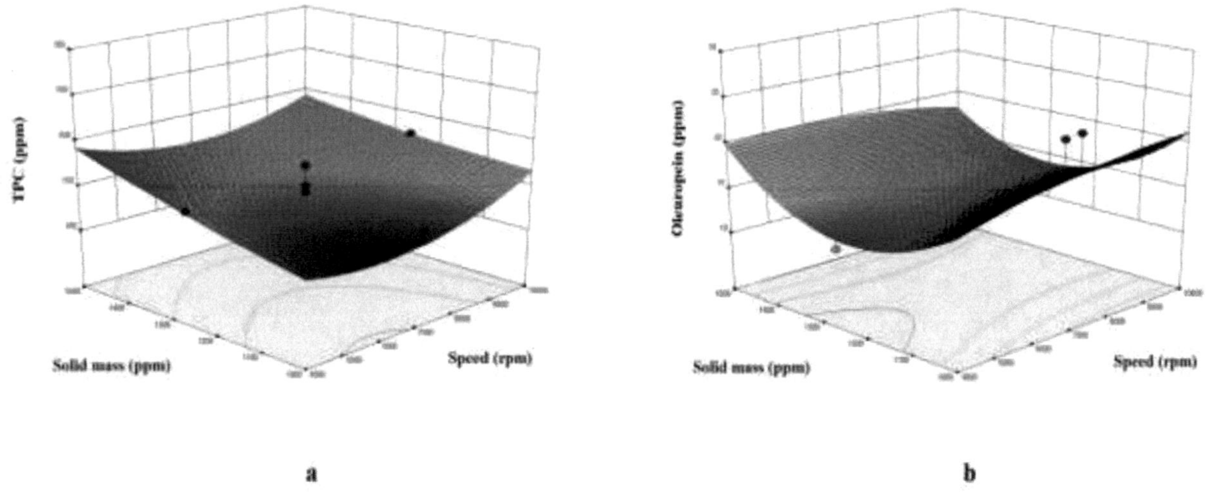

Figure 2. Response surface plots for the TPC (a) and oleuropein (b) of enriched olive oil as a function of solid mass to mixing speed (time=60 sec)

Expectedly, yield of TPC increased by extract quantity as seen in Figure 2a. This is in agreement with the results of Franco et al., who found out that increasing potato peel extract concentration would have improved the control of oil oxidation (Franco *et al.*, 2016). Mei et al. also suggested higher concentrations of the rambutan extract into sunflower oil in order to improve the stability of the oil

(Mei *et al.*, 2014). On the other hand, oleuropein yield had a tendency to decrease until a point, then it started to increase (Figure 2b). This is in agreement with the results of Ballard et al. (Ballard *et al.*, 2010)and Şahin (Şahin, 2015). Ballard et al. observed a 35.8% reduction in TPC when extracting phenolic antioxidants from peanut skins in the mass of the skins from 1.5 to 3.5 g. Şahin showed almost 22 and 13% decrease in the TPC and TFC yields of mandarin leaf exracts obtained by microwave–assisted extraction.

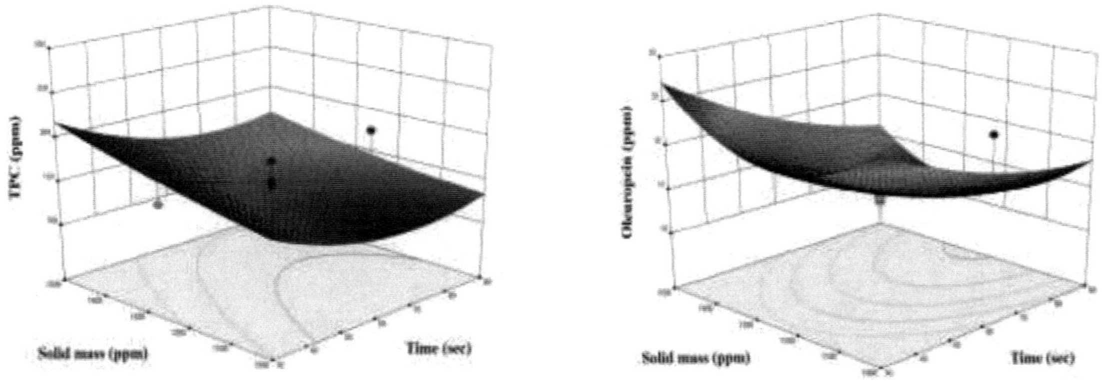

a b

Figure 3. Response surface plots for the TPC (a) and oleuropein (b) of enriched olive oil as a function of solid mass to mixing time (mixing speed=7000 rpm)

Time had a negative effect on the enrichment of olive oil for both variables (Figures.3a and b). This result is most probably due to the degradation of the components caused by overexposure to the heating of the long mixing time (Chan *et al.*, n.d.).

As seen in Figures.4a and b, enrichments by TPC and oleuropein were both increased with mixing speed of homogenizer. Diffusion of the target compounds into the oil is expected to be increased with stirring (Ho and Hsieh, 2001). Moreover, agitation also contributes to the favourable extraction with its additional energy and promoting the homogeneity (Georgiev *et al.*, 2009; Bilgin and Şahin, 2013).

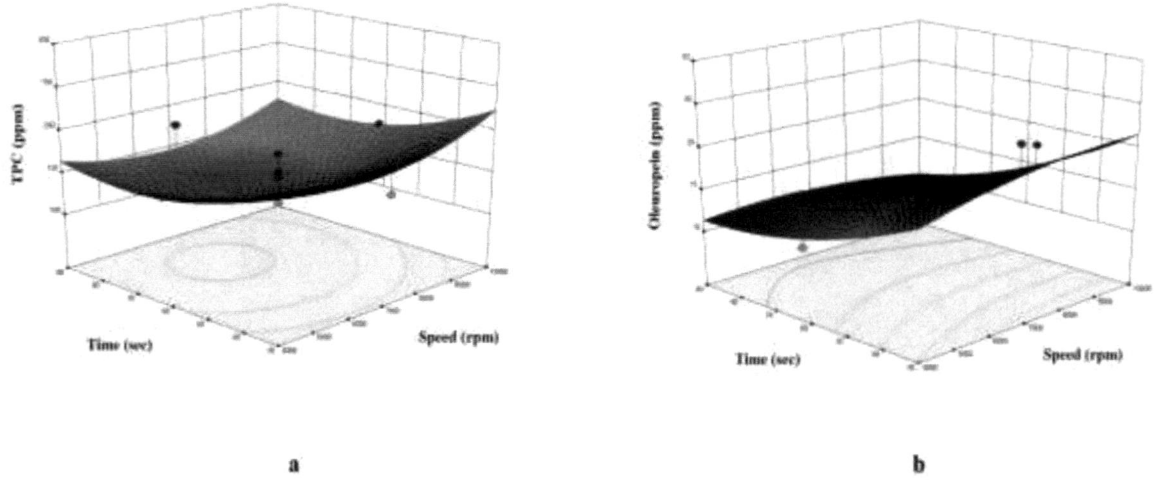

Figure 4. Response surface plots for the TPC (a) and oleuropein (b) of enriched olive oil as a function of mixing time to speed (solid mass=1250 ppm)

Efficiency of the proposed method at the optimum conditions

The results of the present study suggests that 10000 rpm of mixing speed, 30 sec of time and 1500 ppm of olive leaf extract should be employed as optimal operating conditions in order to enrich the greatest TPC (247.168 ppm) and oleuropein (26.471 ppm) in olive oil. Table 3 summarizes the results of enrichment process applied at regarding quality parameters such as TPC, oleuropein, α-tocopherol, antioxidant activity (AA), carotene (TCC) and chlorophyll contents, PV and IT.

Table 3. Changes of olive oil stability parameters before and after their enrichment with the olive leaf extract at the optimum conditions*

	TPC	**Oleuropein**	**α–Tocopherol**	**AA**	**TCC**	**Chlorophyll**	**PV**	**IT**
	(ppm)	**(ppm)**	**(ppm)**	**(%)**	**(mg-β carotene/kg-oil)**	**(mg-Pheophytin/kg-oil)**	**(meq-O_2/kg-oil)**	**(h)**
Pure olive oil	132± 3.19	6.46±0.19	114.38±0.54	13.45±1.15	2.78±0.07	n.d**	9.09±0.40	3.41±0.17
Enriched olive oil	240± 4.29	22.08±1.30	127.83±0.75	34.09±1.05	7.89±0.52	28.17±0.22	7.07±0.54	4.98±0.20

*Data are expressed as the mean (n=3) ± S.D.
**n.d: not detected

Conclusions

A simple, fast and inexpensive method, homogenizer assisted extraction has been applied in order to enrich the virgin olive oil with olive leaf extract. Olive leaf extract has proved that its phenolic content and especially major compound, oleuropein significantly contribute to the antioxidant capacity of the extract. Addition of this natural antioxidant to olive oil has enhanced the stability of the oil as well as increasing its added-value with high quality parameters, substantially. Addition of 0.15 % natural antioxidant gave rise to increase of the stability of the oil (≈46%).

Acknowledgement

The authors wish to thank to The Scientific & Technological research Council of Turkey (TÜBİTAK) for the support of this Project. Project number is 114M728.

References

Ballard, T.S., Mallikarjunan, P., Zhou, K. & O'Keefe, S. (2010). Microwave-assisted extraction of phenolic antioxidant compounds from peanut skins. *Food Chemistry*, **120**, 1185–1192.

Bilgin, M. & Şahin, S. (2013). Effects of geographical origin and extraction methods on total phenolic yield of olive tree (Olea europaea) leaves. *Journal of the Taiwan Institute of Chemical Engineers*, **44**, 8–12.

Chan, E.W.C., Lim, Y.Y., Wong, S.K., Lim, K.K., Tan, S.P., Lianto, F.S. & Yong, M.Y. (n.d.). Effects of different drying methods on the antioxidant properties of leaves and tea of ginger species. *Food Chemistry*, **113**, 166–172.

Franco, D., Pateiro, M., Amado, I.R., López Pedrouso, M., Zapata, C., Vázquez, J.A. & Lorenzo, J.M. (2016). Antioxidant ability of potato (*Solanum tuberosum*) peel extracts to inhibit soybean oil oxidation. *European Journal of Lipid Science and Technology*, n/a–n/a.

Georgiev, M.I., Weber, J. & Maciuk, A. (2009). Bioprocessing of plant cell cultures for mass production of targeted compounds. *Applied Microbiology and Biotechnology*, **83**, 809–823.

Ho, W.-H. & Hsieh, S.-J. (2001). Solid phase microextraction associated with microwave assisted extraction of organochlorine pesticides in medicinal plants. *Analytica Chimica Acta*, **428**, 111–120.

Mei, W., Ismail, A., Esa, N., Akowuah, G., Wai, H. & Seng, Y. (2014). The Effectiveness of Rambutan (Nephelium lappaceum L.) Extract in Stabilization of Sunflower Oil under Accelerated Conditions. *Antioxidants*, **3**, 371–386.

Şahin, S. (2015). A novel technology for extraction of phenolic antioxidants from mandarin (Citrus deliciosa Tenore) leaves: Solvent-free microwave extraction. *Korean Journal of Chemical Engineering*, **32**, 950–957.

Improving the Oxidative Stability of Corn Oil By Natural Antioxidants through Response Surface Methodology

*Selin Şahin, Bedia Güvenilir and Mehmet Bilgin**

Istanbul University, Engineering Faculty, Department of Chemical Engineering, 34320 Avcilar, Istanbul, Turkey

*Corresponding Author E-mail: mbilgin@istanbul.edu.tr
Phone: +90-212-4737070 (17662), Fax: +90-212-4737180

Introduction

Commercially available corn oil was enriched in polyphenols by adding olive leaf and lemon balm extracts, respectively. After extracting the dried and ground leaves with the assistance of homogenizer, the dried extract was partially dissolved into the corn oil to increase the quality and shell-life of the oil enriched by the substances in the plants by means of solid- liquid extraction method. A face central composite design (FCCD) through response surface methodology (RSM) was used to investigate the effects of enrichment conditions (extract content, time and mixing speed) on the responses, total phenolic content (TPC) and antioxidant activity (AA) of the enriched oil. Oxidative stability of the enriched oil was evaluated by means of Rancimat method. Peroxide value (PV), total carotenoid content (TCC) and chlorophyll quantity were also measured, respectively. The quality parameters of the enriched corn oil were compared to those of the pure oil. The last but not the least, the best natural antioxidant source to compensate for the oxidative stability of the corn oil was also determined.

Results and Discussions

The enrichment process was carried out by means of solid-liquid extraction, in which solid olive leaf and lemon balm extracts was partially dissolved into the corn oil, respectively. In order to optimize the process conditions, possible interactions among the operating parameters such as mixing time, speed of the homogenizer and the quantity of the extract should be considered depending on the preliminary experiments. A face central composite design (FCCD) was performed with three variables to explore the effect of variables on the response (Table 1). 20 experimental runs were chosen randomly by Design Expert 9.0.6 as seen in Table 2.

Table 1. Values of the independent variables and their coded forms with their symbols employed in RSM for optimization of enrichment.

Independent Variables	**Units**	**Symbol of the variables**	**Coded Levels**		
			-1	**0**	**1**
Mixing speed	rpm	X_1	4000	7000	10000
Time	sec	X_2	30	60	90
Solid mass	ppm	X_3	1000	1250	1500

Table 2. FCCD of the independent variables (X_1, X_2, X_3) and experimental results for the TPC of the enriched corn oil by olive leaf and lemon balm extracts, respectively*

Run number	Independent variables			TPC (ppm)	
	X_1 (rpm)	X_2 (sec)	X_3 (ppm)	Olive leaf	Lemon balm
1	10000	30	1500	158.01±2.05	44.05±0.05
2	4000	30	1500	101.12±2.10	33.33±0.76
3	7000	30	1250	174.44±3.36	35.00±0.63
4	4000	30	1000	107.33±3.45	21.12±0.45
5	7000	60	1500	170.02±2.05	25.41±0.56
6	7000	60	1000	164.41±1.09	27.06±0.01
7	10000	90	1500	120.06±1.11	31.11±0.02
8	7000	90	1250	199.00±3.33	30.00±0.12
9	7000	60	1250	151.43±2.89	31.32±0.01
10	10000	30	1000	131.00±0.89	31.12±0.44
11	4000	60	1250	137.33±0.55	22.18±0.55
12	7000	60	1250	160.19±4.44	31.06±0.66
13	10000	60	1250	157.26±5.00	33.31±0.70
14	7000	60	1250	163.33±4.89	30.05±0.12
15	7000	60	1250	155.34±4.00	30.42±0.86
16	4000	90	1500	128.88±2.15	37.02±0.90
17	10000	90	1000	97.06±0.09	21.00±0.02
18	7000	60	1250	158.22±0.65	31.33±0.78
19	4000	90	1000	100.06±0.76	25.42±0.65
20	7000	60	1250	152.22±1.56	32.06±0.88

*Data are expressed as the mean (n=9) ± S.D.

Optimization of enrichment conditions for corn oil

In order to explore the potential relationships between the the variables, three dimensional (3D) response surface plots (Figures 1 to 3) are constrcuted according to the quadratic equations 1 and 2.

$$Y_1 = 166.65 + 9.00X_1 - 2.70X_2 + 7.80X_3 - 11.50X_1X_2 + 3.50X_1X_3 + 3.75X_2X_3 - 34.86X_1^2 + 4.64X_2^2 - 14.86X_3^2 \quad (1)$$

$$Y_2 = 29.90 + 2.13X_1 - 2.13X_2 + 4.53X_3 - 3.83X_1X_2 + \text{-}1.325\text{E-}015X_1X_3 - 0.33X_2X_3 - 1.24X_1^2 + 4.09X_2^2 - 2.58X_3^2 \quad (2)$$

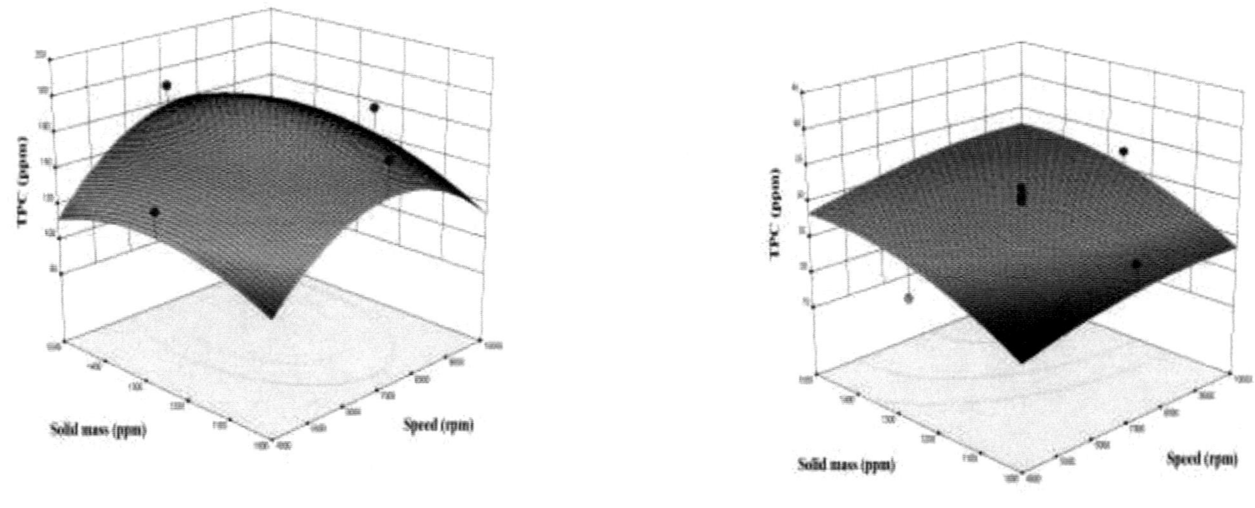

a b

Fig 1. Response surface plots for the TPC of enriched corn oil with olive leaf (a) and lemon balm (b) as a function of solid mass to mixing speed (time=60 sec)

As seen in Figure 1, enrichments by olive leaf and lemon balm extracts were both increased with mixing speed of homogenizer. Diffusion of the target compounds into the oil is expected to be increased with stirring (Ho and Hsieh, 2001). Moreover, agitation also contributes to the favourable extraction with its additional energy and promoting the homogeneity (Georgiev *et al.*, 2009; Bilgin and Şahin, 2013). However, the yield had a tendency to increase until a point, then it started to decrease as for olive leaf extract (Figure 1a). The same trend was also observed by Achat et al., while improving the nutritional value of olive oil by enriching it in phenolic compounds from olive leaves by ultrasonic maceration (Achat *et al.*, 2012). This result is most probably due to the degradation of the components caused by overexposure to the heating of the high mixing speed.

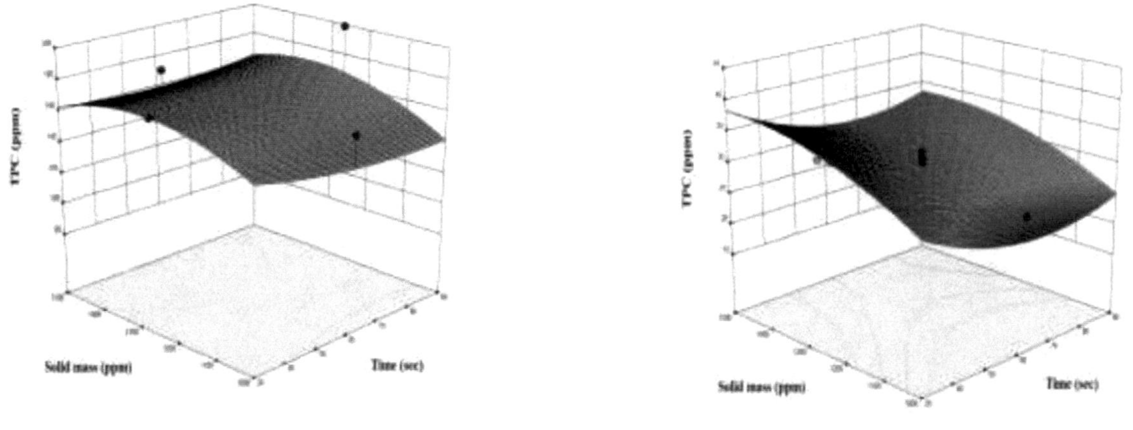

a b

Fig 2. Response surface plots for the TPC of enriched corn oil with olive leaf (a) and lemon balm (b) as a function of solid mass to mixing time (mixing speed=7000 rpm)

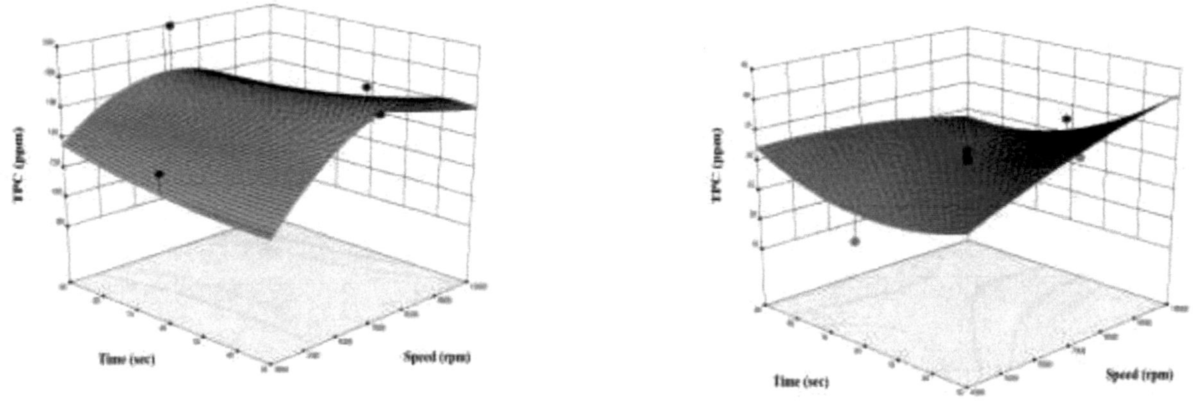

a b

Fig 3. Response surface plots for the TPC of enriched corn oil with olive leaf (a) and lemon balm (b) as a function of mixing time to speed (solid mass=1250 ppm)

As seen in figures 2 and 3, time affected both of the system slightly since it required less time due to the application of an assisting technique such as stirring (Bilgin *et al.*, 2013). Expectedly, both of the yields increased by extract quantity. This is in agreement with the results of Mei et al., who suggested higher concentrations of the rambutan extract into sunflower oil in order to improve the stability of the oil (Mei *et al.*, 2014). Franco et al. also found out that increasing potato peel extract concentration would have improved the control of oil oxidation (Franco *et al.*, 2016).

Efficiency of the proposed method at the optimum conditions

The results of the present study suggests that 7907 rpm of mixing speed, 30 sec of time and 1293 ppm of olive leaf extract should be employed as optimal operating conditions in order to enrich the greatest TPC (177.432 ppm) in corn oil. For the highest TPC (43.144 ppm) of lemon balm extract, 10000 rpm of mixing speed, 30 sec of time and 1486 ppm of the sample should be employed. Table 3 summarizes the results of enrichment process applied at regarding quality parameters such as phenolic, carotene and chlorophyll contents, antioxidant activity, peroxide value and shell life of the oil.

Table 3. Changes of corn oil stability parameters before and after their enrichment with the olive leaf and balm extract at the optimum conditions

Additive	TPC (ppm)	AA (%)	TCC (mg-β carotene/kg-oil)	Chlorophyll (mg-feofitin/kg-oil)	PV (meq-O_2/kg-oil)	IT (h)
Pure	18.00±0.09	1.72±0.00	n.d.*	n.d.	10.30±0.25	1.04±0.01
Olive leaf extract	172.57±0.53	23.85±0.13	3.64±0.01	3.32±0.00	4.70±0.13	1.22±0.05
Lemon balm extract	47.33±0.03	10±0.08	1.56±0.01	n.d.	9.87±0.10	1.23±0.03

*n.d: not detected

Conclusions

Commercially available corn oil was enriched in polyphenols by adding olive leaf and lemon balm extracts, respectively. After extracting the dried and ground plants with the assistance of homogenizer, the dried extracts were partially dissolved into the corn oil to increase the quality and shell-life of the oil enriched by the substances in the plants by means of solid- liquid extraction method. Total phenolic content increased 9.5 and 2.5 times over the pure corn oil. The antioxidant activities of enriched oil by olive leaf and lemon balm extracts have been found almost 20 and 10 times higher than that of untreated oil, respectively. Olive leaf extract enhanced the quality better than lemon balm extract with respect to carotenoid and chlorophyll contents and peroxide value. Addition of 0.12-0.15 % natural antioxidants gave rise to increase of the stability of the oil (≈18%).

Acknowledgement

The authors thank the Research Fund of Istanbul University for financial support for this research project (Project No: 50084).

References

Achat, S., Tomao, V., Madani, K., Chibane, M., Elmaataoui, M., Dangles, O. & Chemat, F. (2012). Direct enrichment of olive oil in oleuropein by ultrasound-assisted maceration at laboratory and pilot plant scale. *Ultrasonics Sonochemistry*, **19**, 777–786.

Bilgin, M., Sahin, S., Dramur, M.U. & Sevgili, L.M. (2013). Obtaining scarlet sage (Salvia coccinea) extract through homogenizer- and ultrasound-assisted extraction methods. *Chemical Engineering Communications*, **200**, 1179–1209.

Bilgin, M. & Şahin, S. (2013). Effects of geographical origin and extraction methods on total phenolic yield of olive tree (Olea europaea) leaves. *Journal of the Taiwan Institute of Chemical Engineers*, **44**, 8–12.

Franco, D., Pateiro, M., Amado, I.R., López Pedrouso, M., Zapata, C., Vázquez, J.A. & Lorenzo, J.M. (2016). Antioxidant ability of potato (*Solanum tuberosum*) peel extracts to inhibit soybean oil oxidation. *European Journal of Lipid Science and Technology*, n/a–n/a.

Georgiev, M.I., Weber, J. & Maciuk, A. (2009). Bioprocessing of plant cell cultures for mass production of targeted compounds. *Applied Microbiology and Biotechnology*, **83**, 809–823.

Ho, W.-H. & Hsieh, S.-J. (2001). Solid phase microextraction associated with microwave assisted extraction of organochlorine pesticides in medicinal plants. *Analytica Chimica Acta*, **428**, 111–120.

Mei, W., Ismail, A., Esa, N., Akowuah, G., Wai, H. & Seng, Y. (2014). The Effectiveness of Rambutan (Nephelium lappaceum L.) Extract in Stabilization of Sunflower Oil under Accelerated Conditions. *Antioxidants*, **3**, 371–386.

458834 Pre-Formulation Tool to Rapidly Predict the Stability of Monoclonal Antibodies

Abstract:

Pre-formulation tool to rapidly predict the stability of monoclonal antibodies

S. Hedberg*, J.Y.Y Heng*, D.R. Williams* and J. Liddell**

ABSTRACT

Protein-protein molecular interactions are known to be involved in protein solution aggregation behaviour and are a common issue for the manufacturing of therapeutic proteins such as mAbs. Much effort has been employed to gain a better understanding of aggregation. However, the mechanisms leading to protein aggregation are still not fully understood. The osmotic second virial coefficient (B_{22}) is a fundamental physiochemical property that describes protein-protein interactions solution, which can be a useful tool for predicting the aggregation propensity of proteins.

One way of predicting aggregation propensity is self-interaction chromatography (SIC), which has demonstrated recent promise as a tool for the better understanding of phase behaviour of proteins. Another technique, cross-interaction chromatography (CIC), has shown to be an even more high-throughput technique than its predecessor with possibly the same capabilities.

The first part of this paper is a scale-down SIC study of therapeutic mAbs from laboratory scale macro-columns to micro-scale columns, which enables the use of micrograms or milligrams of mAb for B_{22} determinations in order to obtain a complete formulation screen.

The second part of this work presents an extensive formulation study of a mAbs system, varying pH and salt, as well as the presence of different stabilisers and different external factors known to induce aggregation. The B_{22} and B_{23} values determined from the formulation study are then correlated with aggregation kinetics obtained from size-exclusion chromatography and dynamic light scattering. It was shown that over all test conditions good correlations could especially be found between B_{22} and aggregation rate.

The third part of the study includes investigation of conformational changes of the monoclonal antibodies. The B_{22} data previously obtained were directly correlated with the conformational changes observed in the mAb to create a predictive tool that can detect any early conformational changes.

In summary this work intends to gain a better understanding of the fundamental mechanisms of aggregation and to develop a more predictive capability by studying the chromatogram output resulting from the self- and cross interactions between the same or different proteins.

REFERENCES

Hedberg, S. H. M., Heng, J. Y. Y., Williams, D. R. & Liddell, J. M. (2016). Micro Scale Self-Interaction Chromatography of Proteins: A Mab Case-Study. *Journal of Chromatography A,* 1434, 57-63.

Quigley, A. & Williams, D. R. (2015). The Second Virial Coefficient as a Predictor of Protein Aggregation Propensity: A Self-Interaction Chromatography Study. *European Journal of Pharmaceutics and Biopharmaceutics,* 96, 282-290.

460658 Clinical Consequences of Heart Treatment Following the TACT Study

Abstract:

The TACT Study published in November 2013 by NIH shows significant impact on heart disease using the chelation method. In this case, using the clinical treatment and the annual exam based on Cleveland Heart Clinic blood tests on the basis for the MDVIP program sponsored by the Procter & Gamble Co. The program has been carried out over a 5 year period and the results are reported here for a single patient. The test shows the impact on the heart, the blood sugar, the myeloperoxidase,the LDL,the HDL, the TSH, and the blood pressure.

461228 Optimisation of Spray Drying Process Parameters for Production of Skim Milk Powder Using a Quality By Design Approach

Abstract:

Spray drying is a key unit operation in the manufacturing of dairy powders. During the spray drying process, a liquid feedstock is transformed into a dried particulate form through atomising the feed into the hot gas within a drying chamber. An inefficient spray drying operation of dairy products may lead to non-compliant product quality (e.g., stickiness, wall deposition and caking), nutritional losses due to Maillard reaction and substantial economic losses. Therefore, improvement of product quality and minimising the potential problems during the spray drying process are the major concerns of dairy powder manufacturers. Optimisation of spray drying processes for dairy powders can involve a high level of trial and error. In addition, optimisation of the spray drying process through empirical tests are usually expensive, time-consuming and sometimes unreliable. Therefore, there has been an appeal for a shift from empirical efforts toward rational design of spray-dried dairy powders based on a better understanding of the interplays between the process parameters of the spray drying process and dairy powders quality.

Quality by Design (QbD) represents a systematic approach to building quality into process development and manufacturing practices from the outset. QbD is a new paradigm in the food industry, inspired by the Process Analytical Technology (PAT) concept introduced originally for the pharmaceutical industry in 2004 by the Food and Drug Administration (FDA, 2004; Munir et al., 2015). The hypothesis behind this paradigm shift is that the quality of the food products can and should be incorporated into process design and not by post-production quality measurements. Process optimization can be implemented by the use of a combination of QbD tools such as Design of Experiments (DoE) and Response Surface Methodology (RSM). DoE is a structured and systematic approach for determination of critical quality attributes that could affect the process or product in order to ensure their reliability and robustness. Likewise, RSM is applied to describe correlations between independent variables (input) and response variables (output), enabling improved understanding of the influence of different excipients and their interactions, as well as improved prediction of optimized process conditions leading to the desired profile. The knowledge of the interplay between process parameters and quality attributes will enable rational process design to achieve the desired outcome.

The physical properties of dairy powders are governed by the size, distribution and shape of the individual particles making up the powder (Dodds, 2013). The shape, structure, and properties of the particles are heavily influenced by the spray drying operational conditions. Evaluating and tuning of drying parameters based on a more reliable accurate process maps is helpful in dealing with several critical issues in powder handling and product quality due to the undesired changes in particle properties. This study is aimed at optimising the spray drying process parameters for production of skim milk powder using the DoE and RSM as QbD tools. The influence of the spray drying process parameters including inlet air temperature, feed flow rate and the total solid content of feed on outlet air temperature, yield, particle size, moisture content and bulk density were investigated. Spray drying experiments are performed using a laboratory-scale spray dryer fitted with an ultrasonic nozzle (Büchi 290, Switzerland). The spray-dried skim milk powders are characterized by Laser Diffraction Particle Size Analyzer, Scanning Electron Microscopy (SEM) and Thermogravimetric Analysis (TGA). In order to determine the optimum level of spray dryer system parameters leading to the best response, graphical and numerical optimisation procedures are carried out. The experimental design matrix, data analysis, and optimisation procedure are performed using Minitab Statistical Software (Version 17). This study shows the applicability of QbD approach for the enhanced production of spray-dried dairy powders. Use of DoE and RSM results in reduced process variability, higher percentage yields, less treatment time with minimum costs.

References

Dodds, J., (2013). 13 - Techniques to analyse particle size of food powders A2 - Bhandari, Bhesh, in: Bansal, N., Zhang, M., Schuck, P. (Eds.), *Handbook of Food Powders*. Woodhead Publishing, pp. 309-338.

FDA, United States Food and Drug Administration, (2004). Guidance for industry: PAT - A framework for innovative pharmaceutical development, manufacturing and quality assurance. Rockville, USA: U.S. Department of Health and Human Services.

Munir, M.T., Yu, W., Young, B.R., Wilson, D.I., I2C, (2015). The current status of process analytical technologies in the dairy industry. Trends in Food Science & Technology 43(2), 205-218.

CHARACTERIZATION OF SWEET POTATO (*Ipomoea batatas* L.) STARCH FROM TWO CLONES AND EVALUATION OF ITS PROPERTIES FOR INDUSTRIAL USES

Ibeth Cepeda; Erika Diaz; Paula García; Maria Tavera; Carmen Pérez Cervera; Arlet P. Franco
DANM/ Desarrollo y Aplicación de Nuevos Materiales, Escuela de Ingenierías y Arquitectura, Universidad Pontificia Bolivariana, Montería, Colombia

Abstract

Sweet potato starch composition and physicochemical properties depend on cultivar variety. Starches, obtained from sweet potato varieties *Tainung* and *199062-1* was characterized by physicochemical analysis and pasting properties were also determined. *Tainung* and *199062-1* presented 23.21% of amylose whereas *199062-1* clone 22.92%. Gelatinization temperature was 78.30 °C for *Tainung* and 78.50 °C for *199061-1.* Starches presented peak viscosity of 1994 cps for *199062-1* and 2206 cps for *Tainung* and setback ratio of 1.30 for *199062-1* clone and 1.33 for *Tainung.* Sweet potato starches from *Tainung* and *199062-1* offers several advantages for industrial uses due to remarkable pasting behavior. Moreover, are considered opaque, suggesting food and pharmaceutical uses as thickening and stabilizing agents.

Introduction

Sweet potato (*Ipomoea batatas* L.) is a dicotyledonous plant belonging to the family *Convolvulaceae*, with approximately 50 genera and over 1000 species (Woolfe, 1992). Artificial selection of sweet potatoes varieties, as well as natural hybrids and mutations resulted in these large cultivars varieties, which differs in composition, physicochemical properties as color, dry matter, root texture, starch content and its parameters as gelatinization temperature, viscosity, water retention and granule size (Zhang; C.G. and Oates, 1999).

Starch obtained from root and tubers are considered more uniformly associated than cereal starches, as a consequence of one stage gelatinization process (Howling, 1980). Starch contributes to the textural properties of foods products and it is widely used for food and industrial applications as thickener, colloidal stabilizer, and gelling, bulking and water retention agent (Singh et al., 2003). Moreover, limitations as low resistance to shear strength and thermal food processing conditions and high tendency towards retro-gradation limits its usage in some industrial applications (Singh et al., 2007), since it determines quality, acceptability, nutritional value and shelf life of processed foods products (Wang and Copeland, 2013).

Retro-gradation process describes the changes that occurs on cooling and storage of starch products, where amylose and amylopectin molecules reassociate and form a different and ordered structure (Hermansson and Svegmark, 1996; Liu, 2005; Wang et al., 2015). In an initial phase, two chains may associate and, under favorable conditions, a crystalline orders appears and physical phase separation occurs. Retro-gradation is important for starch industrial uses, in some applications, it can be desired but it also promotes instability in starch pastes (Liu, 2005).

Amylose/amylopectin ratio characterizes gelatinization process, which occurs when starch granules are heated in presence of water, and it is defined as the collapse of molecular order within the starch granule, producing irreversible changes (Liu, 2005). In this process the starch granules swell and form gel particles. The swollen granules are composed by amylopectin, whereas amylose forms the continuous gel phase outside the granules (Hermansson and Svegmark, 1996).

Those parameters depend on rheological or pasting behavior of starch, which is determined by amylose content, granule parameters as size distribution, volume fraction, shape, interactions between granules and continuous phase viscosity, which depend on botanical source (Bertolini, 2010; Howling, 1980). Native starches generally present amylose content between 20 and 30 % (Hermansson and Svegmark, 1996; Singh et al., 2003; Swinkels, 1985; Wang et al., 2015)

Due to increases on demand of sweet potato products, since those are widely known for its functional and nutraceutical properties owing to antioxidant activities, as a consequence of outstanding content of anthocyanins, carotenoids, flavonoids and other phenolic compounds, previously reported (Teow et al., 2007) and increase of diseases associated to starch from cereal products ingestion, is mandatory to determine the sweet potato starch composition, physicochemical properties and thus its possible industrial uses.

In order to characterize sweet potato starch from two varieties, known as *Tainung* and *199062-1* clones, physicochemical properties of its varieties were determined and consequently suggested for possible industrial uses.

Material and Methods

Samples

Twenty kg of each variety of sweet potatoes were obtained from the indigenous community ASOBARI (Retiro de los Indios Association of Sweet Potato Producers) located in the Retiro de los Indios (Cereté-Colombia). Raw materials were washed with current water, manually peeled with commercial knives, sliced with a vegetables processor CA301 (SAMMIC, España) and immersed during 30 minutes into a sodium bisulfite solution 1500 ppm (1:3 w/v).

The obtained samples were liquefied using an industrial blender LC15 (JAVAR®, Bogotá D.C., Colombia) and immersed into a sodium bisulfite solution 1500 ppm (1:1 v/v), filtered and dried in an oven Thermolab (DIES, Itagüí, Colombia) at 50 °C for 24 hours. Sweet potato starch powder was attained pulverizing the granular material through a 100-mesh sieve. Samples were packed and sealed in polyethylene bags (Figure 1).

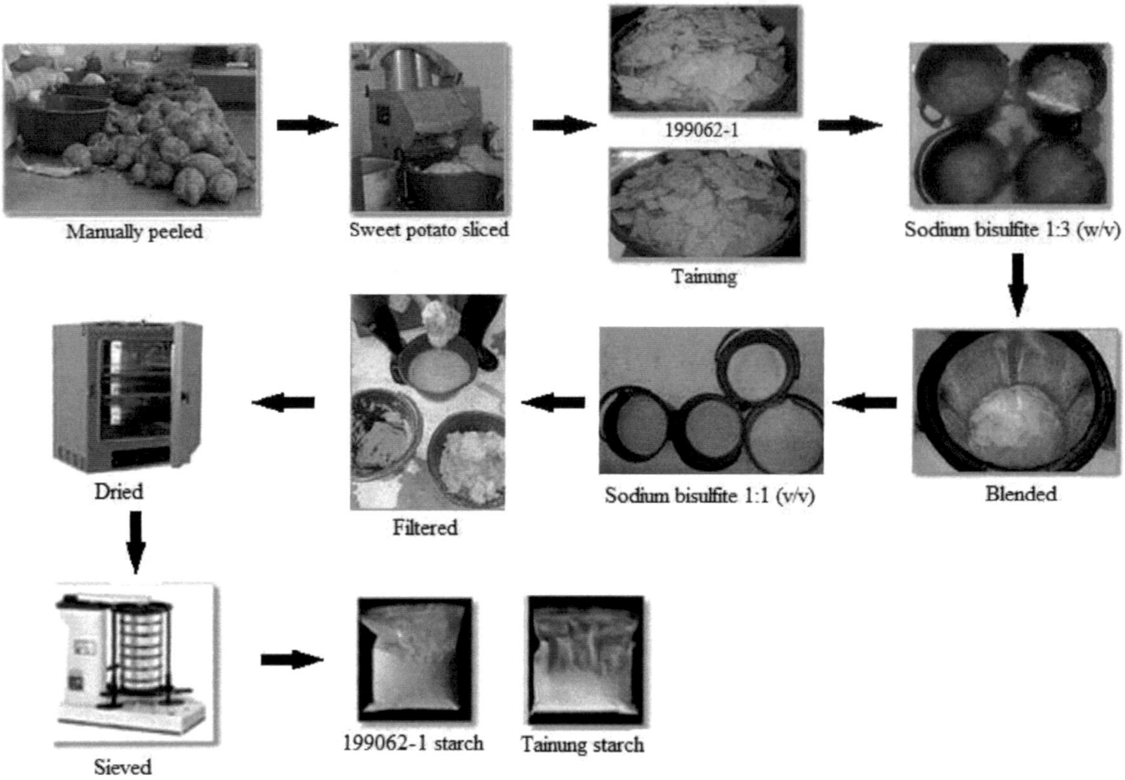

Figure 1. Sweet potato starch extraction process

Starch analysis

Sweet potato starch was analyzed for moisture content according AOAC (2005) by an oven Thermolab (DIES, Itagüí, Colombia). PADES (2015) guide was used to determine amylose and amylopectin content by a spectrophotometer Spectroquant® Pharo 300 (Merck KGaA, Darmstadt, Alemania).

Starch color including L^*(lightness), a^*(redness-greenness) and b^*(yellowness-blueness) were determined by using a portable colorimeter CR-10 PLUS Color Reader (Konica Minolta, Japan).

Granule size was calculated according PADES (2015) by using a sieve equipment CEL 313 (C.I M&M Instrumentos Técnicos, Medellín, Colombia). Gelatinization temperature was determined according Grace (1977), using a magnetic hot plate (Fisher Scientific Isotemp®, China).

Pasting properties were determined using a rheometer (Anton Paar, Austria). Peak viscosity, gel breakdown, starch setback and cooking parameters were obtained from amylograph curves. Starch solubility, swelling volume and water absorption were calculated according Aristizabal and Sanchez (2007). Paste clarity was evaluated as transmittance according Craig et al. (1989) at 650 nm by a spectrophotometer GENESYS™ 10S UV-Vis (Thermo SCIENTIFIC, USA).

Statistical analysis

All analysis were carried out in duplicates by using the software JMP from SAS (Statistical Discovery™). Results were subjected to analysis of variance (ANOVA) and differences were considered significant at $p < 0.05$.

Results and discussion

Moisture content was 4.09% for *Tainung* and 3.17% for *199062-1* clone ($p \leq 0.05$). These values are lower than the previously reported by García and Raleigh (1998), Hernandez-Medina et al. (2008) and Aprianita et al. (2009). Starch granule size of *199062-1* was larger than *Tainung*, with 30.15 μm and 16.80 μm respectively. These values are in agreement with previously reported by Zhang and Oates (1999) for six sweet potato varieties.

In case of amylose content, *Tainung* g presented 23.21% whereas *199062-1* clone 22.92% ($p \leq 0.05$). These values are according to results previously reported for sweet potato by Collado et al. (1999), Aprianita et al. (2009), García and Raleigh (1998), Hernandez-Medina et al. (2008), Takeda et al. (1986) and Zobel (1988), with values between 12.90% and 29.70%. Amylose content values corresponding with founded for corn, rice, tapioca, wheat and potato (Jane et al., 1999; Kim et al., 1995; Zobel, 1988); therefore, is an important alternative to be explored in order to develop bakery and snack based on sweet potato.

Color parameters as lightness *L* was similar for both clones ($p > 0.05$), whereas for parameters *a* and *b*, *Tainung* clone presented higher values of redness-greenness and yellowness-blueness than *199062-1* ($p \leq 0.05$). Sweet potato starch color depends on raw material characteristic as can be seen in Figure 2.

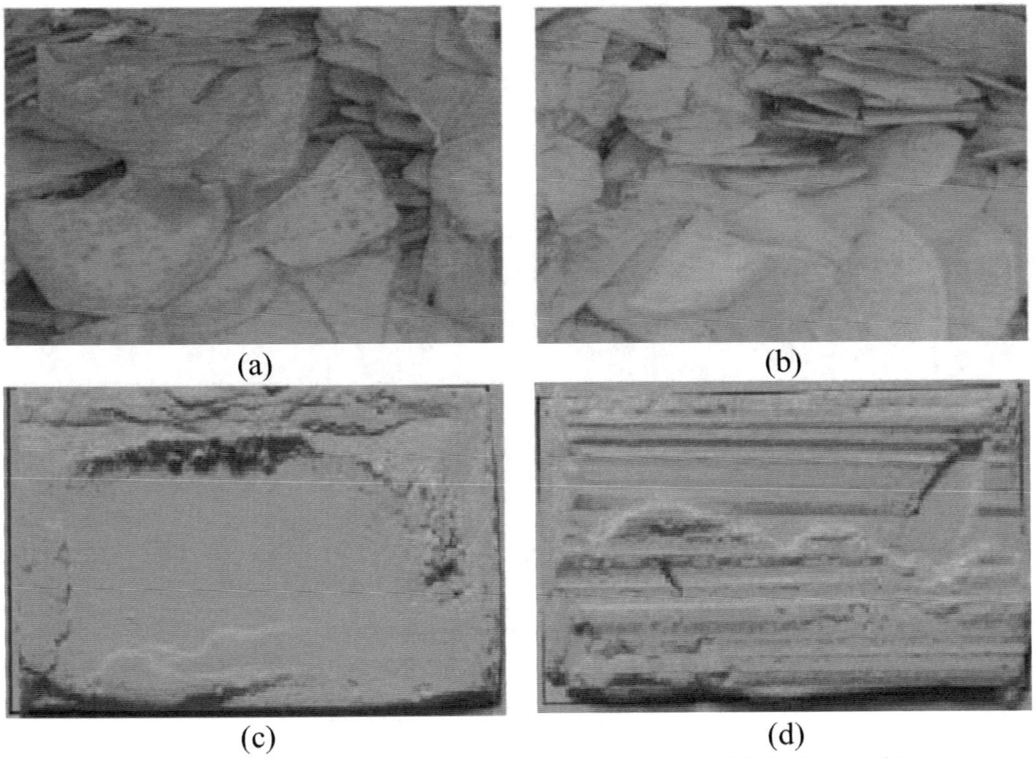

Figure 2. Sweet potato pulp color for *Tainung* (a) and *199062-1* clones (b), sweet potato starch color for *Tainung* (c) and *199062-1* (d) clones.

Gelatinization temperature for both clones were similar, with 78.30 °C for *Tainung* and 78.50 °C for *199061-1* ($p > 0.05$). These values are similar to the previously values reported for sweet potato by Zhang and Oates (1999) and it has shown a higher value than the one reported by Aprianita et al. (2009), Collado et al. (1999), García and Raleigh (1998) and Zobel (1988). Found values for gelatinization temperature for sweet potato are greater than presented for another starch sources as wheat (Jane et al., 1999), tapioca (Hernandez-Medina et al., 2008; Jane et al., 1999), potato from 42 genotypes (Kim et al., 1995) and rice (Jane et al., 1999).

Gelatinization temperature is an important parameter, since it affects pasting properties of sweet potato starches. Zhang and Oates (1999) accomplished lower gelatinization temperature for samples of sweet potato starches presenting less amylopectin content, therefore they reported that gelatinization temperature is governed by amylose/amylopectin ratio.

In case of pasting properties behavior (Figure 3), initial viscosity for both clones was similar, with values of 9.90 and 10.24 cps for *199062-1* clone and *Tainung* respectively. Peak viscosity was 1994 cps for *199062-1* and 2206 cps for *Tainung*. These values are lower than reported by Collado and Corke (1997) for sweet potato starch from 14 genotypes.

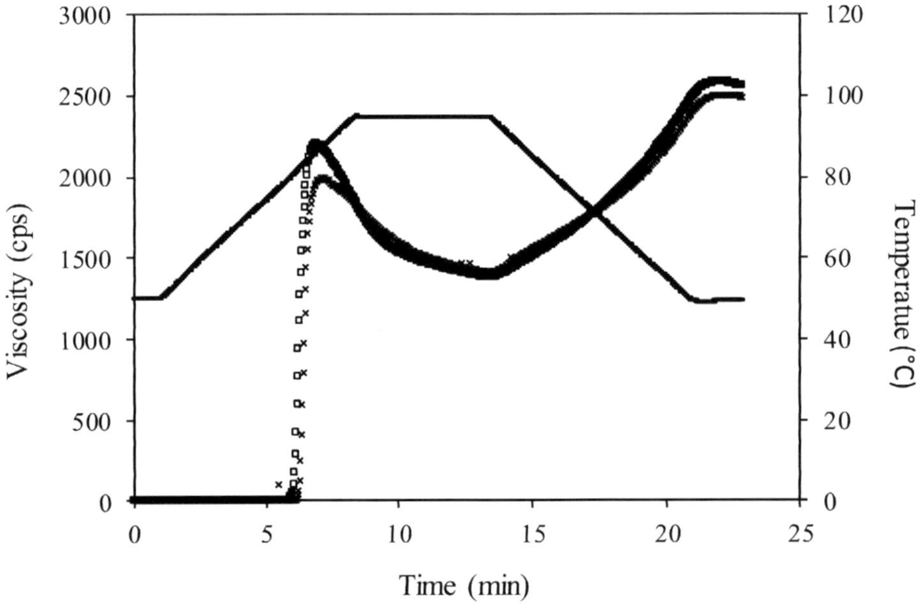

Figure 3. Pasting properties of sweet potato starch for *Tainung* (□) and *199062-1* varieties (×), under temperature changes (—)

Gel breakdown and starch setback were 586 and 495 cps for *199062-1* clone and 830 and 352 cps for *Tainung* ($p \leq 0.05$). Setback ratio was 1.30 for *199062-1* clone and 1.33 for *Tainung*. These values were lower than previously reported by Collado and Corke (1997), which indicates that starches from these sweet potato varieties could be an interesting alternative for food processing industry. In that, starch presenting lower setback values has an inferior tendency for retrogradation (Zaidul et al., 2007).

Cooking parameters values for *199062-1* and *Tainung* were 1.2 min for and 1.0 respectively ($p > 0.05$), which are in agreement with previously reported values by Collado and Corke (1997). Samples presented starch solubility of 1.08% for *Tainung* and 1.30% for *199062-1* ($p > 0.05$), swelling volume for *199062-1* was noticeable higher than *Tainung*, with 14.53 g/g insoluble solids and 5.34 g/g insoluble solids respectively ($p \leq 0.05$). These values are lower than previously reported by Collado and Corke (1997).

Water absorption for *199062-1* was higher than *Tainung*, with 1.88 g of water/g of sample and 1.47 g of water/g of sample respectively ($p \leq 0.05$). For paste clarity, *Tainung* and *199062-1* sweet potato starches showed similar transmittance values (%T), 15.63% and 14.56% respectively ($p > 0.05$). These starches are considered as opaque, since transmittance was lower than 40% (Aristizabal and Sanchez, 2007). These results were inferior to reported by Hernandez et al. (2008) with 24.64%T and Achille et al. (2007) with 20.7%T.

Conclusions

Sweet potato starches from *Tainung* and *199062-1* clones present a similar starch content and physicochemical properties when it was compared with starches from another botanical sources. Gelatinization temperature above 78 °C and pasting properties offer advantages for industrial uses. According to the reduced values of paste clarity and remarkable capacity of forming strong and elastic gels, suggest that it can be employed as thickening and stabilizing agents in sauce, soup, pastry, ice cream and jelly industries.

Acknowledgements

The authors would like to acknowledge financial support from CIDI (Research center for development and innovation) of Universidad Pontificia Bolivariana – Montería.

References

1. Achille, T.F.; Georges, A.N.G.; Alphonse, K. (2007), "Contribution to light transmittance modelling in starch media", *African Journal of biotechnology,* 6, pp. 569-575.
2. AOAC. (2005), *Official methods of analysis*, 18th ed. Washington DC. Association of Official Analytical Chemists.
3. Aprianita, A.; Purwandari, U.; Watson, B.; Vasiljevic, T (2009), "Physico-chemical properties of flours and starches from selected commercial tubers available in Australia, *International Food Research Journal,* 16, pp. 507-520.
4. Aristizabal, J.; Sanchez, T. (2007), *Guía técnica para producción y análisis de almidón de yuca.* Organización de las Naciones Unidas para la Agricultura y la Alimentación (FAO). Roma, 153 pp.
5. Bertolini, A. (2010), "Trends in starch applications", In *Starches Characterization, properties and applications,* A. Bertolini, ed. USA, CRC Press.
6. Collado, L.S.; Corke, H. (1997), "Properties of starch noodles as affected by Sweetpotato genotype", *Cereal Chem.*, 74 (2), pp. 182-187.
7. Collado, L.S.; Mabesa R.C.; Corke, H. (1999), "Genetic variation in the physical properties of sweet potato starch", *Journal Agric. Food Chemistry,* 47, pp. 4195-4201.
8. Craig, S.A.S; Maningat, C.C.; Seib, P.A.; Hoseney, R.C. (1989), "Starch paste clarity", *Cereal Chem.,* 66 (3), pp.173-182.
9. Fredriksson, H.; Silverio, J.; Anderson, R.; Eliasson, AC.; Aman, P. (1998), "The influence of amylose and amylopectin characteristics on gelatinization and

retrogradation properties of different starches", *Carbohydrate Polymers,* 35, pp. 119-134.

10. García, A.M.; Raleigh, W.M.W. (1998), "Physicochemical characterization of starch from peruvian sweetpotato selections", *Starch*, 50 (8), pp. 331-337.
11. Guízar, A. (2009), *Obtención y caracterización física y química del almidón de "Camote de Cerro" (Dioscorea spp.)*. Tesis para obtener el grado de Maestro en ciencas en producción agrícola sustentable. CIIDIR, México, 2009.
12. Grace, M.R. (1977), *Elaboración de la yuca*. Organización de las Naciones Unidas para la Agricultura y la Alimentación (FAO), Roma, 116 pp.
13. Hermansson, AM.; Svegmark, K. (1996), "Developments in the understanding of starch functionality", *Trends Food Sci. Technol.,* 7, pp. 345-353.
14. Hernández-Medina, M; Torruco-Uco, J; Chel-Guerrero, L; Betancur-Ancona, D. (2008), "Caracterización fisicoquímica de almidones de tubérculos cultivados en Yucatán, México", *Ciência e Tecnología de Alimentos*, 28 (3), pp. 718-726.
15. Howling, D. (1980), "The influence of starch structure on its rheological properties", *Food Chemistry,* 6, pp. 51-61.
16. Jane, J; Chen, Y.; Lee, L. F.; McPherson, A. E.; Wong, K. S.; Radosavljevic, M.; Kasemsuwan, and T. (1999), "Effects of Amylopectin Branch Chain Length and Amylose Content on the Gelatinization and Pasting Properties of Starch", C*ereal Chem.*, 17 (5), pp. 629-637.
17. Kim, Y.S.; Wiesenborn, D. P.; Orr, P.H.; Grant, L.A. (1995), "Screening Potato Starch for Novel Properties Using Differential Scanning Calorimetry", *Journal of Food Science,* 60 (5), pp. 1060-1065.
18. Liu, Q. (2005), "Understanding starches and their role in foods". In *Food Carbohydrates*, Steve W. Cui. ed. CRC Press, USA.
19. PADES (2015), Grupo de investigación en Procesos Agroindustriales y Desarrollo Sostenible. Universidad de Sucre, Sincelejo.
20. Singh, N.; Singh, J Kaur, L.; Singh Sodhi, N.; Singh, B. (2003), "Morphological, thermal and rheological properties of starches from different botanical sources", *Food Chemistry*, 81, pp. 219-231.
21. Singh, J.; Kaur, L.; McCarthy, O.J. (2007), "Factors influencing the physico-chemical, morphological, thermal and rheological properties of some chemically modified starches for food applications – A review", *Food Hydrocolloids*, 21, pp. 1-22.
22. Swinkels, JJM. (1985), "Composition and properties of commercial native starches", *Starch*, 37 (1), pp. 1-5.
23. Takeda, Y.; Tokunaga, N.; Takeda, C.; Hizukuri, S. (1986), "Physicochemical properties of sweet potato starches", *Starch,* 38 (10), pp. 345-350.
24. Teow, C.C.; Truong, V.; McFeeters, R.F.; Thompson, R.L., Pecota, V.; Yencho, G.C. (2007), "Antioxidant activities, phenolic and β-carotene contents of sweet potato genotypes with varing flesh colours", *Food Chemistry*, 103, pp. 829-838.
25. Wang, SJ.; Li, C.; Copeland, L.; Wang, S. (2015), "Starch retrogradation: A comprehensive review", *Comprehensive Reviews in Food Science and Food Safety,* 14, pp. 568-585.
26. Woolfe, J.A. (1992), *Sweet potato an untapped food resource*. Cambridge University Press. Cambridge, 643 pp.

27. Zhang; C.G. and Oates. (1999), "Relationship between α-amylase degradation and physico-chemical properties of sweet potato starches", *Food Chemistry*, 65, pp. 157-167.
28. Zaidul, ISM.; Nik Norulaini, AN.; Mohd. Omar, AK.; Yamauchi, H.; Noda, T. (2007), "RVA analysis of mixtures of wheat flour and potato, sweet potato, yam, and cassava starches", *Carbohydrate Polymers*, 69, pp. 784-791.
29. Zobel, H. F. (1988), "Molecules to Granules: A Comprehensive Starch Review", *Starch*, 40 (2), pp. 44-50.

464684 Overcoming Metabolic Toxicity for Improved Isoprenoid Production

Abstract:

Isoprenoids are a large class of secondary metabolites with more than 55,000 currently known, many of which have important industrial roles. While two pathways exist for the production of these compounds, the relatively-newly discovered microbial methylerythritol phosphate (MEP) pathway has failed to produce comparable results to the more extensively studied MVA pathway. However, the MEP pathway has distinct advantages including higher theoretical yield, and a redox balance that enables anaerobic production. Here we discuss a methodology to improve isoprenoid pathway production through the MEP pathway. We target toxicity of intermediate metabolites as a key factor limiting MEP pathway flux. We then probe knockout strains and evaluate their effect on cell health, growth rate, and MEP-pathway flux; arriving at an optimized microbial chassis which will enable future work on the MEP pathway.

463826 Insights intoÂ Ligninolytic Production By Native White Rot Fungi in Malaysia and Its Potential in Industrial Applications

Abstract:

The tropical and humid climate in Malaysia shall permits exploration in advanced research in fungus field. However, far too little attention was given to identify local sources of lignin degrader microorganisms in Malaysian soil. Thus, the fungi ligninolytic potential is still largely unexplored and many novel ligninolytic enzymes may await discovery. In this study, 104 white rot fungi strains isolated from soils were screened using different decolorization assays to identify their ligninolytic potential. Results suggested that different dyes will give an effect on the level of decolorization rate from strains isolated. Selected potent strains were further evaluated for kraft lignin biodegradation and at least one of ligninolytic enzymes preferably lignin peroxidase responsible for its loss of lignin. The highest of lignin loss (43.29±0.15%) were obtained from *Mortierella* sp. WICC F33 strains with cocktail of enzymes (lignin peroxidase, manganese peroxidase and laccase) produced at 134.46±12.94 U L^{-1}, 41.01±1.61 U L^{-1} and 1.43±0.19 U L^{-1}, respectively. Laccase may not noticeable for many isolates; however its role as a mediator enabling the oxidation of non-phenolic lignin model compounds. From this study, three strains of *Mortierella* sp. WICC F32, 32 and 37, *Gongronella* sp. WICC F34, *Cunninghamella* sp. WICC F35, *Fomitopsis* sp. WICC F36, *Phanerochaete* sp. WICC F38 and *Cerrena*sp. WICC F39 were identified and deposited to The National Center for Biotechnology Information (NCBI) for further characterization.

Keywords: White rot fungi, Lignin peroxidase, Manganese peroxidase, Laccase

465516 Optimization of Growth Medium and Fed-Batch Cultivation Using Different Feeding Strategies for Production of *Acinetobacter* Sp. : A Biological Phosphorus Removal

Abstract:

Acinetobacter sp. has been reported previously in their role to enhance the removal of biological phosphorus and heavy metal when introduced in waste water treatment. This is an importance characteristics to be used in remediating the waste water instead of depend on naturally present of microbes. Therefore, high densities of *Acinetobacter sp.* to function in the bioremediation treatment are needed. The objective of this study is to optimize the production medium and bioprocessing condition for high biomass production of *Acinetobacter sp.* in a semi-industrial scale 16-L bioreactor. In this research, six different production media were screened to determine the most potent formulation for cell growth. The cell mass production of 2.54 g L^{-1} was obtained from shake flask cultures of the best medium with sucrose as the carbon source. This medium was then optimized using statistical approach of response surface methodology. Growth kinetic study of statistical optimized medium gives maximal cell mass of 5.58 gL^{-1}. During batch cultures of *Acinetobacter* sp. in the bioreactor 16-L under un-controlled and control pH, the cell mass production is 7.91 gL^{-1} and 5.70 gL^{-1} respectively indicating by pH drop from 7.0 to 6.0. Unfortunately, cultivation under un-controlled pH and dissolved oxygen cell mass was increased up to 156.06 % after 16-18 hours of cultivation. In conclusion, the growth of *Acinetobacter sp.* showing the cell mass production rate of 0.23 g L^{-1} h^{-1} under un-controlled pH when compared with controlled pH cultivation only 0.12 g L^{-1} h^{-1}. Nonetheless, the cell mass was increased when fed-batch cultivations was conducted under controlled dissolved oxygen at 30 % with uncontrolled pH condition. The cell mass production was increased up to 73.19 % (51.60 gL^{-1}) when constantly fed with full medium composition of optimized medium when compared feeding with mono-carbon solution.

KEYWORD: wastewater treatment, phosphate removal, medium optimization, *processing condition, bioreactor, Acinetobacter sp.*

NUTRITIONAL CHARACTERISATION OF PHYTO-BIOFILM STARCH FOR FOOD APPLICATION

Noor Azwani Zainol, UTM, Johor, Malaysia, University of Nottingham, Loughborough, United Kingdom
Siti Nor Azlina Abd Rashid, UTM, Johor, Malaysia
Farahiyah Aman Nor, UTM, Johor, Malaysia
Harisun Yaakob, UTM, Johor, Malaysia
Ramlan Aziz, UTM, Johor, Malaysia

Abstract

Phyto-biofilm offer a novel approach for consumer to improve nutrient intake from vegetables. If vegetables consumed more frequently and in a larger amount, they provide significant contribution of dietary nutrient especially in children. As consumer becoming increasingly aware of the health benefits of vegetables consumption, this research identifies an innovative way to increase the intake of important nutrients in order to prevent chronic disease and have a balance diet. In this study several formulation containing a mixture of starches were prepared however only the best formulation of biofilm was incorporated with spinach dried extract (1%). Biofilm was prepared by casting method. The phyto-biofilm obtained was test for microbial and nutritional content. The dietary biomarker components in spinach dried extract was analysed separately. Lipid results was 305.8 ± 0.2 mg and protein contents was 550 ± 0.6 mg. Results from microbial test indicated that this phyto-biofilm is of a good quality ready to eat food. Dietary contribution of biomarker components of α-Linolenic acid (ALA) and vitamin E is reported in a range of (14-21%) and (6-8%), respectively. In addition to nutritional characterisation, physical characteristic of the Phyto-biofilm is underway.

Introduction

Edible biofilm is a thin layer material made of edible material such as cellulose, starches, other polysaccharides and proteins. These edible biofilms may be formed as either food coatings or film wraps and also pouches. Food coating are applied in horticultural products to prevent them from fungus as well as microorganism and also to extend shelf-life of the products [1]. Other application currently being studied is the used of coating for the reduction of oil incorporated during frying [2][3][4]. On the other hand, monolayer or multilayer films can serve as a substitute for traditional wrapping; as such used in sushi wrapping and onigiri [5]. In addition biofilm wraps or pouches are expected to play a significant role as a carrier to provide nutritional value through diet and can be useful in encapsulating vitamins, nutritional supplements or additives [6]. One of the most features studies in the edible biofilms are the availability of these products to transport and release active compounds [7] which also focus on promotes health benefits for the consumers.

A green vegetable, such as spinach, is reported to contribute a significant amount of essential nutrients [8] while also being a good source of minerals [9]. Phytonutrients, as well as other bioactive components, are stored in lipids which are found in the chloroplast thylakoid membrane; this is the photosynthetic membrane of a green leaf. The eating of spinach is reported to provide a significant contribution of micronutrients namely iron, vitamin A and riboflavin as it is rich in them, these levels could be increased even further if consumed more frequently and in larger amounts, especially in children [10]. Spinach also reported to be rich in antioxidant activity and α-linolenic acid (ALA) which contributed to a healthy effect when consumed [11]. However, it is not very practical to eat large amounts of spinach in a person's daily diet; thus, it is important to identify alternative but acceptable ways for increasing the intake of such nutrients in order to prevent chronic diseases and to have a balanced diet.

Although there are many published work on benefits of eating spinach, incorporation of spinach dried extract in the phyto-biofilm wraps is a new approach to increase the intake of nutrient through diet. Besides, it is also to attract especially the kids to frequently consumed vegetables in a palatable way. This study presents both nutritional data and biomarker compounds of bio-film starch incorporated with spinach dried extract. Safety guideline of this biofilm as ready to eat food also reported.

Methodology

2. Methods

2.1 Extraction of spinach leaf

The stem of the spinach leaves were cut off and weighed before being mixed with 0.3 M (M = molar) of the sucrose solution, at a ratio of 1:3 w/w, they were blended using a household blender for 30 seconds. Sucrose were used as osmotic agents, of which are commonly used in the isolation work of spinach extraction using buffer solutions [12]. Once blended, the slurry was filtered using cheese cloth – the remaining supernatant was then poured into a 50 mL centrifuge tube and centrifuged at 3,500 g, 5°C, for 15 minutes. Finally, the supernatant was discarded and the pellet which remained was collected for further analysis. The pellets were dried with food dehydrator (Excalibur) at 38°C for one day.

2.2 Film preparation

A starch biofilm was prepared using casting method. A film was formed using mixtures of starch (7.0g), glycerol (2.5g) and water (90.5g). The solution was gently stir with a magnetic stir bar until all the starch dissolved. The solution was then was added with 1% of dried spinach and stir using glass rod for about 20 minutes with slow heating until the temperature reaches 80°C - 90°C.

2.3 Analysis

2.3.1 Crude lipid

The method used for lipid extraction was adapted from Bligh and Dyer [13]. The fresh pellet was added to a mixture of chloroform/methanol at a ratio of 2:1, it was then vortexed for 1 minute before 0.3 mL of a saline solution (0.9% NaCl) was added. NaCl was added to facilitate the partitioning of the lipids into the organic phase [14]. This mixture was then vortexed and left for 3 minutes before being centrifuged at 3,500 g, 5°C, for 15 minutes. The bottom layer, which contained the lipids, was transferred into a pre-weigh bottle and subsequently left in a nitrogen dryer until it was dry. After which, the lipids were weighed. The total amount of lipid was calculated using **Equation 1** below.

Equation 1:

$$Lipid\ content\ (\%) = \frac{weight\ of\ lipid\ (g)}{weight\ of\ dried\ pellet\ (g)} \times 100\%$$

2.3.2 Protein

The Pierce®BCA Protein Assay Kit was used to measure the protein content of the starch-based biofilm with dried spinach leaf extract. Bovine Serum Albumin (BSA) standards (containing BSA at 2 mg/mL in 0.9% saline and 0.05% sodium azide) were freshly prepared to obtain a set of standards for proteins. Calibration curves were then made using six different concentrations (0 to 2,000 µg/mL) of BSA. The debris from the dried spinach leaf extract obtained were then dried for 30 minutes at 60°C. The dried samples were then grounded using pestle and mortar before being added to a 1 mL solution of 2% sodium dodecyl sulphate (SDS). At this point, 100 µl of the solution was then taken and 2 mL of standard working reagent was then added. The standard working reagent was then prepared by adding 50 mL of Reagent A (in this case: sodium carbonate, sodium bicarbonate, bicinchoninic acid and sodium tartrate in 0.1 M sodium hydroxide – NaOH) with 2 mL of Reagent B (containing 4% cupric sulphate). The samples were incubated at 60°C for 30 minutes, before being vortexed and then centrifuged at 2,000 g for 3 minutes (this step was repeated twice). After which, the supernatant were collected and diluted 100 fold with 2% of the SDS solution. The samples were then incubated with a set of protein standards for 30 minutes at 37°C. The incubated samples were left to cool for 10 minutes and absorbance was measured at 562 nm against a blank. The concentration of the protein were determined against a standard curve produced using BSA.

2.3.3 Fatty acid

The composition of the fatty acids of the dried spinach leaf extract were determined using gas chromatography-mass spectroscopy (GC/MS), using the Pal System CST Analytical Autosampler and DSQ and TRACE GC Ultra (Thermo Electron Corporation).The lipids extracted were solubilised in 2 mL of chloroform and 0.2 mL of trimethylsulfonium hydroxide. The mixture was left to stand for at least 15 minutes for complete conversion of the fatty acids into fatty acid methyl esters (FAMEs); at which point, 200 µl was then injected into the injector (split flow 50 mL/min-1) at 250°C.The lipids were separated by polyethylene glycol (BP20 ID 0.22 mm x 25 m) gas chromatography column (Milton Keynes, UK), at 30 mL/min nitrogen. The oven temperatures were controlled at 120°C (1 minute), they were then increased by 5°C/minute until they reached 250°C. The retention time of each peak was compared with the FAME standards (Supelco® 37-Component FAME Mix). Furthermore, identification of the peak was also verified by comparing the mass spectrum with the standard library using the Thermo ScientificTM XcaliburTM software program. The internal standard stock solution used in this preparation was a combination of methyl pentadecanoate and glyceryl triheptadecanoate; specifically, 0.1 g of methyl pentadecanoate and 0.1 g of glyceryl triheptadecanoate were weighed and transferred to a 10 mL volumetric flask, where they were then dissolved with chloroform. The solutions were stored at -20°C and were used within three months. The peak areas were normalised based upon the methyl pentadecanoate and the percentage of fatty acid, as calculated using the formula below. The derivatisation efficiency was checked by comparing the glyceryl triheptadecanoate and methyl pentadecanoate peak area ratios (designed as 1:1). The percentage fatty acid component was calculated using **Equation 2**, as below.

Equation 2:

$$\% \, individual\, fatty\, acid = \frac{individual\, fatty\, acid\, peak\, area}{total\, fatty\, acids\, peak\, area} \times 100\%$$

2.3.4 Vitamin E

Before being vortexed, 1 mg of vitamin E lipid extract was diluted with 2 mL of methanol. The sample was filtered using a syringe filter (0.45 μm) before being centrifuged at 2,000 g for 5 minutes, at 4°C. After which, 200 μL of the sample was transferred into HPLC vials and stored at -20°C before being injecting into the HPLC systems. The HPLC analyses for vitamin E were carried out on a Waters® 2695 Separation Module that was equipped with a Waters® 996 Photodiode Array Detector (PDA) and a JASCO Intelligent Fluorescence Detector (FP-920) using a reverse phase Nova-Pak C18 Column (3.9 x 150 mm, 4 μm, Waters).The sentry guard-column used was Nova-Pak C18, 4 μm, 3.9 x 20 mm. A dual gradient mobile phase was prepared, this was made of acetonitrile: methanol: isopropanol: aqueous acetic acid (1%) (at: 45:45:5:5, v/v) for solvent A, and acetonitrile: methanol: isopropanol (at: 25:70:5, v/v) for solvent B. The gradient started with 0% A to 60% A from 0 to 10 minutes and held there for 12 minutes before being returned to the initial conditions. The flow rate was 0.8 mL/min. The 10 μL samples were injected and detection was performed at the excitation wavelength of 298 nm and the emission wavelength of 328 nm. Calibration curves were then plotted using six different concentrations (4-100 μg/mL) of pure standards of vitamin E (for:α, β,γ and δ forms).

2.3.5 Microbial analysis

The microbiological media were prepared in accordance with the manufacturer's instructions. The media and media supplements prepared, and the organisms targeted, included: Plate count agar (Oxoid CM0325) for general bacteria count, *Bacillus cereus* agar base (Oxoid CM0617) for *Bacillus cereus* and other pathogenic Bacillus, Listeria enrichment broth base (Oxoid, CM0862) and Fraser broth base (Oxoid, CM0895) for Listeria and *Rose Bengal Chloramphenicol* agar for yeasts and moulds. Violet Red Bile Glucose agar (from Lab M™) was used for *Enterobacteriaceae*. Analyses were then carried out using standard methods. For Listeria, 25 g samples were diluted 1/10 in Listeria enrichment broth base broths and emulsified in a stomacher for 2 minutes. The primary enrichment was then incubated at 30°C for 24 hours, after which 0.1 mL of the incubated was sub-cultured in 10 mL of Fraser broth and incubated at 37°C for 48 hours. The secondary enrichment (Fraser) broth culture was sub-cultured after 48 hours onto Oxford agar plates. The plates were examined for the presence of typical colonies after incubation. For other tests, 1 g of the sample was dissolved in 9 mL of sterile distilled water (dH_2O). With the exception of the test for *Enterobacteriaceae,* where 1 mL of the sample was plotted using the pour plate method; specifically, 0.1 mL of the sample was spread on the plates of the other media prepared with standard incubation times observed.

3.0 Results and Discussion

3.1 Film Appearance

In general, starch-based biofilm obtained were translucent and colorless. In addition they were easy to handle when dried off. After drying, the biofilm was stored in a close container at room temperature. The biofilms which is ready were kept for further analysis.

3.2 Nutritional composition of bio-film starch based incorporated with spinach dried leaf

Table 1 shows the main analyses of the nutrients found in the bio-film starch based incorporated with spinach dried leaf. Lipids and proteins are the two dominant components in the spinach leaf; to illustrate, lipids were estimated at 300 mg/gram (approximately 30%) and proteins were estimated at 550 mg/gram (approximately 50%). Other research also supports this generated data; specifically lipid content falls within a range of 30% to 40%, while the protein content ranges between 50 and 60% [15]. Total fatty acid component is reported as 291 mg. As expected, α-linolenic acid (ALA) was identified as the highest amount of unsaturated fatty acid in the spinach dried leaf extract compared to the other unsaturated fatty acids chain. This result was supports by other researcher [16] which indicated that spinach is rich in linolenic acid; as such ALA can refer as a biomarker components. The highest amount of ALA presumably comes from the glycolipid components which significantly presents in the spinach leaf.

Table 1. Composition of nutrient in starch-based biofilm with spinach dried leaf extract

Component	**Amount per gram**
Total lipid	305.8 ± 0.2 mg
Total protein	550 ± 0.6 mg
Fatty acid component α-linolenic acid	291 mg 228 ± 6.6 mg
Vitamin E	0.24 ± 0.1 mg

The analyses conducted within this research returned unexpectedly low concentrations of vitamin E. Specifically, only 75% of what was reported in the literature for vitamin E content were found from the dried spinach leaf extract. With regards to vitamin E, this could be as a result of lipid oxidation which may have occurred during storage to produce lipid free radicals. In order to inhibit the peroxidation of the lipid, vitamin E as an antioxidant will interact with these produced lipid free radicals. As a consequence of lipid peroxidation, it will contribute to a reduction of vitamin E content, even though vitamin E was reported to act in the reverse of this to form its normal component after interacting with the lipid radicals [17]. In addition to the oxidation process, the inefficient separation of the solute, during the HPLC analyses, may also have contributed to the lower value of vitamin E.

The dietary contribution calculation for both ALA and vitamin E were based on the Reference Nutrient Intake (RNI). These values used as a guide on a daily basis in order to maintain a healthy diet. Although both of ALA and vitamin E relatively exist in a small amount, yet they provide significant percentage values of the RNI; to indicate ALA is in a range of 14-21% while vitamin E is in a range of 6-8%.

3.3 Microbial analysis

Five tests were conducted in order to determine viable values of bacteria in the dried spinach leaf extract. The following tests were therefore completed: plate count agar (PCA) for the bacteria count; a test for Bacillus; a test for Enterobacteriaceae; a test for yeast and mould; and, finally a test for Listeria. The results were compared to a set of standards that were outlined by [17], these standards can be used to determine the quality of the extract as being satisfactory, acceptable, unsatisfactory or even potentially hazardous. Four terms were used to describe the grade of the microbiological quality, including: satisfactory which is indicative of good microbiology quality, acceptable which is an index reflecting a borderline limit of microbiological quality, unsatisfactory which is indicative that further sampling may be necessary and, finally, unacceptable or potentially hazardous which is indicative that urgent attention is needed to locate the source of the problem – in which case, a detailed risk assessment would be recommended. With regards to this research, a viable number of bacteria were counted on each of the performed tests as being satisfactory (see Table 2). The results herein indicate that the dried spinach leaf extract were of good quality. Although growth was observed in the agar, it is normal that some contamination during the processing stage would be encountered. To illustrate, some of the media demonstrated growth of the tested microbial. It is suggested that this was due to contamination during and/or after preparation of the materials. Practically, cross contamination in food preparation environments can occur during the processing [18]. In practice, it is impossible to totally prevent contamination from the environment and by the food handlers; however, it is possible to minimise the risk by practising good hygiene.

Table 2: Summary of microbial analyses

Test	Cfu/g	Conclusion
General count on plate count agar (PCA)	1.6×10^4	Satisfactory
Test for Bacillus	2.0×10^2	Satisfactory
Test for Enterobacteriaceae	5.0×10^1	Satisfactory
Yeasts and mould	1.0×10^4	Satisfactory
Listeria	Not detected	Satisfactory

4.0 Conclusion

A new approach to increase intake of nutrient in daily diet is a major issue for future research. This research give an information on the edible biofilms which can used as food wrappers; besides increase the nutrient intake. In summary, phyto-biofilm incorporated with spinach extract showed a good nutrition property, good quality based microbial test for ready to eat food and may have potential to be commercial.

5.0 Acknowledgements

The authors acknowledge the financial assistance provided by Islamic Development Bank (IDB), Jeddah during my studies in Nottingham University, UK and Institute Bioproduct Development (IBD), University Technology Malaysia (UTM) for technical assistance.

References

1. M Maqbool, A. Ali, S. Ramachandran, D.R. Smith, P.G. Alderson. Control of postharvest anthracnose of banana using a new edible composite coating. Crop Protection, 29 (10) (2010), pp. 1136–1141
2. S. Albert, G.S. Mittal. Comparative evaluation of edible coatings to reduce fat uptake in a deep-fat fried cereal product. Food Research International, 35 (2002), pp. 445–458
3. A.H. Khalil. Quality of French fried potatoes as influenced by coating with hydrocolloids. Food Chemistry, 66 (1999), pp. 201–208
4. P. Mallikarjuna, M.S. Chinnan, V.M. Balasubramaniam, R.D. Phillips. Edible coatings for deep-fat frying of starchy products. Food Science and Technology, 30 (1997), pp. 709–714
5. Debord, M.D., Mchugh, T.H., Olsen, C.W.: WO2006029128A2 (2006).
6. Gilleland, G.M., Turner, J.L., Patton, P.A., Harrison, M.D.: US20026375981 (2002).

7. Victor Falguera, J. Pablo Quintero, A. Jimenez, J.Aldemar Munoz & A. Ibarz. (2011). Edible films and coatings: Structures, active functions and trends in their use. Trends in Food Science & technology 22 (2011) pg. 292-303
8. Hanif R., Iqbal, Z., Iqbal, M.,Hanif, S. & Rasheed, M. (2006) Use of vegetables as nutritional food: Role in human health. Journal of Agricultural and Biol Science, 1, 18-22.
9. Vaijayanthi Kanabur & Lalitha Reddy, R. P. (2014) Bioactive Components of Spinach and Their Effect on Some Patho Physiological Conditions: A Review. Int. J. Cur. Res. Rev., 06, 156.
10. Mieke F., M., van Jaarsveld, P.J., Laubscher, R., 2007. The contribution of dark-greenleafy vegetables to total micronutrient intake of two- to five-year-old childrenin a rural setting. Water SA 33 (3 (special issue)) 407–412.
11. Castenmiller, J.J.M. (2000). Spinach as a source of carotenoids, folate and antioxidant activity.ISBN9789058081605, pp183
12. Honda S. I., Hongladarom, T. & Laties, G. G. (1966) A new isolation medium for plant organelles. Journal of Experimental Botany, 17, 460-472.
13. Folch, J., Lees, M. & Sloane, G. S. H. (1975) A simple method of the isolation and purification of total lipids from animal tissues. The Journal of Biological Chemistry, 226 (1), 497-509.
14. Hincha, D. K., Oliver, E. A. & Crowe, J. H. (1998).The effects of chloroplast lipids on the stability of liposomes during freezing and drying. Biochimica et Biophysica Acta, 1368, 150-160
15. Abdulnabi, A. A., Emhemed, A. H., Husssein G. D. & Biacs P. A. (1996) Determination of antioxidant vitamins in tomatoes. Food Chemistry, 60, 207-212.
16. Jacob and Miller Jacob, J. S. and Miller, K. R. (1986) the effects of galactolipids depletion on the structure of a photosynthetic membrane. J. Cell Bio. 103: 1337-1347.
17. Poincelot, R. P. (1974) an Improved Method for the Isolation of Spinach Chloroplast Envelope Membranes. Plant Physiol., 54, 780-783.
18. Gilbert M., C. Wilhelm & M. Richter (2000) Bio-optical modelling of oxygen evolution using in vivo fluorescence: Comparison of measured and calculated photosynthesis/irradiance (P-I) curves in four representative phytoplankton species. Journal of Plant Physiology, 157, 307-314.
19. Yoda and Uchimura (2006).An outbreak of Campylobacter jejuni food poisoning caused by secondary contamination in cooking practice at a high school. Japanese Journal of Infectious Diseases, 59, 408-409.

466091 A Model for Vitamin E Recovery from By-Product of Edible Oil Refining By Adsorption/Desorption of Anion-Exchange Resin

Abstract:

Introduction

Vitamin E (V_EH) has high antioxidant and anti-aging activities and widely used in food and pharmaceutical fields[1]. The deodorized distillate, a by-product of edible oil refining, is used as a raw material for V_EH recovery. Its typical composition is free fatty acid (FFA) 44wt%, triglyceride 14wt%, sterol 11wt%, and V_EH 1-5wt% and main component FFA has separation characteristics similar to V_EH. In the current industrial V_EH recovery, a multi-stage molecular distillation at high temperature, 100-250°C, is mainly utilized. Its recovery yield is low, about 50%, because of V_EH thermal degradation and many compounds having similar boiling point remained as impurities, 65%[2]. Thus, further purification is necessary, this causes high production cost.

We have proposed a novel low-temperature process by adsorption/desorption using ion-exchanges resins[3]. This process consists of esterification of FFA using cation-exchange resin catalyst to convert into fatty acid ester without adsorption activity and adsorption/desorption of V_EH using anion-exchange resin at 50°C. Under specific condition, V_EH was completely recovered without thermal degradation and only FFA remained as an impurity, 22%. However, the recovery amount of V_EH after desorption step greatly changed with feedstock supply volumes during adsorption step. This was because V_EH and unreacted FFA competitively adsorbed on the anion exchange resin, and FFA with higher acidity was re-exchanged with the adsorbed V_EH to elute it from the resin. Thus, it is important to determine the optimum feedstock supply volume during the adsorption step for giving maximum V_EH recovery amount.

In this study, a model for competitive adsorption of V_EH and FFA on the anion exchange resin was constructed by considering mass transfer into the resin and adsorption by the ion-exchange reactions. The competitive adsorption behavior of V_EH and FFA was simulated by the model in the column reactor. The V_EH recovery amount during desorption step was predicted by the model.

Experiments

Diaion PA306S (porous-type, average particle diameter: 0.32 mm, ion-exchange capacity: 0.94 mmol/g wet basis) was used as most active anion-exchange resin. The model esterified deodorizer distilllate was V_EH (*Î´*-tocopherol) and FFA (oleic acid) dissolved in ethanol. A glass column (i.d. 1.1cm, length 10cm) was packed with the resin (6.7 g, wet basis) and its temperature was kept constant at 50Â°C. The model feedstock was supplied to the bottom of the column. After the adsorption, ethanol was fed to the column to wash out the feedstock and then desorption eluent, the acetic acid in ethanol, was also fed to the top of the column. Under all operations, the flow rate was 1.0 cm^3/min. The feedstock supply volume in the adsorption step was changed in the range of 40-250 cm^3. The concentrations of V_EH and FFA in the effluent from the column were determined using an HPLC system.

Construction of a model

The following assumptions are set for construction of competitive adsorption model. A linear partition equilibrium for concentrations between the bulk liquid phase and the resin liquid phase was established. Three ion-exchange reactions, i)V_EH adsorption onto the resin by ion-exchange with hydroxyl group, ii)FFA adsorption onto the resin with hydroxyl group, iii)V_EH escape from the resin by re-exchange with FFA, occurred on the resin solid surface. The model equations in the column reactor were derived from mass balance of V_EH, FFA and hydroxyl group considering mass transfer by the convection and dispersion in the bulk liquid phase, partition between the bulk liquid phase and the resin liquid phase and reaction on the resin solid surface. There are unknown constants such as the partition coefficients of each component, kinetic constants and equilibrium adsorption constants of each ion-exchange reaction, and resin porosity. The model constants were estimated by fitting with the separate experimental results of competitive adsorption in batchwise system. The dispersion coefficient is calculated by the empirical equation[4)]for the fixed-bed reactor.

Simulation by the kinetic model

The model well described the concentration profiles of V_EH and FFA in the effluent experimentally obtained during the adsorption step with the feedstock supply volume of 250 cm^3. The overshooting of V_EH concentration in the effluent from the column was observed since the adsorbed V_EH was flowed out by the re-exchange with FFA. On the other hand, the breakthrough of FFA was delayed compared with that of V_EH since the re-exchange as well as the adsorption of FFA occurred.

The model can also simulate the concentration profiles of each component along the axis of the column reactor at each feedstock supply volume condition. By integrating the concentration profiles from inlet to outlet of the column reactor, the adsorbed amount of each component on the resin at each condition was calculated under various feedstock supply volumes. The adsorbed amount calculated by the model simulation were compared with the recovery amounts experimentally obtained during the desorption step. Under any condition, the calculated adsorbed amounts were in good agreement with the experimental recovery amount. This meant that the adsorbed V_EH on the resin was perfectly recovered in the desorption step. The V_EH recovery amount increased with feedstock supply volumes in the adsorption step and it became a maximum at 71 cm^3 of feedstock supply volume. Then, V_EH adsorbed amount became the almost zero at the supply volume of more than 150 cm^3. The model well predicts the optimum feedstock supply volumes for giving a maximum V_EH recovery amount.

1)E.Niki *et al., Free Radic. Biol. Med.*,**66**, 3 (2014)

2)S.T.Jiang *et al., Biosyst. Eng.,* **93**, 383 (2006)

3)K. Hiromori *et al., Food Chem.,* **194**, 1 (2016)

4)H.S. Fogler *Elements of Chemical Reaction Engineering,* 3rd Edition, Prentice-Hall, Upper Saddle River, NJ(1999).

466112 Synbiotics Effect of Probiotic Bacteria Isolated from Mother's Milk and Different Types of Prebiotic for Short Chain Fatty Acids Production

Abstract:

Due to health promoting effects reported for probiotics, these strains are increasingly incorporated into functional food and pharmaceutical markets. The aim of this research is to study the growth and metabolism of probiotic strains isolated from mother's milk in different types of prebiotic as a carbon source. The aim of this research to study the synbiotic effects between the probiotics bacteria and prebiotics to the production of SCFA such as acid, lactic acid, butyric acid, propionic acid and succinic acid. The probiotic bacteria isolated from mother milk used in this research were *Lactobacillus salivarius* and *Lactobacillus delbreuckii bulgaricus* in pure and mixed culture. The growth kinetics and metabolite profiling of SCFA were studied in different types of carbon sources such as inulin, FOS, gum Arabic and growth in glucose as a control. Between the four different types of carbon sources, the best growth profile was obtained by using inulin in mixed culture of *L. bulgaricus* and *L. salivarius* with maximum cell mass production of 8.23 g L^{-1} and followed by FOS and glucose at the value of 7.67 g $L^{-1}$and 6.43 g L^{-1} respectively, whereas specific growth rate for this culture was 0.33 h^{-1}. Furthermore, growth of *L. salivarius* in glucose gave the highest succinic acid production of 3.99 g L^{-1} compared to other types of prebiotics and SCFA. However for cultivation of *L. bulgaricus* was produced the high amount of lactic acid that is 2.39g L^{-1} after 24 h cultivation when glucose as the carbon source. Besides that, growth of mixed culture of *L. bulgaricus* and *L. salivarius* gave high succinic acid production almost 2.62 g L^{-1}in FOS after 24 h cultivation. Generally in this studied, it's shown that growth of mixed culture of *L. bulgaricus* and *L. salivarius* in FOS also gave the highest amount of total SCFA that was 6.47g L^{-1}. As a conclusion, the correlation of the response would help to determine the choice of a good prebiotics for the probiotics strain in order to enhance the beneficial effect to human health.

466197 Encapsulation of *Lactobacillus Bulgaricus* in Carbohydrate-Protein-Coconut Oil Mixtures Provides Protection during Spray Drying Survival, in Simulated Gastrointestinal Conditions and Storage

Abstract:

Encapsulation allows live probiotic bacteria to be protected from its environment, thereby maximizing the probiotic bacteria viability during processing and during storage at 25 Â°C. More importantly the encapsulated product should survive passage through the upper digestive tract in the large numbers, and have beneficial effects in the intestine host. Ideally, probiotics bacteria should be metabolically stable during processing, storage stability, survive passage through the upper digestive tract in the large numbers for colonization and proliferation in the large intestine, and have beneficial effects when in the intestine host. This study would help to understand the effect of feed composition towards viability of *L. bulgaricus* by using carbohydrate â€" protein - coconut oil mixtures. The understanding of the formulation is useful in designing a better formulation of the spray dried encapsulation *L. bulgaricus* as intermediate product for nutraceutical industry. The optimization step was carried out in order to determine the optimal feed composition of spray dried encapsulated *Lactobacillus bulgaricus*. Statistical analysis was carried out by using standard mixture design called simplex lattice, using three factorial variables; the quantity (%) of gum Arabic, gelatine and coconut oil as a mixture of carbohydrate â€" protein - coconut oil as the protective barrier. The optimal feed composition that results in the viability beyond the therapeutic minimum dose was consisted of 30 % gum Arabic, 16.95 % gelatine and 53.05 % coconut oil after spray drying with the viability of the cells are Â within the accepted level which is more than 10^8 cfu/g. Furthermore, the mixture provide a good protection under the simulated gastrointestinal environment with the percent of viability more than 21.75 % with a viable cell was reduced only by 2 log cycles after 6 weeks storage at 25 °C. The mixture of gum Arabic - gelatine - coconut oil shows more protecting to protect against damaging of oxidative stress during storage.

466303 Effect of Dissolved Oxygen on the Cell Growth and Spore Production of *Bacillus Thuringiensis* in Semi-Industrial Scale Bioreactor

Abstract:

In controlling the insect pest in agriculture and insect vectors that lead to human disease, there had been application of chemical insecticides. The use of these chemical pesticides had led to several problems including environmental problem and give effects on human health such as cancer and several immune system disorders. Chemical agents are also probable to lead to emergence of insect resistance and have a long residual effect with toxicity to non-target organisms. Therefore, as an alternative of biocontrol agents, *Bacillus thuringiensis*-based biopesticides had been used as biopesticide in controlling the agricultural pest. Large quantities of bio insecticides possess high insecticidal activity, produced by fermentation, are required for practical applications of bio insecticides. For industrial production, they prefer medium that have lowest cost and give high quality of insecticidal function. To investigate the effects of Dissolved Oxygen Control (DOC) on the cell growth and spore production of *B. thuringiensis*. DOC was controlled at three different levels of uncontrolled DOC, 10, and 30 % of air-saturation by manipulating agitation speed and airflow in a series of batch cultivation in the semi-industrial scale bioreactor. For uncontrolled DO which was started at 100 % at the beginning of cultivation and was not controlled throughout the cultivation. This cultivation showing a significant loss in cell mass due to limitation of DO at 3 to 6 hour of cultivation, this was attributed to cell lysis and the spore formation was only 6.5×10^7 cfu mL^{-1} after 45 hours cultivation when there was no cell growth observed at this phase. No big differences on the cell growth and spore production were observed when the DOC was controlled at 10 %. Unfortunately, when DOC was controlled at 30 %, the highest cell mass was 20.93 g L^{-1} after 16 hours cultivation and then start to enter death phase after 24 hours cultivation with the specific growth rate of 0.07 h^{-1}. The cell mass production was increased up to 103.96 % when compared with growth without controlling the DO. However, the sporulation productivities was increased when the DOC was controlled at 30 % to 2.15-fold about 115.38 %.

466381 Bioprocess Optimization for High Biomass Production of Bacillus Firmus

Abstract:

Bacillus firmus are natural occurring bacteria in the soil environment. *B. firmus* are economically important in agricultural sector whereby the application as biocontrol agent against disease caused by nematodes. *B. firmus* categories as gram positive bacteria rod shape. Research conducted proved that *Bacillus sp*. is able to secrete extracellular enzyme such as protease in which act as a nematicidal agent to control the population of parasite nematodes. The life cycle of *B. firmus* divided into two phase which consists of growth of vegetative cells and endospores formation. Production of high cell mass for different kind of microorganisms, requires extensive study by developing a suitable medium and cultivation strategy. This study will focusing on the development of production medium started with screening of different production media to select the best medium which supports the higher cell growth. The best medium consist of glucose, 20.0 g L^{-1}; peptone from casein 5.0 g L^{-1}; yeast extract 2.0 g L^{-1}, potassium dihydrogen phosphate, 5.0 g L^{-1}; and magnesium sulphate heptahydrate, 2.0 g L^{-1} gave highest cell mass of 6.27 g L^{-1} in shake flask culture. Hence, the medium optimization was carried out by using two approaches: One Factor at Time (OFAT) and statistical optimization. The produced biomass by using un-optimized and optimized medium both yielded 6.82 g L^{-1} and 10.22 g L^{-1}, respectively. As step for bioprocess industrialization, cultivations were carried out using batch mode in 16-L bioreactor under controlled and uncontrolled pH conditions. The maximal biomass produced was 11.91 g L^{-1} in uncontrolled pH culture. Furthermore, cells were cultivated in fed-batch culture using different feeding strategies. Feeding of complete medium using constant rate has increased the biomass production up to 14.89 g L^{-1}. Furthermore, feeding of complete medium components by increasing the feeding rate at 1.027 g L^{-1} h^{-1} for every 10 hours under uncontrolled pH cultivation of *B. firmus*, the cell mass was increased up to 28.07 % with the maximal cell mass achieved was 19.16 g L^{-1}.

466708 Metabolic Engineering of *Corynebacterium Glutamicum* for Biosynthesis of 5-Aminovaleric Acid

Abstract:

Production of bio-based polymers attracts industrial and scientific interests due to growing concerns over petroleum shortage, climate change, and human and animal health. The platform chemicals used for the synthesis of bio-polymers have been produced from renewable feedstocks using metabolically engineered strains. In particular, C5 platform chemicals such as cadaverine, 5-aminovaleric acid, and glutaric acid can be directly produced from lysine, which is readily available from a commerical fermentation process. Among these chemicals, 5-aminovaleric acid (5-AVA) is synthesized from lysine by two-step enzymatic reactions of lysine 2-monooxygenase (DavB) and delta-aminovaleramidase (DavA), and can be used for the synthesis of enineering platics such as nylon 5,6 or 5,6. In this study, we developed metabolically engineered strains of *Corynebacterium glutamicum* for enhanced production of 5-AVA from biosugar. The DavBA expression cassettes under the control of native or synthetic promoters with different strengths were introduced into a lysine-hyper producing *C. glutamicum*. The engineered *C. glutamicum* exhibited higher titre of 5-AVA production (10.4 g/L) than the highest value ever reported and was further evaluated in fed-batch fermentation.

The Establishment of Herbal Processing Technology for the Production of *Orthosiphon Stamineus* Standardised Extracts

Zarani Mat Taher[1], Chua Lee Suan[1], Ngadiran Sulaiman[1], Ware Ismail[1], Abubakar Umar Isah[1] and Muna Mohamad[1], Ramlan Aziz[1]

1. Institute Bioproduct Development, Universiti Teknologi Malaysia, Skudai, Malaysia,

Introduction

Orthosipon stameneus(OS), or locally known as misai kucing in Malaysia, is a family of *Lamiacea*. This medicinal plant is widely found in South East Asia, particularly in Malaysia and Indonesia due to tropical climate and received heavy rainfall over the years. The leaves of *OS* are used as traditional medicine to treat kidney, gouts, bladder inflammation and diabetes due to its anti-allergic, antihypertensive, anti-inflammatory and diuretic properties (Adam *et al*., 2009; Majid and Basheer, 2010; Ameer *et al*., 2012). The flowers are in bell-like shape with long exerted filaments, making the flower look like cat's whiskers.

Akowuah *et al*.,(2004) reported that the separation of bioactive compounds in *OS* extract revealed three methoxylated flavones namely sinensetin (SEN), eupatorin (EUP) and 3'-hydroxyl-5,6,7,4' – tetramethoxyflavone (TMF) rosmarinic acid (RA) , which are caffeic acid derivatives.

Extraction is a process to separate the desired substances from a material. The herbal extraction process involves the separation of bioactive or medicinally active components of plant tissues from inert compounds by using specific solvents (Handa *et al*., 2008). According to Wong (2008), a standardized extract is an herbal extract that contains a specified amount of a certain active compound. Standardization of herbal extract is very important to produce a product with known chemical composition for clinical trials as it provides the information of the presence and quantity of plant constituents (Ismail, 2011). Consequently, the confidence and acceptance level of herbs by consumers due to the medical advancement will be increased. Although *OS* have been studied previously, however, there is still lack of standardised procedure especially in larger scale. This work was undertaken to up-scaling of laboratory findings for the preparation of standardised extract of *Orthosiphon stamineus (OS)*.Therefore, the main objective of this study is to develop a processing method for *Orthosiphon stamineus* in pilot scale with consistent quality of their phytochemical compound and bioactivity.

Material and method

Plant materials

Dried leaves of OS are obtained from local supplier in Malaysia. The plant materials are then be crushed and grounded by commercial blender into fine powder (<0.25 mm). All the chemical and standard were purchased in analytical grade (Sigma-Aldrich (St. Louis, MO, USA).

Small scale solvent and aqueous extraction of plant materials

The small-scale extractions of the plant material are carrying out by using reflux method by deionised water and 50% ethanol respectively. The effects of three processing parameters including sample to solvent ratio, temperature and extraction time are studied. The extracts are then filtered and concentrated by drying under vacuum at 75°C via rotary evaporator (Heidolph, Laborota 4000, Germany). The crude extracts are stored at refrigerator at -4°C until further analysis.

Design of experiment (DOE) of Orthosipon stamineus

Response Surface Methodology is used to optimize and analyse the effects of the independent variables (Bezerra *et al.*, 2008) such as extraction temperature, extraction time and sample to solvent ratio on the yield of crude extracts and concentration of rosmarinic acid (RA). Box Behnken Design (BBD) is used to establish the experiments with three factors at three level (-1, 0 and +1). There are 17 run of experiments with 5 central points.

HPLC analysis for marker compounds

HPLC analysis was carried out to determine the concentration of rosmarinic acid (RA) and eupatorin(EUP). The mixture of methanol-tetrahydrofuran–water (45v/v: 5v/v: 50v/v) as mobile phase. The peaks are detected at 340nm.

Statistical analysis

The statistical analysis including multiple regressions analysis and Pareto analysis of variance (ANOVA) are performed using Design expert software (Version 8.0.5.0, Stat- ease Inc., USA).

Results and discussion

Optimization of Orthosiphon stamineus solvent extraction

For solvent extraction, quadratic model is highly significant for the extraction process as the p-value was less than 0.0001 for quadratic and cubic model. The ANOVA results (data do not show) for yield and RA showed the F-value of 16.11 and 6.02, respectively, which indicated that the model is significant. The regression coefficient, R2 of yield and RA were 0.95 and 0.89, respectively, which means that adjustment of the quadratic model to the experimental data is satisfactory.

The relationships between the responses of the extraction process with the variables (temperature, time and ratio) were described by second order equation. Equation 1.0 and 2.0 show the final equations in term of polynomial equation, for yield of crude extract and RA concentration, respectively.

$$\textbf{Yield (\%)} = 11.897+2.55A+0.593B+0.699C+0.281AB+0.640AC+0.70BC-1.062A2+0.715B2-0.785C2$$
(Equation 1.0)

$$\textbf{RA (\%)} = 417.831-30.641A-25.877B+28.774C-30.745AB+36.113AC-27.682BC+1.871A2+60.966B2+21.087C2$$
(Equation 2.0)

The effect of the extraction variables on the yield and concentration of RA can be represented by 3D surface plot. The response surface plots show the relative effects of any two variables while the rest is kept constant. The effect of the extraction variables on the yield and concentration of RA can be represented by 3D surface plot. The response surface plots show the relative effects of any two variables while the rest is kept constant. Figure 1 shows the effect of extraction variables (temperature, time and ratio) on the crude extract yield and concentration of RA, respectively. For the yield of crude extract of OS will increases with the increase of solvent to solid ratio, extraction temperature and extraction time. Figure 1 shows a decreasing in concentration of RA when the extraction temperature increased. This might be due to the degradation of RA in high temperature. The concentration of RA was found to be increased when the extraction time was increased

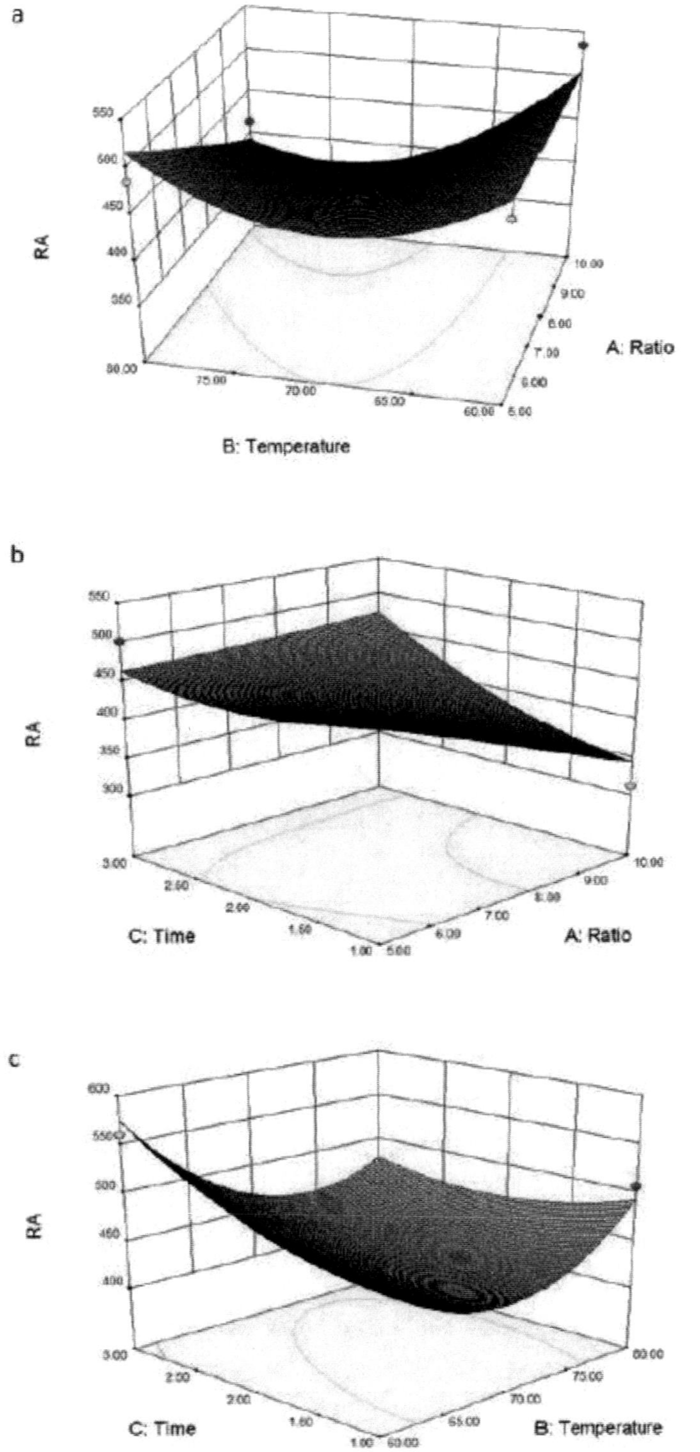

Figure 2: Response surface plot of concentration of rosmarinic acid versus extraction variables for solvent extraction of *Orthosiphon stamineus(OS)*

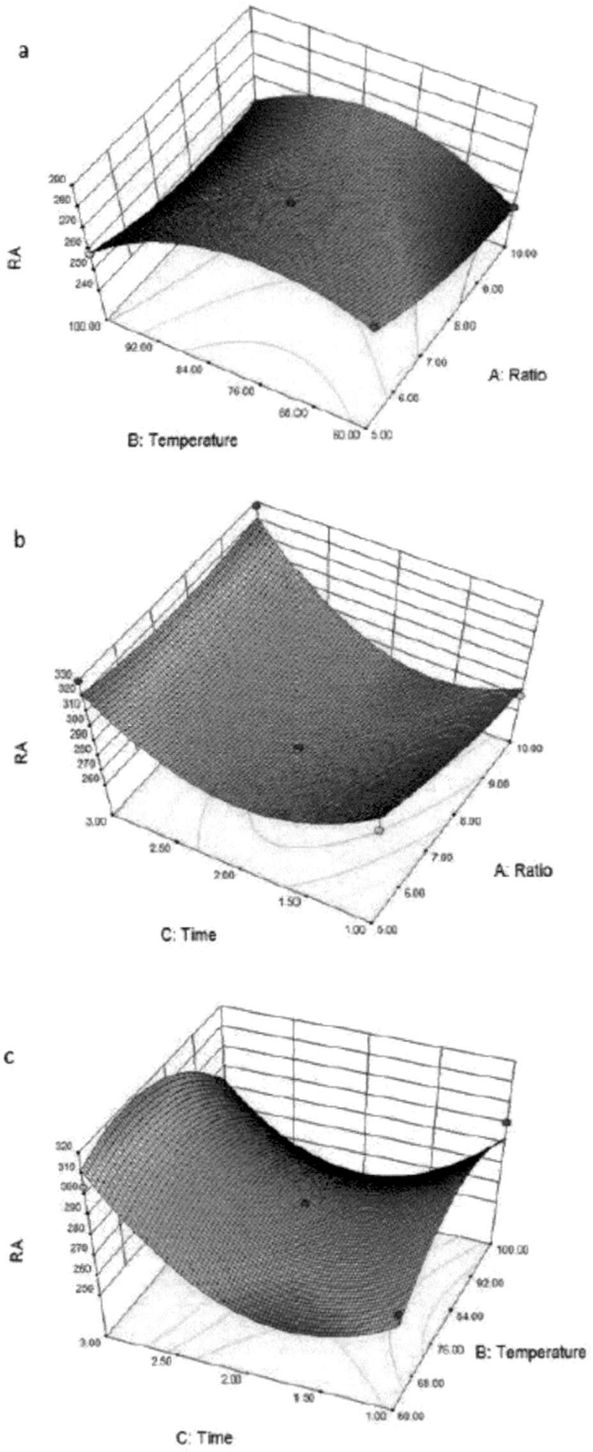

Figure 2 : Response surface plot of concentration of rosmarinic acid versus extraction variables for aquoues extraction of *Orthosiphon stamineus*

Optimization of Orthosiphon stamineus aqueous extraction

The quadratic model is highly significant for aqueous extraction process as the p-value was less than 0.0001 for quadratic and cubic model. The model is significant as the F-value of yield and RA showed 9.59 and 17.64, respectively. Equation 3.0 and 4.0 show the relationships between the responses of the extraction process with the variables (temperature, time and ratio) in term of polynomial equation, for yield of crude extract and RA concentration, respectively. Based on the analysis using RSM, the crude yield of extract increases with the increase of solvent to solid ratio, extraction temperature and extraction time. However, a decreasing in concentration of RA when the extraction temperature increased. This might be due to the degradation of RA in high temperature. The concentration of RA was found to be increased when the extraction time was increased.

$$\textbf{Yield}\ (\%)= 17.258++2.4875A +0.5184B -0.6375C +0.3963AB -0.4125AC +0.0725BC -2.2695A2 -0.3638B2 -2.0888C2 \quad \text{(Equation 3.0)}$$

$$\textbf{RA}\ (\%)= 270.086-10.772A -3.0117B +12.3598C +6.2648AB +10.7826AC -9.712BC +5.6921A2 -18.0879B2 +33.4679C2 \quad \text{(Equation 4.0)}$$

Biological activities of Orthosiphon stamineus

Anti-inflammatory refers to the property of a substance or treatment that reduces inflammation or swelling. Their properties drugs make up about half of analgesics, remedying pain by reducing inflammation as opposed to opioids, which affect the central nervous system. However, anti- radical scavenging is a molecule that inhibits the oxidation of other molecules. Oxidation is a chemical reaction that transfers electrons or hydrogen from a substance to an oxidizing agent. Their reactions can produce free radicals. Table 1.0 and 2.0 shows the result of anti-radical scavenging and anti-inflammatory of *Orthosiphon stamineus* . Overall, fractionation improved the anti-inflammatory activities for the plant fractions. Even though 80% EtOH of OS exhibited the highest scavenging activity but its anti-inflammatory activity was less than 40% EtOH fraction. Further investigation is required in order to determine the bioactive compounds in the fractions for these biological activities.

Table 1.0: Anti-radical scavenging of ***Orthosiphon stamineus***

Fraction	**MK (50% inhibition) in ppm**
Vitamin C (for compare)	5.7
Crude extract	2424
20% EtOH	1576
40% EtOH	1348
60% EtOH	766
80% EtOH	377*
100% EtOH	1196

Table 2: Anti-inflammatory of *Orthosiphon stamineus*

Fraction (1g/mL)	**MK (% inhibition) in ppm**
Crude extract	77
20% EtOH	75
40% EtOH	100
60% EtOH	90
80% EtOH	88
100% EtOH	84

Conclusion

The effects of solid-liquid ratio, particle size, and temperature and extraction time on the responses of solid-liquid extraction process were successfully studied. The suitable condition for solid liquid extraction of OS for solvent were 60°C extraction temperature with ratio of solvent to solid of 10:1 (ml:g), and 2.52 hour of extraction time, respectively. Extraction by using aqueous were 81.50 °C extraction temperature with ratio of solvent 8.9:1(ml:g), and 3 hour of extraction time. Among all fractions, 80% EtOH of OS exhibited the highest radical scavenging activity.

References

1. Adam, Y., Somchit, M. N., Sulaiman, M. R., Nasaruddin, A. A., Zuraini, A., Bustamam, A. A. & Zakaria, Z. A. 2009. Diuretic properties of *Orthosiphon stamineus* Benth. Journal of Ethnopharmacology, 124, 154-158.

2. Akowuah, G. A., Ismail, Z., Norhayati, I. & Sadikun, A. 2005. The effects of different extraction solvents of varying polarities on polyphenols of Orthosiphon stamineus and evaluation of the free radical-scavenging activity. Food Chemistry, 93, 311-317.

3. Ameer, O. Z., Salman, I. M., Asmawi, M. Z., Ibraheem, Z. O. & Yam, M. F. 2012. *Orthosiphon stamineus*: Traditional Uses, Phytochemistry, Pharmacology, and Toxicology. J Med Food, 15, 678-90.

4. Bezerra, M. A., Santelli, R. E., Oliveira, E. P., Villar, L. S. & Escaleira, L. A. 2008. Response surface methodology (RSM) as a tool for optimization in analytical chemistry. Talanta, 76, 965-977.

5. Handa, S. S., Khanuja, S. P. S., Longo, G. & Rakesh, D. D. 2008. Extraction Technologies for Medicinal and Aromatic Plants, Earth, Environmental and Marine Sciences and Technologies

6. Ismail, Z. (2011). Herb industry and marketing view. Presented in: Current Herbal Development Seminar, 10 – 12 October 2011, Putrajaya.

7. Tiwari, P., Kumar, B., Kaur, M., Kaur, G. & Kaur, H. 2011. Phytochemical screening and Extraction: A Review. Internationale Pharmaceutica Sciencia, 1.

466390 Enzymatic Production of Prebiotic Fructo Oligosteviol Glycosides

Abstract:

Fructo-oligosaccharides are low-caloric sweeteners with 30-50% of the sweetness of sucrose. They are also used as functional food ingredients due to their prebiotic properties, i.e. they stimulate the growth and activity of lactobacilli and bifidobacteria in the digestive tract. Such compounds are normally extracted from chicory, but they can also be produced enzymatically from sucrose using fructosyltransferases. Steviol glycosides are naturally sweet constituents of Stevia rebaudiana, a plant species native to subtropical and tropical regions of western North America and South America. But even highly purified steviol glycosides retain attributes such as bitterness, sweet aftertaste and liquorice flavour, which reflect their degree of glycosylation. Here we describe the enzymatic two-stage conversion of steviol glycosides to prebiotic fructo-oligosteviol glycosides. The steviol glycosides are initially transfructosylated by *Microbacterium saccharopilum* fructosyltransferase and in the second stage the fructosyl chain is elongated by *Aspergillus terreus* fructosyltransferase. The product combines the prebiotic functionality and palatability of fructooligosaccharides with the sweetness of native steviol glycosides.

467063 Rapid Freshness Assesment of Bakery Products Using Discriminant Analysis and FTIR Spectroscopy

Abstract:

Bakery products are one of the most widely consumed staple food all over the world. Since fresh consumption of bakery products is typically preferred, freshness is one of the key determinates of consumer acceptance and a major contributor to the product quality (Heenan et al. 2009). Although freshness is a complex collective quality measure of a product involving sensory properties, such as appearance, odor, taste and texture, instrumental techniques can also be put in practice to devise analytical methods for qualitative assessment of food freshness (MacFie and Hedderley, 1993; Kalua et al. 2006;Sinelli 2007).

A pragmatic approach to classify scones as a model bakery product according to state of freshness was developed using a chemical profile specific analytical tool, Fourier Transform Infrared (FTIR) spectroscopy, and chemometrics. Spectral responses of the ester (C=O), aliphatic chain (C-H) stretch and water (O-H) bend vibrational modes were found to be affected due to molecular transport of fat and moisture over four-day storage period. Multivariate discrimination models were constructed to classify fresh and aged products with respect to the extent of storage on the basis of surface chemical profile information, employing a composite partial least-squares- canonical variate analysis (PLS–CVA) algorithm. The fingerprint, the ester & moisture and the C-H stretch regions of spectra were evaluated separately to find out the best freshness correlated spectral features from discrimination patterns which are statistically determined in PLS-CVA plots. It was shown that the degree of freshness was better correlated with the combined fatty acid ester & moisture and the C-H stretch regions of mid-IR spectrum.

The talk will be primarily on discussion of spectroscopic determination of physicochemical changes in scones due to molecular transport of moisture and fat. Then, use of spectral features to create discriminant models will be introduced. Discrimination of fresh bakery products and storage day based classification of samples on 2D plots will be illustrated to show effectiveness and potential of the proposed analytical approach.

References:

Heenan, S.P., Hamid, N., Dufour, J-P., Harvey, W., Delahunty, C.M. (2009). Consumer freshness perceptions of breads, biscuits and cakes. *Food Quality and Preference*, *20*(5), 380-390

Kalua C.M., Bedgood D.R., Bishop A.G., Prenzler P.D. (2006). Discrimination of storage conditions and freshness in virgin olive oil. *Journal of Agricultural and Food Chemistry, 54(19)*, 7144-7151.

MacFie, H.J.H., Hedderley, D. (1993). Current practice in relating sensory perception to instrumental measurements. *Food Quality and Preference. 4*(1), 41-49

Sinelli, N., Cosio, M.S., Gigliotti, C., Casiraghi, E. (2007). Preliminary study on application of mid infrared spectroscopy for the evaluation of the virgin oil freshness. *Anaytica Chimica Acta. 598*(1),128-134.

467183 Scaling-up Processing Parameter of *Andrograpis Paniculata* Standardised Extracts

Abstract:

Andrographis paniculata (AP) is commonly known as King of Bitters and widely cultivated in tropical areas of Southeast Asia. This work was carried out to develop an efficient processing method of AP standardised extract to retain high quality of biomarker especially in large scale. Response surface methodology (RSM) was selected in designing better processing condition . Conferring to the laboratory preparation, the optimum condition of AP extract were 50:1 (ml:g) for ratio solid to solvent with particles size less than 1.20 mm and 25 minutes for extraction process to completed. It is evident that fractionation of AP extract did not contribute to the enhancement of scavenging activity as the crude extract of AP showed at lowest value of IC_{50}. Thus, fractionation improved the anti-inflammatory activities for the plant fractions. As a result, the best fit model that represents the scaling-up extraction of AP is Sherwood number. In this work, the Sherwood number for AP is 2.03 in lab scale, while 5.17 for pilot scale. The verification of the models was done at production scale and some of the parameters were adjusted to obtain the same dimensionless number.

467191 Direct Homogenization of Beverage Emulsions with the Innovative Counterflow Injection Process

Abstract:

Soft drinks often contain beverage emulsions in order to incorporate water insoluble flavours (flavour emulsion), to imitate cloudiness of natural juice (cloud emulsion) or both combined. Despite the widespread use, physical instabilities in beverages like ringing or sedimentation are still an issue.

The production of beverage emulsions usually needs three steps: Pre-emulsification by high shear agitation, followed by deaeration and two or more homogenization steps using a high pressure homogeniser with pressures up to 400 bar (5800 psi).

In our project emulsions, particular beverage emulsions, are directly homogenised with a counterflow injection plant, a process derived from the polyurethane production. The advantages of this process are: complete separation of oil and aqueous phase prior to the mixing head, only one combined emulsification and homogenization step and lower injection pressures compared to high-pressure homogenization. This leads to lower temperature rise during emulsification, less energy dissipation and extremely variable batch sizes from lab trials of 200 mL to industrial scale of 2 t/h with one plant.

The plant mainly consist of two vessels and two high pressure metering pumps, each for the oil phase and for the aqueous phase as well as a mixhead with two injection nozzles. Prior to homogenization or during idle times, oil and aqueous phase recirculate in the specified ratio at the required pressure for injection. A spool valve in the mixhead shifts from recirculation to injection resulting in direct homogenization of both phases by high-pressure impingement on one another.
First results show that it is possible to produce stable beverage emulsions at moderate pressures of 120 bar to 160 bar (1740 â€“ 2320 psi). Emulsions mainly consist of 12 % oil phases (orange oil and weighting agent), water, sodium octenyl succinate starch, as emulsifier and citric acid. All trials were repeated for comparison with a standard GEA NiroSoavi high-pressure homogenizer.

At 160 bar a Sauter mean diameter (SMD, $D_{3,2}$) of 520 nm was achieved with the counter flow injection plant. The temperature increase of only 4 K compared to 12 K of the high-pressure homogenization shows the efficiency of this new process. Turbidity loss of the emulsions in ready-to-drink beverages at accelerated storage conditions was at the same low level as the standard emulsions made with the high-pressure homogeniser. Further potential will be investigated within the injection process and the layout of the injection nozzles to exceed present emulsion stability by high-pressure homogenization.

467066 The Lemgo D- and z-Value Database for Food – an Instrument for the Dimensioning of Pasteurization Process

Abstract:

The meaning of 'minimal processing' in the food industry is to reduce the usage of material and resources without a loss in quality or security. This principle still hasn't yet arrived in pasteurization processes for food and packaging. In practice most of the time rough estimations of death rates are the basis for the parametrization. Often it depends for the most part on the experience of the chief engineer. Major safety margins are necessary to provide secure products.

For example, in beer pasteurization state of the art is to applicate twenty pasteurization units (PU) for a safe product. This base upon a series of publications from the 50th of the last century [1–3]. Latterly some big breweries tried to reduce the heat impact to their beer to only 8 PU without losing any product security.

On the other side to underdose pasteurization can lead to a loss of food safety and if the worst comes to the worst expensive product recalls and an immense image loss [4].

New products with modern ingredients like vitamin preparations or probiotic not only are more damageable to heat, they also susceptible for microbial spoilage. Thus a precise regulation of the pasteurization process is more important than ever.

In science this problem is known and a lot of research is done to scan the thermal death rates of microorganisms. Several hundreds of publications are available each describing the thermal death of one or a few species in a specific medium. For industrial food engineer or food scientists it is difficult to keep track of all the data. For this a database has been developed at the Institute of Food Technology.NRW (ILT.NRW). The "Lemgo D- and z-value Database for Food" (LDzBase) is an open access database [5]

Because the majority of existing data are given as D- and z-values this was our first choice of data model. A D-value describes the decimal reduction time; this means the time necessary to reduce the microorganism by the factor of 10 at a given temperature. The z-value describes the temperature dependency of the D-value. It is the change in temperature necessary to change the D-value by the factor of 10. By now the database consists of about 6,000 D-values for 140 species at about 2,700 different environmental conditions [6]. The data can facilitate the dimensioning of pasteurization processes. Comparing the different potential spoiling organisms will show those species that needing the most intensive heat exposure for the necessary reduction.

A comparison within species can be used for meta-analysis of the collected data. With 700 D-values concerning Listeria monocytogenes you can get a good overview about the influence of the medium, the microorganism live in. Factors like pH-value, a_w -value or salt concentration can influence the mortality by the factor of 1,000 and more.

In our contribution we will introduce the database and show how to work with the data.

[1] **Baselt, F. C.; Dayharsh, C. A.; Del Vecchio, H. W.:** *Thermal deathtime studies on beer spoilage organisms - III.* A.S.B.C. Proceedings, S. 141–146, 1954.

[2] **Dayharsh, C. A.; Del Vechio, H. W.:** *Thermal deathtime studies on beer spoilage organisms - II.* A.S.B.C. Proceedings, S. 48–52, 1952.

[3] **Del Vecchio, H. W.; Dayharsh, C. A.; Baselt, F. C.:** *Thermal deathtime studies on beer spoilage organisms – I.* A.S.B.C. Proceedings, Bd. 1951, S. 45–50.

[4] **Oliver Daumen, B.:** *Aseptische Abfüllanlagen. Teil 1: KZE-Anlagen in der Praxis.* Brauwelt, 28/29, S. 905–912, 2003.

[5] **Schwarzer, K.:** *Lemgo D- and z-value Database for Food,* http://www.hs-owl.de/fb4/ldzbase/, [Datum des Zugriffs] 09.05.2016.

[6] **Schwarzer, K.; Schneider, J.; Müller, U.; Becker, B.; Wilhelm, P.:** *Lemgo Database for D- and z-values.* Brauwelt International, Nr. 5, S. 252–256, 2010.

467270 Comparison of the Static and Dynamic Freeze-Drying By Means of Modelling

Abstract:

The freeze-drying (lyophilisation) is process, which is used extensively in many areas of the food industry and pharmaceutical industry. High costs lead to the treatment of only high-quality and valuable products e.g. starter cultures. The microbial activities of these cultures are preserved well by freeze drying.

The freeze-drying is slow due to a low partial pressure of the vapor given by the low temperature as well as to increasing heat and mass transfer resistances. Specifically, the drying velocity is reduced with the growing dried layer. Consequently, of the conductive heat transfer to the ice front as well as of the diffusive vapor transfer from the ice front are slowed down.

To produce starter cultures, the biomass is increased, harvested and frozen by dropping into liquid nitrogen. Particles of approx. 3 to 4 mm are formed. This standard process is followed by dynamically freeze drying the pellets. Dynamical freeze drying describes the process of freeze drying in a vacuum contact drying, i.e. drying during continuous movement of the particles in a solid mixer. The formed dry layer on the surface is scraped off, so that at every time ice-crystals are present on the surface and sublimation without diffusion hindrance can proceed. It was shown, that such a freeze-drying can be carried out steadily in the constant drying rate regime. In addition a comparison of required drying times of this â€œdynamic freeze-drying processâ€ with static freeze-drying at optimum drying conditions will be given. Actual heat transfer coefficient were determined in dependency of the dryer loading, drying-temperature and rotary frequency of the mixing element starting at a mean particle size of about 4 mm. A calculation of the whole drying process gives the registered energy which leaves the drying arrangement as steam and accordingly also the required time to remove all 'frozen' water. The obtained drying time is compared with the drying time of the conventional drying.

469011 Production of Branched-Chain Fatty Acid in High Percentage

Abstract:

Branched-chain fatty acids (BCFAs) are important precursors for the production of advanced biofuels with improved cold-flow properties. Previous efforts in engineering type II fatty acid synthase (FAS) for BCFA production suffered from low titers and the co-production of a large amount of straight-chain fatty acids (SCFAs), complicating efficient isolation of BCFA. Synthesis of both SCFAs and BCFAs requires FabH, the only Î²-ketoacyl-(acyl-carrier-protein) synthase that catalyzes the initial condensation reaction between malonyl-ACP and a short-chain acyl-CoA. We demonstrated that replacement of the acetyl-CoA-specific *Escherichia coli* FabH with a branched-chain-acyl-CoA-specific FabH directed the flux to the synthesis of BCFAs, resulting in 81-fold enhancement in BCFA titer compared to a strain containing both acetyl-CoA- and branched-chain-acyl-CoA-specific FabHs. We further identified the incomplete lipoylation of 2-oxoacid dehydrogenases as a key bottleneck in BCFA production, and thus improved the BCFA percentage to 85% of the total free fatty acids (FFAs) by engineering of protein lipoylation pathways. Furthermore, we achieved fine-tuning of BCFA branch positions by engineering the upstream pathway to control the supply of different branched-chain acyl-CoAs, and eventually BCFA can be produced from glucose at 181 mg/L and 72% of total FFA.

469547 Solubility of Ternary Systems Containing Capsaicin + β-Carotene and Lauric Acid + Palmitic Acid in Supercritical Carbon Dioxide

Abstract:

The solubility data of solid mixtures in supercritical fluid (SCF) are rarely reported in literature. Capsaicin and β-carotene are both components of paprika (*Capsicum annuum*), one of the oldest source of natural colorants used in the food and cosmetic industries. β-Carotene is one of several carotenoids found in paprika. Capsaicin is the most important component among various capsaicinoids which causes pungency of paprika and should therefore be removed from coloring compounds (carotenoids).

On the other hand, vegetable oils contain a large number of compounds that can be classified into unsaponifiable and saponifiable fraction. The latter, accounting for about 98% of the oil, is mostly constituted from fatty acids forming esters with glycerol to produce mono-, di-, and tri-glycerides. The fatty acids composition of oils varies considerably according to the vegetable matrix (olive, sunflower, corn, palm, hazelnut, soybean, etc.) and even varies for the same vegetable matrix when considering variety, local factors, climatic condition, harvest time, extraction processes from fruits or seeds, etc.

Supercritical extraction with CO2 seems to be a good alternative. Due to carbon dioxide's relatively low critical temperature (Tc= 304.4 K), it is especially suitable for processing thermo-labile compounds. Information about solubility of solids in supercritical fluids is of great importance because supercritical fluids have several advantages over organic solvents about physical properties. Accurate experimental solubility data are fundamental to test theoretical, empirical or semi-empirical models. Semi-empirical models based on density are common approaches used to correlate experimental solid solubilities in supercritical fluids.

The aim of this work is to present new experimental solid solubilities of the ternary systems: Capsaicin + β-Carotene + Carbon dioxide, and Lauric acid + Palmitic acid+ Carbon dioxide at 313, 323 and 333 K an

pressures up to 30 MPa, based on the static-analytical method. The solubility data were correlated with the Mendez-Santiago and Teja model.

469698 Optimized Supercritical Fluid Extraction of Capsaicinoids from the Capsicum annuum Cultivar (Red Pepper)

Abstract:

The relationship between supercritical CO_2 extraction solubility parameters and the specific surface area of pretreated capsicum annuum cultivar will be examined in order to obtain the optimum extraction efficiency for the major capsaicinoids including capsaicin, dihydrocapsaicin, and nordihydrocapsaicin. Linearly increasing the solubility parameter pressure of supercritical CO_2 isothermally, increases solvent density and subsequently solvent power leading to increased solubility of non-polar capsaicinoid molecules and greater extraction efficiency from the raw cultivar. However, in the target extraction zone, linearly increasing the solubility parameter temperature, to take advantage of capsaicinoid phase change, has a net positive improvement in extraction efficiency at the expense of supercritical fluid density. Diffusion and transfer to the bulk solvent phase is increased but not nearly to the degree that specific surface area through cultivar pretreatment facilitates extraction efficiency. Increasing cultivar specific surface area through milling pretreatment yields an increase in capsaicinoid extraction efficiency via supercritical CO_2 as long as mass transfer is not impeded by high levels of tortuosity. Product and process cost efficiency of commercial extraction is a key economic element in the value chain of natural substance production from not only capsicum annuum cultivars but a wide range of agricultural products found globally.

469904 Comparative Evaluation of 13 Yeast Species in the *Yarrowia* Clade on Lignocellulosic Biomass Hydrolysate and Genetic Engineering of Inhibitor Tolerant Strains for Lipid and Biofuel Production

Abstract:

Yarrowia lipolytica is an oleaginous yeast that has garnered interest for commercial production of single cell oil and other fatty acid-derived chemicals because of its GRAS status and genetic tractability. Three recent peer-reviewed studies have highlighted the possibility of lipid production by this yeast using lignocellulosic biomass hydrolysates, which provide an inexpensive and renewable feedstock for bioprocessing. Prior research on *Saccharomyces cerevisiae* suggests that yeast strains can vary widely in their tolerance to common inhibitors in cellulosic hydrolysates. In the present study, we investigated growth and lipid production by 57 yeast isolates, representing 13 species in the *Yarrowia* clade, on a dilute acid-pretreated switchgrass hydrolysate with *Y. lipolytica* strain W29 as a control. Phenotypic traits, such as inhibitor tolerance, substrate utilization, cell growth, and oleaginicity, varied considerably within the clade. Furthermore, we identified a lesser-known *Yarrowia* species that was superior to W29 in terms of lipid content (49.7% of dry weight) and lipid titer (2.9-fold improvement). The top performing strains, including members of a novel species, were subsequently evaluated for amenability to genetic engineering. An integration vector was constructed with a selectable antibiotic resistance marker and successful transformations were demonstrated by modifying a protocol developed for *Y. lipolytica*. Then, endogenous *Yarrowia* promoters were characterized in the new species by monitoring specific expression of a fluorescent reporter protein in relevant conditions. Using our newly developed genetic tools, we were able to manipulate a set of candidate target genes in a non-model host species and improve promising strains for enhanced conversion of cellulosic sugars to triacylglycerols. In summary, this work describes identification, characterization, and engineering of novel oleaginous strains in the *Yarrowia* clade for lipid and biofuel production from renewable biomass.

472230 Genomic Hot Spots for Strong and Stable Expression Sites in *Yarrowia Lipolytica.*

Abstract:

Integrative expression in *Y. lipolytica* has been accomplished by targeting the zeta site for multiple copy integration to promote strong expression and by site specific integration. While this provides for stable and strong integration, there is little control of copy number and/or expression strength. Therefore, being able to control expression strength and attain strong yet stable expression is essential for metabolic engineering applications. It is well known that in eukaryotes expression strength is genomic loci-dependent as there are several factors that contribute to controlling transcriptional activation and amplification at the genomic scale. Here, we sought to identify loci in *Y. lipolytica* that promote strong expression and study the physical environment that enables such expression levels. To identify these loci, we transformed linear DNA carrying a GFP reporter fluorescence gene under a strong hybrid promoter and allowed *Y. lipolytica* to randomly integrate the DNA into its genome. A library of fluorescent cells was sorted into bins of expression strength using Fluorescent Activated Cell Sorting (FACS). Our results indicate that expression strength varied based on integration locus. A combination of genome sequencing methods was used to identify GFP integration location in the genome. The highly fluorescent colonies were further characterized for expression strength under different carbon and nitrogen conditions to determine if local regulation influences heterologous gene expression. The information gained from these experiments will allow future metabolic engineering efforts to attain maximum expression from genetic cassettes when integrating onto the genome.

475172 Enhanced *N*-Butanol Fermentation By Engineered *Clostridium Tyrobutyricum* Integrated with Gas Stripping

Abstract:

Butanol can be produced by acetone-butanol-ethanol (ABE) fermentation from lignocellulosic biomass such as soybean hull and sugarcane bagasse. Traditional ABE fermentation usually suffers from low butanol titer, yield and productivity, which however, can be boosted by applying external driving forces to redirect the electron and carbon flow towards butanol synthesis instead of byproduct formation. In this work, an engineered strain of ***Clostridium tyrobutyricum*** with ***ack*** knock out overexpressing ***adhE*2** was used for butanol production from various carbon sources including glucose and xylose. Increased butanol production with high yields (>0.30 g/g) was achieved by providing artificial electron carrier such as methyl and benzyl viologen. Fed-batch fermentation in a fibrous-bed bioreactor integrated with gas stripping gave a total butanol production of 24.7 g/L, yield of 0.32 g/g and productivity of 0.10 g/L·h from glucose/xylose mixture. The process produced 12.1 g/L butanol at a yield of 0.25 g/g and productivity of 0.12 g/L·h from soybean hull hydrolysate (SHH), and 10.7 g/L butanol at a yield of 0.31 g/g and productivity of 0.081 g/L·h from sugarcane bagasse hydrolysate. This work demonstrated the feasibility of producing biobutanol from low-cost agricultural residues such as soybean hull and sugarcane bagasse.

476090 The Design and Preparation of Fe_3O_4/TiO_2/ go for Photo-Fenton Catalyst in the Degradation of Tetracycline Hydrochloride

Abstract:

A novel Fe_3O_4/rGO/TiO_2 photocatalyst was synthesized by a simple two-step process, an electrostatic self-assemble and hydrothermal method. The as prepared Fe_3O_4/rGO/TiO_2 exhibited an enhanced photocatalytic activity for the degradation of Tetracycline Hydrochloride(TCH) in aqueous medium using sunlight. The reaction parameters, including the initial solution pH, reaction temperature, concentrations of TCH and H_2O_2, dosage of the catalyst, on the degradation rates of the TCH were systematically investigated. The photocatalyst was highly effective for the degradation of TCH in a wide pH range of 2.0â€"11.0. At optimal conditions, 97% degradation rate of TCH was achieved with 360min. A possible photocatalytic mechanism for TCH by Fe_3O_4/rGO/TiO_2 and H_2O_2 was proposed based on the quenching tests and high efficiency liquid chromatography-mass spectrometry (HPLC-MS) analysis. In addition, leaching test was also carried out and the result suggested that the leached iron from the catalyst was negligible and the catalyst still exhibited high photocatalytic activity after five reaction cycles, which demonstrated that the Fe3O4/rGO/TiO2 was reusable, excellent stable and highly effective.

476563 Simulation of a sugar beet factory using a chemical engineering software to perform energy analysis

Abstract:

Nowadays, it is of the utmost interest to optimize energy consumption within factories in a systematic way to face recent environmental regulations. This study shows a methodology to model food processes to in fine combining Pinch and exergy analysis, using a chemical engineering software (ProSimPlus ® , version 3.5.3.1.). The methodology was applied to a sugar beet factory. A UNIFAC-Larsen thermodynamic model was used to estimate the thermophysical properties of the streams. Each unit operation related to the sugar beet process was considered as a combination of several modules to distinguish the thermodynamic irreversibilities of the process: a unit operation module only simulates one energy change (pressure, temperature or composition). In addition, the numeral unit operations specific to the sugar beet process not existing in the ProSimPlus ® database (i.e. diffusion, mechanical press, filtration and centrifuge) were modeled. To do so, basic and generic unit operation modules and "constraints and recycles" modules (SPEC) were combined. The developed model allows for the simulation of a sugar beet factory, with a deviation range within the industrial error bars.

454184 Modeling and Simulation Efforts in Scale up of Sustained Release Microspheres

Abstract:

Abstract

Generic pharmaceutical industry has to operate in tight timelines of product development and scale up from lab to plant. To meet such needs, Dr Reddys Laboratories use approaches like modeling and simulation for better process understanding and to reduce scale up risks. This presentation provides example cases of such efforts in a microspheres product.

Microencapsulation process development demands understanding of the impacts of various material properties and process parameters to achieve the right product characteristics. Employing appropriate modeling approaches at different scales can aid in developing this understanding. In this work, a first principle based diffusion model which uses Monte Carlo simulation helps link the impact of microsphere product attributes like size distribution, porosity, and polymer properties on the sustained release of drug from microspheres.**[1]** Results from a numerical simulation of this model gauge the risk of each of the product characteristics on the release. This understanding identifies the key product attributes to focus on and manipulate with the process conditions.

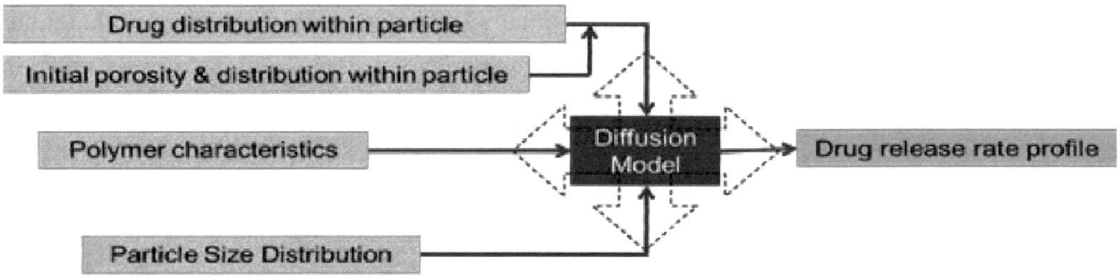

Manufacture of such micro-encapsulated products consists of multiple unit operation steps like phase separation, extraction and drying to get the desired product characteristics. A combination of first principle, and CFD based approaches for the unit operations links how each of the process steps impacts the product attributes. Some of the important linkages made include scale dependent variations of: mixing conditions on particle size using vessel averaged Reynolds and Weber number obtained from CFD simulations; cycle times for extractions by single particle spherical diffusion model; and drying cycle predictions by first principle discrete element model with mixing simulated by randomization of the discrete elements. Custom lab scale experimentation provides the necessary model parameters and enables extension of the understanding towards the plant scale. This provides the necessary framework for a successful scale-up.

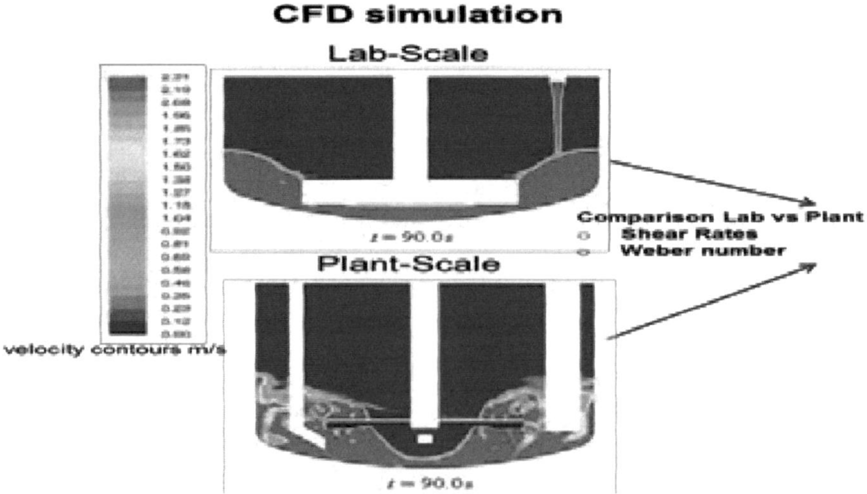

[1] J. Siepmann, N. Faisant, J.P. Benoit, A new mathematical model quantifying drug release from bioerodible microparticles using Monte Carlo simulations, Pharm.Res. 19 (2002) 1885–1893.

455552 Combined Task Bayesian Optimization for Efficient Scale-up and Technology Transfer

Abstract:

Operating condition optimization is crucial to assure product quality and reduce operation cost in any industry. In particular, one of the most challenging problems is to optimize operating conditions of commercial-scale equipment when only a small number of data are available just after scale-up from pilot-scale equipment. Such a situation is quite common because the experimental cost is significantly higher with commercial-scale equipment than pilot-scale equipment. In the present work, we aim to solve this problem and propose a new operating condition optimization method, which is referred to as combined task Bayesian optimization (CTBO). The proposed method is applicable not only to the scale-up problem but also to the problem of technology transfer from a mother plant to a copy plant; that is, when two plants are the same but operating conditions need to be optimized at the copy plant due to various uncertainties.

A conventional method builds a model that relates product qualities with operating conditions, and then optimizes the operating conditions by using the model so that the target product qualities are realized and operation cost is minimized under various constraints. Unfortunately, this method does not work well when only little data is available at a commercial-scale plant, because it is difficult to build an accurate model. Such a situation always occurs just after scale-up from a pilot-scale plant. The optimization performance would be improved if data of the pilot-scale plant could be used jointly with data of the commercial-scale plant. However, combining data of different scales is not straightforward because pilot-scale and commercial-scale plants have different numbers and types of sensors, which are operated under different conditions.

The goal of this research is to develop a new method that can efficiently and accurately derive optimal operating conditions by using data obtained from a commercial-scale plant and a pilot-scale plant even when only little data is available at the commercial-scale plant. The proposed method uses Bayesian optimization (BO).

BO can systematically determine a plan for new operating conditions to be evaluated for further optimization. Thus, without experimental design, BO can find a better solution through fewer experiments than conventional methods. BO uses two important tools: Gaussian process regression (GPR) and an acquisition function. GPR is used as a nonlinear regression method, and the acquisition function is used to determine new operating conditions to be evaluated by taking account of the trade-off between exploration and exploitation. By combining them, BO can efficiently optimize operating conditions.

Transfer learning aims to exploit knowledge from one or more source tasks and to apply the knowledge to the target task. A key idea of the present work is that transfer learning is useful for solving the scale-up problem by regarding the source task and the target task as the pilot-scale plant and the commercial-scale plant, respectively. Among transfer learning algorithms, adaptive transfer learning and frustratingly easy domain adaptation are used in this work. The former is an algorithm established for GPR, and the latter is a famous, simple algorithm.

Conventional transfer learning algorithms assume that the number of input variables in the source task is the same as that in the target task. In practice, however, the number and types of sensors of a pilot-scale plant are different from those of a commercial-scale plant. Thus, to apply transfer learning to the scale-up problem, we introduce a transformation matrix, which transforms data of the pilot-scale plant into data of the commercial-scale plant so that data from both can be used for modeling the commercial-scale plant. To derive the transformation matrix automatically, we propose to use Markov Chain Monte Carlo (MCMC), which is also used to determine GPR hyperparameters.

In the present work, combined task Bayesian optimization (CTBO) is proposed by integrating BO, transfer learning, and the transformation matrix. To validate the effectiveness of the proposed method, CTBO was compared with BO and LW-PLS + jDE (locally weighted partial least squares + self-adaptive differential evolution) through their applications to numerical examples and a pharmaceutical granulation process. Two types of CTBO were tested: CTBO(ATGP), i.e., CTBO with adaptive transfer learning, and CTBO(DAGP), i.e., CTBO with frustratingly easy domain adaptation.

The results of the numerical examples showed that the optimization performance of BO was superior to that of LW-PLS + jDE since BO appropriately determined operating conditions one by one. CTBO further improved the optimization performance by combining data of both the source task and the target task. Similarly, in the pharmaceutical granulation process, CTBO outperformed LW-PLS + jDE and BO. The improvement rate of CTBO(DAGP) compared with LW-PLS + jDE and BO were 48.6% and 15.2%, respectively, when the number of samples on the commercial-scale plant was 10. The results demonstrated that CTBO used data of the pilot-scale plant effectively to optimize the operating conditions of the commercial-scale plant.

456730 Development and Optimization of a Mixed Suspension Mixed Product Removal (MSMPR) Crystallization Process Incorporating Wet-Milling

Abstract:

In recent years, there has been an increased interest in moving pharmaceutical manufacturing from batch to continuous processing. The potential to reduce cost, simplify production and improve product quality have driven significant investment in continuous manufacturing. While crystallization is an important separation and purification unit operation, continuous crystallization processes have not been widely studied within the pharmaceutical industry primarily because of the design complexity and difficulty of robust operation. The MSMPR crystallizer is arguably one of the most promising technologies as its operation is conceptually simpler than for e.g. tubular crystallizers and it is easy to convert batch crystallizers to continuous operation. However, issues such as fouling and encrustation, transfer line blockage, and classification due to limited nucleation are frequently experienced with numerous reports in the literature.

The work presented herein solves the fouling and classification issues encountered for one of Takedaâ€™s APIs by adding wet-milling to the first stage of an MSMPR crystallizer. The API studied was found to be characterized by slow nucleation kinetics resulting in a prolonged wash out period. During this period, large (>1mm) particles were formed that were unable to be transferred leading to classification issues, build up of supersaturation and severe fouling.

The high shear imparted by the wet-mill was found to control particle size and increase both primary and secondary nucleation rates. This reduced supersaturation and therefore fouling and encrustation. The wet-mill was incorporated in the first stage of the cascade which was then experimentally optimized investigating factors such as solvent composition, types and materials of construction of the impeller, wet-mill rotational speed, wet-mill turnover numbers, residence time and start-up methods. Finally, a multi-stage system was developed to optimize productivity.

457164 Developing Process Models of Hydrogen Peroxide Decontamination in Pharmaceutical Manufacturing

Abstract:

Decontamination is a critical technology for providing sterile environment in the production of sterile drug products such as injectables, intravenous bags and eye-drops. In the modern production facilities, these drug products are filled in the equipment such as isolators, which are decontaminated before the operation using vapor hydrogen peroxide. This industry-standard decontamination agent is diffused in the equipment as vapor and mist, kills the microorganisms on the surface, and removed by aeration before the filling can start. The success of this decontamination process is a prerequisite for assuring the quality of the above mentioned sterile drug products. Furthermore, decontamination is a relevant process regarding productivity, because the entire process including conditioning and aeration dominates the production time. Despite of such relevance in quality and productivity, scientific elucidation of decontamination is still in infancy, which is the obstacle to more rational design of decontamination processes.

The objective of our research is to develop a model of decontamination using hydrogen peroxide, which can serve as the basis for process design in pharmaceutical manufacturing. We aim to create a model with linking the design parameters to so-called â€œD-valueâ€, and further to the objective functions such as time required for decontamination or energy consumption. D-value is the representative parameter of decontamination efficacy, and is defined as the time required to kill 90% of surviving microorganisms. As the design parameters, concentration of hydrogen peroxide as well as humidity are considered, and the influence of condensation is also taken into account in the model. The relationship between the design parameters and D-value is to be investigated by experiments using biological indicators with following the standard application procedure in the pharmaceutical industry. In addition to decontamination efficacy, our investigation will cover time and energy consumption required for the decontamination process including conditioning and aeration.

So far we set up experimental apparatus for the decontamination process in the steady-state condition, and measured D-values with varying humidity and hydrogen peroxide concentration. The preliminary results of the experiments suggested that the influence of humidity and concentration on decontamination efficacy is strong as known in the previous studies. A mathematical expression has been firstly created considering the relationship between the varied parameters and the observed D-values. This initial mathematical model will incorporate the influence of condensation on decontamination efficacy by using vapor-liquid equilibrium, diffusion, and mass transfer. Additionally, the model will cover the effect of the concentration of residual hydrogen peroxide on the product. After consideration of the condensation and the residual concentration, the time required for decontamination process will be evaluated, and the time will be linked to the objective functions. In future, we intend to create a design method for the decontamination processes using the process model to be developed in this work.

457244 Comparison of Raman and UV Spectroscopy for the in-Line Quantification of Active Ingredient in Pharmaceutical Semi-Solid and Liquid Formulations

Abstract:

Comparison of Raman and UV spectroscopy for the in-line quantification of active ingredient in pharmaceutical semi-solid and liquid formulations

N. Bostijn[1,*], C. Vervaet[2], T. De Beer[1]

[1]Laboratory of Pharmaceutical Process Analytical Technology, Ghent University, Ottergemsesteenweg 460, 9000 Ghent, Belgium

[2]Laboratory of Pharmaceutical Technology, Ghent University, Ottergemsesteenweg 460, 9000 Ghent, Belgium

Introduction

In-line spectroscopic techniques are increasingly proposed as alternative methods for the quantification of active ingredients in pharmaceuticals. This is due to their advantages over the traditional methods, such as fast and non-destructive measurements without the need of sample pretreatment. One of these spectroscopic techniques is Raman, that is until now, mostly applied for solid dosage forms. UV spectroscopy is a widely used quantification technique, although less frequently implemented inside a process environment for in-line monitoring.

This study compares the feasibility of in-line Raman and UV spectroscopy for the quantification of active ingredients in pharmaceutical formulations. A suspension and gel containing 0.09 % (w/w) and 2 % (w/w) active ingredient, respectively, were used as model formulations. The performance of both spectroscopic techniques was assessed via accuracy profiles [1-3].

MATERIALS AND METHODS

Materials

Two commercial available pharmaceutical formulations were selected from a pharmaceutical company. The suspension contained 0.09 % (w/w) active ingredient and the gel 2 % (w/w). Both formulations were prepared based on confidential information.

Methods

Calibration (80, 90, 95, 100, 105, 110 and 120 % of target) and validation (85, 95, 100, 105 and 115 % of target) samples were created by producing a batch for the lowest concentration. After batch completion, a specific amount of active ingredient was added, corresponding to the subsequent concentration level, followed by the collection of spectra. This procedure was repeated until the highest concentration level was reached and spectra were recorded from each concentration. In total, one calibration batch and three validation batches (three days) were produced for each formulation.

The suspension and gel were prepared using a customized IKA LR2000 modular mini plant reactor system (IKA, Staufen, Germany). The mixing vessel was equipped with a double wall for controlling the temperature of the process. A peristaltic pump (Watson-Marlow, Falmouth, Cornwall, UK) was used for pumping water from a water bath (Type 1032, GFL, Burgwedel, Germany) to the mixing vessel.

In-line Raman spectra were recorded using a Raman Rxn2 spectrometer (Kaiser Optical Systems, Ann Arbor, MI, USA), equipped with a fiber-optic PhAT probe. The laser wavelength was 785 nm. The spectral range of the system was from 150–1890 cm^{-1} with a resolution of 5 cm^{-1}. An exposure time of 15 s with no averaging was used.

An Avaspec-ULS2048L spectrometer (Avantes, Apeldoorn, The Netherlands) was connected by a fiber-optic cable to an immersion probe. The light source was an AvaLight Deuterium-Halogen Lamp. All spectra were acquired in the 200-1100 nm spectral range. The exposure time was 1000 ms and 950 ms for the gel and suspension, respectively, with each spectra the average of 5 scans.

The calibration models of the gel were developed using the mean-centered spectra, after standard normal variate (SNV) correction and first derivative transformation, over the regions of 390.1-404.8, 655.0-666.7, 1340.2-1354.9 and 1570.0-1600.0 cm^{-1} for the Raman spectra, while for the UV spectra the region 280.1-296.9 nm was used.

The Raman spectra of the suspension were mean-centered and SNV corrected, followed by taking the first derivative and selecting the region between 1390.0-1430.2 cm^{-1}. For the UV spectra, the same preprocessing was applied, except for the SNV correction and the selection of another spectral region (310.1-325.6 nm).

Spectroscopic data were correlated with the actual concentration of active ingredient, through partial least square (PLS) regression. The PLS models were created using the SIMCA software (Version 14, Umetrics, Umeå, Sweden).

Accuracy profiles, introduced by the SFSTP [1-3], were used as validation approach for the quantitative analytical methods.

RESULTS AND DISCUSSION

Gel

The RMSEP values of the Raman prediction set (0.026, 0.024 and 0.039 %) were lower compared to the ones of the UV data (0.058, 0.071 and 0.059 %). Accuracy profiles confirm the better performance of Raman. The β-expectation tolerance intervals exceed the 5 % acceptance limits only at the 1.75 and 2.37 % (w/w) concentration levels, whereas for UV, the β-expectation tolerance intervals exceed the 5 % acceptance limits over the whole concentration range.

Suspension

It can be noticed from the accuracy profile of the Raman method that the low concentration levels (0.077 and 0.086 % w/w) were predicted higher, the middle concentration level (0.091 % w/w) was predicted correct and the high concentration levels (0.095 and 0.104 % w/w) predictions were lower, suggesting that all the different concentrations were predicted the same value. This observation, the spectral data and the high RMSEP values (0.0097, 0.01 and 0.098 %) confirm that it was not possible to discriminate between the different concentration levels with Raman spectroscopy. For the UV calibration model, the β-expectation tolerance intervals exceed the 5 % acceptance limits over the whole concentration range, but fall within the 10 % acceptance limits at the 0.0865, 0.091 and 0.0955 % (w/w) concentration levels. RMSEP values (0.0050, 0.0015 and 0.0017 %) were lower than the RMSEP values of the Raman based quantification method.

CONCLUSION

The study showed that Raman spectroscopy performed better than UV spectroscopy for the in-line quantification of 2 % (w/w) active ingredient in a gel, but both techniques were feasible. For the low dosed suspension (0.09 % w/w), Raman spectroscopy failed to discriminate between the different concentration levels and thus the quantification of the active ingredient. UV spectroscopy was adequate for the in-line quantification, however between the acceptance limits of 10 %, instead of 5 %.

REFERENCES

1. P. Hubert et al. Harmonization of strategies for the validation of quantitative analytical procedures A SFSTP proposal-part I. Journal of Pharmaceutical and Biomedical Analysis. 2004; 36: 579–586.
2. P. Hubert et al. Harmonization of strategies for the validation of quantitative analytical procedures A SFSTP proposal-part II. Journal of Pharmaceutical and Biomedical Analysis. 2007; 45: 70–81.
3. P. Hubert et al. Harmonization of strategies for the validation of quantitative analytical procedures A SFSTP proposal-part III. Journal of Pharmaceutical and Biomedical Analysis. 2007; 45: 82–96.

458314 Process Development for Continuous Extrusion and Pelletization of a Heat and Shear-Sensitive Formulation

Abstract:

Hot-melt extrusion is an emerging technology for pharmaceutical manufacturing. Not only does it offer the possibility to establish amorphous systems to increase bioavailability, but also carries attractive implications process-wise [1]. It is an inherently continuous processes, performs several processing steps in one machine and offers the possibility of continuous monitoring and direct product shaping, e.g. to pellets, strands or films. At the same time, intense shear forces and dissipative heat generation are challenging when using some traditional carrier polymers [2].

In this case an immediate release formulation was developed containing a BCS class II with poor solubility. The formulation consists of the API, Eudragit E and Methocel E5. Methocel E5 was added in order to achieve immediate release from the very dense pellet. Besides the shear and heat sensitivity of Methocel E5, a high pressure drop across the extrusion die is obtained due to the high viscosity and the yield point. Therefore, the extrusion process line consists of (1) a hot-melt extruder with a melt pump, (2) and a strand pelletizer. The strand pelletizer is used to directly produce pellets of small size (x_{50} = 1000 µm) for subsequent continuous tableting.

The process development focused on the split-feeding unit, the intake zone, the temperature profile, the die plate temperature and the die plate design. After developing a stable process, a screening DoE was performed to investigate content uniformity, dissolution behavior, pellet size and material degradation. Moreover, the residence time distribution was determined.

References:

[1] J. Breitenbach, "Feste Lösungen durch Schmelzextrusion - Ein integriertes Herstellkonzept," *Pharm.*

Unserer Zeit, vol. 29, no. 1, pp. 46–49, 2000.

[2] A. Meena, T. Parikh, S. S. Gupta, and A. T. M. Serajuddin, "Investigation of thermal and viscoelastic properties of polymers relevant to hot melt extrusion - II : Cellulosic polymers .," *J. Excipients Food Chem.*, vol. 5, no. 1, pp. 46–55, 2014.

458474 Wet Granulation-Induced Polymorphic Transitions in Piracetam: Raman Spectroscopy As a Potential in-Situ Process Analytical Tool

Abstract:

Wet granulation induced polymorphic transitions in Piracetam: Raman spectroscopy as a potential in-situ process analytical tool

Gayathri Kollamaram, Catherine B. Kelly, Denise M. Croker, Gavin M. Walker

University of Limerick, Synthesis and Solid State Pharmaceutical Centre, Department of Chemical and Environmental Sciences, Limerick, Ireland.

Gayathri.Kollamaram@ul.ie

Wet granulation is one of the most popular techniques used during the post processing of active pharmaceutical ingredients (APIs) due to its ability to improve powder flow and compaction properties however the addition of water can affect the crystal structure of certain APIs exhibiting polymorphism. Piracetam is a highly aqueous-soluble drug which can exist in many polymorphic forms including hydrates. The potential therefore exists for process-induced transformations during wet granulation. Process-induced polymorphic transitions in APIs are a major challenge in the pharmaceutical industry as these unpredictable phase transitions can affect processability of the drugs owing to altered physical properties such as compressibility and can result in regulatory non-compliance due to altered bio-availability. Prior knowledge and understanding of the mechanism of these process-induced polymorphic transitions is therefore of the utmost importance. XRD and DSC techniques were used along with Raman to examine the polymorphic transitions of Piracetam during wet granulation so as to determine the effectiveness of Raman spectroscopy as a process analytical technology (PAT) tool for this system.

Piracetam undergoes polymorphic transitions under the influence of both solvent and temperature. FIII of Piracetam is the most stable form at room temperature. FIII transforms to FI when heated to 140°C which transforms to metastable FII upon cooling to room temperature following the Ostwald's step rule of successive transformation. At ambient conditions FII will eventually transform to the more stable FIII. It is known that polymers can have a stabilizing and/or inhibitory effect on the polymorphic transitions in APIs so different excipients with a range of hydrophilicities, aqueous solubilities and molecular weights were examined. Excipients used were Lactose monohydrate (LMH), microcrystalline cellulose (MCC), Hydroxy propyl methyl cellulose (HPMC), Hydroxy propyl cellulose (HPC) and Polyethylene glycol (PEG). LMH was chosen as the excipient representing low molecular weight; MCC, HPMC and HPC represent the excipients with varying degree of hydrophilicity while PEG is a water soluble polymer.

Piracetam was wet granulated with excipients with drug to excipient ratio of 3:2 as Piracetam is a high dosage drug. The granules produced were dried overnight at room temperature and the dried granules were then analysed at pre-determined times points using XRD, Raman and DSC. Raman was also employed to investigate the use of this technique as an in-situ analysis method. Principle component analysis (PCA) and direct cluster analysis (DCA) were used to process the Raman spectra as shown in (figure 1). Prior to processing all spectra were normalised and the first derivative was taken. XRD indicated that PIR granulated without any excipients transformed from FIII to Piracetam monohydrate (PIR-MH) in the presence of water and then recrystallized to FII upon drying. FII so obtained gradually recrystallised to the stable FIII at room temperature within two weeks. The Raman spectra gained from the granulation and drying of PIR were analysed using PCA and DCA and showed clustering into three distinct spectra which were confirmed to be those of FII, FIII and PIR-MH using XRD. Raman was therefore used in conjunction with XRD to inspect the progress of the transformation both from FIII to PIR-MH and from FII to FIII (Figure 2).

Granules of PIR containing LMH, MCC, HPMC and HPC exhibited similar transformations as pure PIR during granulation, namely that FIII transformed to monohydrate in the presence of water which recrystallized to FII upon drying (Figure 3). The kinetics of solid state transformation of FII to FIII at room temperature was retarded in the presence of excipients in comparison to that of pure PIR and by different degrees depending on the polymer. This was attributed to the interaction of excipients with the active functional groups in the drug and by the hydrophilicity and molecular weight of the excipients. However PEG had a different effect on these polymorphic transitions. Piracetam remained in its stable FIII throughout the processes of wetting and drying. Unlike the other excipients used in the study PEG is freely soluble in water and therefore it is possible for PEG to competitively inhibit the interaction between PIR and water. It is suggested that PEG inhibited the solution-mediated polymorphic transition of FIII to FII through the intermediate monohydrate.

Through the use of the complimentary techniques of XRD, DSC and Raman spectroscopy the polymorphic transformations exhibited by Piracetam during wet-granulation have been investigated. The importance of detecting the monohydrate form of the API as a precursor to the production of the metastable FII, the effectiveness of Raman spectroscopy in this detection, and the importance of understanding the excipients typically used during wet granulation have been demonstrated.

Description: Description: C:\Users\Gayathri.Kollamaram\Desktop\Exp\Piracetam\Images\PCA.tif

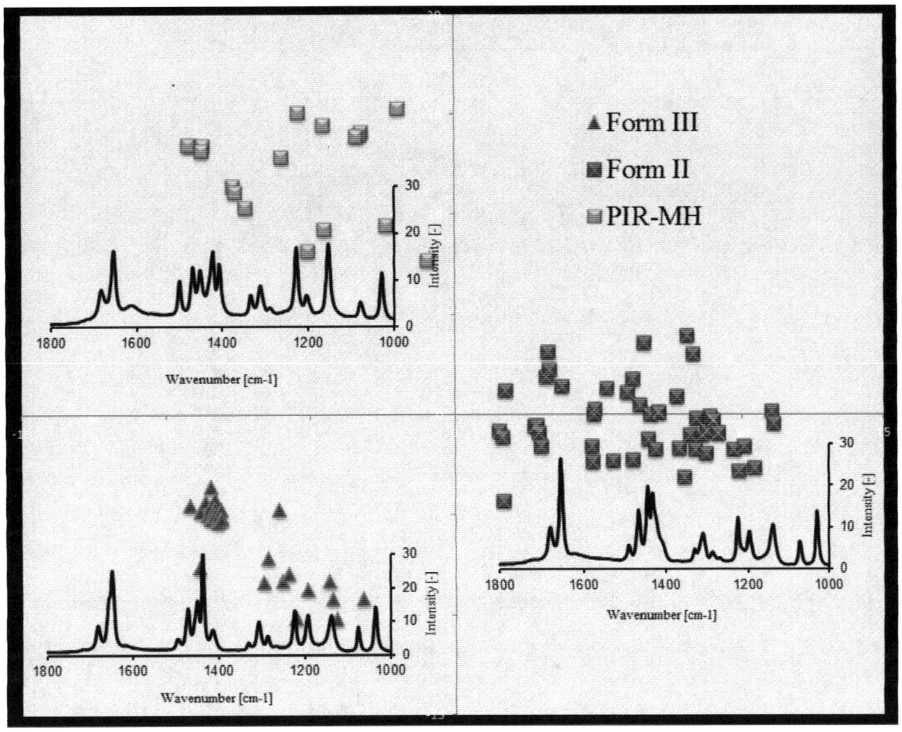

Fig 1: Prinicpal component analysis of the Raman spectra gained from the studies of Piracetam in water indicated clustering of the spectra into three distinct forms.

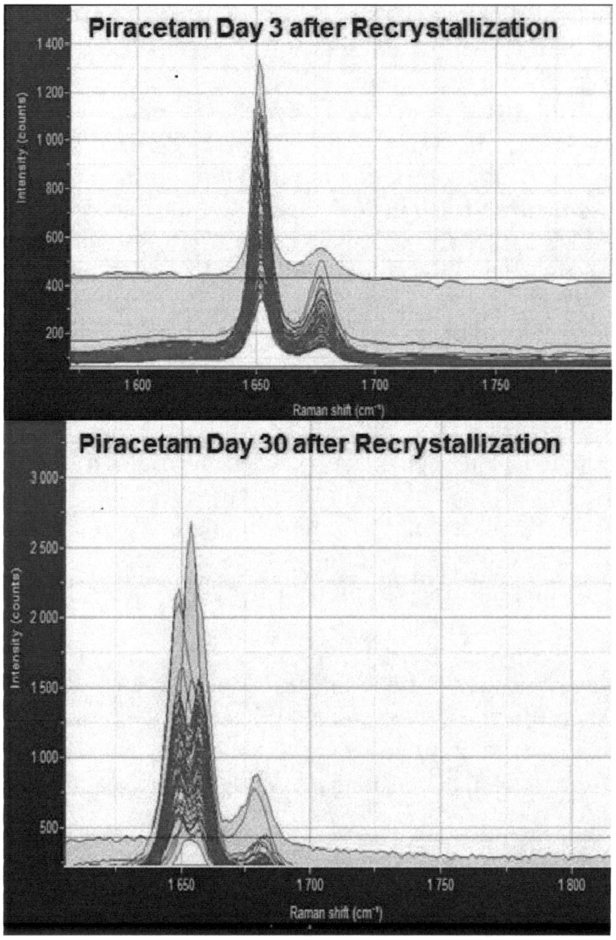

Fig 2: Raman spectra collected from an area on the surface of the API on Day 3 and 30 after recrystallisation from water illustrating the transformation of Piracetam FII to FIII during thirty days at ambient temperature.

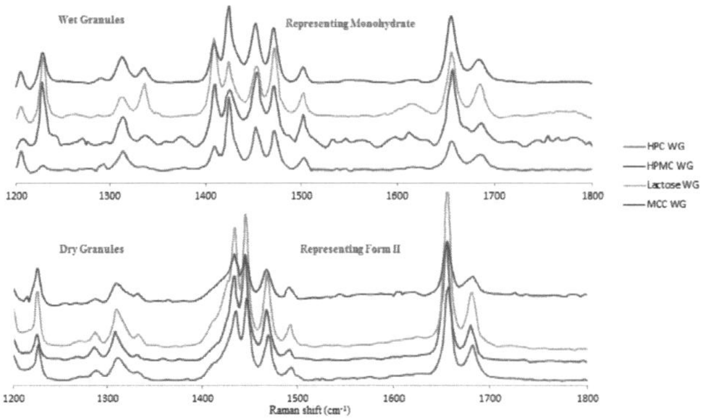

Fig 3: Raman spectra showing transformation of Piracetam from monohydrate to FII upon drying.

459850 Dissolvable Microcarriers for Efficient Cell Production and Recovery

Abstract:

Cell scale-up on microcarriers is an increasingly important technique for bioprocess applications. However, the harvest, separation, and concentration of final product (e.g. cell, virus, or antibody) can be time and resource consuming. To address these issues, Corning has developed a new generation of microcarriers that can be quickly dissolved, allowing easy collection of grown cells or secreted products without the need for microcarrier separation. Here we present unique properties and potential applications for dissolvable microcarriers. Human mesenchymal stem cells (hMSCs) and Vero were grown in spinner flask cultures or bioreactors on dissolvable microcarriers. For all cell types evaluated, growth performance on dissolvable microcarriers was equivalent to other commercially available microcarriers, and high quality cells were recovered without the need for microcarrier separation.

461030 The Effect of Particle Engineering on the Processing and the in Vitro Performance of Inhalation Blends in Dry Powder Inhalers

Abstract:

Introduction

In order to use API (active pharmaceutical ingredient) particles intended to target the tiny airways of the deep lung in dry powder inhalers (DPIs), the range of an aerodynamic diameter of 1 µm to 5 µm is highly preferred. Particles in this size regimen are typically cohesive and possess poor flow properties [1], leading to difficulties concerning volumetrically dosing. To overcome this flowability problem, it is a general practice to formulate carrier-based formulations wherein the API particles are attached to the surface of a larger carrier particle (50 µm – 200 µm).

Materials and Methods

Spray dried and micronized salbutamol sulphate (SS, Selectchemie, Zurich, Switzerland) were chosen as model API. Spray dried active was prepared on a Nano Spray Dryer B-90 (Buechi Labortechnik AG, Flawil, Switzerland). The spray drying conditions were chosen according to our previous work [2]. Micronization was done with an air jet mill (Spiral Jet Mill 50 AS, Hosokawa Alpine AG, Augsburg, Germany) at an injection pressure of 6 bar and a pressure inside the micronizer chamber of 3 bar. Modification of the carrier was done through wet decantation, which was supposed to decrease the fine particles on the carrier material and to modify the carrier surface [3]. Spray-dried (SDSS) and micronized API (MSS) was blended with α-lactose monohydrate (LAC_BD) and α-lactose monohydrate decanted (LAC_AD), so four adhesive mixtures with 2% API were prepared via sandwich method in a tumble blender TC2 (Willy A. Bachofen Maschinenfabrik, Muttenz, Switzerland). The mixing time was 60 min at 60 rpm. Both carriers, the APIs as well as the adhesives mixtures were extensively characterized (e.g. particle size, particle shape, flow properties, surface topography) to compare the different particulate properties.

Subsequent capsule filling was performed with different process setting (Setting 1: 3.4mm dosator, 2.5mm dosing chamber, 5mm powder layer; Setting 2: 3.4mm dosator, 2.5mm dosing chamber, 10mm powder layer) at a filling rate of 2500 capsules per hour (cph) on a dosator nozzle capsule filling machine (Labby, MG2, Bologna, Italy) with a target fill weight of 20 to 25mg. To evaluate the performance of the different mixtures, in vitro lung deposition experiments were carried out with a next generation impactor (NGI, Copley Scientific, Nottingham, United Kingdom), the emitted dose (ED) and fine particle fraction (FPF) were calculated based on the specification of the European pharmacopoeia. The inhalation device used for these experiments was the Aerolizer®/Cyclohaler®, a capsule inhaler.

Results and Discussion

Figure 1 shows SEM images of the spray dried (Fig. 1a) and micronized salbutamol (Fig. 1b) particles. SEM images show that both techniques were able to produce particles that have suitable size for inhalation but with very different shape and morphology. Spray dried particles are spherical, whereas micronized particles are needle shaped. Moreover, spray dried particles are amorphous while micronized particles were largely crystalline. The mean particle size (x50) determined via laser diffraction for spray dried and micronized particles as well as the blends can be found in table 1.

Figure 1: SEM images of spray dried (SDSS) and micronized (MSS) salbutamol sulphate particles

Table 1: Particle size and distribution of engineered actives and adhesive mixtures

	x_{50} [µm]	Span [x_{90}-x_{10}/x_{50}]
MSS	1.99	2.21
SDSS	2.91	1.68
MSS+LAC_BD	183.26	1.05
MSS+LAC_AD	173.45	1.06
SDSS+LAC_BD	193.11	1.09
SDSS+LAC_AD	185.48	1.04

Table 2: Fill weight and fill weight variability (RSD)

	Setting 1 [mg]	RSD	Setting 2 [mg]	RSD
MSS+LAC_BD	19.74	6.57	27.22	2.00
MSS+LAC_AD	24.07	2.37	25.08	1.97
SDSS+LAC_BD	20.85	4.19	27.19	2.33
SDSS+LAC_AD	24.74	2.21	27.33	2.00

Table 3: In Vitro performance: Fine particle fraction(FPF), Emitted dose (ED) and stage reached in the impactor

	Setting 1			Setting 2		
	FPF [%]	ED [µg]	max. Stage	FPF [%]	ED [µg]	max. Stage
MSS+LAC_BD	19.20	829.88	6	22.41	1131.95	7
MSS+LAC_AD	32.53	939.91	7	32.53	939.91	6
SDSS+LAC_BD	6.49	1035.02	5	5.07	1114.91	4
SDSS+LAC_AD	2.39	1261.05	4	3.35	1162.72	5

The FPF for micronized powder blends increased significantly, whilst spray-dried blends showed a decrease for decanted mixtures in comparison to the undecanted. Also the attainable stage in the NGI got improved through micronization of the API, but showed no correlation to the wet decantation of the powder blends. The different powder bed height settings in capsule filling (table 2) led to distinctions in attainable stage and FPF but do not seem to be influencing factors on the emitted dose. In summary, micronization of the API has much more influence on the ability for reaching the deep lung (API detachment from carrier) than wet decantation of the carrier material has. The obtained data are highly useful, to improve the understanding of the relationship between carrier morphology and carrier type, drug detachment and capsule filling efficiency and will help to generate DPI formulations with the desirable performance.

[1] Pilcer, G., Wauthoz, N. and Amighi, K. (2012) *"Lactose characteristics and the generation of the aerosol"* Adv. Drug Deliv. Rev., vol. 64, no. 3, pp. 233–256.

[2] Littringer, E., Zellnitz, S., Hammernik, K., Adamer, V., Friedl, H. and Urbanetz, N.A. (2013) *"Spray Drying of Aqueous Salbutamol Sulfate Solutions Using the Nano Spray Dryer B-90—The Impact of Process Parameters on Particle Size"* Dry. Technol., vol. 31, no. 12, pp. 1346–1353.

[3] Faulhammer, E., Zellnitz, S., Wahl, V., Khinast, J.G., Paudel, A. (2015) "*Carrier-based dry powder inhalation: Impact of carrier modification on capsule filling processability and in vitro aerodynamic performance" Int J Pharm; 491: pp231–242.*

461690 Feasibility Analysis of Flowsheet Models in Continuous Pharmaceutical Manufacturing Processes Considering the Effects of Noise

Abstract:

Continuous manufacturing has gained increasing attention in the pharmaceutical industry due to its potential in reducing cost and producing consistently qualified products [1]. As it is becoming more time-consuming to discover and develop a new drug, and together with the intensified competition from generic companies, continuous manufacturing stands out as a promising alternative for pharmaceutical companies to improve production efficiency and maximize profits within the productsâ€™ patent life [2, 3]. The development of continuous pharmaceutical manufacturing has been facilitated by the improved continuous equipment (e.g. feeding, mixing, tableting, etc. [1]) and advanced process analytical technology (PAT) tools for process understanding and process control [4]. However, in order to systematically generalize the process knowledge and optimize the process design, we need to use the computer-aided process systems engineering (PSE) approach [3].

Recently, extensive modeling work has been done to simulate the dynamic behaviors of unit operations, based on which integrated flowsheet models can be built to predict the process output under a certain set of input values [3, 5]. Such models can not only correlate the materials properties (e.g. flow rate, powder bulk density, mixture concentration, etc.) and operating conditions (e.g. blade speed, turret speed, etc.) to the steady-state process response (mass hold-up, product properties), but also dynamically trace the variation in any input factors and its propagation through the subsequent unit operations. Such information can be useful for risk assessment and sensitivity analysis of the process. Following the requirement from Quality by Design (QbD) [4], it is naturally to consider how to apply the predictability of flowsheet models to the quantification and optimization of the design space of the manufacturing process, which is known as â€œfeasibility analysis problemsâ€ in the PSE community.

Feasibility analysis is to identify the feasible region where the process can meet all operating, quality, and product constraints under uncertainty [6, 7]. This is important to help understand how much variation the process can withstand while maintaining operability. However, the feasibility analysis problem can be very difficult to solve because of the complexity of the flowsheet model, which may include hundreds of variables and constraints. Moreover, the computational cost for the simulation can be high, which requires a highly efficient sampling strategy for the analysis. In this presentation, we focus on using surrogate-based approaches (e.g. Kriging, Radial Basis Function, etc.) to efficiently approximate the computational expensive flowsheet model and predict the feasible region under uncertainty. More specifically, we apply this approach not only to deterministic models, but also to models that contain noise in the output prediction. The objective is to provide a framework of feasibility analysis for flowsheet models that contain various sources of uncertainty.

References

[1] Schaber S D, Gerogiorgis D I, Ramachandran R, et al. Economic analysis of integrated continuous and batch pharmaceutical manufacturing: a case study[J]. Industrial & Engineering Chemistry Research, 2011, 50(17): 10083-10092.

[2] Plumb K. Continuous processing in the pharmaceutical industry: changing the mind set[J]. Chemical Engineering Research and Design, 2005, 83(6): 730-738.

[3] Boukouvala F, Ierapetritou M G. Surrogate-based optimization of expensive flowsheet modeling for continuous pharmaceutical manufacturing[J]. Journal of Pharmaceutical Innovation, 2013, 8(2): 131-145.

[4] Lawrence X Y. Pharmaceutical quality by design: product and process development, understanding, and control[J]. Pharmaceutical research, 2008, 25(4): 781-791.

[5] Rogers A J, Inamdar C, Ierapetritou M G. An integrated approach to simulation of pharmaceutical processes for solid drug manufacture[J]. Industrial & Engineering Chemistry Research, 2013, 53(13): 5128-5147.

[6] Rogers A, Ierapetritou M. Feasibility and flexibility analysis of black-box processes Part 1: Surrogate-based feasibility analysis[J]. Chemical Engineering Science, 2015, 137: 986-1004.

[7] Grossmann I E, Calfa B A, Garcia-Herreros P. Evolution of concepts and models for quantifying resiliency and flexibility of chemical processes[J]. Computers & Chemical Engineering, 2014, 70: 22-34.

464170 Characterization of Compaction Behaviour of Powders in Solid-Dosage Manufacturing Using Finite Element Analysis

Abstract:

Powder compaction is an essential step in pharmaceutical solid-dosage manufacturing, irrespective of the chosen path for powder pre-processing. The compaction changes the structure of material from a loose arrangement of particles with various shapes and sizes to a condensed structure that behaves like a continuum. An optimised compaction process is necessary to ensure production of high-quality tablets. Experimentally determining the effects of involved parameters and subsequent optimization of the compaction process is very difficult and expensive for this process, as tablets contain the expensive drug substance. Therfore, numerical simulation of the compaction process based on Finite Element Analysis (FEA) is suggested. The FEA is used for calculating forces, deformations, stresses and strains throughout a bonded structure, which is later used to ensure the formation of compact material with uniform density, free from cracks or chips. Hence, these simulations provide a detailed and cost-effective means of understanding and predicting the compaction profiles of the formulation material based on the processing parameters, such as imposed pressure, relative density of the powder and die shape.

In this study, the mechanical properties of a range of pharmaceutical formulations were approximated using the modified Drucker-Prager Cap (DPC) plasticity model and applied for FEA. The DPC model parameters were varied depending on the local relative density using an external subroutine. Additionally, the effect of wall friction on the mechanical behaviour of different formulation compacts during compaction is explored. This numerically predicted density distribution is compared with experimental measurements. The verification results are important for establishing a predictive capability of this model. The results from this study contribute to a better understanding of the impact of powder properties and process settings on the tabletting process and final properties of the produced tablet.

464286 Gender-Dependent PBPK Modeling

Abstract:

Personalized medicine has the potential to improve disease prevention and treatment by tailoring therapies to the unique physiological and environmental factors that describe a subpopulation or even an individual patient. Personalized medicine aims at providing patients with "the right drug at the right dose at the right time" (1) to maximize therapeutic efficiency and minimize adverse events. Anatomical differences (height, body weight, body composition), ethnicity, genotype, disease state, age, and gender, amongst other factors can lead to significant variations in the bioavailability of a drug due to differences in the rates and/or extent of absorption, distribution, metabolism and elimination (ADME) following administration. The success of personized medicine hinges on the industry's capacity to understand the role of these physiological and genetic factors that drive differences in ADME and ultimately the action of a drug. Identifying factors with the potential to impart the greatest effect on bioavailability is a critical step to explaining clinical observations and understanding the usefulness of a particular treatment. One such tool to enable this mechanistic understanding *in silico* is physiologically-based pharmacokinetic (PBPK) modeling.

Unlike traditional pharmacokinetics which calculates key parameters from existing experimental data, PBPK models have the ability to extrapolate bioavailability predictions to new populations or disease states, and for new administration routes or formulation, in addition to explaining physiological differences such as age or gender. PBPK models consist of a system of mathematical equations to describe the ADME processes across the various organs and tissues of the human body (2). These mechanistic models, in comparison to traditional compartmental modeling, expand the usefulness of pharmacokinetics to drug discovery, development and regulation for knowledge-based decision making. These models leverage drug physical chemical properties and body-specific physiological parameters to predict the amount of drug that is absorbed and reaches systemic circulation (3).

Physiological differences between genders are of particular interest given that women respond differently to therapies than males, both in regards to the effectiveness of a particular treatment as well as in the extent of observed adverse effects. These observations are well cited in the literature, indicating that physiological differences between genders can lead to vast differences in pharmacokinetic parameters (AUC, Cmax, clearance, volume of distribution) and potentially variations in pharmacodynamic responses (4-6). Differences in pharmacokinetics and pharmacodynamics has been observed clinically across many systems of the body including the central nervous system, cardiovascular system, energy metabolism, and the immune system (4). The clinical relevance of the differences in pharmacokinetic responses varies with the physicochemical properties of the drug, female physiology, and mechanism of action of the drug. The vast differences in drug behavior across a multitude of bodily systems suggests that sex-dependent pharmacokinetics are highly complex as the potential for differences span across all ADME processes.

Gender dependent responses are not always consistently reported in clinical studies given the strong influence of sex hormones during development, the menstrual cycle and with the use of oral contraceptives to drive higher variability in the pharmacokinetics for women as compared to men. Thus, incorporating gender differences is an important consideration and the next logical step in the expansion of PKPD modeling for more accurate predictions of clinical performance to tease out differences between males and females in well-defined and controlled virtual clinical studies. This expansion begins with an understanding of how the ADME processes vary in males and females in order to accurately predict bioavailability in both genders. In general, males and females are believed to have similar metabolic and signaling pathways with sex-dependent regulation of transcription, translation, and receptor signaling by sex hormones (7). Therefore, the focus of this work is to establish gender-specific PBPK model parameters for male and females that can be used for prediction of absorption and ultimately bioavailability of an oral solid dosage form. The opportunity for gender-dependent pharmacokinetics of an orally administered drug begins with absorption from the gastrointestinal tract. Sex hormones influence absorption from the gastrointestinal tract on several levels including gender-dependent physiology (passive diffusion), expression and activity of intestinal transporters (active transport), and expression and activity of gastric enzymes and cytochrome P450 isozymes (gut metabolism) in addition to gastrointestinal pH and transit time (8).

The properties of the drug (or formulation) for a solid oral dosage form dictate the extent of the observed differences in pharmacokinetics between genders. In a preliminary case study, we chose to study the effect of formulation properties, mainly pH-dependent solubility or *in vivo* dissolution, on the rate and extent of absorption in males and females. The purpose of this study was to establish an understanding of the gender-dependent sensitivity of absorption to changes in *in vivo* release. Using GastroPlus™ software, the Advanced Compartmental Absorption and Transit model parameters were adjusted to reflect male and female physiology. Gender-dependent absorption was predicted given dissolution profiles as input to the software with varying rates and extent of release.

1. Hamburg MA, Collins FS. The path to personalized medicine. The New England Journal of Medicine. 2010:301-4.

2. Yang RS, Mayeno AN, Liao KH, Reardon KF, Reisfeild B. Integration of PBPK and reaction network modelling: Predictive xenobiotic metabolomics. ALTEX 22. 2005(2).

3. Rowland M, Peck C, Tucker G. Physiologically-based pharmacokinetics in drug development and regulatory science. Annu Rev Pharmacol Toxicol. 2011;51:45-73.

4. Soldin OP, Chung SH, Mattison DR. Sex differences in drug disposition. J Biomed Biotechnol. 2011;2011:187103.

5. Ibarra M, Vazquez M, Fagiolino P, Derendorf H. Sex related differences on valproic acid pharmacokinetics after oral single dose. J Pharmacokinet Pharmacodyn. 2013;40(4):479-86.

6. Chen ML, Lee SC, Ng MJ, Schuirmann DJ, Lesko LJ, Williams RL. Pharmacokinetic analysis of bioequivalence trials: implications for sex-related issues in clinical pharmacology and biopharmaceutics. Clin Pharmacol Ther. 2000;68(5):510-21.

7. Regitz-Zagrosek V. Sex and Gender Differences in Pharmacology. Heidelberg: Springer; 2012. 599 p.

8. Freire AC, Basit AW, Choudhary R, Piong CW, Merchant HA. Does sex matter? The influence of gender on gastrointestinal physiology and drug delivery. Int J Pharm. 2011;415(1-2):15-28.

464607 Model-Based Process Analysis in Continuous Manufacturing of Pharmaceuticals: Calibration of a Fluid Bed Dryer Model

Abstract:

The goal of this work is to investigate the dominating phenomena in the fluid bed dryer in a continuous pharmaceutical wet granulation line, which are acting on the Particle Size Distribution (PSD) and moisture content of the product. Knowledge about these phenomena is built with model-based analysis, using a semi-mechanistic model. A thorough and novel experimental data collection campaign is done to create sufficient data for the calibration and validation of this model. Along with that, this work provides clarity into the application of a model for researching the particle property dynamics in a fluid bed dryer, and to produce concrete criteria for the success of every step in the application of the model. In this way, model-based process analysis for the transition to continuous manufacturing in the pharmaceutical industry, becomes more concrete.

The interaction and evolution of PSD and moisture content is investigated in a ConsiGma 25™ semi-continuous fluid bed dryer, which is fed with wet granules from a twin-screw granulator through pneumatic transport in a tube. In the dryer unit, the continuous flow of granules is divided over its six separate drying segments, in which the granules are batch dried. This partitioning into parallel and sequential batch drying processes approaches a continuous mode over the entire dryer.

In total, 49 different drying experiments were performed with wet granules that were produced under constant granulation settings. For the dryer, the following process settings were varied: the dryer inlet air flow rate was varied between 360 and 440 m^3/h and the inlet air temperature between 40 and 60°C. The filling time of a dryer cell, governing the cell load, was varied between 60 and 180 seconds. For each of the combinations of the previous settings and one center point of the settings, experiments were conducted for, at least, five different drying times, ranging from the filling time plus 20 seconds to maximally six times the filling time. As such, the drying times in the experiments span from 80 to 1080 seconds. The high-resolution PSD of the dried granules was collected using a particle size analyser, QICPIC (Sympatec), with a dispensing system, Gradis (Sympatec). The sample is collected after drying and directly sieved into five fractions. The moisture content of each sieve fraction, as well as the total sample, is determined by means of the Loss-On-Drying method (Mettler LP16 moisture analyzer).

An obvious observation can be made that the moisture content of the granules of the total sample decreases when drying time increases. Two phases can be distinguished during drying: (1) a fast phase, where evaporation of the surface water occurs, and (2) a slower drying phase in which water from the core needs to migrate to the surface to get evaporated (Mortier et al., 2012). It can be noticed that smaller granules have a lower moisture content compared to the larger granules, caused by the larger specific surface of the small granules. This phenomenon can only be noticed when shorter drying times are applied. When the drying time is increased, the moisture content of the different size fractions does not deviate significantly, indicating that drying is finished. The experimental data also indicates that most of the breakage occurs during the transport of the granules to the dryer and the loading into the dryer segments. The extent of this breakage was significantly different between the two formulations used in the experiments.

Through the application of a 2D Population Balance Model (PBM), the evolution of PSD and moisture content is simulated. The PBM describes the granules in the system according to a two-dimensional grid of discrete classes of particle size and moisture content, where the last one is given as the wet radius of the particle. Breakage and loss of moisture content of the granules are solved sequentially within each time step, approaching a simultaneous simulation of these phenomena (Mortier, 2014). Breakage is solved using the fixed-pivot discretization scheme. The change in wet radius is modelled according to a drying curve with two regimes, separated by the point where the wet radius equals the particle radius. When the wet radius is smaller, it is assumed that the moisture in the particle resides in its pores, resulting in a lower drying rate. The evolution of the moisture content is solved using the high resolution finite volume scheme (Mortier et al., 2013).

The three steps in the application of the model are model selection, calibration and validation. Model selection denotes selecting the breakage function, which describes the breakage regime in the system. Once a suitable breakage function is found, its empirical components, as well as the drying rates, are calibrated to the system for some of the different process conditions, using the experimental data. The other part of the process conditions are reserved for validation of the model. The methodology of these three steps, including the success criteria, will be presented. The study is done based on criteria from the statistics of comparison of distributions, applying the moments of a PSD to assess calibration and validation success. Each moment contains different information on the PSD. Therefore, incorporating various moments into a success criterion results in an all-round assessment of the PBM prediction. This way, the concrete and statistically sound comparison of PSD is brought into the calibration and validation steps of a PBM, important steps that are often overlooked in practice (Mortier, 2014).

References:

Mortier, S. T. F. C., De Beer, T., Gernaey, K. V., Vercruysse, J., Fonteyne, M., Remon, J. P., Vervaet, C., Nopens, I. (2012). Mechanistic modelling of the drying behaviour of single pharmaceutical granules. *European Journal of Pharmaceutics and Biopharmaceutics,* Vol. 80(3), 682–9.

Mortier, S. T. F. C., Gernaey, K. V., De Beer, T., Nopens, I. (2013). Development of a Population Balance Model of a pharmaceutical drying process and testing of solution methods. *Computers & Chemical Engineering,* Vol. 50, 39–53.

Mortier, S.T.F.C. (2014). *Modelling the drying behaviour of wet granules in the context of fully continuous pharmaceutical tablet manufacturing.* Ghent University. Faculty of Bioscience Engineering, Ghent, Belgium.

466116 Development of a Model Predictive Tool for Continuous Manufacturing of Solid Oral Dosage Forms Via Roller Compaction

Abstract:

Continuous Manufacturing (CM) is currently receiving significant attention in the pharmaceutical industry. In the past, common manufacturing processes for oral solid dosage (OSD) forms relied on batch processing followed by final product testing to ensure the desired quality has been achieved. This approach is considered inefficient due to increased material consumption and prolonged processing times. Therefore, it is crucial to take a step towards model predictive CM to enable prediction of the final tablet critical quality attributes (CQAs). To develop a quality by design (QbD) framework, a compaction and compression integrated work stream has been developed to enable prediction of compaction and compression behaviour of various powders and their blends. The blends consist of varying amounts of active pharmaceutical ingredient (API), 1 wt% lubricant and as remainder microcrystalline cellulose (MCC). These are either directly compressed or compressed after dry granulation via roller compaction. This work develops a real time control tool to predict final tablet CQAs based on powders with varying properties. This tool will ultimately provide control of roller compaction and tablet press settings to produce desired tablet properties. For the development of this tool a detailed understanding of the impact of powder properties on the compression and compaction behaviour was established. Three different model APIs were used in this study to establish the production limits and to define the optimised process settings of the roller compactor and tablet press. Each API displays a different morphological structure (spherical, monocline and needle shaped) that produces varying flow, compaction and compression characteristics. Experimental work was carried out using various (10 – 40 wt%) API loadings to establish the optimal compaction and compression conditions. The resulting experimental data was analysed to identify the key process parameters, which were linked to the model predictive tool. A top fed roller compactor was used under ambient conditions with varying screw speed (12 – 24 rpm), roller speed (3 - 5 rpm) and roller pressure (10 - 20 bar) for

ribbon production. The ribbon density, which is a key factor of roller compaction, was analysed with a GeoPyc 1360 envelope density analyser. The ribbons were granulated in a conical mill (Comil) and analysed for particle size distribution using sieving and a Microtrac Particle Size Analyser. The granule produced was compacted into tablets using various die sizes (6, 8, 10 and 12 mm) and a single punch Manesty F3 tablet press. Displacement measurements were assessed using commercial and custom built compaction testing equipment to determine compaction behaviour of powders and their blends. The tablets were tested for properties, including weight consistency, hardness, thickness, diameter, friability, dissolution and disintegration time.

466270 Application of Model-Predictive Control to a Feeding Blending Unit

Abstract:

Continuous pharmaceutical manufacturing is pursued by the FDA [1][2]. By continuously monitoring the process and by actively taking actions in case of deviating process parameters, the product quality will be improved while the production costs decrease. The proper operation of the continuous production line requires the development and implementation of sophisticated control strategies. On the one hand, the control strategy has to ensure that intermediates of bad quality are discharged. On the other hand, this missing material has to be compensated by means of adjusting the production speed of the upstream and/or downstream unit operation. This operation mode poses the requirement of adjusting the production speed of individual unit operations.

The paper at hand focuses on the design of a model predictive control (MPC) strategy [3][4] for a feeding blending unit (FBU), see Figure 1. The considered setup utilizes two feeders and a blender in order to create a blend of a given composition. The controller aims at tracking a prescribed blender outlet mass flow. Simultaneously, the mass hold-up in the blender, as well as the blender speed should remain within given bounds. A mathematical model of the plant, which is based on physical relations, is given. Model parameters have to be identified from measurement data. Based on the presented plant model, a model predictive control strategy is designed.

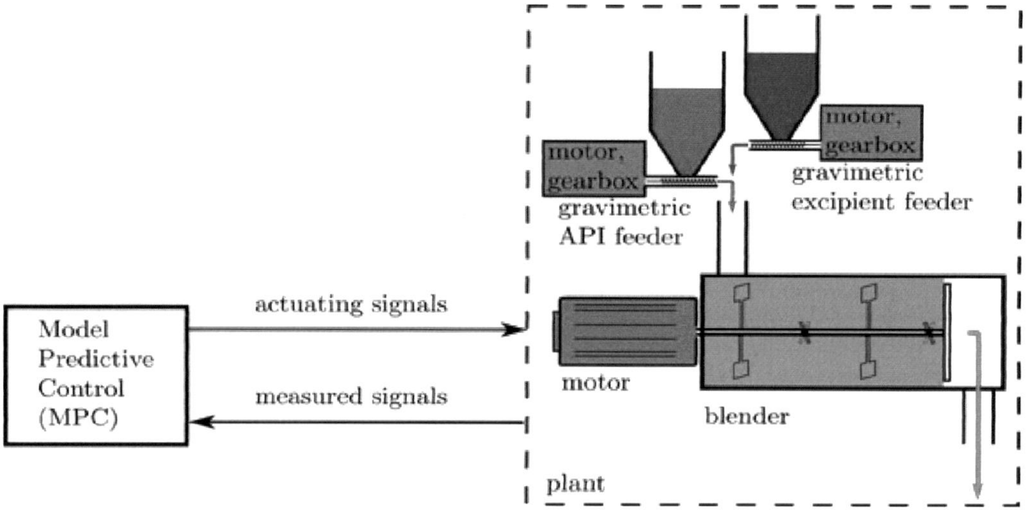

Figure 1: Considered setup

The presented plant model is based upon the idea of dividing the blender into several compartments. For each compartment, the mass balance equation is solved. The inlet mass flow of the compartments is known: for the first compartment, it is equal to the sum of the feeder mass flows, and for the following compartments, the inlet mass flow is equal to the outlet mass flow of the preceding one. The outlet mass flow of one compartment is computed from the hold-up in the compartment and the mean residence time of the material in this compartment. The mean residence time depends on the hold-up and on the blender speed. A simple relation, which relates the mean residence time to the blender speed is fitted from measured data. By considering the mass hold-ups of the individual compartments as state variables, a state space realization [5] of the plant is derived. The number of state variables equals the number of compartments.

Furthermore, the input/output behavior of the blender regarding concentration is modeled by means of a third order low pass filter with additional dead-time. The time constants and the dead time have to be identified from measurements.

A model predictive control strategy, which is based on the presented plant model, is proposed. The formulation of the objective function is discussed. Constraints on the state and the actuating signal are taken into account by the suggested control strategy. Several test scenarios, which investigate the fluctuations of feeder mass flows on the outlet of the blender are set up.

Simulation studies demonstrate the advantages of the proposed concept. The tuning of the controller is very intuitive, multi-input multi-output systems with constraints can be handled naturally.

[1] U.S. Department of Health and Human Services, "Guidance for Industry - PAT - A Framework for Innovative Pharmaceutical Development, Manufacturing, and Quality Assurance," 2004.

[2] S. Chatterjee, "FDA Perspective on Continuous Manufacturing," *IFPAC Annual Meeting*, 2012. [Online]. Available: http://www.fda.gov/downloads/AboutFDA/CentersOffices/OfficeofMedicalProductsandTobacco/CDER/UCM341197.pdf. [Accessed: 21-Mar-2016].

[3] J. B. Rawlings and D. Q. Mayne, *Model Predictive Control: Theory and Design*. Nob Hill Publishing, 2015.

[4] J. Maciejowski, *Predictive Control with Constraints*. Prentice Hall, 2001.

[5] B. Friedland, *Control System Design: An Introduction to State-Space Methods*. New York: McGraw-Hill, 1986.

466304 Systematic Technoeconomic Comparison of Separation Solvents for Green Continuous Pharmaceutical Manufacturing of Diphenhydramine

Abstract:

Continuous pharmaceutical manufacturing (CPM) has been shown to provide significant cost savings and enhanced material efficiencies compared to the currently used, yet inefficient, batch methods [1]. With the highest research and development costs of all industrial sectors, and historically increasing commercialisation costs [2], technological innovation is essential for pharmaceutical firms to remain profitable and sustainable [3]. For the benefits of CPM to be realised by industry, demonstration of its feasibility and viability via systematic process evaluation is paramount [4].

Continuous flow synthesis is the key to all candidate CPM processes, which has been demonstrated for several active pharmaceutical ingredients (APIs) to date. Several promising API candidates have already been identified as suitable for CPM implementation, including ibuprofen and artemisinin, which have already been examined in recent publications [5][6]. Systematic flowsheet synthesis, process modelling and simulation are an essential method for rapid assessment of potential processes and have already been widely applied for CPM case studies [7].

Diphenhydramine, a first-generation antihistamine API with hypnotic and anti-depressant uses [8] has been shown to be a promising candidate API for CPM implementation. Our previous work developed a process model on the continuous upstream processing of diphenhydramine [9] on the basis of a published continuous flow synthesis route which demonstrates that this economically and societally important API can be synthesised in plug flow reactors with molten ammonium salts [10]. This method achieves solvent and waste minimisation, achieving low total costs, enhanced material efficiency, and reduced hazards of chemical waste disposal.

The present paper illustrates a comprehensive technoeconomic analysis of various separation solvents for the continuous purification of diphenhydramine in this novel CPM process. This work provides a systematic methodological framework toward assessing liquid-liquid extraction (LLE) performance and material efficiency of the CPM process using different separation solvents. The inefficiencies incurred by each solvent at varying feed ratios on plant capacity, unit operation sizes and the corresponding capital, operating and total costs are also calculated here and compared to the CPM process with a batch separation train.

It is shown that whilst there is significant variation in process equipment sizes, costs and material efficiencies for CPM with different separation solvents, there is a clear trade-off between greatest cost savings, improvements in material efficiency, solvent toxicity and green, sustainable CPM operation. Original results indicate small computed reactor volumes and considerable heating requirements for the plant capacity. Chloroform emerges as a potent extraction agent in comparison to other candidate solvents, allowing the highest material efficiency (environmental factor of 3.43, total cost savings of 49.5%), but its high toxicity prohibits CPM use. Methyl-cyclohexane is the next strongest performer (environmental factor of 31.06, total cost savings of 37.3%) whose chemical properties render it significantly more acceptable for CPM implementation.

The present study exhibits the potential for technological innovation available *via* CPM in an effort to facilitate the transition from batch production methods in the pharmaceutical industry; by demonstrating the technoeconomic feasibility and economic viability for CPM of diphenhydramine via a rigorous comparison of a wide range of continuous separation options, we highlight the importance of systematic process systems engineering methods toward facilitating promising transitions from batch to CPM processes.

LITERATURE REFERENCES

1. Plumb, K., 2005. Continuous processing in the pharmaceutical industry - Changing the mind set, *Chem. Eng. Res. Des.* **83**(A6): 730–738.

2. Morgan, S., Grootendorst, P., Lexchin, J., Cunningham, C., Greyson, D., 2011. The cost of drug development: A systematic review, *Health Policy* 100(1): 4–17.

3. Behr, A., Brehme, V.A., Ewers, C.L.J., Gron, H., Kimmel, T., Kuppers, S., Symietz, I., 2004. New developments in chemical engineering for the production of drug substances, *Eng. Life Sci.*, **4**(1): 15–24.

4. Teoh, S.K., Rathi, C., Sharratt, P., 2016. Practical assessment methodology for converting fine chemicals processes from batch to continuous, *Org. Process Res. Dev.*, **20**(2): 414-431.

5. H. G. Jolliffe and D. I. Gerogiorgis, D.I., 2015. Process modelling and simulation for continuous pharmaceutical manufacturing of ibuprofen, *Chem. Eng. Res. Des.* **97**: 175–191.

6. H. G. Jolliffe and D. I. Gerogiorgis, D.I., 2016. Process modelling and simulation for continuous pharmaceutical manufacturing of artemisinin, *Chem. Eng. Res. Des.* (in press, available online).

7. K. V Gernaey, K.V., Cervera-Padrell, A.E., Woodley, J.M., 2012. Development of continuous pharmaceutical production processes supported by process systems engineering methods and tools, *Future Med. Chem.* **4**(11): 1371–1374.

8. Richardson, G.S., Roehrs, T.A., Rosenthal, L., Koshorek, G., Roth, T., 2002. Tolerance to daytime sedative effects of H1 antihistamines, *J. Clin. Psychopharmacol.* **22**(5): 511–515.

9. Diab, S., Gerogiorgis, D.I., 2015. Process modeling and simulation for Continuous Pharmaceutical Manufacturing (CPM) of diphenhydramine, *Proceedings of the AIChE Annual Meeting,* Salt Lake City, UT, USA.

10. Snead, D.R., Jamison, T.F., 2013. End-to-end continuous flow synthesis and purification of diphenhydramine hydrochloride featuring atom economy, in-line separation, and flow of molten ammonium salts, *Chem. Sci.* **4**(7): 2822–2827.

466733 Continuous Wet Granulation in a Twin Screw Extruder: Effect of Process Variables on Residence Time Distribution and Granule Characteristics

Abstract:

Wet granulation using a twin screw extruder is a promising continuous processing technique for granulation of pharmaceutical formulations. Although granulating with more consistency than a traditional high shear batch mixer,[1] the understanding of the granulation mechanisms in twin screw granulation is still not complete.[2] Furthermore, the residence time is very important, especially when multiple unit operations are connected in continuous processes.

Here we analyse the dependence of the residence time and granule size distribution (GSD) in a twin screw granulator on process parameters (screw speed, powder feed rate, liquid-to-solid ratio) and screw configuration (conveying, kneading, and cutting elements). The formulation was granulated in a 12 mm twin screw extruder (Three-Tec, Switzerland) with a length-to-diameter ratio of 40, using water as the granulating liquid. GSD was measured using a laser diffraction technique, while the residence time distribution was measured by adding a small dose of dye into the powder feeding port during steady state operation, followed by the sampling cup method described elsewhere.[3] Parameter values were chosen using a d-optimal factorial design to identify the most important parameters, and the results were rationalised to elucidate the granulation mechanisms in the different screw regions/screw elements.

[1] E.I. Keleb, A. Vermeire, C. Vervaet, J.P. Remon, Extrusion granulation and high shear granulation of different grades of lactose and highly dosed drugs: A comparative study, Drug Development and Industrial Pharmacy. 30 (2004) 679â€"691.

[2] M.R. Thompson, Twin screw granulation - review of current progress, Drug Development and Industrial Pharmacy.

(2014) 1â€"9. doi:10.3109/03639045.2014.983931.

[3] A.S. El Hagrasy, J.R. Hennenkamp, M.D. Burke, J.J. Cartwright, J.D. Litster, Twin screw wet granulation: Influence of formulation parameters on granule properties and growth behavior, Powder Technology. 238 (2013) 108â €"115.

467254 Virtual High-Throughput Screening Pipeline: Limitation Characterization and Application to Different Targets

Abstract:

Drug leads comprise a small portion of all known compounds. In an effort to identify promising leads, thousands to hundred thousands of compounds are experimentally tested for biological activity in high-throughput screening (HTS) experiments. However, typical hit-rates are only a few percent at best (Dobson, 2004). To increase efficiency, researchers are using previously obtained data and computational techniques to identify probable leads to focus resources on.

Previously, we developed an iterative virtual HTS (vHTS) pipeline. Using previously experimental data as input, the pipeline will develop quantitative structure-activity relationships (QSAR) models capable of classifying a compoundâ€™s class (active or inactive) and predicting its quantitative activity (Chen and Visco, 2016). Applied to a compound database, we can screen for probable leads. We tested our pipeline using NCBIâ€™s PubChem Bioassay 825 dataset, targeting a receptor implicated in viral diseases. We experimentally confirmed predictions from 2 iterations, with a hit rate of 19% in the 1st iteration and 75% in the 2nd.

To expand on our work, we will examine three key parameters to determine the robustness and applicability of our pipeline. Two parameters are related to the experimental datasets used to train our models: size and distribution. Size refers to the number of entries in the dataset while distribution refers to the classification of each entry and the number of members in each class. These two parameters together dictate the amount and quality of information available for training our models. Determining limitations for these two parameters will further direct experimental efforts towards fruitful pursuits as well as increase prediction confidence. The third examined parameter is pipeline iteration. By iterating, training set data can be updated with the addition of data from the validation experiments. However, there are diminishing returns with additional iterations. Identifying when it happens will increase efficiency.

In this poster, we will present work towards determining the aforementioned limitations. From PubChemâ€™s Bioassay database, we have identified datasets that are useful to determining the effect of each parameter. While controlling for two parameters, we examine the effect of the third by apply our pipeline to these datasets and experimentally validate predictions. Based on the observed changes in prediction accuracy and efficiency, we have a clearer idea what limitations exist and what effect a particular parameter has. We present the results of our work and aim to show our success is repeatable and our pipeline is robust enough to handle different datasets targeting different receptors while also investigating its limitations.

Chen J.J.F., Visco D.P. Jr., Developing an in silico pipeline for faster drug candidate discovery: Virtual high throughput screening with the Signature molecular descriptor using support vector machine models, *Chemical Engineering Science*, 2 March 2016. http://dx.doi.org/10.1016/j.ces.2016.02.037.

Dobson, C.M., 2004. Chemical space and biology. Nature 432, 824-828.

466960 Real-Time Monitoring and Ejection of out-of-Spec Tablets Via NIR Chemical Imaging

Abstract:

Near-infrared chemical imaging (NIR-CI) using high-speed cameras based on the push-broom acquisition principle is a strongly evolving technology in food, waste or mineral sorting. Yet, in the pharmaceutical industry, it is mainly used in lab settings.

Here we present a NIR-CI system, which is capable of real-time monitoring of approx. 500.000 tablets per hour and forwarding aggregated data (mean API content, standard deviation (SD) and number of analyzed tablets) to a process control system (PCS). Alternatively, the obtained chemical information can be used to identify and eject single out-of-spec tablets. The system consists of a push-broom spectrometer (EVK Helios G2 CLASS), a software for image analysis and pneumatic valves for ejection of tablets.

The camera (312 pixels spatial-, 248 pixels spectral resolution, 500 Hz measurement rate) generates 49 MB/s of spectral raw data. To reduce the data, chemometric models developed in Simca 13 are transferred to the spectrometer. They are executed in real-time on the onboard FPGA (Field Programmable Gate Array). The resulting video stream (0.29 MB/s) is further analyzed via openCV, an image processing library. For each detected tablet the mean and SD of API content, predicted via PLS, are calculated and averaged for one second. The results (3×4 byte/s) are written to an OPC-DA server, a standard interface to PCS, enabling a reaction to measured deviations. For ejection of out-of-specs tablets the video stream is analyzed via a DSP/TI System. Objects, which have to be ejected are forwarded to an IO subsystem, which controls the pneumatic valves.

Thereby the overall quality of tablet manufacturing can be raised by removal of potential out-of-spec tablets from the product stream at full manufacturing speed.

467823 Multivariate Statistical Process Control for the Continuous Manufacturing of Solid Oral Dosage Forms

Abstract:

Nowadays a commodity in the industrial software industry, the application of principal components analysis (PCA) to monitor the state of a process with Multivariate Statistical Process Control (MSPC) charts is far from a venture. Its implementation in a commercial plant with a contemporary control and/or automation system requires a modest effort, given a stable process and the availability of a historian data base with abundant data representative of the normal operating conditions (NOC) for the plant. This work presents the development of an MSPC monitoring scheme for a continuous drug product manufacturing process, using information available at the R&D stage. At the time when the process is still being developed, the data representative of the expected NOC may not be abundant. Additionally, the product development efforts may have other experimental priorities, for example running experimental trials at different mass rates and/or configuration of the feeders, mixer, and/or tablet press equipment.

The approach presented is based on the assumption that, regardless of the target set point for the main operating conditions (e.g. mass rate), the common cause variability in the process will remain the same. Hence the covariance structure for the variables in the system can be identified by a PCA model on a matrix resulting from the concatenation of data from multiple â€œtrial runsâ€ that were pre-centered around their mean values. An analysis of the loadings vectors reveals variable correlations that matches the expectations from a process engineering standpoint; and the corresponding movement in the scores space reveals directions of expected normal (and tolerated) variability, as well as patterns that can be matched with the startup and the shutdown of the system.

Our case study presents the analysis of a model built with the data collected from the system of screw feeders that dispense material to the unit, as well as the analysis of the model obtained with data collected from the tablet press. Furthermore, a method is presented to determine the lag time between a measurement vector from the feeders, and a measurement vector from the tablet press; such that a combined lagged model can be built. The combined model eases the holistic identification of contributions in the case of a disturbance, and the use of the three models (one per system and the combined) in tandem provides a complete monitoring approach for the system as a whole.

468412 Modelling Approaches to Spray Drying Process Development, Scale-up and Troubleshooting

Abstract:

Modelling approaches to spray drying process development, scale-up and troubleshooting

P.C. Valente, Í. Duarte, F.Neves & M.N. Temtem

[1] R&D Drug Product Development, Hovione Farmaciência S.A, Sete Casas, 2674-506 Loures, Portugal;

Initiatives such as Quality by Design in the pharmaceutical industry have greatly increased the emphasis on physical and mathematical modelling of the various processes and the focus on science-based development [1]. The shift from the more traditional approach of the Quality by Testing is driven by the rational that better understanding of the underlying physical mechanisms leading up to the final dosage form leads to improved control of the quality of the final product and to accelerated product and process development. Spray drying of drug product intermediates is one of the pharmaceutical processes being shifted from Quality by Testing to Quality by Design and it is also illustrative of the various models that need to be developed in order to reach that goal. For example, when a spray drying process is employed to obtain an amorphous solid dispersion, an adequate understanding of the miscibility/phase separation between polymer and API can accelerate the screening of polymer candidates and optimized process conditions to ensure a homogeneous molecular dispersion between polymer and API. However, due to the complexity of the underlying physical mechanisms together with the fact that multiple physical phenomena are simultaneously at play, the task of creating an adequate model can be quite daunting. To overcome these difficulties it is crucial to have a pragmatic approach to modelling and to adequately formulate the problem statement so that no more complexity than absolutely necessary is fed into the model.

In this work we discuss five case studies where different levels of modelling are used, from statistical to first principles, to address product and process development needs as well as troubleshooting. In the first and second case-studies two models based on first principles are applied to the formulation of amorphous solid dispersions and to forecast the impact of particle design on the dissolution behavior. In the third and fourth case-studies two different levels of modelling, statistical and mechanistic, respectively, are applied to predict particle size of a spray dried powder based on the nozzle geometry, atomization and process parameters, showing that there is an unavoidable compromise between model accuracy and model generality. This case-study is also used to emphasize that the choice of model closely depends on the problem statement. A fifth case-study of troubleshooting a pharmaceutical process using computer fluid dynamics is also presented (Fig. 1).

Finally, the adequateness of the model complexity for tackling a given problem is discussed and two illustrative examples of overkilling or oversimplifying a model are given.

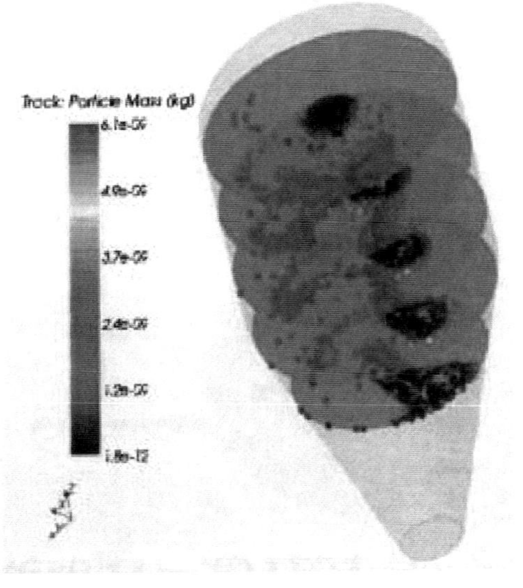

Figure 1: Troubleshooting example of a spray drying process using computer fluid dynamics. (Left) Gas velocity contours, (center) droplet spatial distribution within the drying chamber and (right) gas streamlines in the gas disperser and drying chamber of a spray drier.

References

[1] Lawrence X. Y. "Pharmaceutical Quality by Design: Product and Process Development, Understanding and Control", Pharm. Res., 25 (4) 2008.

468611 Solid-State Reactivity of Drugs with Potential Reactive Excipient Impurities: In silico and Experimental Studies

Abstract:

Demonstrating the chemical (long term) stability of drugs and drug formulations is a regulatory requirement, and guidelines for the evaluation of chemical stability are available. However, the most common formulation of small molecule drugs is the solid oral dosage form, and experimental procedures for the measurement of chemical degradation in the solid state are not well established, the predictive power of existing procedures is limited, and the required experimental effort is considerable.
One issue that is commonly encountered in the development of solid state drug formulations is the degradation of the active pharmaceutical ingredient (API) through a reaction with small molecule impurities present in excipients [1,2]. Here, we present first results of our attempts to use a combination of computational chemistry and experiments to investigate the impact of the solid state, its structure and dynamics, on chemical degradation in such systems.
We studied this type of reaction using a set of small molecular weight drugs that can potentially react with formaldehyde, a commonly observed impurity in polymeric excipients. Comparing the reactivities of these compounds with formaldehyde in solution to the corresponding reactivities of solid state (crystalline powder) samples, we found that the correlation of relative reactivities of different compounds in solution versus the solid state is limited. We performed an analysis of the solid state samples based on the structures of crystal bulk and surfaces using molecular modelling and Molecular Dynamics (MD) simulations, and defined a number of descriptors that can be calculated from first principles. These descriptors were used to explain the different reactivities observed for the studied drug compounds concentrating on the effects of limited accessibility of reactive sites in the crystalline state and relative lattice and surface stabilities and the related degree of disorder. Our results provide insights into the impact of solid state on relative reactivities and pave the way for the development of theoretical methods that can be used to optimize formulations towards improved stability.

References
[1] Hotha, KK, Roychowdhury, S, Subramanian, V. (2016) "Drug-Excipient Interactions: Case Studies and Overview of Drug Degradation Pathways." American Journal of Analytical Chemistry, 7, 107-140.
[2] Wu, Y, Levons, J, Narang, AS, Raghavan, K, Rao, VM. (2011) "Reactive impurities in excipients: profiling, identification and mitigation of drug-excipient incompatibility." AAPS PharmSciTech. 12, 4, 1248-1263.

467448 Prediction of Formulation Effects of Continuously Formulated Solid Oral Dosage Forms on Bioavailability Using PBPK Modeling

Abstract:

The *Quality by Design* (QbD) paradigm posits that in depth understanding of the relationship of a pharmaceutical product's *Critical Quality Attributes* (CQAs), ingredient (including API) *Critical Material Attributes (CMAs)*, and *Critical manufacturing Process Parameters* (CPPs), to the *in vivo* characteristics of the product as defined by the appropriate *Quality Target Product Profile* (QTPP) and *Clinically Relevant Specifications* (CRS), including absorption/bioavailability, will lead to better products more rapidly and economically for the benefit of the patient [1]. As stated, the QbD paradigm essentially requires establishing reliable quantitative relations between the "input" to the system (CMAs and CPPs) and the surrogate outputs (CQAs), which are presumed to be predictive of the "real outputs", (CRS). Ideally, the relationships between CMAs, ingredient ratios, and CPPs to highly predictive CQAs would be established *a priori* without extensive experimentation, especially *in vivo*, or substantial time/cost investments. For establishing such quantitative relations, we need to a) accurately, effectively and efficiently tune CQAs, ideally using predictive relationships to the ingredient proportions, CMAs and CPPs; and b) accurately, effectively and efficiently predict CRS from the CQAs (and perhaps, also using information from CMAs and CPPs).

Meeting this target has remained elusive. In recent years, two major advances have made a significant impact. First, *Continuous Manufacturing* (CM) has clearly demonstrated itself as a superior alternative to batch processing allowing for tight control of product specifications, thus making it possible to target specific CQA. Second, physiology-based mathematical models of dissolution and absorption are becoming increasingly more reliable and dependable for connecting *in silico* CQA and CRS. However, despite previous studies linking *in silico* CQA and CRS, we still lack a clear demonstration of an integrated framework that considers the overall "supply chain" of drug manufacturing, starting from the API, manufacturing process parameters, ingredient proportions and CMAs, and that eventually leads to *in vivo* predictions of the biorelevant profile**.** The lack of such integrated formalism a) creates sub-optimal designs and delivery systems; and b) does not take full advantage of the opportunities offered by CM. We propose to develop, implement, and demonstrate a systematic computational framework for connecting CAQ with CRS *in silico* in the context of continuous solid oral dose manufacturing processes.

More specifically, we explore computational methods for linking formulation-dependent dissolution characteristics to *in vivo* bioavailability. We achieve this aim in two main steps: (1) using physicochemical and processing parameters we predict *in vivo* dissolution from formulation and process parameters; and (2) using the predicted dissolution profile along with physicochemical and physiological properties we predict *in vivo* bioavailability. These two steps are accomplished by making use of DDDPlus^{TM} and GastroPlus^{TM} respectively. DDDPlus^{TM} simulates the dissolution curve of a drug product based on parameters such as diffusivity, solubility, particle size, and pKa, along with tablet properties such as composition, compression force, and tablet dimensions. The second step consists of feeding the output dissolution curve from DDDPlus^{TM} into a GastroPlus^{TM} model. GastroPlus^{TM}is based on the ACAT (Advanced Compartmental Absorption and Transit model) which has the ability to simulate the expected systemic bioavailability of a drug in the form of a plasma versus time graph [2, 3].

We hypothesize that since formulation regulates the parameters inputted into the dissolution model, for which the output profile serves as an input into the absorption model, it should have a direct impact on simulated bioavailability. An important part of this work will thus consists of quantifying the extent to which formulation can affect bioavailability. To aid in this endeavor, a parameter sensitivity analysis (PSA) will be performed to establish the parameters within the manufacturing design space that have the highest impact [4]. Furthermore, the significance of process variability on bioavailability will be studied in order to determine optimal operation range and to map out a design space for a given target bioperformance.

The advantage of mathematical modeling is the ease with which useful information can be produced compared to more labor-intensive and costly experiments. We seek to demonstrate the tremendous potential of PBPK models in guiding pharmaceutical processing development for oral tablet manufacturing within a Quality by Design framework. Future work also involves utilizing this mechanistic system to advance personalized medicine. By varying physiological properties, the model can be made to account for age, sex, ethnicity, extent of disease, and circadian rhythms, an aspect which could significantly enhance the customizing of drug formulation for specific populations.

1. Kesisoglou, F. and A. Mitra, *Application of Absorption Modeling in Rational Design of Drug Product Under Quality-by-Design Paradigm.* AAPS J, 2015. **17**(5): p. 1224-36.

2. Huang, W., S.L. Lee, and L.X. Yu, *Mechanistic approaches to predicting oral drug absorption.* AAPS J, 2009. **11**(2): p. 217-24.

3. *GastroPlus User Manual - Simulation Software for Drug Discovery and Development*, Cognigen, Editor. 2015, Simulations Plus.

4. Zhang, X., et al., *Utility of physiologically based absorption modeling in implementing Quality by Design in drug development.* AAPS J, 2011. **13**(1): p. 59-71.

467752 Predicting the *In Vivo* Performance of Amorphous Solid Dispersions Based on Molecular Descriptors and Statistical Analysis

Abstract:

Over the last decade, amorphous solid dispersions (ASDs) have become an increasingly popular strategy to address the solubility constraints of new chemical entities. The main advantage of ASDs is the potential to create and maintain a supersaturated state, once the drug is dissolved in the the GI fluids. While supersaturated, drug concentration in solution may reach values well above its intrinsic solubility, meaning that more drug would be available to be absorbed.

To retard drug precipitation, the selection of the stabilizing polymer (POL) is crucial. From a screening perspective, it is of interest to early evaluate different polymers in terms of their supersaturation potential and precipitation inhibition effect, in order to reduce time, risk and costs during formulation development.

Current screening methodologies to predict ASDs performance in solution generally consist in the implementation of miniaturized kinetic solubility experiments based on pH- or solvent-shift methods, using smaller volumes apparatus or in 96-well plates [1]. An alternative is the use of computational tools particularly the existing physiologically-based pharmacokinetic (PBPK) models, such as GastroPlus™ or Simcyp®. However these models need to be combined with accurate *in vitro / in vivo*dissolution experiments as input parameters. These data is typically obtained at advanced stages of formulation development. Thus, from an early screening perspective, there is lack of a theoretical basis for the ranking of the best polymers with precipitation inhibition effect.

Today, there is a significant number of research papers demonstrating the *in vivo* bioavailability of ASDs when compared with the reference products [2]. Taking advantage of the data available in the literature,

*** **Formatting Issue - Best Available Paper/Graphic** ***

the purpose of this work was to develop a statistical model, based on multivariate data analysis tools that could help on guiding ASD formulation design to obtain the desired *in vivo*performance.A database with 37 observations (ASDs) and 35 XY variables was constructed as schematically shown in Figure 1. The X variables include simple molecular descriptors of the APIs and POLs (*e.g.* MW, Log P, T_m, T_g, number of H-bond donors/acceptors, etc) and interactions thereof, while the Y variables correspond to experimental data (*e.g.* C_{max}, t_{max}, $AUC_{in\ vivo}$) obtained from the literature. Principal component analysis (PCA) and partial least square (PLS) models were developed using SIMCA-P+ 13.0 (Umetrics, Sweden). All variables from the dataset were mean centered and scaled to unit variance before effective analysis, in order to give variables equal weight.

Figure 1. Schematic representation of the dataset.

First, an outlier's identification stage was conducted using PCA to handle variability in data from disparate sources. An amorphous dispersion was identified as an outlier, which was then removed from the dataset.

After outlier's identification, a two-component PLS model with multiple responses was developed - Log ($AUC_{in\ vivo,\ ASD}/AUC_{in\ vivo,\ REF}$) and Log ($C_{max\ in\ vivo,\ ASD}/C_{max\ in\ vivo,\ REF}$). These Y-variables were identified as being correlated from the PCA analysis (data not shown). Figure 2 shows the correlations between the observed Log ($AUC_{in\ vivo,\ ratio}$) and Log ($C_{max\ in\ vivo,\ ratio}$) and the respective predicted values, together with the predicted test set. These correlations were obtained after variable selection to maximize model performance/minimize prediction error.

A

B

Figure 2 – Observed data *versus* predicted data by the PLS model: A – Log AUC_{ratio} response; B - Log $C_{max,\ ratio}$; training set (red circles); prediction set (blue circles).

The R^2 and Q^2 values for each Y-variable were around 0.7 and 0.5, respectively. Given the existing uncontrolled variability in the data - *in vivo*data obtained disparate sources and different animal models - the obtained model is considered adequate for interpretation purposes. Figure 3 shows the loading plot together with the most important variables (VIP) plot of the model. The loading plot shows the relationships between the inputs and output variables simultaneously, while the VIP shows the variables by descending order of influence in the model.

A

B

Figure 3– PLS loading plot (A) and VIP plot (B) of model with two PCs, based on the training set.

As can be seen from Figure 3, the most important variables for the model included general API-related molecular descriptors, POL-related molecular descriptors and API-POL interaction variables mainly

*** **Formatting Issue - Best Available Paper/Graphic** ***

associated with the capacity of the POL to establish hydrogen bonding with the APIs. These results confirm the key role of polymers for the kinetic stabilization of drugs in solution.

Figure 4 shows two scatter plots of two of the most important variables for the model. The size of each point/observation corresponds to the $AUC_{in\ vivo}$ and $C_{max,in\ vivo}$gains which were categorized in three different levels (low, medium and high).

A

B

Figure 4 – Scatter plots of two of the most important variables for the model. The size of the points/observations represent the $AUC_{in\ vivo}$ gain (A) and the $C_{max,in\ vivo}$gain (B).

The importance of hydrogen bonding for improving the *in vivo* performance of ASDs was in line with the observation that polymers with higher solubility parameters also tend to contribute for higher AUCs. In general, cellulose-based polymers (*i.e.* HPMCAS, HPMC, HPMCP) seem to provide better precipitation inhibition across different classes of drugs, when compared with other polymer families.

To conclude, chemometric tools based on multivariate data analysis can be applied to assess correlations between molecular descriptors of the formulation ingredients and performance related output variables, namely $AUC_{in\ vivo}$ and $C_{max,\ in\ vivo}$. These models are useful not only for narrowing the scope of formulations variables to be further explored, but also to rapidly identify suitable systems with synergistic interactions for subsequent clinical evaluation.<span">

References:

[1] Bevernage J. *et al.*, Evaluation of gastrointestinal drug supersaturation and precipitation: Strategies and issues, International Journal of Pharmaceutics, 2013, 453 (1), pp. 25-35;

[2] Kawabata Y. et al., Formulation design for poorly water-soluble drugs based on biopharmaceutics classification system: Basic approaches and practical applications, International Journal of Pharmaceutics, 2011, 420 (1), pp. 1-10;<span">

468110 Behaviour of Magnesium Stearate in Continuous Feeding

Abstract:

Behaviour of magnesium stearate in continuous feeding

Raghu V. G. Peddapatla*[1,2], Caroline A. Blackshields[1,2], Michael F. Cronin[1], Abina M. Crean[1,2,3]

[1]School of Pharmacy, University College Cork, Ireland

[2]Pharmaceutical Manufacturing Technology Centre, Ireland

[3]Synthesis and Solid State Pharmaceutical Centre, Ireland

*School of Pharmacy, University College Cork, Ireland, Tel: + 353 (21) 4901683,

email: r.peddapatla@umail.ucc.ie

INTRODUCTION

In a pharmaceutical continuous process line, feeding powders constantly and accurately into downstream processing units is challenging [1]. Consistent feeding is particularly challenging for cohesive materials such as the powder lubricant, magnesium stearate (MgSt). Even small changes in the feedrate of powders being fed for a short period of time, could result in products falling off the specification limits[2]. Hence, the initial feeding raw material step is important and must be monitored continuously to achieve consistency in feeding. Feeding of pharmaceutical excipients into continuous process lines is most commonly accomplished through Loss-In-Weight (LIW) or Loss-In-Volume (LIV) feeders which operate in gravimetric or volumetric mode [1,3].

In this study the continuous feeding of magnesium stearate alone and as a component of a tablet blend was studied. Magnesium stearate was selected as it is the most commonly used powder lubricant in tablet processing, and variability in its properties can comprise tablet ejection compaction, and performance (hardness and dissolution)[4] .

The objectives of this study were

1. To identify how the physical attributes of four different grades of magnesium stearate affect its continuous feeding performance.

2. To identify the effect of continuous feeding rate on the compaction and properties tablets prepared from blends (with and without lubricant).

MATERIALS

Four different grades of magnesium stearate obtained from two different suppliers, hereafter referred to as samples 1-4.

METHODS

Magnesium Stearate characterisation

Physical properties such as powder flow, bulk density, % compressibility are analysed using Brookfield powder flow tester (PFT) and Laser Light Scattering (Malvern 3000) for particle size distribution

Blends Preparation

Two blends are prepared in double cone Erweka blender blended for 30 minutes @ 30 rpm. Magnesium stearate is added to one of the blend and blended for extra 1minute @ 30rpm Blend 1 composed of 90% microcrystalline cellulose PH200 and 10% metoclopramide HCl. Blend 2 composed of 89.5% microcrystalline cellulose PH200 and 10% metoclopramide HCl and 0.5% magnesium stearate.

Feeder setup and evaluation

Loss in weight feeder are designed to monitor the changes in the bulk density of the powder in the feeder hopper[5]. Powders are fed through a K-Tron MT12 feeder using self-cleaning screws with fine square screen in gravimetric mode. Magnesium stearate samples are fed using Coarse Concave Screws (CCS) with Fine Square Screen (FSqS) setup, with two different set points of 0.15kg/hr. and 0.25kg/hr. Both blends are fed at 3 different set points 0.2238kg/hr., 0.5594kg/hr. and 1.0069kg/hr. with Fine Concave Screw (FCS) & Fine Square Screen (FSqS).

Mass flow performance was measured independently using a K-Sample Test System for Gravimetric Feeders. The relative standard deviation (RSD %) was determined from data collected at 30 sec intervals over a 30 minute duration.

RESULTS AND DISCUSSION

Continuous feeding performance

Feed factor is the gravimetric speed equivalent for 100% screw speed with a specific screw and screen setting. It gives an indication of the max. feed rates that can be achieved for a powder with specific equipment set up. Table 1 shows the measured bulk density and feed factors for the 4 grades of magnesium stearate studied.

Table 1

Sample	Bulk Density (g/cm^3)	Feed Factor (kg/hr)
MgSt Sample-1	0.255	1.9486
MgSt Sample-2	0.186	1.2784
MgSt Sample-3	0.203	1.3676
MgSt Sample-4	0.154	0.7635

Physical properties of magnesium stearate impact on gravimetric powder feeding behaviour. Samples feed factors are directly proportional to their bulk densities.

A measure of feed performance is the variability in quantity feed which is expressed as the relative standard deviation (RSD %). In this study weights where determined at 30 sec intervals for a duration of 30 minutes. The results for magnesium stearate alone and tablet blends are shown in Figure 1.

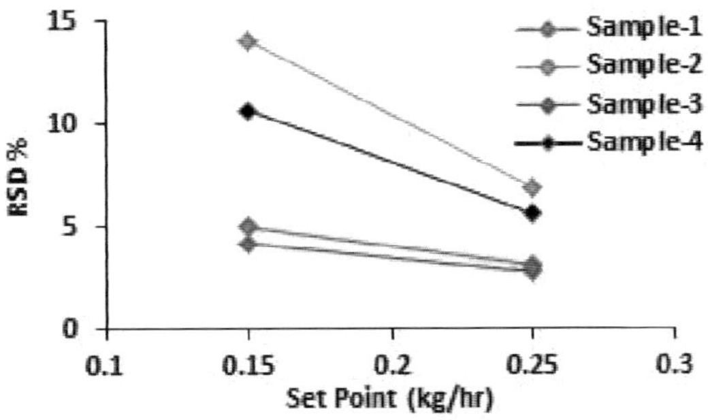

Figure 1: Feeding performance of magnesium stearate samples (left) and blends (right)

Magnesium stearate samples showed less variability at higher set points and greater variability at lower set points (Figure 1 left). Brookfield powder flow testing results showed magnesium stearate samples to exhibit very cohesive to cohesive flow behaviour and no relationship was seen between the feed variability and the measured flow function coefficient.

Blend 2 (which contained magnesium stearate) showed a lower % RSD compared to blend 1 at 3 different set points (Figure 1 right). Two blends fed at the lowest fed rate, 0.2238kg/hr, showed the highest variability.

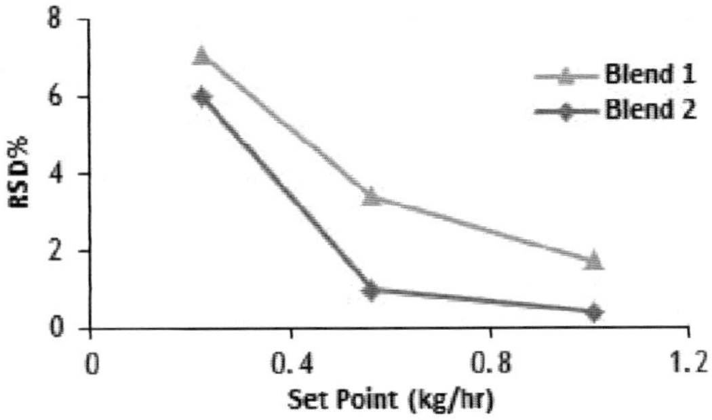

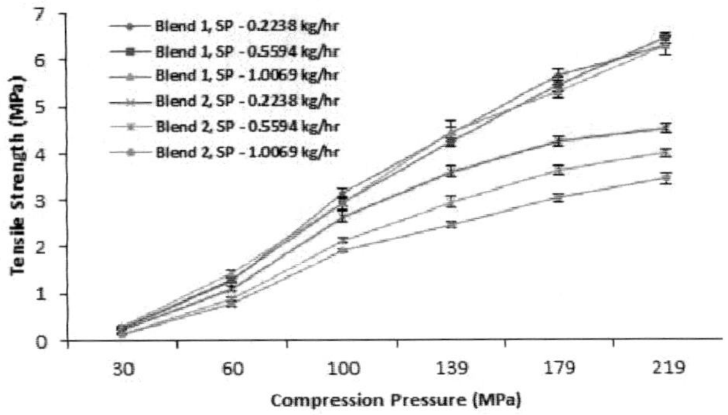

Figure 2: Tensile Strength profiles of both blends fed through the feeder at a range of fed rate set points.

There was no significant difference observed in tensile strength profiles fed at different set points for control blend 1, without magnesium stearate. Whereas the blend 2, with magnesium stearate, showed decrease in tablet tensile strength profiles upon feeding at higher set points (Figure 2). This may be due to increased mixing between the screws, as screws rotate with greater speeds upon increase in set point and would increase blending of the blend.

CONCLUSIONS

The bulk density magnesium stearate effected its feed factor - the higher the bulk density, the higher the feed factor achieved. While all magnesium stearate samples showed higher feeding variability at a lower feed rate, the difference was more pronounced for magnesium stearate samples 1 and 4. The physical properties responsible for differences in magnesium stearate feeding performance are currently being investigated.

The feeding rate of the blend containing magnesium stearate was seen to undermine its compaction properties. The higher the feeding rate the lower the tensile strength of the tablet produced. Future work will look at feeding rate impact on tablet dissolution.

ACKNOWLEDGEMENTS

Research funded by Pharmaceutical Manufacturing Technology Centre (PMTC), an Enterprise Ireland, Irish Development Agency funded centre and Synthesis and Solid State Pharmaceutical Centre (SSPC) and Science Foundation Ireland (SFI) under grant number 12/RC/2275. PH200 was donated by FMC Corporation. K-Tron for support in use of MT12 micro feeder.

BIBLIOGRAPHY

1. Engisch WE, Muzzio FJ. Method for characterization of loss-in-weight feeder equipment. *Powder Technol*. 2012;228:395-403. doi:10.1016/j.powtec.2012.05.058.

2. Pernenkil L, Cooney CL. A review on the continuous blending of powders. *Chem Eng Sci*. 2006;61(2):720-742. doi:10.1016/j.ces.2005.06.016.

3. Simonaho S, Ketolainen J, Ervasti T, Toiviainen M, Korhonen O. European Journal of Pharmaceutical Sciences Continuous manufacturing of tablets with PROMIS-line — Introduction and case studies from continuous feeding , blending and tableting. *PHASCI*. 2016. doi:10.1016/j.ejps.2016.02.006.

4. Gupta A, Hamad ML, Tawakkul M, Sayeed V a, Khan M a. Difference in the lubrication efficiency of bovine and vegetable-derived magnesium stearate during tabletting. *AAPS PharmSciTech*. 2009;10(2):500-504. doi:10.1208/s12249-009-9229-y.

5. Engisch WE, Muzzio FJ. Feedrate deviations caused by hopper re fi ll of loss-in-weight feeders. *Powder Technol*. 2015;283:389-400. doi:10.1016/j.powtec.2015.06.001.

468802 Comparing and Contrasting the Solid-State Esterification of APIs with Excipient: Computational and Experimental Assessment

Abstract:

Chemical stability of pharmaceutical products is one of the key quality attributes. There are limited studies that attempt to construct a molecular level definition of a drug product and predict how solid state factors influence the chemical stability of solid pharmaceutical products. In this study, we aim to combine computational and experimental approaches in an effort to better understand drugâ€"excipient degradation reactions in the solid state.
For our studies, we chose carvedilol (CAR) and codeine phosphate (COP) in formulation with citric acid (CA) as our model systems. Both CAR and COP possess a secondary alcohol and are known to undergo esterification in presence of CA [1,2]. To identify effects that are purely based on differences in the crystal structure, all calculations and experiments were carried out with two different polymorphic forms of, each, CAR and COP. For the morphologically important crystal faces, Molecular Dynamics (MD) simulations were used to calculate a range of descriptors which we expect to account for relevant structural, energetic and kinetic properties of the API crystal surfaces [3].
For the experimental assessment of the solid state stability, we first recrystallized the different crystal forms of CAR and COP and fractionated different sizes by sieving in order to obtain crystals with defined characteristics such as particle size and surface area. We then produced mixtures of the model APIs with citric acid and stored the samples at accelerated conditions. The accelerated storage conditions were chosen in a way that all crystal forms are stable, i.e. no polymorphic transformation and no loss of crystal water would occur. A different set of degradation experiments was performed where disorder was introduced to our APIâ€"excipient systems by ball milling.
By comparing results from our computational and experimental assessments, we gain insights into which material properties crucially determine the stability of pharmaceuticals in formulation with excipients in the solid state.

References

[1] Larsen, J, Cornett, C, Jaroszewski, JW, Hansen, SH. (2009) Reaction between drug substances and pharmaceutical excipients: formation of citric acid esters and amides of carvedilol in the solid state J. Pharm. Biomed. Anal., vol. 15, pp. 11-17.
[2] Silver, B and Sundholm, EG. (1987) Solid-State Esterification of Codeine Phosphate by the Acid Constituent of Effervescent Tablets J. Pharm Sci., vol. 76, pp. 53â€"55.
[3] Brunsteiner, M, Gressl, Khinast, JG Paudel, A 1 C, Davis, AL, Landis, M, Pencheva, K, Scrivens, G, Wood, GPF (2015) Stability by Design (SbD): Utilizing material science principles to predict physical and chemical stability of pharmaceutical solids AIChE 2015.

469699 The Effects of Formulation Factors on the Aerosolization Performance of Metered Dose Inhalers

Abstract:

Introduction: Metered dose inhalers (MDIs) are complex drug-device combination products widely used as portable delivery systems to treat pulmonary disorders, such as asthma and chronic obstructive pulmonary disease (COPD). A typical MDI consists of a canister containing the formulation, a metering valve, and an actuator-mouthpiece.[1] The formulation can be either a solution [active pharmaceutical ingredient (API) dissolved in the liquid propellant] or a suspension (API particles dispersed in the liquid propellant) along with inactive ingredients (e.g., co-solvents and surfactants).[2] The product performance of MDIs depends on a multitude of factors including, but not limited to, the physicochemical properties of API, device geometry (e.g., valve metering chamber volume, actuator nozzle orifice diameter, actuator sump depth, and actuator orifice jet length), and nature and amount of inactive ingredient(s).[3] Under the Quality by Design (QbD) paradigm, systematic investigations are necessary to understand how changes in critical quality attributes (CQAs) of formulation, device and manufacturing process influence the product performance.[4] Although much is known about the effects of changes in device geometry on the aerosolization performance of MDIs,[5],[6] the effects of changes in formulation factors on MDI product performance are not clearly defined. Therefore, the purpose of this work is to provide a better understanding of the effects of different levels of inactive ingredients and API particle size distribution (PSD) on key aerosolization performance parameters of MDI products. The systematic approach applied in this work can be utilized as a QbD tool to develop mathematical models and design spaces, allowing the manufacture of formulations with desired *in vitro* product performance parameters through a proper combination of formulation factors.

Methods: Three commercial MDI products (albuterol sulfate and mometasone furoate suspension formulations, and beclomethasone dipropionate solution formulation) were chosen for this study. Reverse engineering was utilized to determine the qualitative and quantitative composition of the formulations. The aerosolization performance was characterized by measuring the delivered dose (DD) and aerodynamic particle size distribution (APSD) performance. MDI model systems were then developed and used as anchor point for developing a statistical experimental design to study the effects of formulation factors (inactive ingredients concentrations and API PSD) and simulate design spaces of product performance. The purpose of defining the MDI model systems was not to achieve *in vitro* bioequivalence to the commercial MDI products, but to develop systems similar to the commercial MDIs with respect to formulation composition and key aerosolization performance parameters. A reduced factorial statistical design of experiments (DoE) approach was then used to develop a MDI batch manufacturing plan to vary the levels of inactive ingredients (ethanol and oleic acid) and API PSD (D_{50}, the value of the particle diameter at 50% in the cumulative distribution, that is, the median diameter of the PSD). The following ranges were studied:

API	D_{50} (μm)	Ethanol (% w/w)	Oleic Acid (% w/w)
Albuterol Sulfate	1.4 - 2.5	7 - 20	0.005 - 0.1
Mometasone Furoate	1.1 - 2.0	0.45 - 3.6	0.001 - 0.025
Beclomethasone Dipropionate	N/A	7 - 9	0 - 2

API D_{50} was varied for the suspension formulations only; micronized API was prepared from the same mother batch by sizing down to the desired API D_{50} using jet mill process. While the commercial solution MDI chosen for this work does not contain oleic acid, such an inactive ingredient was included in the DoE plan to study its effects in the solution formulation in comparison to the suspension formulations. The MDI batches were manufactured under non-GMP conditions, characterized and the data was statistically analyzed – a formulation factor (ethanol and oleic acid concentrations, or API D_{50}) was considered to have a statistically significant effect on the aerosolization performance parameter if p value < 0.05. Multivariate mathematical models and design spaces were then developed to predict the MDI aerosolization performance with respect to DD and APSD according to the different levels of formulation factors. The potential effects of different propellants [hydrofluoroalkane (HFA)-134a in the albuterol sulfate and beclomethasone dipropionate MDIs, and HFA-227 in the mometasone furoate MDIs] were not studied in this work.

Results: For the albuterol sulfate suspension MDI formulations studied, the effect of ethanol on the DD was statistically significant – as the ethanol concentration increased from 7 to 20% w/w, the DD decreased by 17%. Both ethanol and API D_{50} showed statistically significant effects on the APSD – as the ethanol concentration and API D_{50} increased from, respectively, 7 to 20% w/w and 1.4 to 2.5 μm, the fine particle dose less than 5 μm (FPD<5) decreased by 40%. The effect of oleic acid on the DD and APSD was not statistically significant. For the beclomethasone dipropionate solution MDI formulations studied, the effect of ethanol on both DD and APSD was not statistically significant within the range studied. However, the effect of oleic acid on the DD and APSD was statistically significant – as the oleic acid concentration increased from 0 to 2% w/w, the DD decreased by 11% and the FPD<5 decreased by 65%. For the mometasone furoate suspension MDI formulations studied, the DD was significantly affected only by ethanol – as the ethanol concentration increased from 0.45 to 3.6% w/w, the DD increased by 7%. All three formulation factors (ethanol and oleic acid concentrations, and API D_{50}) showed statistically significant effects on the APSD, but the strongest effects were provided by the ethanol and API D_{50} – for these, a change of approximately 40% in FPD<5 was obtained from the lowest to the highest factor level.

Conclusions: The changes in API PSD had statistically significant effects on the APSD performance of suspension MDI formulations studied, but not on DD. The changes in concentrations of inactive ingredients (ethanol and oleic acid) showed, in some cases, statistically significant effects on DD and APSD performance of suspension and solution MDI formulations studied. However, several cases without effects were also found, despite some large changes in concentrations of inactive ingredients studied. The possible effects of varying these must hence be studied on a case-by-case basis. The outcomes of this study allowed defining design spaces for DD and APSD performance according to the different levels of formulation factors (ethanol and oleic acid concentrations, and API D_{50}). The systematic approach utilized in this work can contribute as a QbD tool to evaluate the extent to which the formulation factors govern the aerosolization performance of MDI products, helping to design MDI formulations with desired *in vitro* product performance parameters.

Keywords: Metered Dose Inhalers (MDIs); Quality by Design (QbD); Factorial Design of Experiments (DoE).

References:

[1] Stein, S.W., *et al.*, Advances in metered dose inhaler technology: Hardware development. *AAPS PharmSciTech*, 2013. **15**(2): p. 326-338.

[2] Bell, J. and S. Newman, The rejuvenated pressurised metered dose inhaler. *Expert Opinion on Drug Delivery*, 2007. **4**(3): p. 215-234.

[3] Smyth, H.D.C., The influence of formulation variables on the performance of alternative propellant-driven metered dose inhalers. *Advanced Drug Delivery Reviews*, 2003. **55**(7): p. 807-828.

[4] Lionberger, R.A., *et al.*, Quality by design: Concepts for ANDAs. *The AAPS Journal*, 2008. **10**(2): p. 268-276.

[5] Smyth, H., *et al.*, Spray pattern analysis for metered dose inhalers I: Orifice size, particle size, and droplet motion correlations. *Drug Development and Industrial Pharmacy*, 2006. **32**(9): p. 1033-1041.

[6] Smyth, H., *et al.*, Spray pattern analysis for metered dose inhalers: Effect of actuator design. *Pharmaceutical Research*, 2006. **23**(7): p. 1591-1596.

469725 Exploiting Solvent and Impurity Effects to Modify Crystal Shape and Bulk Powder Properties of API and Intermediate Molecules

Abstract:

Crystallization design is mainly focused on designing API manufacturing process to deliver drug molecules with desired particles size and shape, as well as bulk powder properties to allow downstream formulation process to be more robust and less expensive. However, similar focus on upstream steps of the API process could benefit from improved physical properties of the API or intermediates that lead to better filtration, impurity rejection, cycle time, and throughput on scale. This work will present examples showcasing merits of exploiting solvent and impurity effect to improve process robustness and productivity.

469947 Continuous Manufacturing of Co-Crystals in Twin Screw Extruder: A Case Study

Abstract:

Co-crystals are crystalline materials, comprised of two or more components that are solids at room temperature and held together by noncovalent forces. A pharmaceutical co-crystal consists of an active pharmaceutical ingredient and a pharmaceutically acceptable co-former. Nowadays, pharmaceutical co-crystals are becoming an integral part in pharmaceutical strategies as they can improve the physicochemical properties of the API.Wet granulation is a widely used process in the pharmaceutical industry and its main goal is to produce granules from fine particles of active pharmaceutical ingredients (API) and excipients using a granulation liquid. In a twin screw extruder, the process is undergone under the high shear forces exerted by the screws on the particles.

The main aim of this study is to investigate and develop novel processes for continuous pharmaceutical processing of co-crystals. Wet granulation was chosen as a manufacturing method since it combined synthesis and granulation in a single continuous operation step. A 12 mm twin screw extruder (Three-Tec, Switzerland) with a length-to-diameter ratio of 40 was used for the manufacturing of the granules.A novel Theophylline-Aspirin (THP-ASP) model system was selected for the study. The effects of different process parameters (feed rate, screw speed, liquid-to-solid ratio) and formulation parameters (API and co-former ratio) on the final granules properties and the extent of conversion were investigated. Particle size distribution was analysed using Microtrac particle size analyser. The solid state form of the final granules was characterised with various techniques, including DSC, and PXRD.

470423 Fault-Tolerant Control Design and Risk MAP Based Resiliency Analysis of Continuous Solid Dose Manufacturing

Abstract:

The recent development and adoption of continuous manufacturing technologies in pharmaceutical industries[1-3] has also created the demand for more efficient process monitoring and control techniques. This is particularly the case with solid dose manufacturing processes, in which unit operations, such as blending, milling, *etc*., that in a classic batch processing are performed sequentially, are now integrated with a higher degree of automation, and operated at steady state with less manual intervention.[4-8] It has also become apparent that fast and reliable control strategy design is a requirement for successful continuous manufacturing technology. Hence, in this study, a systematic framework for fault-tolerant control design and resiliency analysis is proposed for continuous solid dose manufacturing processes based on risk mapping, accessing, and planning (Risk MAP).[9]

First, a risk mapping was designed to characterize the frequency and severity of the potential faults during the design and operation stages, *e.g*., model-plant mismatch, calibration errors, failing/missing sensors, process disturbances, *etc*. The three levels of hierarchical control strategies design, which ranges from single proportional-integral-derivate (PID) control loop to plant-wide advanced model predictive control (MPC), were then assessed subject to the potential faults by using control key performance indices (KPIs), such as, time to product (T2P), magnitude to product (M2P), and other control measures, such as Morariâ€™s resiliency index (MRI).[10] Mitigation planning for severe faults were also suggested, *e.g*., to activate or decommission a specific control level, or to divert the off-spec products.

The proposed systematic framework for fault-tolerant control design was demonstrated in a pilot plant for continuous dry granulation process[11] and was shown to be efficient and robust in handling potential faults and achieving a robust processing line for solid dose.

References

1. Ierapetritou M, Muzzio F, Reklaitis G. Perspectives on the continuous manufacturing of powder-based pharmaceutical processes. *AIChE Journal*.2016;62:1846-1862.
2. Lee SL, Oâ€™Connor TF, Yang X, Cruz CN, Chatterjee S, Madurawe RD, Moore CMV, Xu LX, Woodcock J. Modernizing pharmaceutical manufacturing: from batch to continuous production. *Journal of Pharmaceutical Innovation*. 2015; 10:191-199.
3. Buchholz S. Future manufacturing approaches in the chemical and pharmaceutical industry. *Chemical Engineering and Processing*. 2010; 49:993-995.
4. Vervaet C, Remon JP. Continuous granulation in the pharmaceutical industry. Chemical Engineering Science. 2005; 60:3949-3957.
5. Teng Y, Qiu Z, Wen H. Systematical approach of formulation and process development using roller compaction. *European Journal of Pharmaceutics and Biopharmaceutics*. 2009; 73:219-229.
6. Singh R, Ierapetritou M, Ramachandran R. An engineering study on the enhanced control and operation of continuous manufacturing of pharmaceutical tablets via roller compaction. *International Journal of Pharmaceutics*. 2012; 438:307-326.
7. Singh R, Sen M, Ierapetritou M, Ramachandran R. Integrated moving horizon-based dynamic real-time optimization and hybrid MPC-PID control of a direct compaction continuous tablet manufacturing process. *Journal of Pharmaceutical Innovation*. 2015; 10:233-253.
8. Hsu SH, Reklaitis GV, Venkatasubramanian V. Modeling and control of roller compaction for pharmaceutical manufacturing. Part II: control system design. *Journal of Pharmaceutical Innovation*. 2010; 5:24-36.
9. http://www.fema.gov/media-library/assets/documents/18274.
10. Morari M. Design of resilient processing plants III a general framework for the assessment of dynamic resilience. *Chemical Engineering Science*. 1983;38(11): 1881-1891.
11. Hsu SH, Reklaitis GV, Venkatasubramanian V. Modeling and control of roller compaction for pharmaceutical manufacturing. Part I: process dynamics and control framework. *Journal of Pharmaceutical Innovation*. 2010; 5:14-23.

470558 A Comprehensive Statistics-Based Approach to Proactive Process Analysis Including Multivariate Analysis Methods

Abstract:

Following the initial design and validation of a pharmaceutical drug product manufacturing process, the process will undergo further monitoring. For example, those process parameters that are most likely to impact the critical quality attributes, may be monitored over a series of batches. The resulting data can be used to 1) satisfy continued process verification, as part of stage 3 PPQ, or other regulatory needs and 2) ensure the process is in control and increase robustness by proactively improving and optimizing the process (i.e. proactive process analysis).

The approach to process monitoring for continuous improvement purposes can be varied. An example of a comprehensive systematic, statistics-based approach to proactive process analysis is presented here. The approach includes using statistical process control to visualize and identify trends before an impact to quality occurs. Furthermore, the use of multivariate analysis methods, such as principal component analysis (PCA) and partial least squares (PLS), to identify interactions between process parameters and point of cause, in support of root cause investigations, is illustrated.

470746 The Utilization of Amino Acid Consumption Rate Assessment As a Tool for CHO Clonal Stability Evaluation.

Abstract:

<span">CHO cells are commonly utilized for the production of therapeutics.To ensure product reproducibility and quality, assessment of clonal stability is important for both regulatory and economic reasons. We are continuously striving to improve methods for assessing CHO cell bank stability. One new tool we are currently evaluating is specific amino acid (AA) consumption rate and how it relates to CHO clonal stability. By calculating AA consumption data generated at various PDL ages we are able to observe trends between the AA consumption rates per cell per day of multiple clones throughout the process. Clones which exhibit increased AA consumption as they age can potentially result in total depletion of one or more amino acids within the culture. As a result false conclusions may be drawn that a clone is unstable when in reality the feed strategy may need to be further assessed in order to maintain a high level of productivity.

<span"> For this experiment we ran selected clones in an AMBR250 using a platform process. Time point samples were analyzed for AA content using a Hitachi L-8900 amino acid analyzer and titers were analyzed using RP-HPLC. These data were used to calculate specific productivity (Q_P) and specific amino acid consumption (Q_{AA}). The data presented shows that of three high producing clones, C10 exhibits increased AA consumption and decreased titer as the PDL increases. C81 AA consumption and titer production remain constant, while C88 AA consumption levels decrease but titer remains constant across PDL ages. Based on this information, C81 and C88 are better candidates for establishment of stable master and working cell bank. It also points to the potential to prevent the loss of $Q_P$in C10 by exploring additional feeds for late PDL cells.

470932 Powder Tester Comparison to Predict Discharge Rate from Hoppers and Filling Tablet Press Dies

Abstract:

To enhance the productivity in the manufacturing of solid dosage forms it is important to ensure that powders flow from hoppers and into tablet press dies at appropriate rates to enable high throughput and reduce variation to minimize product defects. This paper discusses the results of different flow testers to describe flowability of materials and how they relate to solid dosage form production.

The manufacturing of tablets requires fine powders to flow through a number of hoppers, feeders and finally into dies of a tablet press. The authors evaluated the flowability of fine powders with commonly used powder characterization testers in the pharmaceutical industry namely shear testers, compressibility cells, permeability cells, and funnel, orifice plate and avalanche testers.

For this study the authors selected 10 cellulose ether excipients which are investigated for flowability assessment. The flowablity measured by the different flow testers is compared with the limiting flow rate based on Johanson's work [1]. It was found that the discharge rate or flowability is best described by the permeability and the bulk density. The flow function or the flow factor could not describe the discharge rate appropriately. The Carr index describes the compressibility of a powder but cannot predict the discharge rate. However, the Carr index can be used to evaluate flowability trends if the variation of Carr index measurement is considered.

The impact of flowability of a material on tableting was evaluated experimentally with a Piccola bilayer press. It was found that the standard deviation of tablet weight, compaction force and radial tensile strength can be described as a function of flowability. With increasing flowability, the variability in all three metrics was reduced. Thus, powder flowability can be applied to understand variation in the tableting process. Consequently in this study the flowablity of all cellulose ether excipients was measured and it was shown that Methocel™ DC2 has the best balance between flow properties and the above mentioned tableting metrics.

[1] Johanson, J. R., "Two-Phase Flow Effects in Solids Processing and Handling," ***Chemical Engineering***, pp. 77–86 (Jan. 1, 1979)

470797 PAT Application in Expedited Development of Multi-Step Chemical Syntheses of Active Compounds

Abstract: "PAT application in Expedited Development of Multi-Step Chemical Syntheses of Active Compounds"

Rapid kinetic methods that allow to acquire knowledge of reaction mechanism and rate early-on in the research and development process are becoming widely utilized. They are used to determine the range of operability space in order to ensure a safe, reliable and successful scale-up of complex molecules. One of the limitations to taking such a modern approach to development and scale-up in general is the need for a large number of experiments and an extensive amount of analytical data. In addition, tools are needed for non-experts to interpret, evaluate, and convert the data into chemical reaction knowledge and understanding. PAT, and in particular in situ ATR-FTIR is being increasingly recognized as a key technology to automatically collect a large number of analytical data from chemical reactions in situ and in real time. Since more information is retrieved from each experiment, a smaller number is required to provide concentration dependence of reaction components as well rate and equilibrium constants. This high-throughput approach to data collection can then be complemented with simplified means of representing and analyzing the data, among others, multi-dimensional visuals and Reaction Progress Kinetic Analysis. Literature examples of combining real time ATR-FTIR, simplified kinetic analysis now abound.
The presentation will describe several examples, including the concise scalable synthesis of a chiral dipeptide acid, substructure of the HCV protease inhibitor Faldaprevir. An extensive utilization of PAT early-on provided sufficient process knowledge to complete three steps in one operation in a much shorter time, and afforded a greener and robust process.
scheme 1, 2).

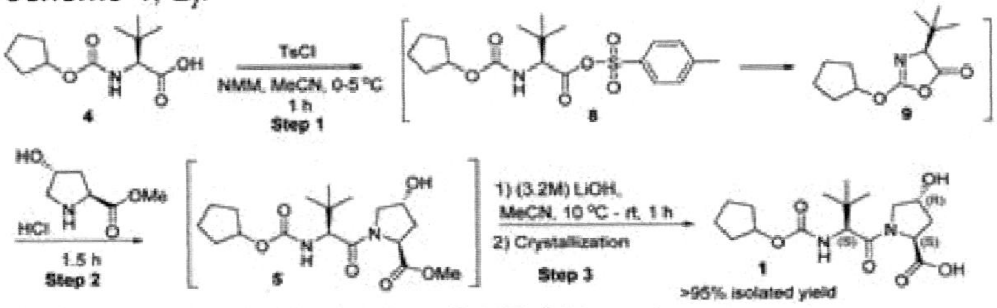

Scheme 1: Scale-up process for Faldaprevir

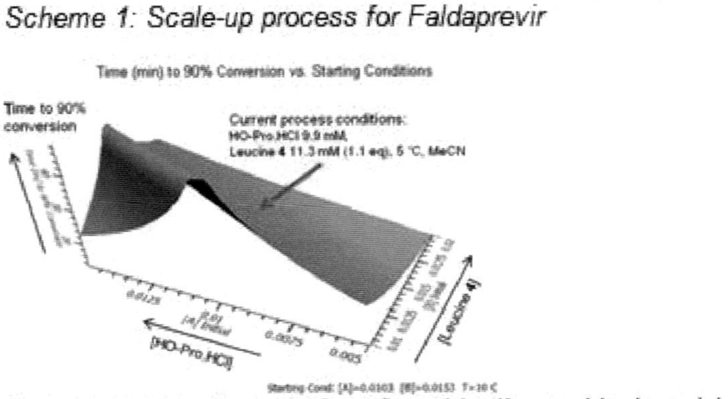

Scheme 2: Response surface from kinetic empirical model

470864 Process Innovation and Optimization of a Commercial Cell Culture Process

Abstract:

Reduction of cadence and optimization of process yield are both required to maximize commercial biopharma manufacturing plant output. Often, these are mutually exclusive and concessions must be made to one or both areas to improve overall plant efficiency or maintain product quality. This abstract describes innovative optimization approaches on two different mammalian cell culture unit operations to improve both cadence and yield.

In the first example, the mammalian cell culture process described requires control of inoculum expansion at a low final viable cell density target, above which viability drop and growth lag are observed in the subsequent stage. The inoculum expansion and seed bioreactor stages require a precise control of transfer conditions to maintain the culture within certain viable cell density range. A flexible and scalable inoculum expansion process with robust control has been implemented. The process is resilient and supports a wide variety of plant cadence targets. In the second example the mammalian cell culture process described showed no increase in final output when production bioreactor titer improved by 20%. Data driven analysis of the production process showed a negative correlation between production bioreactor titer and primary recovery step yield. Multi Variate Analysis (MVA) was performed to identify drivers for lower step yields and process optimization.

471020 Design of Nonlinear Observer for State Estimation of a Bioreactor

Abstract:

Measurement or estimation of process states is critical for process monitoring, advanced process control, and process optimization[1,2]. For chemical process whose state information cannot be measured directly, techniques such as state estimation need to be developed. Model-based state estimation is one of the most widely applied methods for estimation of unmeasured states basing on a high-fidelity process model. However, for some complex process, it requires extensive knowledge and effort to build a high-fidelity model to describe the whole system. Certain unmodeled disturbance or unknown inputs will generate model-plant mismatch, and in some case the mismatch is significant. For a case of bioreactor, there are several situations that can generate model-plant mismatch which can be treated as unknown input or disturbance, such as the effect of nutrient limitation, death of cells due to bust of air bubbles, oxygen delivery at high cell density, carbon dioxide stripping, and accumulation of inhibitive compound[3]. There are several different attempts to address the model-plant mismatch. One of the solutions is to apply parameter estimation for each run of the experiments. However, the adaptive parameter estimation requires extensive measurements of states and cannot be used for state estimation for a new experiment until a new set of parameters is updated. Filter and observer are mathematical methods to extract state information from corrupted measurements, which offer an alternative method for state estimation in the presence of model-plant mismatch.

Unknown input observer was developed to estimate the state in presence of certain types of faults and disturbance[3]. Comparing to other observers, unknown input observer can eliminate the effect of certain disturbance or fault despite of their size when their possible influences are estimated in the process model. To estimate the process state in presence of process disturbance or unknown inputs, a new design of nonlinear unknown input observer is proposed and it is applied for state estimation of a bioreactor. Design of such observer is provided and sufficient and necessary conditions of the observer are discussed.

Experimental study of batch and fed-batch operation of a bioreactor was performed, which used Saccharomyces cerevisiae strain mutant SM14 to produce Î²-carotene. Fed-batch experiment was based on the feeding of glucose and ethanol. Based on the mathematical modeling of the process which is demonstrated in our previous study[5], unknown input observer for the bioreactor was developed. Model-plant mismatch was observed between the original process model and measurements when changing the initial conditions or operating mode of the experiments. State estimation of the process from the designed observer was demonstrated and compared to the experiment measurements. Results indicated that the state estimation from unknown input observer has smaller estimation error compared to the original process model, especially for the fed-batch experiment.

1. Heidarinejad M, Liu J, Christofides PD. State-estimation-based economic model predictive control of nonlinear systems. *Systems & Control Letters.* 2012;61(9):926-935.

2. Ricker NL. Model predictive control with state estimation. *Industrial & engineering chemistry research.* 1990;29(3):374-382.

3. Ozturk S, Hu W-S. Cell culture technology for pharmaceutical and cell-based therapies: CRC Press; 2005.

4.Chen W, Saif M. Unknown input observer design for a class of nonlinear systems: an LMI approach. Paper presented at: American Control Conference, 20062006.

5. M. Carolina Ordoñez, Jonathan P.Raftery, Tejasvi Jaladi , M. Nazmul Karim. Modelling of batch kinetics of aerobic carotenoid production using Saccharomyces cerevisiae. *submitted.*

471193 Computational Fluid Dynamics (CFD) Guided Scale up of Cell Culture Bioreactors from Bench to Pilot to Commercial Scale

Abstract:

Computational fluid dynamics is a valuable tool for better understanding bioreactors and medium preparation tanks used for mammalian cell cultivation. Knowledge generated by CFD, such as on flow characteristics or on impeller mixing, aid in better transfer and scale up of cell culture processes.

This study used single phase simulations to predict power input and mixing time of production bioreactors at different scales. Impeller power numbers generated by CFD were compared across sites and assessed against historical vendor values to align cross-site scale up parameters. Approaches to applying CFD to large scale cell culture bioreactors will be discussed.

472590 Chemical Engineering Principles in Oral Solid Dosage Process Development-Where Are We?

Abstract:

While solid processing is a common engineerâ€™s domain across industries, there are still gaps in the pharmaceutical industry in applying basic chemical engineering principles, from API development to drug product development, especially equipment design and process control.Â General gaps and misconceptions in these areas gathered from field experience will be presented along with case studies. While the barrier to apply equipment modification or to upgrade process control could be complicated and cumbersome in the heavily regulated industry, the benefit of such change cannot be ignored due to the fact that there were (and are) significant amount of development work caused by the lack of basic principles in designing the current system (as simple of sufficient heat transfer). The other element of barrier is rather cultural, i.e., no desire to change as the current system is workable. In summary, opportunities along with challenges in the pharmaceutical process development are waiting for more chemical engineers to explore.

472435 Characterization of Mixing and Heat Transfer in Continuous Lab Scale Reactors

Abstract:

With proper design and optimization, continuous reactors can provide advantages of heat transfer, mixing, and productivity in a smaller footprint. However, batch reactors have the advantage of a flexibility, including slow or variable feed-times. Continuous plug flow reactors with a single feed point provide stoichiometric feed ratio's upfront. This places higher demand on quality of mixing and heat transfer at the point of mixing for continuous plug flow reactors especially if there are possible side reactions. Chemical systems e.g. Bourne reactions were used to assess the quality of mixing in glass Advanced Flow Reactors (AFR manufactured by Corning) to determine the quality of mixing, based on selectivity from parallel competitive reactions as a function of process flow rates. The quality of mixing is compared to lab-scale batch reactors (EasyMax manufactured by Mettler). Heat Transfer through AFR plates will independently be measured. The residence time distribution through the AFR will also be discussed. The methods described herein are generally applicable for characterizing lab-scale continuous reactors.

472802 Screening a Library of Commercially Available Compounds for Drugs with Novel Anti-Biofilm Activity

Abstract:

It is now well established that bacterial infections are often associated with biofilm phenotypes that demonstrate increased resistance to common antimicrobials. Further, due to the collective attrition of new-antibiotic development programs by the pharmaceutical industries, drug re-purposing has become an attractive alternative. In this work, we have screened 1,280 existing commercially available drugs in the Prestwick Chemical Library, some with previously unknown antimicrobial activity, against an organism that has been implicated as a causative pathogen in bacterial infections, such as burn and wound infections: *Staphylococcus aureus* and *Pseudomonas aeruginosa*. From a preliminary screen of the entire Prestwick Chemical Library at a fixed concentration of 10mM, we found 104 drugs that showed activity against planktonic *S. aureus*, and not surprisingly these were mostly antimicrobials and antiseptics. The activity of 18 selected non-antimicrobials observed in the preliminary screen was confirmed in dose response experiments. Finally, a selection of nine drugs with little or no reported antimicrobial activity was tested against pre-formed biofilms of *S. aureus*. Three of these drugs retained their antimicrobial activity against pre-formed biofilms making them attractive candidates for repurposing as novel anti-biofilm therapies. On the other hand *P. aeruginosa* biofilms were resistant to most of the drugs in the library. However, combinations of some of the drugs in the library with sub-inhibitory concentrations of antibiotics and surfactants were effective against the biofilms. Together, our data show that the anti-biofilm activity makes these drugs attractive candidates for re-purposing as novel anti-biofilm therapies, especially in combination with traditional antimicrobials, and warrant further attention to determine their modes of action.

475290 Controlling amorphization versus crystallization of carbamazepine nanoparticles using a supercritical-assisted spray drying process

Abstract:

Spray drying is now regarded as one of the most commonly used technologies in the pharmaceutical industry for API (Active Pharmaceutical Ingredient) particle production in the micron-size range[1]. It is a method of producing dry powders from a solution or a suspension by rapidly evaporating solvent with a hot air stream. This is one of the preferred technologies for material processing and scale-up in the pharmaceutical and food industries because it is a fast, continuous, and a one-step process. However, the solid-state nature of the spray-dried materials is often amorphous due to rapid solidification and is restricted to the micron size scale range. Although amorphous materials provide huge improvements in the APIs apparent solubility, a major limitation of these materials is that the drug is thermodynamically unstable and tends to recrystallize during storage (depending on the stability provided by the polymer matrix in the solid dispersion) where the outcome might be a crystalline form with a much lower solubility.

Pharmaceutical crystalline materials provide a higher potential for enhanced solid state stability of poorly soluble APIs compared to amorphous materials, although providing a poorer solubilities and dissolution rates. Interestingly, producing poorly soluble APIs as nanocrystals provides considerable improvements in their solubility, dissolution rate as well as solid-state stability. One of the most successful top-down methods to obtain APIs nanocrystals or solid nanocrystalline dispersions with the desired size distribution consists of wet milling APIs down to a nano size range, adding excipients to these nanosuspensions to inhibit crystal growth due to Ostwald ripening and spray drying them later on[2]. However, milling limits the control of the API crystal habit and other surface properties, can take long times to achieve the required size, can introduce impurities from abrasion of the milling media, and can lead to thermal/mechanical degradation of temperature-sensitive materials.

In this work a supercritical CO_2-assisted spray drying technology has been used to produce and control the solid-state form carbamazepine (e.g. amorphous, crystalline) in a nano-sized range. This process involves contacting the API solution with supercritical CO_2 in the nozzle and immediately spraying it through the nozzle exit for solvent extraction at atmospheric pressure[3]. By restricting the high pressure exclusively to where it is indispensable, it is possible to control the precipitation of the API before the nozzle (anti-solvent nucleation/crystallization) or after the nozzle (atomization/spray drying) while providing processing flexibility. This also facilitates the adaptation of this technology onto existing spray drying equipments.

Solutions of CBZ (carbamazepine) dissolved in methanol ([CBZ] = 50mg/ml) with or without selected additives (e.g. ethyl cellulose, sodium stearate, maltitol, Kollidon VA64) were spray dried using supercritical CO_2. In this process the carbamazepine solutions are fed to the nozzle exit, while immediately and continuously feeding the solutions or suspensions through the nozzle to obtain the final powders. Depending on the type of spray drying mechanism being used (atomization or antisolvent-assisted atomization) by changing process variables (e.g. pressure, temperature, supercritical antisolvent/solvent ratio) has provided control on the size and solid-state outcome (e.g. amorphous, crystalline) of the carbamazepine nanoparticles. The inclusion of small amounts of selected additives (0.1-10% w/w) in the carbamazepine-methanol solutions has provided an additional step to controlling the formation of nanoparticles in the amorphous form or favoring the crystallization of a particular polymorph.

1. Vehring R. *Pharm. Res.* 25 **(2008)** 999-1022.

1. Paredes *et al. Drug Dev. Ind. Pharm.* (**2016**), DOI: 10.3109/03639045.2016.1151036
2. Padrela *et al. J. Supercrit. Fluids* 86 (**2014**) 129-136.

This work has been supported by Science Foundation Ireland, Grant number: 12/RC/2275.

475659 Evaluation of the Therapeutic Potential of FOXC1 Knockdown for Treatment of Basal-Like Breast Cancers

Abstract:

Basal-like breast cancer (BLBC) accounts for 15% of breast cancer diagnoses, is characterized as one of the most aggressive forms of breast cancer, and exhibits the worst prognosis. The cancer tends to metastasize to the brain and lungs during early stages [1]. Due to the BLBC triple negative phenotype (estrogen receptor-, progesterone receptor-, human epidermal growth factor 2-), chemotherapy remains the only treatment despite its poor success rate and propensity for relapse. The aggressive growth and mobility of BLBC contribute to its difficulty of treatment and result in low survivability [2-3]. This mortality indicates the urgency for new therapeutic approaches to replace or supplement the limited chemotherapeutic treatment options. Since the forkhead-box transcription factor C1 (FOXC1) was isolated as a unique biomarker for BLBC [4], alternative therapeutic approaches focusing on inhibition of FOXC1 have emerged. FOXC1 overexpression in BLBC was linked to excessive cell proliferation and aggressive mobility. Knockdown of FOXC1 using RNA interference, in contrast, was shown to decrease proliferation and mobility of BLBC cell lines *in vitro*. This suggests that the elimination or knockdown of FOXC1 is a potential therapeutic approach to the treatment of BLBC. This project evaluates this therapeutic potential both *in vitro* and *in vivo* by various gene therapy methods.

Knockdown of protein products from aberrant or overexpressed genes can be achieved by delivering repair machinery to act at the DNA level, or RNA interference machinery to elicit degradation of mRNA products of targeted genes. The CRISPR/Cas9 system is a cutting-edge method that has the ability to produce double stranded cuts in DNA at targeted locations in the eukaryotic genome in order to produce controlled deletions or insertions of genetic code. RNAi was used to knockdown FOXC1 in 4T1 cells (mouse BLBC). Knockdown (>95%) was verified via western blot and proliferation studies showed a statistically significant ($p<0.013$) reduction in cell growth up to 96 hours post transfection. The CRISPR/Cas9 system was used to produce clone 4T1 cells exhibiting knockout of the FOXC1 gene (verified via western blot). Clone proliferation studies *in vitro* showed similar results to knockdown studies. These clones were used to produce tumors in mice models that were compared to 4T1 tumors in control specimens. Knockout tumors exhibited no reduction in metastatic activity compared to controls and showed no statistical reduction in growth rate. The study indicates that therapies targeting FOXC1 are possible and even successful *in vitro*. However, the inherent complexity of physiological systems convolutes the effect of any positive therapeutic response *in vivo*.

1. Smid M, et al. *Cancer Res*. 2008; 68:3108-3114

2. Seewaldt VL, Scott V. *N Engl J Med* 2007; 356: e12

3. Sarrio D, et al. *Cancer Res* 2008; 68:989-997.

4. Ray P, et al. *Cancer Res* 2010; 70:3870-3876

469512 A Generalizable Crispr-Cas9 System for Rapid and Scalable Gene Target Discovery and Optimization of Metabolism

Abstract:

Metabolic engineering enabled the enhanced bio-renewable production of numerous drop-in fuel replacements and specialty chemicals. Often, these traits are achieved by rewiring the native metabolism in many hosts as well as fine-tuning the expression of local pathways. However, traditional strain engineering faces many limitations in its scope: DNA incorporation to create new strains is time-consuming and there are only a limited number of markers available to multiplex this process. The speed of genetic modification juxtaposed with a large number of required design-build-test cycles required to optimize a strain thus limits strain engineering. To bypass these limitations, we first develop a plasmid-based CRISPR-Cas9 system in yeast to specifically program gene expression in the absence of â€œhard-wiredâ€ genetic modification. Second, we utilize this system to rapidly prototype metabolic pathways via expression perturbation in order to identify key bottlenecks limiting flux and then multiplex these gene targets in a manner that is rapidly scalable and combinatorial. We generalize this method by demonstrating several pathway case studies of interest enabling us to screen and optimize gene targets across yeast metabolism for varied phenotypes. Finally, this method for target identification scales with transformation, thus allowing for rapid identification of gene targets using a combinatorial approach.

470247 Improving Yields of Glucose-Derived Products Using the Wood-Ljungdahl Pathway: A Case Study of 3-Hydroxybutyrate Production in *Clostridium Ljungdahlii*

Abstract:

Clostridium ljungdahlii has recently emerged as an attractive candidate for the bioconversion of synthesis gas (CO, CO2, H2) to a variety of fuels and chemicals through the Wood-Ljungdahl (WL) pathway. Here we examine another use of the WL pathway; namely the augmentation of fermentation yields on heterotrophic substrates by using reducing equivalents produced in glycolysis to fix the CO_2 released, thus resulting in complete carbon conversion. As a proof of concept, we focused on 3-hydroxybutyric acid (3HB), a chiral compound that serves as feedstock in a variety of applications. First, we varied promoters, codon usage, and ribosome binding site spacing to optimize 3HB production, achieving titers of 1 g/L. Second, we implemented a CRISPRi approach to reduce flux to unwanted byproducts. Finally, we quantified the activity of the WL pathway during fermentation using isotopic tracers. The results demonstrate the ability to enhance fermentation yields above those that can be achieved in typical engineering hosts such as *E. coli* or *S. cerevisiae*.

457539 A Cell-Free Framework for Rapid Biosynthetic Pathway Prototyping and Enzyme Discovery

Abstract:

Speeding up design-build-test (DBT) cycles is a fundamental challenge facing biochemical engineering. To address this challenge, we report a new cell-free protein synthesis driven metabolic engineering (CFPS-ME) framework for rapid biosynthetic pathway prototyping. In our framework, cell-free cocktails for synthesizing target small molecules are assembled in a mix-and-match fashion from crude cell lysates either containing selectively enriched pathway enzymes from heterologous overexpression or directly producing pathway enzymes in lysates by CFPS. As a model, we apply our approach to *n*-butanol biosynthesis showing that *Escherichia coli* lysates support a highly active 17-step CoA-dependent *n*-butanol pathway *in vitro*. Further, we leverage robotic liquid-handling systems to access combinatorial design space exceeding typical pipelines pursued in cells. The elevated degree of flexibility in the cell-free environment allows us to manipulate physiochemical conditions, access enzymatic nodes, discover new enzymes, and prototype enzyme sets with linear DNA templates to study pathway performance. We anticipate that CFPS-ME will facilitate efforts to define, manipulate, and understand metabolic pathways for accelerated DBT cycles without the need to reengineer organisms.

464040 Construction of a Synergetic Carbon Utilization Mechanism for Biosynthesis of Glucose-Based Compounds

Abstract:

Conventional glucose utilization in *Escherichia coli* depends on glycolysis and pentose phosphate pathway to achieve catabolism. During this process, important catabolites such as acetyl-CoA and pyruvate contribute to cell growth and product synthesis. Unconventional utilization of glucose involves applying glucose as a C6 building block for production of glucose-based compounds. This non-catabolic usage of glucose conflicts with the catabolism which naturally leads to breakdown of glucose for biomass. To reserve glucose as a building block, blocking catabolic pathways leads to cell growth retardation and thus low productivity. To address this conflict, we introduce a second carbon source glycerol and design a synergetic carbon utilization mechanism to strengthen the connection between glucose and glycerol utilization. This new mechanism couples glucose uptake and catabolism of glycerol via the phosphoenolpyruvate (PEP) as a driving force for glucose transport. We validate the mechanism by introducing glucose-based trehalose biosynthesis model. Before introducing the mechanism, the titer of trehalose is only 1.22 g l^{-1} by consuming 7.39 g l^{-1} glucose in 48 h. After enhancement and optimization of the mechanism, the titer of trehalose is 3.67 g l^{-1} in 48 h by consuming 5.86 g l^{-1} glucose. The conversion efficiency of glucose to trehalose is improved from 0.16 g trehalose/g glucose to 0.63 g trehalose/g glucose. After extension of cultivation time to 96h, 8.20 g l^{-1} trehalose is produced in shake flasks. Remarkably, the conversion efficiency of glucose to trehalose reaches 0.86 g trehalose/g glucose, which represents 91% of the theoretic maximum. Hence, this synergetic carbon utilization mechanism, which is established and demonstrated for the first time, can be applied for non-catabolic use of glucose as C6 building block for synthesis of glucose-based compounds. It also provides a novel strategy for industrial microbial production of trehalose.

469461 Genetic Refactoring of *Escherichia coli* with a Formaldehyde-Inducible Promoter for Synthetic Methylotrophy

Abstract:

The discovery of large natural gas reserves within the United States have led to increased interest in converting natural gas into liquid fuels for use in existing transportation infrastructure. Methanol production from methane has increased recently through chemical and bioconversion processes, and its increasing supply, declining price, and low contamination risk make it a promising substrate for the production of chemicals and biofuels. Initial attempts to generate a strain of the model organism *Escherichia coli* capable of efficiently utilizing methanol as a substrate have been met with various metabolic bottlenecks.

Formaldehyde is a cytotoxic compound and the product of the first step of methanol assimilation, catalyzed by methanol dehydrogenase (MDH). Improper pathway balancing and gene regulation can lead to formaldehyde accumulation and low levels of methanol assimilation in engineered methylotrophic *E. coli* strains. Utilizing an *E. coli* formaldehyde-inducible promoter to drive expression of key methanol assimilation genes, including 3-hexulose-6-phosphate synthase (HPS) and 6-phospho-3-hexuloisomerase (PHI) in the ribulose monophosphate (RuMP) pathway, emulates native methylotrophic regulation and avoids the need to add costly inducers. It also reduces the metabolic burden that high expression promoters can place on the cell, and instead allows for expression levels driven directly in response to cell needs.

We genetically refactored methylotrophic *E. coli* strains by strategically placing formaldehyde-responsive promoters before key genes of interest. Refactored strains were evaluated by assessing MDH and HPS activities, transcriptional expression levels, biomass yields on methanol, and methanol consumption rates. A promoter engineering approach was used to elucidate transcription factor binding sites and generate customized promoters for placement. A formaldehyde-responsive promoter library was generated with error-prone PCR and evaluated using a GFP reporter and fluorescence-activated cell sorting (FACS). Utilizing key metabolic intermediates as inducers is a promising approach for automated pathway balancing and increased product yields.

This work was supported by the US DOE ARPA-E agency through contract no. DE-AR0000432.

473002 Design, Construction, and Validation of Modular Chassis Cells for Efficient Combinatorial Biosynthesis of Chemicals

Abstract:

Transforming biology into engineering practice has a great potential to shape the industrialization of biology that will drive rapid development of novel microbial manufacturing platforms. These platforms will be capable of producing a vast number of industrially relevant chemicals at scale from alternative renewable feedstocks or wastes (e.g., biomass residues, biogas methane, syngas, CO2) without harming the environment. The challenge is to develop novel microbial manufacturing platforms to produce targeted chemicals with high efficiency in a rapid, predictable, and reproducible fashion. The conventional approach involves engineering each desirable cell in many trial-and-error optimization cycles. To engineer new desirable cells to produce other target chemicals, the entire process must be repeated, which is laborious and expensive. To address these bottlenecks, we have developed the MODCELL (Modular Cell) tool to design modular cells that can metabolically and genetically couple with a diverse class of exchangeable production modules for combinatorial biosynthesis of novel chemicals in a plug-and-play fashion requiring minimum iterative strain optimization cycles. We will present the design, construction, and validation of an *E. coli* modular cell for a combinatorial biosynthesis of novel bioesters that can be used as fragrances, flavors, solvents, and biodiesels from biomass-derived sugars and waste streams.

461773 Systematic Optimization of Protein Secretory Pathways in *Saccharomyces Cerevisiae* to Increase Production of Hepatitis B Vaccine

Abstract:

Systematic Optimization of Protein Secretory Pathways in *Saccharomyces cerevisiae* to Increase Production of Hepatitis B Vaccine

Hunter Flick, Jiayuan Sheng, Xueyang Feng*

Department of Biological Systems Engineering, Virginia Polytechnic Institute and State University, Blacksburg, VA 24060

Hepatitis B is a major disease that threats over 40% of the population in the world, especially in developing countries. Currently, one of the effective vaccines to treat Hepatitis B is the Hepatitis B Small Antigen (HBsAg), which is produced by the recombinant yeast *Saccharomyces cerevisiae*. However, the price for the Hepatitis B vaccine is still too high (~$20/dose) for people in developing countries to afford. In this study, we aim to improve the production of HBsAg by systematically engineering the protein secretory pathways. In general, we screened 194 *S. cerevisiae* strains with single gene knocked out in four major steps of the secretory pathway: ER degradation (47 genes), protein folding (67 genes), unfolded protein response (39 genes), and translocation and exocytosis (40 genes). From these screenings, we found that the deletion of YPT32 (YGL210W), SBH1 (YER087C-A), and HSP42 (YDR171W) led to the most dramatic increase of HBsAg production, with 1.92, 1.66, and 1.62-fold increases over the wild type *S. cerevisiae* strain, respectively. During the screening, we also found that the deletion of IRE1 (YHR079C) led to dramatic decrease of HBsAg production. We next overexpressed IRE1 and found that the HBsAg production was increased by 1.29-fold. We then tested the combination of two strategies: gene knockout and gene overexpression and found that deletion of YPT32 with the overexpression of IRE1 led to a 2.12-fold increase in HBsAg production over the wild type which indicates potential synergistic effect among different genes in protein secretory pathways. In summary, our discoveries promoted the bio-manufacturing of a broad scope vaccines in a cost-effective ways, allowing the vaccine to be more accessible throughout the world.

465652 Engineering the Type III Secretion Apparatus for Increased Protein Export

Abstract:

Bacterial hosts are attractive for heterologous protein production because they are robust, easy to manipulate, and inexpensive to culture. Protein production traditionally occurs in the cytoplasm; product recovery requires cell lysis and several downstream purification steps, often including refolding. To avoid these challenges, it is desirable to develop a system to export proteins directly to the extracellular space. We use synthetic biology and protein engineering techniques to optimize the *Salmonella enterica* type III secretion system (T3SS) to export high titers of heterologous protein cargo. The type III secretion apparatus is a multimeric protein needle complex that shuttles proteins from the cytoplasm to the extracellular space through a polymeric needle composed of a single protein, PrgI. Mutational studies in homologous T3SSs indicated that single point mutations in the needle protein alter secretion titer, and we hypothesized that we could replicate this behavior in the *S. enterica* T3SS. We mapped a set of point mutations from these studies to PrgI and identified a hypersecreting mutant. We are now applying protein engineering techniques to create additional hypersecreting mutants of PrgI and increase bulk secretion titer. Removing bottlenecks in protein export advances development of the T3SS as a general platform for protein production.

472556 Tunable Expression and Display of Proteins on Bacterial Spore Surface

Abstract:

Bacteria endospores are the most robust and durable type of cells found in nature, attributed to the multiple layers of proteins that encase the spore and form a protective shield. Studies have shown that a few proteins on the outmost layer of the spore can be used as anchors to display proteins. Compared with soluble proteins or proteins displayed on other platforms such as cell surface, spore surface display of proteins offer a few advantages. First, spores are highly robust systems and can survive various extreme environments, such as extreme heat or cold, toxic chemicals, and radiations. Proteins displayed on the spore surface can be also protected from the harsh environments. Second, these proteins are part of the spore and can be easily produced through sporulation. No further purification is needed and this significantly reduces the production costs. Third, for enzymes used in industrial processes, spore based enzymes can be reused as they are part of the spore and easy to separate from the reaction mixture. Moreover, the worn enzymes on the spore can be re-generated through a germination-sporulation cycle. Altogether spores are an attractive platform for protein display. However, the assembly of spores is a delicate and complex system. Currently protein display on spore surface relies on a few anchor proteins and the expressions of them are all under native regulatory elements. No expression system is available to display multiple enzymes and modulate their expression level in response to external signals. In this work, we designed tunable expression systems that integrate the regulatory elements for tight sporulation control of a few anchor proteins and binding sites for non-sporulation transcription factors including LacI and TetR. Using GFP and mCherry as reporters, it was shown that the designed systems allow for functional expression and display of heterologous proteins in the range of 0.2 to 15 times of those subject to the native regulatory control. The development of the tunable expression and display systems enable the fine tuning of the display of multiple proteins on the spore surface, which is critical to investigate the synergy between proteins and optimize the overall efficiency.

471362 The Effect of *N*-Glycosylation on Biophysical Properties of an Anthrax Toxin Bioscavenger

Abstract:

Glycoprotein based therapeutics have significantly expanded due to their outstanding potential for the treatment of multiple diseases in recent decades. However, protein pharmaceuticals display a series of fundamental physiochemical instability problems during their production, purification and storage that can adversely impact their final therapeutic efficacies. In general, glycans have an effect on increasing the overall stability of glycoproteins, because *N*-glycosylation is one of the most common posttranslational modifications to occur in protein biosynthesis. Manipulation of the glycosylation parameters through glycoengineering either *in vitro* or *in vivo* methods could become a promising approach to improve the stability of protein pharmaceuticals. Molecular farming of tobacco plants provides a viable option over traditional microbial and mammalian cell culture technologies for production of recombinant glycoproteins. However, the effect of *N*-glycosylation on the stability of Fc-fusion proteins is poorly understood, therefore current work is focused on the production, purification and stability of differentially glycosylated recombinant anthrax toxin receptor Fc fusion protein (CMG2-Fc). This chimeric fusion protein, designed to protect against the deadly anthrax toxins, is composed of the von Willebrand factor A (VWA) domain of human capillary morphogenesis 2 (CMG2), an effective anthrax toxin receptor, and the Fc region of human immunoglobulin G (IgG). In this work *Agrobacterium tumefaciens* harboring genes for three *N*-glycosylated variants of CMG2-Fc (Aglycosylated, ER-retained and Apo-targeted) under the control of a constitutive CaMV 35S promoter are delivered into *Nicotiana benthamiana* plants via vacuum agroinfiltration. A viral RNA gene silencing suppressor (P19) is co-expressed to improve CMG2-Fc protein production. Purification of target protein molecule from the leafy biomass was achieved by Protein A affinity chromatography. The purity and quantity of the target protein was assessed using SDS-PAGE, western blot, ELISA and MS/MS. Site specific glycoform analyses were performed using UPLC system coupled to a triple quadrupole mass spectrometer. Biophysical properties of the differentially glycosylated CMG2-Fc protein variants are assessed, including thermal stability, aggregation, and secondary structure. Finally, anthrax toxin binding efficacy of the CMG2-Fc protein variants are evaluated.

465622 Glyco-Engineering of CHO Cells for Improved Protein Quality

Abstract:

Glycosylation affects the quality of glycoproteins by influencing protein folding, stability, bioactivity, and therapeutic efficacy. Unfortunately, heterogeneous glycans and incomplete glycosylation are still the biggest issues in protein quality control. Owing to recent advances in gene editing technologies and the availability of CHO genome sequences, stable genetic engineering has become a remarkable approach to regulate glycosylation. In this study, CHO host cells were engineered using genome editing tools via modulating the enzymes involved in glycosylation pathway to produce proteins with enriched sialic acid and non-fucosylated glycan. Specifically, CHO DG44 host cell was first genetically engineered using overexpression and CRISPR/Cas9 simultaneously, followed by the human IgG1 producing cell line construction. The glycan structure of the production cell line was evaluated to confirm the increased sialic acid contents and absent core fucose structure. Finally, the engineered CHO DG44 cell was characterized using proteomics to develop an in-depth understanding of cell engineering on the expression of host cell proteins involved in glycosylation.

469361 *Agrobacterium* Mediated Transient Expression of Butyrylcholinesterase, a Biodefense Agent, in *Nicotiana Benthamiana* Plants

Abstract:

Organophosphates (OP) are highly toxic inhibitors of the acetylcholine-hydrolyzing enzymes. The resulting accumulation of acetylcholine can lead to respiratory collapse and death. Current therapies are based on elevating the serum levels of OP bioscavengers. The most advanced candidate so far is human butyrylcholinesterase (hBChE). It is a 340 kDa blood protein made in the liver, which is capable of hydrolyzing different choline based esters.Â The major limitation of this therapy is the high cost, with plasma-derived hBChE costing more than $10,000/treatment. Limitations like cost and availability necessitate an alternative expression platform capable of large scale, low-cost production of a fully active and efficacious recombinant hBChE. The development of an effective recombinant BChE (rBChE) is a pressing national security concern in terms of protecting the nationâ€™s warfighters and civilian population from the threat of attack with OP agents. We describe the use of viral amplicon-based gene expression system based on *Tobacco mosaic virus* (TMV) to express a functional rhBChE in *Nicotiana benthamiana* plants using a transient agroinfiltration method. A viral gene silencing suppresser (P19) was used to improve the production of the target molecule. In this work, upstream bioprocessing was optimized. The production was monitored at different post infiltration time points using whole plants and harvested leave tissue as a production platform. At each time point, the production level was monitored using SDS-PAGE, western blot and an enzymatic assay (Ellman assay). The maximum production level of functional plant rBChE in whole plants reached 22 0.4 mg/kg fresh weight (FW) 6 days post infiltration, while it reached 41 0.7 mg/kg FW in harvested leaf tissue 12 days post infiltration. The concentration of the agrobacterium harboring the target gene was varied in detached leaves from OD_{600} 0.25 to 1.0, and the optimal OD_{600} was found to be 0.5 giving a yield of 32 0.1 mg/kg FW.

471266 Modulating Cell-Matrix Interactions Is an Effective Method to Control Cell-Cell Junctions and Thus Improve the Predictive Capability of the Caco-2 in Vitro Model of Drug Permeability

Abstract:

Oral bioavailability is a key parameter in the development of any therapeutic compound, be it a biologic or a small molecule drug. The traditional preclinical method to screen for the oral bioavailability of a drug in vitro is the Caco-2 assay in which human colorectal adenocarcinoma cells (Caco-2) are cultured as a monolayer on top of collagen I-coated transwell inserts. While the Caco-2 assay has been a very successful in vitro predictor for many drug compounds, it has long been known to be ineffective at evaluating drugs that are transported via the paracellular route. Previous studies have demonstrated that paracellular transport is regulated by the presentation and organization of tight junction proteins at cell-cell contact points. The development and maintenance of tight junctions is influenced by focal adhesions at cell-matrix contact points, and tight junctions and focal adhesions are connected through the actin cytoskeleton. Therefore, we hypothesized that regulation of focal adhesions through matrix engineering could be leveraged to regulate tight junction formation and hence enable significant paracellular transport across Caco-2 monolayers. Protein engineering was used to synthesize a recombinant engineered extracellular matrix (eECM) based on an elastin-like amino acid sequence. Two independent matrix parameters, stiffness and density of integrin-engaging RGD ligands, were used to alter focal adhesion formation. As expected, the number of mature focal adhesions, as quantified by talin and vinculin staining, was significantly reduced on more compliant matrices and on matrices with fewer RGD ligands. Interestingly, the average Caco-2 cell area on the less adhesive eECM substrates was significantly less than that for cells grown on collagen I, and the eECM-cultured Caco-2 cells had similar cell area to primary native intestinal tissue epithelial cells. Furthermore, Caco-2 cells cultured on softer, lower integrin-engaging eECM exhibited a more small intestinal phenotype compared to collagen I-cultured Caco-2 cells, with increased expression of the tight junction marker claudin-2 and the small intestinal markers sucrase-isomaltase and aminopeptidase N. In drug permeability assays, the eECM cultured Caco-2 cells displayed a significant increase in paracellular transport of the model drugs inulin-FITC, ranitidine, famotidine, and nadolol. Notably, the use of eECM in this assay was found to not affect the permeability of dexamethasone and naproxen, drugs that are transported via transcellular routes. Our results demonstrate that the substitution of an eECM material for collagen I would be an easily implementable strategy for improving the physiological accuracy of the Caco-2 assay in preclinical screens for oral absorption of pharmaceutical drugs.

470786 Establishing Mechanically Active Synthetic Mucosal Interface in a Multi-Well Plate

Abstract:

Introduction: Human gastro-intestinal tract (GIT) accounts for more than 75% of the body surface that comes directly in contact with the external environment. This makes GIT the prime location for host-microbiome interactions, since 70-80% of immune cells are present along the tract. GIT epithelium consists of a monolayer of polarized cells arranged into microscopic features called villi and crypts, which increases the GIT surface area tremendously. Recent studies suggest that peristaltic motion is important for epithelial cell differentiation, polarization and structural arrangement into villi and crypts. Some progress has been made in recent times, there is still a growing need to develop model systems to integrate microbiome, epithelial cells, and immune cells in a single platform along with providing sufficient level of biological and mechanical complexity. To this end, we developed a mechanically actuated 3D tissue culture model system that can achieve the aforementioned integration of different cell types with multiple layer complexity in an easily adaptable well plate platform. Pneumatically powered mechanical actuator applies magnetic force that cause cyclic stretching of PDMS membrane supporting cell culture and can simulate peristaltic motion. Actuation frequency is easily controllable using Arduino software. In this study we demonstrate how this platform was used to mechanically stimulate colon carcinoma cells (HT-29, ATCC® HTB-38™) leading to polarization, vertical growth and 3D morphogenesis of these cells.

Materials and Methods: The in-vitro platform consists of an "insert" made of three parts, a PDMS membrane (100 μm thick, 100 μm pore size, and 400 μm pore separation), a medical grade stainless steel ball embedded in a PDMS peg, and a PDMS ring. The PDMS peg was bound to the PDMS membrane and the PDMS ring bound on the outer periphery of the membrane using plasma treatment (Harrick Plasma). The PDMS membrane was coated with collagen on the apical side and HT-29 cells were seeded. The 24 well plate was then placed on the mechanical actuator (at 37°C, and 5% CO_2) with mechanical actuation rate of 1 cycle/minute and cell culture medium was changed every day.

Results and Discussion: Cyclic stretching of HT-29 cells grown on top of a stretchable PDMS membrane cause vertical growth and formation of 3D morphological structures in HT-29 cells. **C**onfocal fluorescence image confirms 3D growth of HT-29 cells. Commercial transwell inserts were used previously that allows for 2D culture of epithelial cells that were then exposed to pathogenic bacteria on the apical side and human peripheral blood mononuclear cells (PBMC) on the basal side. Preliminary results showed immune-modulatory effects of epithelial cells in terms of cytokine responses from PBMCs measured using ELISA (R&D systems). However, 2D cultures are not good representatives of the in-vivo physiology. Therefore, we replaced the 2D transwell system with aforementioned in-vitro platform. Currently, this system is being used to study bacterial interactions with functionally active epithelial barrier and immune cells.

Conclusions: A high throughput in-vitro platform was developed that allows for mechanical stimulation and polarization of epithelial cells via magnetic field. The system allows for compartmentalized co-culture of bacterial and mammalian cells to study the role of microbial invasion, epithelium, and cytokines in pathophysiology of human GIT.

471233 Recapitulating Liver Fibrosis in Vitro By Recreating the Fibrotic Milieu

Abstract:

Liver damage as a consequence of chronic stress can lead to fibrosis causing restructuring of the liver microenvironment (LME). The increase in the liver stiffness is a strong indication of the progression of liver diseases. However, the effect of increased stiffness on different liver cells has not been established. Clinical and animal studies have contributed significantly to our understanding of liver fibrosis biology but there exists a need for in vitro models to investigate the essential biochemical mechanistic pathways that associated with liver fibrosis microenvironment. Such models need to incorporate the dynamic changes in LME including the change in liver stiffness. In our study, we investigated the effect of substrate stiffness on the functions of cultured primary hepatocytes and liver sinusoidal endothelial cells (LSECs) utilizing a polydimethyl siloxane (PDMS) based substrate with tunable stiffness. We employed a soft substrate to represent the healthy liver tissue stiffness and a stiff substrate to represent diseased liver tissue and compared the cellular properties with the cells cultured on tissue culture polystyrene surfaces (TCPS). We observed that hepatocytes cultured on the soft substrate displayed a consistently more differentiated phenotype for a prolonged duration as compared to the hepatocytes that were cultured on the stiff substrate and TCPS. We further demonstrated that hepatocyte cultured on soft substrate showed better maintenance of hepatocyte-specific functions of urea and albumin synthesis. Similarly, cytochrome P450 (CYP) activity on soft substrate displayed a multi-fold increase in CYP activity, as compared to TCPS. In LSECs, we observed a stiffness mediated dedifferentiation in the cells, validated by defenestration of cells and overexpression of vascular endothelial cell markers. Together, these results indicate that the substrate stiffness plays a significant role in modulating hepatic cell behavior. Establishing mechanosensitivity of hepatic cells in liver fibrosis could result in the development of anti-fibrosis therapeutic approaches that target the mechanotransduction pathways of the cells to achieve functional restoration.

468907 Micropatterned Hydrogels to Promote Alignment in Co-Cultures Systems for an *in Vitro N*euromuscular Model

Abstract:

Introduction:

In vitro models are of immeasurable value to researchers and have become the gold standard in most tissue engineering and drug discovery applications. However, many of the currently used models do not fully capture the complex cell-cell interactions and 3D environment of multicellular organisms. The phenotype of cells can be vastly influenced by chemical and mechanical interactions within their microenvironment, therefore, it is critical to mimic native tissue environment when designing an *in vitro* models. **[1]**.

Primary cardiac muscle cells, cardiomyocytes (CMs), seeded on and inside of hydrogels, often lose their propensity to spontaneously contract, a visual cue of endogenous CM phenotyping, after only a few days **[2]**. However, the beating characteristics and phenotype of CM can be maintained *in vitro* for longer periods of time when CMs are subjected to external mechanical **[3]** or electrical stimuli **[4]**. Additionally, the presence of cardiac fibroblast in CM co-culture systems has been shown to prolong CM beating **[5]**. We hypothesize that in an *in vitro* model, the intrinsic cardiac neurons that help modulate heart rate **[6]** also may play a role in maintaining CM phenotype.

To test this hypothesis, we aim to construct a 3D co-culture system by encapsulating cardiac neurons and CMs in a methacrylated gelatin (GelMA) hydrogel. To align both cell types and mimic the architecture of native heart tissue, we have designed a two-step photolithography technique to encapsulate multiple cell types in a micropatterned GelMA hydrogel. This novel technique can be employed to study the interactions between the nervous system and the myocardium in a biomimetic environment.

Methods:

Encapsulated cells are aligned in the 3D hydrogel constructs through a multi-step photolithography technique modified from a single step technique developed by Nikkhah et al **[7]**. GelMA hydrogels are synthesized by photocrosslinking of pregel solutions using free radical chemistry. GelMA is produced by reacting methacrylic anhydride to the anime groups in gelatin as previous reported by Nichol et al **[8]**. Crosslinking is mediated via exposing the pregel solution: 10% (w/v) GelMA solutions containing 0.1% (w/v) free radical initiator (Irgacure 2959) to UV light (300-400nm, intensity 6.9mW/cm^2) for up to 75 seconds.

As a proof of concept, fluorescent beads were used as analogs for trypsinized cells. To pattern these beads, first a solution of a pregel-bead mixture is exposed to UV light though a photomask to form patterned lines of crosslinked GelMA (200 um widths with 200 um separation). Next, in order to create a contiguous hydrogel with distinctly aligned beads, the uncrosslinked suspension is washed away and a second pregel-bead solution is crosslinked in between partially patterned hydrogel by irradiating the entire sample. Hydrogels are imaged using fluorescent microscopy and the degree of gel segmentation in 3D is analyzed via intensity profiles of z-stacks taken at 10x on a Zeiss Axio Observer with NIH ImageJ software.

Results and Discussion:

We have successfully fabricated patterned hydrogels of two alternating type of fluorescent beads. Analysis of patterned hydrogels demonstrates two distinct patterns inside 3D GelMA hydrogels showed only a small percentage of leakage between line sections. However, this method is relatively efficient for spatially containing two individual cell suspensions. We have further plans to validate this process for a number of different line segment and spacing measurements.

Utilizing this technique, muscle cells will be patterned prior to neurons for our *in vitro*neuromuscular model. Future work will combine the neurons found in the heart and CMs into our patterned hydrogel to investigate if their interactions have an effect on neurite sprouting and growth as well as CM phenotype.

References:

1. Bhana, B., et al. (2010). Biotechnology and Bioengineering **105**(6): 1148-1160.
2. Bursac, N., et al. (1999). The American journal of physiology **277**(2 Pt 2): 44.
3. Radisic, M., et al. (2004). Proceedings of the National Academy of Sciences **101**(52): 18129-18134.
4. Zimmermann, W. H. H., et al. (2002). Circulation research **90**(2): 223-230.
5. LaFramboise, W. A., et al. (2007). American Journal of Physiology - Cell Physiology **292**(5).
6. Zarzoso, M., et al. (2013). Cardiovascular Research **99**(3): 566-575.
7. Nikkhah, M., et al. (2012). Biomaterials **33**(35): 9009-9018.
8. Nichol, J. W., et al. (2010). Biomaterials **31**(21): 5536-5544.

469318 Multi-Zonal Three-Dimensional in Vitro Culture Model of Growth Plate Cartilage Using Alginate Hydrogel Scaffolds

Abstract:

Organ regeneration using engineered tissues has the potential to revolutionize personalized patient care. However, precisely controlling the rate and direction of tissue growth remains a significant challenge for tissue engineers. Advanced tissue culture models that enable investigation of the mechanisms underlying anisotropic growth are necessary to overcome this challenge. During skeletal development, cartilaginous growth plates at epiphyses drive polarized bone growth, or endochondral ossification. This process is controlled by the precise regulation of chondrocyte differentiation via long- and short-range signaling mechanisms within the growth plateâ€™s architecture. As a result, the growth plate is organized in morphologically unique zones, each containing chondrocytes with unique phenotypes. While much has been described about the process of development of the growth plate, there is still a gap in the understanding of the interplay between specific signaling mechanisms and resulting architecture. To date, not enough is known about these intricacies to effectively modulate chondrocyte maturation in existing *in vitro* cartilage culture models. The design of an *in vitro* culture model of growth plate cartilage will be utilized to elucidate the structure and function of the regulatory network that coordinates chondrocyte maturation and leads to the formation of zonal architecture. This project focuses on the utilization of alginate hydrogels in the design of a novel, multi-zonal, three-dimensional (3-D) *in vitro* culture system aiming to recapitulate the supportive extracellular matrix and diffusional gradients of signaling factors found in growth plate cartilage. In order to test the hypothesis that the activation of the hedgehog (HH) signaling pathway is necessary for zonal arrangement, and in particular the formation of the columnar structures found in the proliferative zone of growth plate cartilage, radially layered alginate bead constructs are employed. We first show that an alginate bead culture method is supportive of cell viability and conducive to the presentation of factors known to modulate HH signaling, such as parathyroid hormone (PTH/PTHrP) and Indian Hedgehog (IHH), and the subsequent response of primary mouse growth plate chondrocytes to these factors. This response was observed with flow cytometry, ddPCR and western blot to monitor HH signaling. In order to present these factors in a gradient, a layered bead construct was designed and optimized for the controlled response of chondrocytes to signaling factors. HEK293T cells transfected with a plasmid to produce PTH were encapsulated in an innermost alginate bead, while growth plate chondrocytes were encapsulated in an alginate layer around this inner bead, receiving PTH in a radially diffusing fashion. IHH was added to the media to set up a bidirectional diffusional gradient. The response of the chondrocytes to this gradient and the resulting zonal arrangement of the growth plate chondrocytes was monitored by the above methods along with the use of Fluorescent In Situ Hybridization (FISH) to visualize the characteristic, radially differential expression of genes specific to each of the defined zones. This scaffold design has proven to be a powerful tool in investigating the role of these specific factors on zonal development in the growth plate. With the knowledge gained from these experiments, we begin to form a better picture of the mechanisms behind growth plate cartilage maturation, which will lead to the development of better therapeutic techniques for patients suffering from growth plate cartilage disease or injury.

469666 High-Throughput Drug Screening Using 3D Micro Organotypic Liver Models

Abstract:

There are thousands of chemicals whose effects on human and animal populations are unknown or only partially understood. Specifically, the combinatorial effects of these chemicals to the liver are unknown. Traditional *in vitro* and *in vivo* studies, while informative, can be laborious and time-consuming. High-throughput screening (HTS) methodologies can rapidly test individual and cocktails of chemicals across a wide range of concentrations. This approach can identify chemical mixtures that are toxic to the liver. To date, HTS assays on liver cells have primarily been conducted on 2D monolayers or co-cultures. These analyses are useful, but do not capture the complex interplay that exists between hepatic cells and their *in vivo*microenvironment.

When the principal liver cell types, hepatocytes, are cultured in standard monolayers, they begin to lose their liver-specific functions within 24 hours. Moreover, these models fail to recapitulate the native stratified architecture of the liver. We have designed 3D liver models that contain all hepatic cell types and maintain the cellular compositions over extended time periods. The combination of HTS methodologies and these 3D liver models has the potential to transform current testing of pharmaceuticals and environmental toxicants.

We have assembled a novel *m*icro liver *o*rganotypic *c*ulture *m*odel assembled in a 96-well plate (denoted as mOCM) that can be used to rapidly screen individual and mixtures of chemicals. A novel aspect of the mOCMs is the use of automated procedures. An automatic cell dispenser was used to coat the substrates with collagen, seed the hepatic cell types, change medium, administer chemicals, and conduct assays within the 96-well plates. The mOCMs contained a biopolymeric membrane composed of collagen (type 1) and hyaluronic acid that separated the hepatocytes from the liver sinusoidal endothelial cells (LSECs) and Kupffer cells (KCs). This membrane mimicked the height and composition of the Space of Disse that separates the hepatocytes from the LSECs and KCs *in vivo*.Acetaminophen

(APAP), troglitazone and perfluorooctanoic acid (PFOA) were administered to the mOCMs across a range of concentrations (sub-lethal to highly toxic). Hepatocyte monolayer (HM) and collagen sandwich (CS) cultures served as 2D comparisons. The mOCMs exhibited up to 43 ± 6%, 33 ± 8% and 69 ± 3% decreases in viability when treated with 10 mM APAP, 50 mM troglitazone and 0.5 mM PFOA, respectively. When these concentrations were increased to 40 mM APAP, 200 mM troglitazone and 2 mM PFOA, cell death increased up to 79 ± 2%, 67 ± 12% and 84 ± 4%. In contrast, the HM and CS cultures only exhibited significant cell death at the highest concentrations of PFOA and APAP. The increased sensitivity of the mOCMs compared to the HM and CS cultures clearly demonstrate that these cultures maintain biotransformation capabilities of these chemicals.

Changes in viability, necrosis and apoptosis were simultaneously measured in the cultures using a high-throughput assay (ApoTox-Glo^{TM} Triplex Assay). This HTS assay simultaneously provided information on cell death, apoptosis and necrosis by simply measuring the luminescence and fluorescence from each culture. In response to 40 mM APAP, there was a 6.9-fold increase in necrosis relative to controls in the mOCMs. The corresponding change was only 1.6-fold in HM and CS cultures. In response to 200 mM troglitazone, necrosis increased by 2.9, 6.4 and 16.1-fold in the HM, CS and mOCMs, respectively. Apoptosis increased up to 3.0-fold and 2.1-fold in the 2D cultures and mOCMs, respectively, relative to controls. *In vivo*, the expected mechanism of cell death for both APAP and troglitazone is necrosis. Therefore, these results indicate that not only are the mOCMs more sensitive than traditional 2D cultures, but the mechanism of cell death is more representative of the hepatic response *in vivo*.

Ongoing efforts are focused on the use of HTS assays to investigate changes in hepatic cellular functions, protein expression and functional markers in the presence of individual and cocktails of chemicals. By seamlessly combining HTS methodologies with a liver mOCM we seek to provide a platform for medium to high throughput testing of toxicants.

458646 Wnt-YAP Interactions during Neural Tissue Patterning of Human Induced Pluripotent Stem Cells

Abstract:

WNT-YAP INTERACTIONS DURING NEURAL TISSUE PATTERNING OF HUMAN INDUCED PLURIPOTENT STEM CELLS

Julie Bejoy, Liqing Song, Yan Li

Department of Chemical and Biomedical Engineering, FAMU-FSU College of Engineering, Florida Agricultural and Mechanical University, Tallahassee, Florida, USA

Abstract

Human induced pluripotent stem cells (hiPSCs) have special ability to self-assemble into mini-organ like structures (e.g., organoids, mini-brains). In this process, Wnt signaling impacts regional patterning and positional identity of hiPSC-derived neural progenitors. One important function of Wnt signaling is to regulate Yes-associated protein (YAP) expression (nuclear or cytoplasmic), the pivotal regulator of cell proliferation and differentiation during organ growth and tissue generation. However, the crosstalk between Wnt and YAP expression during neural patterning of hiPSCs has not been well investigated. The objective of this study is to reveal the capability of Wnt signaling in the regulation of YAP expression to modulate 3-D neural spheroid formation from hiPSCs. Human iPSK3 cells were induced toward neural lineages through embryoid body formation. Wnt signaling was activated using CHIR99021, which was found to induce nuclear localization of YAP. CHIR99021 treatment upregulated the expression of HOXB4, the marker for hindbrain/spinal cord, while in the absence of Wnt activation, the cells maintained rostral forebrain neural identity (expression of TBR1 and BRN2) (by RT-PCR analysis and immunocytochemistry). Modulation of YAP expression with cytochalasin D also influenced neural cell identity, indicating bi-directional interactions of Wnt signaling and YAP expression. The perturbation of neural patterning was also evaluated by incorporating the neural spheroids with microparticles fabricated with triblock copolymer polycaprolactone-poly(dimethylsiloxane)-polycaprolactone. This study should advance our understanding on the biological processes regulated by Wnt signaling and YAP activity during neural tissue patterning and brain organoid engineering. The results have the significance in neurological disease modeling, drug screening, neural tissue regeneration, as well as studying virus infection.

456522 3-D Spheroid Model of Human Adipose-Derived Stem Cells for Superior Adipogenic Differentiation

Abstract:

Development of *in vitro* models of adipogenesis is critical to find therapeutics for obesity and related comorbidities. Here we present the 3-D spheroid model using human adipose-derived stem cells (hASCs) and their subsequent adipogenic differentiation. Our work describes construction of a 3-D spheroid culture system atop positively charged elastin-like polypeptide-polyethyleneimine (ELP-PEI) coating to study the differentiation and maintenance process of hASC adipocyte spheroids when subjected to fatty acid treatment. Measurements taken from spheroid micrographs indicated spheroid size increased during the maintenance phase, though the spheroids treated with linoleic acid or oleic acid had smaller size compared to those cultured in the control maintenance medium. Increased spheroid size correlated with increased triglyceride accumulation and the linoleic acid or oleic acid treatments stimulated triglyceride accumulation in hASC cultures. The CD36 expression, representing cell competency for consuming extracellular fatty acids, was consistently found in hASC 3-D spheroids. Overall, we observed that, the hASC spheroids undergo slower differentiation along the adipogenic lineage and attainment of physiological functionality compared to the 3T3-L1 mouse spheroids. Taken together, we have shown the feasibility of adipogenic differentiation of patient-derived hASC spheroids, which with further development, may help elucidate key features in the adipogenesis process.

456713 Mass Transfer Characterization of a Biphasic Continuous Reaction System for Improved Process Understanding and Control

Abstract:

Continuous processing, while commonplace in other chemical industries, remains a novel and emerging technology in the pharmaceutical industry and presents potential advantages in manufacturing footprint, quality, and enhanced control of mass and heat transfer limitations associated with scale-up. Recent developments in the regulatory environment and pressure to control costs have encouraged growth of continuous manufacture in the pharmaceutical industry. Despite numerous advantages, continuous processing contains challenges in areas such as understanding laboratory and pilot plant equipment capabilities for improved process control. One fundamental element to process development is the characterization of continuous equipment to enable process understanding in terms of engineering first principles. With process understanding, effective control strategies can be implemented to deliver quality product at the lowest possible cost.

A key asset in the continuous portfolio at GlaxoSmithKline involves process development of a partially miscible chemical reaction mixture, in which the reaction is limited by mass transfer. To support process understanding, control, and scale-up from the laboratory to pilot plant facilities, a protocol was developed to characterize the mass transfer of the system in terms of fundamental engineering first principles. This presentation will detail the characterization of a partially miscible mass transfer limited reaction for continuous processing including observations of equipment specific scale-up parameters.

461095 The Development of Integrated Continuous Drug Substance Processes

Abstract:

As pressure grows on pharmaceutical companies to deliver increasing levels of process performance, reduced capital and manufacturing costs the industry is looking more and more to continuous processing to help tackle some or all of these challenges.

Within this presentation case studies will be presented where unique continuous processing hybrid platforms have been utilised for the development and operation of continuous processes. These processes included heterogeneous and homogenous chemistries, work-up/distillations, crystallization and isolation. Specifically within the case studies presented details will be outlined, via PAT integration, engineering kinetic and statistical models, experimental and lab rig design, required to enable the design, optimisation and operation of these platforms over multiple week long periods. During the presentation the design considerations over CSTRs, PFRs, hybrid or novel reactor design configurations will be presented and how such design selections can impact or tailor the design properties of the intermediates and importantly the final API solid state characteristics.

464864 Process Intensification through Continuous Spherical Crystallization Using an Oscillatory Baffled Crystallizer

Abstract:

Of current interest in the pharmaceutical industry is the development of continuous processes either within or for the manufacture of final drug products. These processes include, but are not limited to, drug synthesis, crystallization, formulation and tableting/encapsulating. Crystallization is the key link in going from drug synthesis to final formulation, and can often be the bottleneck in downstream process efficiency. For these reasons, there has been increased importance in the development and control of continuous crystallization systems of active pharmaceutical ingredients to produce crystals of good physical, processing, and biopharmaceutical properties. Continuous spherical crystallization (CSC) is a process intensification technique that can address many of the present flaws of traditional crystallization systems.

Peña & Nagy[1] studied and showed the benefits of CSC as a process intensification technique whereby both internal (primary crystals) and external (agglomerates) properties can be controlled experimentally through a decoupled CSC approach in a two-stage mixed suspension mixed product removal (MSMPR) crystallizer; providing the means by which both biopharmaceutical (bioavailability, dissolution) and manufacturing (flowability, filtration, drying) properties can be simultaneously adapted to meet desired quality specifications. Tahara et al.[2] also presented the value of the CSC technique using a one-stage MSMPR to make spherical API particles with a recycle stream; allowing for reduced anti-solvent consumption given many crystallization systems require low solution to anti-solvent ratios. Aside from these recent studies, previous CSC of a model drug compound was first carried out by Kawashima et al.[3]

Herein a novel concept and method in the area of process intensification though continuous spherical crystallization is proposed, which is based on performing crystallization/spherical agglomeration in oscillatory baffled crystallizer (OBC) system. OBCs are similar to plug flow crystallizers (PFCs) in that they are both

tubular crystallizers, however, the OBC has periodically spaced orifice baffles with oscillatory motion overlapped on net flow.[4] This allows for operations similar to PFCs without the high flowrates, high residence times and long crystallizer length while maintaining good mixing. The system allows for both coupled and decoupled nucleation, growth and agglomeration mechanisms. Decoupling through spatially distributed solution, solvent, anti-solvent and bridging liquid addition will offer more degrees of freedom for the control of each mechanism. This allows for products tailored to biopharmaceutical benefit and efficacy (e.g. bioavailability, dissolution, morphology) and processing efficiency (e.g. filtering, drying, & friability), while continuously producing particles with desired properties in an efficient manner and productivity at the manufacturing level.

In this work, a model oscillatory baffled crystallizer (OBC) consisting of seven segments was evaluated for both coupled and decoupled spherical crystallization systems. For coupled systems nucleation, growth and agglomeration occur from the onset as the solution contains bridging liquid. For spatially distributed (decoupled) systems nucleation and growth occur in the first four segments due to solution and anti-solvent addition, while bridging liquid is added at the end of the fourth segment making the final three segments agglomeration zones. The OBC will be evaluated for feasibility, control of crystallization mechanisms and final product properties.

1. R. Peña; Z. K. Nagy. Process Intensification through Continuous Spherical Crystallization Using a Two-Stage Mixed Suspension Mixed Product Removal (MSMPR) System. *Crystal Growth & Design*, 2015, 15(9), 4225-4236.
2. K. Tahara; M. O'Mahony; A. S. Myerson. Continuous Spherical Crystallization of Albuterol Sulfate with Solvent Recycle System. *Crystal Growth Design*, 2015, 15, 5149–5156.
3. Kawashima, Y.; Kurachi, Y.; Takenaka, H. Preparation of spherical wax matrices of sulfamethoxazole by wet spherical agglomeration technique using a CMSMPR agglomerator. *Powder Technology*, 1982, 32, 155–161.
4. S. Lawton; G. Steele; P. Shering; L. Zhao; I. Laird; X.-W. Ni. Continuous crystallization of pharmaceuticals using a continuous oscillatory baffled crystallizer. *Organic Process Research & Development*, 2009, 13(12), 1357– 1363.

466160 Designing for Continuous: A Step Change in Performance

Abstract:

Traditionally the majority of active pharmaceutical compounds have been manufactured at large plant scales using batch methodologies. Continuous manufacturing is becoming an increasing attractive alternative however due to apparent benefits such as smaller manufacturing footprints, increased process safety, lower manufacturing costs and increased supply-chain flexibility. Realising all of the above benefits using conventional continuous equipment is however quite challenging and sometimes not achievable. In this project, a new approach to the design of continuous manufacturing platforms is being presented which is not restricted to using such conventional equipment but which is based around the bespoke design and manufacture of all unit operations, optimised to fit the specific process being examined. Processes are quantified in detail using methods such as computational fluid dynamics (CFD) leading to tailored equipment designs which can be manufactured with high precision using metallic 3D printing.

The first case study presented in this paper examines a biphasic reaction between an amine and an acid chloride in the presence of an aqueous sodium hydroxide solution. In this type of biphasic reaction, the main design challenge is centred around achieving a sufficiently large interfacial surface area between the two phases. Conventionally, an agitated vessel operating in either batch or continuous mode would be used in order to maximise this area. However when tested in this case study, adequate dispersion between the immiscible phases required very high energy input levels of at least $1kW/m^3$, making such an approach difficult to scale up. Plug flow reactors (PFRs) are the most common continuous alternative to agitated vessels. These typically operate under Taylor flow conditions (where alternating aqueous and organic slugs move sequentially through the reactor), giving only limited contact between the phases. In order to overcome the limitations of these conventional approaches, a bespoke unit operation based upon the specific needs of the reaction is designed in this paper which is capable of achieving extremely high dispersion rates between the two phases under the very low Reynolds numbers typically encountered in the pharmaceutical industry ($Re < 5$). Due to its optimised design, this unit is able to significantly outperform both conventional methods tested,

achieving yields of 98% with residence times of less than 0.067 minutes (the batch agitated reactor achieved a 91% yield with a 6 minute residence time while the flow reactor achieved a 49% yield with a residence time of 7 minutes).

In order to demonstrate the suitability of applying this design approach to the entire manufacturing process, a case study is also presented in which two reactions are coupled together, followed by an extraction and wash step and a liquid-liquid separation. All unit operations within this platform are designed using fundamental engineering principles combined with numerical analyses (based upon the requirements of the specific process), prior to being 3D printed and combined to create a bespoke, sealed, hard-piped manufacturing system. The optimised design of the units enables extremely high throughput rates to be achieved corresponding to 580,000 kg/L/day (compared to an approximate batch productivity of 525kg/L/day).

466468 Continuous Drying of Powders and Slurries in an Extruder without Changing Product Particle Size Distribution

Abstract:

Academia, industry and regulators have increasingly demonstrated and addressed the advantages of continuous manufacturing during recent years. Slowly, this is leading to a change in mindset for pharmaceutical manufacturing, moving towards advanced continuous processing based on real-time quality assurance and away from batch processing.

Greater interest is being shown in the pharmaceutical industry for linking primary and secondary manufacturing into a single continuous manufacturing line. This demand is driven by the need of a cost-efficient process development and upscaling phase, which occurs during phase two and three of drug product development. Within the framework of continuous manufacturing, several techniques are commonly used for producing drugs. These include filtration, drying, milling/blending, granulation, and tableting, among others.

Drying is one of the most challenging process steps, if linking primary and secondary production. Not only does drying consume a vast amount of energy, but moreover, particle properties can significantly change during drying. Since the particle structure is carefully considered during continuous crystallization, it is highly desirable to avoid any changes in the particle properties while drying. Two counteracting forces, attrition and agglomeration, constantly occur during the drying of wet powder by shear movement.

When considering continuous manufacturing, not only must the product quality be considered, but also the processability. Therefore, the residence time distribution must be taken into account. Ideally, the residence time distribution should be narrow and no material should be stuck to walls and in dead spaces. Using an extruder helps to overcome many of the challenges normally faced during drying. Agglomeration can be avoided and the residence time distribution can be very narrow if using a twin-screw extruder with forced-feed and self-cleaning screws.

This paper will highlight successful efforts for drying wet powders or slurries, with high solid content, in a twin-screw extruder (Leistritz 27mm twin-screw). This work is focused on the processability of a test substance with a particle size below 100 µm. The screw configuration was optimized to best suit the needs of the drying process. Different process settings where tested and some showed very promising results. Products with different starting moisture contents were also tested. The percent reduction in moisture observed following the drying procedure was 99% with a residual moisture below 0.1% and there was no agglomeration observed during drying.

466532 Continuous Co-Crystallization of Novel Theophylline and 4-Aminobenzoic Acid Co-Crystals from Twin Screw Extrusion

Abstract:

Recently, increased demand for greener and more environmentally sustainable processes has focused the attention on mechano-chemical process. Mechano-chemical synthesis is one in which solid reactants are vigorously ground (milling or grinding) together with minimal or even no solvent [1]. Mechano-chemical synthesis is environmentally friendly, easy, cheap, fast reaction rates and high yields; these advantageous properties made mechano-chemical synthesis a method of choice for co-crystal formation [2 & 3]. Pharmaceutical co-crystals are solids that are crystalline materials comprise of two or more components held together by non-covalent forces [4]. In recent years co-crystals are being studied intensively due to the potential for improved pharmaceutical properties such as increased solubility, bioavailability, chemical stability and hygroscopicity of active pharmaceutical ingredients [1].

High-shear granulation and twin screw extrusion, have recently been used for co-crystal preparation. These methods are environmentally friendly as they avoid, or minimise, the use of organic solvents [2 & 5]. Twin screw extruder mediated co-crystallisation is a promising processing technique for continuous co-crystallisation. Although twin screw extruder mediated co-crystallisation has been reported with various primary experimental parameters (screw geometry, temperature, feed rate and the RPM of the screws) that can affect co-crystal conversion, information on the impact of processing parameters processing temperature (*eg.* at eutectic temperature and room temperature), shear dependent crystallization, residence time and addition of organic solvents is less available. Therefore to enhance understanding of continuous co-crystallisation process in the present study the effect of key extrusion processing parameters that influence extent of conversion to the co-crystals has been studied. A Theophylline-4-Aminobenzoic acid novel co-crystal forming system was selected for the study. Appropriate stoichiometric ratios of

Theophylline-4-Aminobenzoic acid were prepared. Co-crystallization of novel co-crystal system was carried out in a co-rotating 12 mm Hybrid Mini-Extruder (Three-Tec GmbH, Switzerland) with a length-to-diameter ratio of 40:1. The extrudates were mainly characterised by powder X-ray diffraction (PXRPD), differential scanning calorimetry (DSC), and *in vitro* drug release studies. As shown parameters had a significant influence on co-crystallization process, consequently they are important to consider in co-crystallization process. Finally twin screw extruder mediated co-crystallisation can be considered easy, cheap, scalable, and environmentally friendly for continues co-crystallization.

Key words: continuous co-crystallisation, twin screw extrusion, Theophylline-4-Aminobenzoic acid.

References

1. S.F. Chow, M. Chen, L. Shi, A.H.L. Chow, C.C. Sun, Simultaneously improving the mechanical properties, dissolution performance, and hygroscopicity of ibuprofen and flurbiprofen by cocrystallization with nicotinamide, Pharm. Res. 29 (2012) 1854–1865.

2. D. Daurio, C. Medina, R. Saw, K. Nagapudi, F. Alvarez-Nunez, Application of twin screw extrusion in the manufacture of cocrystals, Part I: four case studies, Pharmaceutics 3 (2011) 582–600.

3. James, S.L., Adams, C.J., Bolm, C., Braga, D., Collier, P., Friˇsˇci ´ c, T., Grepioni, F., Harris, K.D.M., Hyett, G., Jones, W., Krebs, A., Mack, J., Maini, L., Orpen, A.G., Parkin, I.P., Shearouse, W.C., Steed, J.W., Waddell, D.C., 2012. Mechanochemistry: opportunities for new and cleaner synthesis. Chem. Soc. Rev. 41, 413–447.

4. FDA Guidance (draft), December 2011: Regulatory Classification of Pharmaceutical Co-Crystals.

5. R.S. Dhumal, A.L. Kelly, P. York, P.D. Coates, A. Paradkar, Cocrystallization and simultaneous agglomeration using hot melt extrusion, Pharm. Res. 27 (2010) 2725–2733.

466998 A Facilitated Diffusion Mechanism Establishes the Drosophila Dorsal Gradient

Abstract:

During the first three hours of Drosophila embryogenesis, the transcription factor Dorsal (homologous to NF- κB) patterns the dorsal-ventral axis. Dorsal is present in a nuclear concentration gradient with the highest concentration at the ventral midline and a steady decay to about 40% of the embryo's circumference. The Dorsal gradient is initialized on the ventral side of the embryo via Toll signaling, which phosphorylates the inhibitor protein Cactus (homologous to IκB), marking it for degradation. In the absence of Cactus, Dorsal can enter the nuclei and activate expression of target genes in a concentration-dependent manner. Recent work measuring the dynamics of the Dorsal gradient has shown that, while Dorsal nuclear levels are initially uniform, Dorsal protein diffuses towards the ventral midline, which causes the overall accumulation of a Dorsal gradient peak. To explain this accumulation phenomenon, which seems to act against standard Fickean diffusion, we hypothesize a facilitated diffusion, or "shuttling" mechanism. The Dorsal/Cactus system has each of the features required for facilitated diffusion: (1) Dorsal binds to a "carrier" molecule (Cactus) that protects it from capture; (2) the Dorsal/Cactus complex is diffusible; and (3) the complex is broken in a spatially-dependent manner.

After using a photoactivatable GFP to show that nucleus-to-nucleus diffusion of Dorsal occurs within the embryo, we extended a previous computational model of Dorsal/Cactus interactions. Our model generically predicts that shuttling occurs in the embryo, and that three experimental perturbations would reveal a shuttling-specific phenotype. First, embryos with a half dose of Dorsal develop gradients that are wider and flatter than normal. Second, slowing Dorsal or Cactus diffusion causes the gradient to widen. And third, expanding the spatial domain where Toll signaling causes Cactus degradation results in a widening or even splitting of the Dorsal gradient. We performed each of these model-guided experiments and observed the predicted phenotypes.

Our results suggest that Cactus plays multiple roles in dorsal-ventral axis specification. Besides its predominant role, which is to maintain Dorsal in the cytoplasm, a secondary, but important role is to shuttle Dorsal towards the ventral midline. Given that this mechanism has been found in other, independent systems, we suggest it may be more prevalent than previously thought.

470148 Nuclear Rheology Associated with Early Development and Aging

Abstract:

The nucleus contains the genome and all of the regulatory proteins required for gene expression. However, gene expression changes drastically over the lifetime of an organism, particularly during early development and during phenotypic switches in a cellâ€™s lifetime. In addition to containing the code for the cell, the genome within the nucleus is a complex, self-assembled polymeric structure with unique rheological properties. The genome of metazoan cells is surrounded by an intermediate filament network known as the nucleoskeleton, which also changes drastically in composition during development and aging. Using spectroscopy, imaging, micromanipulation and computational techniques, we measure the mechanics of the nucleoskeleton and the nuclear interior at various length scales. We are particularly interested in the role that force and cytokine treatment play in altering nuclear mechanics and gene expression in primary human cells. Motor activity from the cytoskeleton transduced through the nucleoskeleton impacts the driving force for nuclear and subnuclear movement, and altered chromatin condensation shifts the resistance and propagation of forces. We also quantify nuclear stiffness in a broad spectrum of cell types: cells with less regulated gene expression patterns, including stem cells and cancer cells, have much softer nuclei whereas aged cells have stiffer nuclei. While the mechanisms directing stiffness are still being elucidated, we have quantified dramatic downstream impacts of nuclear stiffness on cellular migration. Generally, nuclear architecture and mechanics impacts cell fate directly by altering cell stiffness and indirectly by modulating gene expression. These results have broad implications in cell biology, inhibition of cancer metastasis, and for applications in cellular therapies.

470269 Quantification and Modeling of BMP Signaling during Zebrafish Embryo Development

Abstract:

Bone Morphogenetic Proteins (BMPs) act in developmental pattern formation as a paradigm of extracellular information that is passed from an extracellular morphogen to cells that process the information and differentiate into distinct cell types based on the morphogen level. Numerous extracellular modulators and feedback regulators establish and control the BMP signaling distribution along the dorsal-ventral (DV) embryonic axis in vertebrates to induce space and time-dependent patterns of gene expression. To identify how the dynamic pattern is regulated during development, we have developed a seamless data-to-model integration and optimization strategy. First, the nuclear intensities of fluorescent stained Phosphorylated-Smad5 (P-Smad) are acquired for each nuclei in each embryo from staged populations to provide a quantitative time-course for the BMP signaling gradient. Next, the nuclei are segmented to yield quantitative point-clouds of P-Smad level at each nuclei. The individual point clouds are registered to similarly staged embryos using a process called Coherent Point Drift (CPD) and the registered populations provide rigorous quantification and comparison of phenotype. To delineate the mechanism of BMP signal inhibition by the secreted binding proteins Chordin (Chd), and Noggin (Nog) a mathematical model was developed and optimized against the population data for wild type and combinations of the Chd mutants with and without morpholino knockdown of Noggin and Follistatin. We found a model that reproduced the experimentally observed patterning in WT embryos as well as a number of tested mutants. Models consistent with experimental behavior require that Chordin has a greater range than Noggin, that Noggin is not as freely diffusible, and that the BMP ligands are freely diffusible.

471629 Quantitative Synaptic Aging Characterization in Caenorhabditis Elegans As a Neuronal Health Metric

Abstract:

The multicellular organism *C. elegans* has been widely studied as a model for multiple fundamental biological processes, ranging from gene silencing to apoptosis, metabolism, and neurogenesis. In particular, *C. elegans* has had a large impact in aging research. Using this model organism, it was first discovered that aging is a genetically controlled process. Aging is becoming a major challenge in todayâ€™s society. Lifespan has been dramatically increased; yet serious age-associated diseases significantly impact the quality of life at older age. Better understanding the aging process should shed light on disease prevention and on environmental and life-long variables that could significantly improve quality of life at older age.

In this study, we aim to better understand age-induced neuronal decline in *C. elegans*. In particular, we focus on morphological and functional characterization of synaptic connections. We apply quantitative high-throughput experimental platforms, which integrate microfluidics, computer vision, and automation, to obtain large data sets of multi-parametric characterization of synaptic sites. Morphological characterization is based on imaging a set of representative synaptic-localized fluorescent fusion proteins. Functional characterization is based on quantitative drug resistance and optogenetic stimulation. Through the combination of these quantitative platforms, we focus on elucidating the main changes that occur at synaptic connections in populations of animals at various points during their lifespan, and potential genetic, environmental, or activity-driven prevention strategies for age-induced neuronal decline.

460333 Spatiotemporal Analysis Reveals Divergent Effects of Activating Mutations on Developmental Signaling

Abstract:

According to the Centers for Disease Control (CDC), approximately one in every six children in the U.S. is born with a developmental disability. A large class of such abnormalities, believed to affect ~1/1000 human births, results from activating mutations in components of the highly conserved RAS/ERK signaling pathway. Known as RASopathies, these disorders are characterized by a broad spectrum of phenotypes, including craniofacial malformations, heart defects, and neurocognitive delay. While many such defects have been successfully mimicked by introducing disease-causing mutations into model organisms, much less is understood about effects of these mutations on developmental dynamics and the emergence of the observed phenotypes. To explore these effects, we used a combination of imaging and genetic experiments with multiple activating mutations introduced into the early *Drosophila* embryo, a powerful model for studies of the RAS/ERK signaling with high spatiotemporal resolution. We focused on a panel of activating mutations in MEK, an ERK kinase, and analyzed their effects on the first wave of ERK activation in the blastoderm embryo. Our results revealed that the effects of activating mutations are highly dependent on cell context: they cause ectopic ERK activation in cells that do not receive endogenous signals and reduce ERK activation in stimulated cells. Importantly, the observed reduction of ERK activation leads to a pronounced effect on downstream gene expression and causes loss-of-function morphological phenotypes. Our results can be compactly summarized using a testable mathematical model, whereby the reduced ERK activation is a result of a negative feedback loop induced by premature pathway activation. In addition to describing the data, the model makes several predictions, all of which are supported by the results of additional genetic, imaging, and biochemical experiments. Furthermore, the divergent effects on activating mutations on ERK signaling are conserved in zebrafish, suggesting the generality of the proposed model, in which mutations that are activating at the level of the molecule can cause loss-of-function phenotypes at the systems level. Our results provide a first step in the quantitative understanding of large class of developmental abnormalities and present rational guidelines for thinking about their origins and potential treatment.

463213 Incorporation of Non-Canonical Amino Acids into the Developing Murine Proteome

Abstract:

Analysis of the developing proteome has been complicated by a lack of tools that can be easily employed to label and identify newly synthesized proteins within complex biological mixtures. Here, we demonstrate that the methionine analogs azidohomoalanine and homopropargylglycine can be globally incorporated into the proteome of mice through facile intraperitoneal injections. These analogs contain bio-orthogonal chemical handles to which fluorescent and affinity tags can be conjugated to facilitate identification of newly synthesized proteins within specific time windows during development.

The methionine (Met) analog azidohomoalanine (AHA) carries an azide functional group and is incorporated into sites normally occupied by Met during protein synthesis. Similarly, the Met analog homopropargylglycine (HPG) carries an alkyne functional group and is also incorporated at Met sites during protein synthesis. Both azide and alkyne handles can be covalently conjugated to affinity tags or fluorphores via copper-catalyzed azide-alkyne cycloaddition (CuAAC), a widely used "click" chemistry reaction. This enables fluorescent detection of the AHA and HPG containing proteins via SDS-PAGE size-separation and enrichment of the AHA and HPG containing proteins for downstream mass spectrometry identification from complex mixtures such as soluble cell lysate or solubilized membrane fractions and extracellular matrix.

In this work we show these non-canonical amino acids are incorporated into various tissues in juvenile mice and in a concentration dependent manner. Additionally, administration of these methionine analogs to pregnant dams during a critical stage of murine development, E10.5 – E12.5 when many tissues are assembling, leads to robust incorporation of AHA and HPG into the embryo and does not overtly disrupt development as assessed by proteomic analysis and normal parturition and growth of pups. Fluorescence imaging as well as intensity profiles of fluorescently labeled cellular fractions from AHA and HPG treated mice reveal unique banding in multiple tissues and robust incorporation of non-canonical amino acids in embryos. Furthermore, fluorescence imaging reveals unique banding in cytoplasmic, nuclear, membrane, and extracellular matrix fractions of developing tissues, suggesting AHA is incorporated into newly synthesized proteins of all types. Comparison of the intensity profiles of the total protein content of AHA and control samples indicates that protein expression was not noticeably altered as a result of AHA administration.

Proteomic analysis of control and AHA-treated embryos showed that less than 10% of identified proteins have significantly different levels of expression ($p > 0.05$). When considering a more stringent threshold ($p > 0.01$), the number of differentially expressed proteins dropped significantly to less than 3%. One differentially expressed protein, Pdia6, a member of the protein disulfide isomerase family, is expressed during times of ER stress and is thought to help attenuate the unfolded protein response via mechanisms independent of disulfide bond forming activity. Overall, it is not unexpected that components involved in protein folding show a difference in expression between the two treatments, given the differences in side chain chemistry between Met and AHA. It is important to note that overall AHA treatment has little effect on the identity and quantity of proteins. This corroborates our observation that AHA treatment does not overtly disrupt embryogenesis or subsequent development after parturition. This successful demonstration that AHA and HPG can be directly administered *in vivo* for labeling of newly synthesized proteins will enable innumerous future studies that seek to characterize protein dynamics during growth, disease and repair.

473115 Cellular Signaling Dynamics Regulate Stem Cell Mediated Neurogenesis in the Adult and Aged Mammalian Brain

Abstract:

Adult neural stem cells (NSC) continuously contribute new neurons in specific regions of the brain - a process known as adult neurogenesis - to modulate learning and memory. As a result, they represent therapeutic targets due to their potential roles in age-related cognitive decline, in disease pathologies, and as endogenous sources for cell replacement therapies. The adult NSC microenvironment is highly dynamic, with signaling molecules likely varying in concentration and duration at timescales including rapid electrophysiological activity, circadian rhythm, environmental cues, and organismal aging. Wnt signaling activates Î²-catenin signaling, which we (Nature Neuroscience, 2012) and others have shown regulates the differentiation of adult hippocampal neural stem cells into neurons *in vivo* and *in vitro*, and can contribute to age-associated declines in neurogenesis. To address the question of how dynamics in signaling impacts cell function, our lab has developed a tunable optogenetic system to modulate Î²-catenin signaling via Cry2 oligomerization of the LRP6 intracellular domain (Nature Methods, 2013). Similar to Wnt3a activation of the canonical Wnt pathway, blue light illumination of Cry2-LRP6c expressing NSCs induces robust neuronal differentiation. This raises the question: do stem cells differentiate simply when the integral of a signal during a given temporal window exceeds a key threshold, or do dynamics in signal presentation matter? Continuous illumination at different light intensities *in vitro* resulted in a progressive, saturable increase in neuronal differentiation from 5% to 60%. However, variation in signaling intensity over time yielded different results. Specifically, we observed that initial pulses of light for variable duration, or oscilllating illumination at frequencies >12 hours, yielded considerably less neuronal differentiation than in cells that received the same overall signal dosage but with continuous illumination. Furthermore, this signal stimulation followed by signal loss led to increased apoptosis, indicating exposure to the differentiation signal also rendered cells dependent upon it for survival, potentially offering a mechanism for removal of incompletely or poorly differentiated stem cells from tissue. Moreover, investigation of candidates emerging from

RNAseq has provided insights into the molecular mechanisms underlying how dynamic signaling regulates two alternate cell fates. In sum, these results harness optogenetics to demonstrate that not only the overall dosage of a signal but temporal dynamics in its presentation can exert a strong impact on stem cell behavior, work that offers further insights into the biology and translational potential of adult neurogenesis.

461505 Upgrading Waste for Food, Fuel and Chemicals Production Via Metabolic Engineering

Abstract:

Waste is rapidly becoming a major problem of the modern world. Disposing of waste in an environmentally acceptable manner presents major issues and in many cases it incurs major costs to communities all over the world. On the other hand, waste, in its various forms, (gases, liquids or solids) could be a valuable feedstock if it could be converted to biofuels and other products. This is possible for some types of waste that is available in aggregated form and at low or negative cost. Various waste-processing technologies have been developed in recent years including industrial biotechnology. In this task, the latter is aided by critical advancements in Metabolic Engineering that transforms microbes into little chemical factories capable of converting feedstocks to a variety of products. In this talk, I will present a technological platform based on native or engineered microbes that convert waste gases and municipal wastes to lipids for biofuel production, specialty products, food and bioingredients. The platform is centered around volatile fatty acids that can be generated from waste streams by various biological and thermochemical methods.

470941 Scaling up and Industrializing the Production of Viral Vectors and Cells for Therapeutic Use

Abstract:

Cell and Gene Therapies have the potential of revolutionizing medicine in the 21st Century. Bioprocessing the therapeutic agents for these therapies still face many challenges during scale up. The utilization of scale down unit operations from bioreactors to downstream purification is very useful in defining the design space and preventing costly scale up errors. Case studies will be presented to demonstrate the challenges and solutions associated with process development of such therapeutics. Optimization for adherent cells cultures in two different platforms â€“ microcarriers and flatware â€“ is presented. The cultures are scaled up to industrial scale for both platforms. Optimal conditions for the production of viral vectors in fixed bed bioreactors are determined through DOEâ€™s and scaled up according to the scaling rules associated with these type of reactors. Purification development and scale up for these vectors are presented.

459696 Snapshot of Sequential SNARE Assembling States Between Membranes Shows That N-Terminal Transient Assembly Initializes Fusion

Abstract:

Many prominent biological processes are driven by protein assembling between membranes. Understanding the mechanisms then entails determining the assembling pathway of the involved proteins. Because the intermediates are by nature transient and located in the intermembrane space, this determination is generally a very difficult, not to say intractable, problem. In this presentation, I will present a new setup with sphere/plane geometry. Using this setup, we have been able to freeze one transient state in which the N-terminal domains of SNARE proteins are assembled. A single camera frame is sufficient to obtain the complete probability of this state with the transmembrane distance. I will show that it forms when membranes are 20 nm apart and stabilizes by further assembling of the SNAREs at 8 nm. This setup that fixes the intermembrane distance, and thereby the transient states, while optically probing the level of molecular assembly by Förster resonance energy transfer (FRET) can be used to characterize any other transient transmembrane complexes.

454158 Modulating Amyloid and Virus Particle Interactions Using Osmolytes

Abstract:

Osmolytes, compounds that modulate cell volume and fluid balance, play an important role in modulating the hydrophobic interaction of proteins and larger biomolecular nanoparticles. It is likely that this modulation is caused by the water caging ability of osmolyte molecules. While in the lab of George Belfort, we were working on methods to modulate the interaction of hydrophobic amyloid proteins and peptides. His work, and the work of others, has shown that amyloid kinetics can be modulated with osmolytes. In my independent lab, we have utilized the ability of osmolytes to change the interaction of hydrophobic molecules by developing osmolyte flocculation technology (OFT) for virus particle purification. OFT has the potential to greatly reduce both development time and cost for future vaccine products. While osmolytes have many uses in protein stabilization, the ability to modulate hydrophobic interactions may open a new avenue of use for these versatile compounds.

467720 Building in Vitro Models of Human Barriers Tissues

Abstract:

We have focused our recent efforts on building a â€œBody-on-a-Chipâ€ as a human surrogate for both drug testing and as model for testing the potential human toxicity of chemicals or chemical mixtures. Such a model consists of barrier tissues (such as skin, GI tract, blood brain barrier, kidney, etc.) and integral body organs (such as liver, heart, fat, etc). These barrier tissues control the uptake and presentation of drugs and chemicals to the rest of the body. Typically cells representing the barrier tissue are supported on a membrane, but ideally a very open porous membrane so that transport limitations are controlled by the cell barrier rather than the supporting membrane. In this talk I will focus on examples and issues in terms of building barrier tissue models and their integration with the whole body model.

AUTHOR INDEX

AUTHOR INDEX

AUTHOR INDEX

AUTHOR INDEX

AUTHOR INDEX

AUTHOR INDEX

AUTHOR INDEX

AUTHOR INDEX

AUTHOR INDEX

AUTHOR INDEX

AUTHOR INDEX

AUTHOR INDEX

AUTHOR INDEX

AUTHOR INDEX

AUTHOR INDEX

AUTHOR INDEX

AUTHOR INDEX

AUTHOR INDEX

AUTHOR INDEX

AUTHOR INDEX

AUTHOR INDEX

AUTHOR INDEX

AUTHOR INDEX